progress in industrial microbiology

volume

19

Series Editor: M.E. Bushell, Department of Microbiology, University of Surrey, Guildford, Surrey GU2 5XH, Great Britain
Previous Series Editors: Vols. 1—13 D.J.D. Hockenhull (published by Churchill Livingstone, Edinburgh, Great Britain)
Vols. 14--16 M.J. Bull

Vol. 14 (1978)	edited by M.J. Bull (1st reprint 1983)
Vol. 15 (1979)	edited by M.J. Bull
Vol. 16 (1982)	edited by M.J. Bull
Vol. 17 (1983)	edited by M.E. Bushell
Vol. 18 (1983)	Microbial Polysaccharides, edited by M.E. Bushell
Vol. 19 (1984)	Modern Applications of Traditional Biotechnologies, edited by M.E. Bushell
Vol. 20 (1984)	Innovations in Biotechnology, edited by E.H. Houwink and R.R. van der Meer

progress in industrial microbiology

volume

19 modern applications of traditional biotechnologies

EDITED BY

M.E. BUSHELL

Department of Microbiology, University of Surrey
Guildford, Surrey GU2 5XH (Great Britain)

ELSEVIER
Amsterdam — Oxford — New York — Tokyo 1984

ELSEVIER SCIENCE PUBLISHERS B.V.
Molenwerf 1
P.O. Box 211, 1000 AE Amsterdam, The Netherlands

Distributors for the United States and Canada:

ELSEVIER SCIENCE PUBLISHING COMPANY INC.
52, Vanderbilt Avenue
New York, NY 10017

ISBN 0-444-42364-8 (Vol. 19)
ISBN 0-444-41666-8 (Series)

Printed in The Netherlands

PREFACE

The availability of enzymatic techniques which allowed true genetic engineering have resulted in the popularization of the term 'biotechnology'. In volume 17, I referred to the stimulation in other areas of industrial microbiology which has resulted. This volume contains re-appraisals of a number of traditional processes which have benefited from modern expertise in microbiology and process technology.

The sciences of alcoholic beverage and fermented food production are found to be practised in the most primitive societies. Indeed, exploitation of substrate-level phosphorylation yeasts seems high on the list of priorities for human communities! The continuous refinement to which such processes have been subjected is exemplified in the production of wine. The quality of the product is assessed according to complex organoleptic criteria whose subtleties defy a chemically precise rationale, but are, nevertheless, objectively assessed and reproducible. Process development for product improvement has been inhibited by the resultant difficulties in assessing progress. The application of modern fermentation technology to wine production in the U.S.A. has received some publicity, but those outside the subject are often surprised to learn of the existence of the many institutes devoted to the study of oenology and viticulture, at least one of which is associated with each of the wine-producing regions throughout the world. Chapter 1 describes developments in the microbiology of wine production and has been written by S. Lafon-Lafourcade and P. Ribereau-Gayon. The latter author is the Director of the Bordeaux Institute of Oenology. The manipulation of yeast physiology in the alcoholic fermentation is described and the influence of the lactic acid bacteria on the process is discussed. A section on microbial spoilage is also included.

The production of beer is considered in Chapter 2 in which current approaches to brewery fermentations are presented by J. MacDonald, P.T.V. Reeve, J.D. Ruddlesden and F.H. White of Bass Brewing Ltd. The modern fermentation process has been derived by the marriage of modern hardware with traditional methodology. A great deal is now known, however, about the influence of fermentation conditions on the detailed chemical composition of the product. Developments in engineering technology have effected improvements in construction materials, ease of reactor cleaning, automation and control equipment and gas recovery and recycling. The most significant advances, however, have been derived from manipulation of the yeast physiology. The influences of economics, market requirements and legislation on the process are also described. Design exercises, as well as process philosophies are considered with respect to both batch and continuous cultures. The article concludes that the recent high level of capital investment and the current overcapacity preclude any major new processing developments within the current decade.

The third major class of alcoholic beverages receives attention in Chapter 3 in which D.R. Berry of Strathclyde University describes the physiology and microbiology of Scotch whisky production. Although such factors as malt preparation and the distillation and maturation stages of the process are important, Berry concludes that the majority of the organoleptic compounds in the final product are produced by the yeast during the fermentation. Laboratory scale fermentations in the author's laboratory have indicated differences between whisky and beer fermentations which include growth limitation by amino nitrogen content in the former compared with oxygen or fatty acids in the latter.

To follow all the alcohol, it seemed appropriate to introduce Chapter 4 at this point. It describes the properties of microorganisms and their enzymes in the maturation of cheese (B.A. Law, National Institute for Research in Dairying, U.K.). The replacement of calf rennet (chymosin), with enzymes of microbial origin had received attention recently with the announcement of a genetically engineered microbially-produced enzyme. The influence of starter bacteria on the process in the fermentation, proteolytic and lipolytic phases are described. The effects of the secondary flora on final conditioning (ripening) are then considered. Some of the negative effects of microbes are also described with respect to the production of toxins and other forms of spoilage.

As a contrast to human foodstuffs, Chapter 5 deals with the application of biotechnology to the animal health industry (W.J. Marsheck, Diamond Shamrock Corporation, U.S.A.). Products from classical fermentation processes, as well as those promised by recombinant DNA technology, are considered. The involvement of microbial metabolites in this area covers antibiotic, growth stimulatory and immunological applications. Marsheck concludes that vaccines derived from recombinant DNA represent a significant product potential; he also points out that novel growth hormones and immunostimulators will probably be developed in this field as a result of advances in gene cloning techniques.

The production of Xanthan from Xanthomonas campestris was developed as a response to a requirement for polysaccharides to supplement or replace those derived from plant sources, such as seaweed, trees, seeds and cereals. Fermentation overcomes variations in supply due to seasonal variations in plant production and harvesting. Chapter 6 (Kennedy and Bradshaw) describes the chemical and physical properties of Xanthan, its production, recovery and commercial application.

Chapter 7 is a contribution from another University of Strathclyde author (B.J.B. Wood) who is well known in the U.K. as the individual responsible for setting up a production plant in Glasgow for the manufacture of Soy-sauce and related products by fermentation. Wood describes the production, biochemistry and microbiology of soyabean-based fermentations. Anxiety felt by microbiologists at the possibilities for mycotoxin production in these largely non-aseptic fermentations may be allayed by studies cited which failed to observe toxin production in all shoyu mould strains tested. The traditional moromi fermentation is very slow (months) and complex. Wood's description of this and the Miso and Tempeh processes uncovers a very large potential for research in this area.

After the chapters covering the various beverages and foodstuffs, we round off our meal in the traditional manner, with tea and coffee. In the final chapter, K.L. Jones and S.E. Jones review the state of knowledge of the fermentations involved in the production of cocoa, coffee and tea. The authors concede the lack of scientific methodology in current processes, but point to the need to remedy this situation when raw materials costs and desire for product uniformity are considered. Areas in which studies could be initiated are made apparent in this survey.

I should like to thank all the authors for providing such splendidly readable material and for reminding those of us in more recently established areas of fermentation technology of our roots.

M.E. Bushell
University of Surrey, 1984.

CONTENTS

DEVELOPMENTS IN THE MICROBIOLOGY OF WINE PRODUCTION

SUZANNE LAFON-LAFOURCADE and PASCAL RIBEREAU-GAYON
Institut d'Oenologie, Université de Bordeaux II et I N R A
33405 TALENCE, FRANCE

CONTENTS

1. INTRODUCTION

The origin of scientific oenology is "Etudes sur le vin" (1866), in which Pasteur analyzes the spoilage of wines. Oenology has continued to progress from that point on.

Traditionally the transformation of must into wine remains a "spontaneous phenomenon". That means it is induced by indigenous microorganisms which are found on grapes and in winery equipment.

Taxonomic and systematic studies on the specific microorganisms of must and wine have multiplied since the beginning of this century. They have led to a good knowledge of the microflora. The development of techniques of analysis has led to the determination of several hundred components of wine. The development of knowledge in biochemistry had led to a better understanding of the theory of fermentations.

Nevertheless many serious microbiological problems have remained. Their solution may have been delayed by the fact that one tried to apply directly to oenology notions acquired in fundamental microbiology.

The orientations of the researches now in progress take into account the specificity of microbiology and biochemistry of wine. First the composition of the medium, notably the concentration of must sugars, affects the growth of yeasts. Also the higher concentration of alcohol and the acid pH affect the growth of lactic acid bacteria. This and the technological operations impose upon the microorganisms a specific growth cycle and a particular metabolism which constitute the characteristics of this biotechnology.

These considerations explain the limited number of microbial species able to multiply in must and wine (yeasts, lactic acid bacteria, acetic acid bacteria) ; they also account for the specific metabolism of these microorganisms.

In the first step,yeasts transform sugars into alcohol (alcoholic fermentation), in the second step, which is general for red wines but occasional for white, malic acid is transformed by lactic acid bacteria (malolactic fermentation). Then, during aging all the microorganisms, mainly the acetic acid bacteria,must be inhibited.

Concerning growth, activity or metabolism of these three microorganisms, new results have recently been obtained. They have technical applications which may improve the way of wine making. Also they open new directions for further researches.

2. ALCOHOLIC FERMENTATION OF GRAPE MUST

2.1. New concepts

The control of the alcoholic fermentation of grape must and its completion are not always easy. One of the most serious problems is that of premature arrest of fermentation,that is to say stuck fermentation, leaving quantities of residual sugar in the wine. This is not compatible with wines of suitable quality. Also, because of the great difficulty to induce a second fermentation in such a medium, stuck wine can be the origin of bacterial development and thus alterations.

The first approach to solve alcoholic fermentation difficulties was an ecological study of yeasts of the vine and the wine. This research was conducted in most viticultural regions of the world. It showed that the grape must, by its acid pH and high concentration of sugar, is a particularly selective medium in which only a restricted number of species could develop. It was shown that *S. cerevisiae* is the main yeast that "makes" the wine. The species *S. bayanus*, with a superior fermentative capacity, is "the finishing" yeast (1).

Until now, only cellular multiplication and the factors which influence it have been considered in the stimulation of yeast fermentation. Consequently, research was focused towards enriching the nutritional capacity of the must by addition of growth factors (ammonium salts, vitamins). Another possibility tested for improved fermentation is the inoculation by yeasts selected for their fermentative capacity. The implantation of these exogenous yeasts in the grape must, rich in indigenous well adapted yeasts, was facilited by the commercialisation of active dry yeasts.

More recently, by the researches in this laboratory,the specificity of the growth cycle of yeast and of their metabolism in grape must was pointed out. It appeared that, in grape must rich in sugar, the viability and fermentative activity of non proliferating yeast populations was as important as total growth to avoid stuck fermentation. The new concept of survival factors was introduced. It was shown that the extinction of populations and the inhibition of their metabolism were linked to the production by yeasts themselves of toxic substances acting in synergy with ethanol.

Research on the stimulation of fermentative activity is thus now directed towards the elimination of these toxic substances.

This new result modifies the way of interpretation and consequently the approach to the problems of grape must fermentation and the solutions resulting. On the other hand, because of the use of spontaneous fermentation conditions, genetic progress has had until now little effect in oenology. A tendancy concerning standard table wines consists of aiming to industrialize their production, giving a constant quality product by the use of selected aromatic yeast able to develop preferentially in a natural medium, without it needing to be sterilized. Research on killer yeasts is in reponse to this need.

2.2. Yeast growth and the origins of fermentation problems

2.2.1. The yeast growth cycle in grape must

Fresh grape must induces particular conditions of growth on the yeasts responsible of a spontaneous fermentation. In an extreme case, one could mention the must very rich in sugar (320 g/l) extracted from over-ripened grapes (2).

One is able to distinguish three principal phases in the yeast growth cycle during fermentation : (a) a limited phase of multiplication that lasts for two to five days ; (b) a stationary phase lasts for about eight days ; (c) a decline phase which may last several weeks (figure 1).

These phases represent the classical microbiological growth cycle but, with wine fermentation, several variations may be noted :

(1) the duration is particularly long ;

(2) the total growth is limited and corresponds to about 4 or 5 generations ;

(3) the cessation of growth does not arise from the exhaustion of sugar in the must ;

(4) a disproportion exists between the principal phases ; the decline phase is able to last 3 or 4 times longer than the phase of multiplication.

The kinetics of fermentation are directly related to the growth cycle. In the example cited, the speed of fermentation is pratically constant and maximal until day eleven (figure 1). At this stage the viable cell population is abundant ($2{,}6 \cdot 10^7$ cells/ml), but it has lost a large part of its metabolic activity. At day 14, the must still contains around 150 g/l of sugar ; the fermentation continues, however, but at a much reduced speed. The fermentation stops around day 40, leaving 100 g/l of sugar not metabolized.

The first conclusion may be derived from this example. For the fermentation to be rapid and complete (that is utilisation of all sugar), it must take place entirely during the proliferation and stationary phases. This is generally with musts containing less than 200 g/l of sugar. For musts with more elevated levels of sugar, the final part of the fermentation is conducted by cells in the decline phase, where their metabolic activity has decreased. One can fear here, a premature cessation of the fermentation.

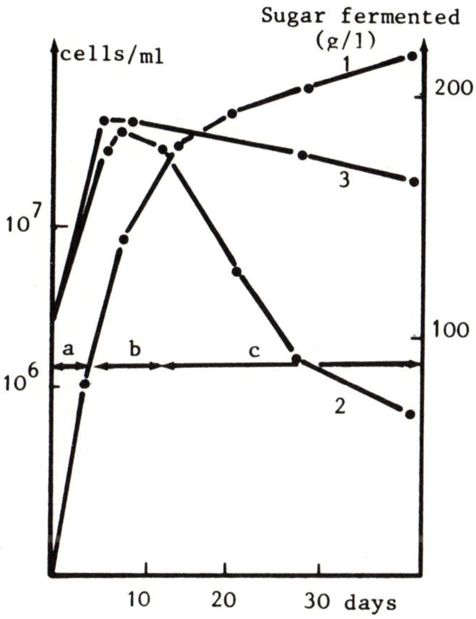

Fig. 1 - Growth cycle and kinetics of fermentation in grape must of high sugar concentration. Symbols : (1) fermentation curve ; (2) viable yeasts (cells per ml) ; (3) total yeasts ; (a) exponential phase ; (b) quasi-stationary phase ; (c) decline phase

(LAFON-LAFOURCADE and al, 1979, Appl. Env. Microbiol. 38,6, 1069)

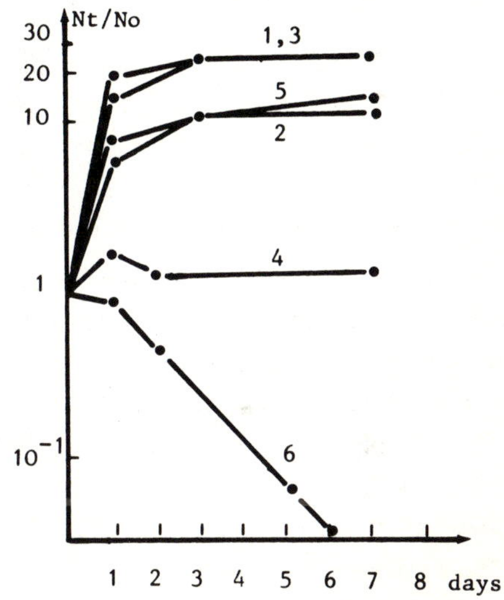

Fig. 2 - Evolution of Saccharomyces cerevisiae inoculated in a phase of proliferation into different media : (1) control unfermented I (ethanol 1,7% vol; sucrose 160 g/l) ; (2) prefermented I ; (3) control unfermented II (ethanol 7% vol; sucrose 65 g/l); (4) prefermented II; (5) control unfermented III (ethanol 9,5% vol ; sucrose 23 g/l) ; (6) prefermented III; (Nt, number of cells per ml at time t; No, number of cells per ml at beginning, around 10^7)

(GENEIX and al, 1983,C.R. Acad. Sc. Paris, 296, série III, 943)

2.2.2. Incidence of products of yeast metabolism

Ethanol is well know as an inhibitor of yeasts. Also some nutritional deficiencies may intervene. However, certain experiments show that products of the yeast, other than ethanol, may be implicated in its inhibition (3).

A synthetic medium was inoculated with *S. cerevisiae* and fermented at 25°C. At different stages, during the fermentation, corresponding to the proliferation (I), stationary (II) and decline phases (III) of the growth cycle, samples were removed and the yeasts sedimented out by centrifugation ; the super natants were kept to represent the prefermented media I,II and III respectively. In addition, control unfermented synthetic media were prepared and sugar and ethanol added to give concentrations corresponding to those in the prefermented media I,II, and III ; these are called non-fermented media I,II and III. The various media were then inoculated with cells of *S. cerevisiae* in proliferation phase ; the extent of growth was recorded.

In each of the control, non-fermented media, one observes an increase in the viable population, more or less according to the concentration of ethanol (figure 2). In the prefermented media I and II, the same trends were noted, but the total growth was respectively two and twenty five times less than their respective non fermentation controls. By contrast, in prefermented medium III, all growth was prevented and the viable population had decreased from 10^7 to 10^5 cells/ml by seven days. In the three control media, all of the sugar was degraded by 7 days ; but it remained at 7 to 10 g/l in the prefermented media. These results clearly show that, among the products formed by the yeasts, ethanol is not the sole inhibitor of yeast growth.

In the conditions of the experiment, nutritional deficiencies were not responsible for the inhibition. In fact, additions of vitamins and activators to the prefermented media were without effect.

In consequence, it is concluded that the inhibitory phenomenon is caused by the formation of toxic substances during the fermentation.

In another series of experiments, the destruction of these possible inhibitory substances was attempted by heating at 65° C for 15 minutes. This treatment was without effect. It seams, therefore, that these substances are not protein in nature. We have examined, equally, fixation on bentonite or charcoal. Treatment of prefermented media with bentonite did not remove the inhibitory substances. However, treatment with charcoal did remove some inhibitory material since the rate of growth is increased (figure 3).

The substances, therefore, must be essentialy of low molecular weight since they were non-adsorbed by the bentonite, but in part eliminated by the charcoal.

Attempts have been made to identify these substances as being among the secondary products of yeast metabolism, namely the higher alcohols, esters and fatty acids.

Yeast growth was little changed by addition of higher alcohols to the control media (figure 4). It was appreciably reduced by the addition of esters and a mixture of esters and higher alcohols. But, above all, the action of the three fatty acids, at doses existing in wines (3 mg/l), is particularly significant. The only reference concerning the anti-yeast action of these acids seems to be work of NORSTRÖM (1964) (4) who showed an unfavourable effect of these acids on beer yeasts. The influence of fatty acids formed by yeast, on the fermentability of media rich in sugars, has important pratical significances.

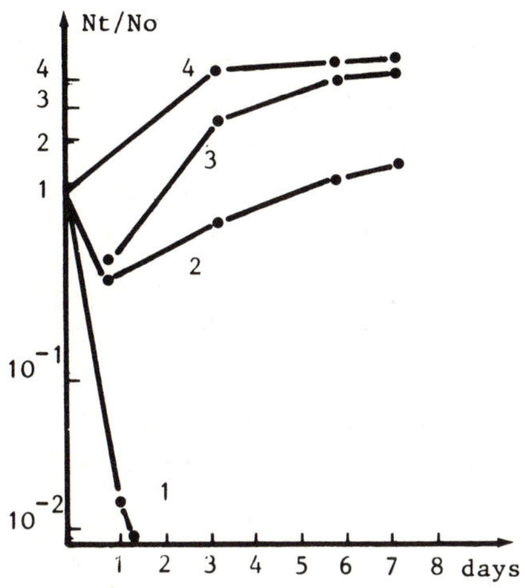

Fig. 3 - Evolution of Saccharomyces cerevisiae inoculated in a phase of proliferation into different media : (1) prefermented III (ethanol 10,5 % vol; sucrose 25 g/l) ; (2) prefermented III, treated with charcoal (2g/l) ; (3) control unfermented III (ethanol 10,5 % vol ; sucrose 25 g/l) ; (4) control unfermented III, treated with charcoal; (Nt, number of cells per ml at time t; No, number of cells per ml at beginning, around 10^7)

(GENEIX and al., 1983, C.R. Acad. Sc. Paris, 296, serie III, 943)

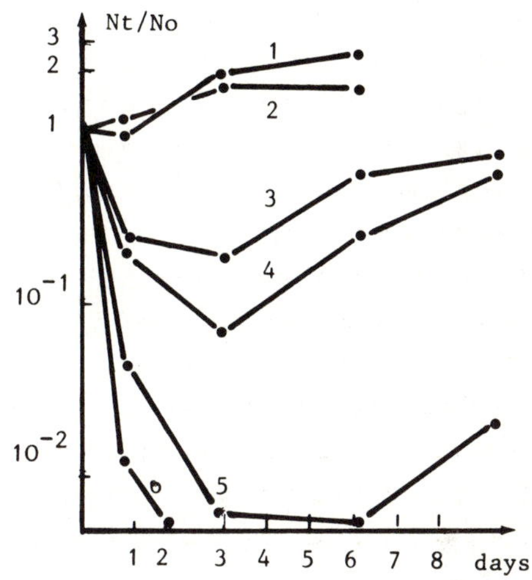

Fig. 4 - Evolution of Saccharomyces cerevisiae inoculated in a phase of proliferation into different media : (1) control unfermented III (ethanol 9,25 % vol ; sucrose 26,5 g/l) ; (2) control unfermented III higher alcohols added ; (3) control unfermented III esters added ; (4) control unfermented III higher alcohols and esters added ; (5) control unfermented III fatty acids added ; (6) prefermented III (ethanol 9,25 % vol ; sucrose 26,5 g/l); (Nt, number of cells per ml at time t ; No, number of cells per ml at beginning, around 10^7).

(GENEIX and al, 1983, C.R. Acad. Sc. Paris, 296, serie III, 943)

2.2.3. Microbial antagonism

The phenomena of microbial antagonism include:a) inhibition of yeasts by substances secreted by the other yeasts or by other microorganisms ; b) competition between yeasts (killer effect) or yeasts and lactic acid bacteria ; c) effect of *Botrytis cinerea* which parasitizes grapes and induces the growth of other microorganisms. *Gluconobacter*, *Acetobacter* are found ; these include *Aspergillus*, *Penicillium*, *Mucor*, *Alternaria*.

The killers (K) are capable of excreting a toxin that is lethal for so-called "sensitive" (S) strains. The so-called "neutral" (N) strains do not produce the toxin and are resistant to it (5). The phenomenon was discovered in yeasts by Bevan and Makover (6). The determinants of the phenotype killer are following : (a) cytoplasmic ; consisting of two types of R N A associated with a proteinaceous capsid which form together a particle of viral type ; (b) nuclear ; the responsible genes are of three types, a gene for maintenance (required for maintenance of the viral particles in the cytoplasm), a gene for expression (required for secretion of the toxin), and a gene conferring resistance to the toxin. The killer character can be introduced into sensitive strains by hybridization. (7) or by cytoduction (8).

Killer yeasts can be found amoung the following genera : *Saccharomyces*, *Debaromyces*, *Torulopsis*, *Candida*, and *Pichia*. The frequency of the occurence of (K) yeasts is a controversial subject. Barre (5) tested 908 wild yeast strains and found 514 (K) strains, 299 (S) strains, and 95 (N) strains.

The K toxin is a glycoprotein (9) (10) which adheres to the cytoplasmic membrane. After a variable time of latency one finds an arrest to the synthesis of macromolecules, an outpouring of coumpounds such as ATP and K^+ ions (11), and an inhibition of the active transport system for amino acids (12) (13). At 5 % glucose concentrations, the optimum pH is 4,5 and the production of the toxin is independant of the nitrogenous nutrients for the yeast (14). The toxin resists heating to 40°C for 1 hour. In spite of some unfavorable factors (elevated sugar concentration, low pH and presence of polyphenols), BARRE (5) could detect a killer effect in grape must when (K) type yeasts accounted for 2 % of the total population. Besides one observes the preservation of the killer effect in a secondary fermentation induced by a sensitive yeast population. Disappearance of the killer effect modifies the metabolism of a strain significantly (14). The killer character can be used as a criterion for strain selection (15). The killer phenomenon permits an explanation of fermentation difficulties observed with the inoculation of must with active dried yeast preparations (16).

The lactic acid bacteria are not generally found on grapes and sometimes these bacteria are added to the must to stimulate the malolactic fermentation (1) (17). When a large inoculation of bacteria is made when the activity of yeasts is weak, for instance before or at the end of alcoholic fermentation, there may be a partial inhibition of the yeasts and a premature arrest of the fermentation leaving residual sugar (table 2) . Thus, in general wine making practice , it is imperative to inoculate lactic acid bacteria only after all the must sugar has been utilized (16).

Certain sweet wines are made from must extracted from grapes particularly rich in sugar concentration, induced by over ripening by special infection with *Botrytis cinerea* ,called "noble rot" (pourriture noble) (1). These musts have a population of acetic acid bacteria which is fairly high (18). On grapes that have undergone "grey rot" (pourriture vulgaire), other microbial species can develop.

When one makes a fermentation with different mixtures of musts from healthy grapes and grapes parasitized by *B. cinerea*, all taken to the same pH and sugar concentration, fermentation is slowest in the mixtures containing the highest proportion of botrytized grapes (figure 5).

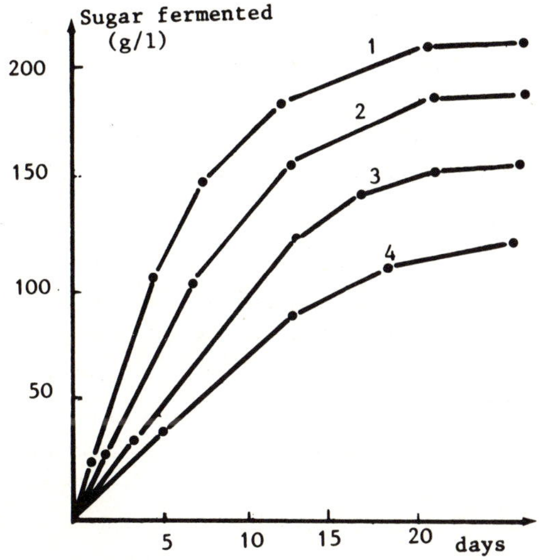

Fig. 5 - Effect of overipening, and growth of *Botrytis cinerea* on fermentability of grape musts.

A must parasited by *B. cinerea* (420 g/l sugar)

B healthy grape must (220 g/l sugar)

(1) 1/4 A + 3/4 B ; (2) 1/2 A + 1/2 B;
(3) 3/4 A + 1/4 B ; (4) A

(S. LAFON-LAFOURCADE and P. RIBEREAU-GAYON, 1979, Connaissance Vigne et vin, 15, 51)

When antifungal substances are precipitated by 80 % ethanol, the precipitate exerts on the alcoholic fermentation an inhibitory action clearly superior to the alcohol it contains (table1) (19). The formation of acetic acid and glycerol (normal products of sugar fermentation) are strongly increased. The precipitate also contains 5 % of proteins, but is essentially composed of polysaccharides rich in mannose.

The growth of other fungi on the grapes has the effect of retarding the commencement of fermentation, diminishing the activity of yeasts and reducing the biomass formed. Also the growth of certain species of acetic acid bacteria, mainly *Gluconobacter oxydans*, caused a decrease of total yeast growth and premature stoppage of alcoholic fermentation (18). The lactic acid bacteria are not generally found on grapes and sometimes these bacteria are added to the must to stimulate the malolactic fermentation (1). When a large inoculation of bacteria is made when the activity of yeasts is weak, for instance before or at the end of alcoholic fermentation, there may be a partial inhibition of the yeasts and a premature arrest of the fermentation leaving residual sugar (table 2). Thus in general wine making practice, it is imperative to inoculate lactic acid bacteria only after all the must sugar has been fermented by yeast (17).

Table 1

Effect, on the alcoholic fermentation, of the addition of the alcoholic precipitate from botrytized grapes must(19)

((1) Results in g/l ; (2) Results in millimoles for 1 mole fermented sugar)

	Fermented sugar (1)	Formed acetic acid (1)	(2)	Formed glycerol (1)	(2)
Control	217	0,35	4,8	4,8	43
+ 2 g alcoholic precipitate	173	0,66	11,4	5,3	59
Control	177	0,41	6,9	3,5	38
+ 4 g alcoholic precipitate	88	0,72	25	3,6	81

Table 2

Inoculation of *L. oenos* at different stages of alcoholic fermentation of must (17)

(Initial sugar content 204 g/l ; pH 3,3 g/l ; initial malic acid 4,7 g/l)

Time of bacterial inoculation	Lenght of incubation (at 25°C)	Hexoses (g/l)	Malic acid (g/l)	Lactic acid (g/l)	Acetic acid (g/l)
(1) before alcoholic fermentation	1 month	25	0,05	0,41	0,70
(2) during the yeast stationary phase (residual sugar : 25 g/l)	1 month	1,63	3,98	0,17	0,36
	2 months	0,09	3,20	0,18	0,38
(3) during the yeast decline phase (residual sugar : 5 g/l)	1 month	3,56	0,08	0,30	0,53
	2 months	0,13	0	0,38	0,69
(4) In wine	1 month	1,51	3,65	0,20	0,36
	2 months	-	0	0,20	0,32

2.2.4. Influence of technological constraints

Certain technological contraints increase the difficulties of fermentation of grape must.

The conditions of anaerobiosis dictated by the holding of large volumes of must may prohibit the biosynthesis of sterols and limit the exchanges of cells with the media (20).

The clarification of white wine musts, before fermentation : (a) removes part of the microflora ; (b) deprives the yeasts of supports and also of traces of oxygen (necessary for sterols synthesis) included in these particles ; (c) eliminates some nutrients contained in removed material (21) ; (d) deprives the medium of particles which can adsorb toxic substances produced on yeast itself.

In addition, each must possesses, from its extraction, a "particular fermentability". These qualities are related to the grape variety, conditions of maturity at harvest and, above all, treatment of the grapes in the winery. Mechanical and fast working presses give a must of mediocre fermentability. The must obtained by gently pressing of whole grapes has good fermentation which is increased further by short skins contact. The interpretation of these observations requires further researches ; they seem to be related to the extraction of sterols from the skins.

2.2.5. Mechanisms of the inhibition of the yeasts in the must

It has been previously shown that the inhibition of the alcoholic fermentation is not caused by nutritional deficiencies ; it is due in fact to the synthesis by the yeast of substances toxic to themselves. Some ethyl esters and fatty acids have been identified among these substances, acting synergistically with the ethanol. To our knowledge this is the first time that this phenomenon has been advanced as an explanation of the problem of fermentation arrest in grape must.

Concerning the mechanism of inhibition at the cell level, Navarro and Durand (22) reported that the accumulation of ethanol in the cell initiates the arrest of growth. But under the conditions of vinification (23), one observes (table 3): (a) a relatively weak variation in the endo cellular concentration of ethanol ; (b) in the presence of steroids (ergosterol, oleanolic acid), a concentration of cellular ethanol two times higher, with a higher concentration of sugar fermented ; (c) an absence of cellular pyruvic acid and acetaldehyde which testifies to the activity of carboxylase and alcohol dehydrogenase.

In contrast, the cellular concentration of glucose (table 3) of the population does not cease to decrease during the proliferation phase ; it is relatively stable during the stationary phase ; then it decreases during the decline phase, falling to zero. The transformation of must sugar to ethanol attests to enzymes functioning even at the end of the decline phase. According to these results, under the conditions of vinification, the slowing of the fermentation would be due, essentially, to a difficulty in sugar penetration. On other hand, further results showed that the toxic action of fatty acids is caused by their adsorption on the yeast cell wall, perhaps creating problems of exchange with the medium.

Table 3

Changes in endocellular hexoses and ethanol contents of *Saccharomyces cerevisiae* during fermentation (23)

		Hexoses (g/l cellular water)	Ethanol (g/l cellular water)	Viable cells (10^6 / ml)
Control grape must				
	day 2	22,5	3,9	3,5
	6	14	6,8	25
	14	3,1	4,3	5
	23	0	4,5	1,2
Addition of ergosterol (50 mg/l)				
	day 2	30	6,0	3
	6	14,2	14,4	26
	14	1,8	8,3	8,6
	23	0	8,5	6

Moreover it is known that sterols, notably ergosterol, are indispensable to the yeast (25). They intervene in the regulation of the physiological state of the cell membrane, affecting the nature of the phospholipids which contribute to membrane structure (26). During fermentation of the must (20), the concentration of sterols in the yeast population decreases rapidly during the phase of proliferation and then more slowly during subsequent phases (table 4). The cessation of growth and the slowing of the fermentation coincide with the depletion of cellular sterols.

All these results tend to suggest that inhibition of yeasts in the must is due mainly to a membrane alteration phenomenon.

Table 4

Sterols content in yeast during fermentation (20)

	Fermented sugar (g/l)	Sterols (% dry yeasts weigh)	Viable cells (10^6/ml)
Control grape must			
day 2	37	1,60	22
5	113	·0,60	13
End of fermentation	170	0,05	0,05
Addition of ergosterol (25 mg/l)			
day 2	36	1,40	20
5	111	1,10	10
End of fermentation	199	0,60	0,5

2.3. Stimulation of the alcoholic fermentation

2.3.1. Influence of nutritional growth activators (ammonium phosphate or sulfate, thiamin)

These stimulate multiplication of yeasts provided they are added to the must before the start of the fermentation so that they can be fully utilized for growth. The population may be doubled and, consequently, the fermentation is active from the start. The efficiency is more or less pronounced depending on the natural concentration of the must in the coumpounds. However, they do not affect survival of non-proliferating yeast populations. Therefore, they do not lead to a more complete fermentation of musts high in sugar concentrations (figure 6) (2).

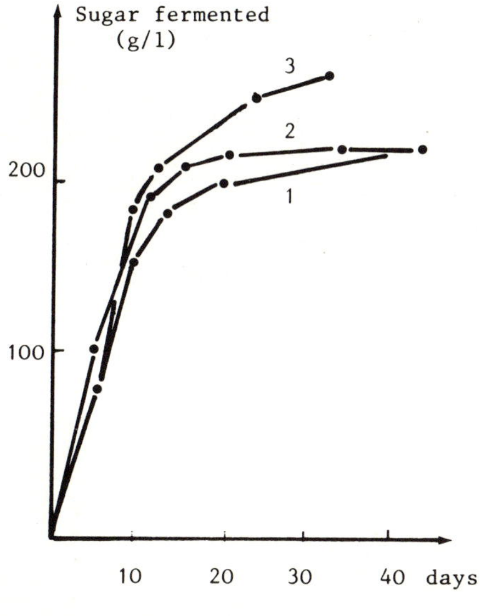

Fig. 6 - Fermentation kinetic in grape must of high sugar concentration, with and without added nutritional growth activators and with selected yeasts. Symbols : (1) *S. cerevisiae* ; (2) *S. cerevisiae* + nutritional growth activators (diammonium phosphate 200 mg/l, thiamine 0,5 mg/l) ; (3) *S. bayanus*.

(LAFON-LAFOURCADE and al, 1979 Appl. Env. Microbiol. 38,6, 1069)

2.3.2. Influence of aeration

Permanent and significant aeration, during the phase of anaerobic proliferation, stimulates yeast growth. This phenomenon has been known for a long time (1) but it is difficult to interpret it theoretically. This activation can only be explained by considerations of energy exchange (27). In effect the sugar concentrations of the must inhibit the synthesis of cytochromes ("Crabtree effect" (28) ; "reverse Pasteur effect" (27)). Therefore, yeast does not assimilate sugar from grape must except through the fermentative pathway, regardless of the conditions of aeration.

The effect of aeration on the cellular concentration of sterols appears to be significant (20) (table 5).

Table 5

Effect of anaerobic and semi-aerobic conditions on the sterol content of the yeast cells and on fermentation (20)

	Fermented sugar (g/l)		Sterols (% dry yeast weigh)	
	anaerobic	semi-aerobic	anaerobic	semi-aerobic
Control grape must				
day 2	37	30	1,6	2,7
9	164	187	0,4	1,2
19	170	246	0,3	1,0
Addition of ergosterol				
day 2	36	27	1,4	2,8
9	169	175	1,0	1,1
19	199	234	0,6	0,8

These results suggest that the oxygen introduced into the must permits the synthesis of some essential metabolites by yeasts, such as the sterols. This hypothesis has been suggested by Kirsop (29) for fermentations of the brewing industry.

Aeration is a current practice in the production of red wines by "pumping over". This technique consists of drainage of the fermenting must into a vessel open to the air. From this vessel the must is pumped back into the top of the fermenter. This method affects the fermentation in three different ways : (a) it favors sterol biosynthesis in yeasts ; (b) it facilitates the diffusion of oleanoic acid in the warm and alcoholic medium ; (c) it assures a better distribution of yeast which are abundant in the pulp into the entire fermenting volume (1). Until now this "pumping over" technique has not been applied to white wine production because of the fear of oxidative reactions, which affect the flavour of the product.

2.3.3. Influence of the cellular concentration of the inoculum and of the use of selected yeast

When a grape must containing 250 g of sugar per liter was inoculated with increasing populations (15 . 10^4 to 18 . 10^7 total cells per ml of *S. cerevisiae*) cultivated aerobically (commercial dry yeast), it was noted that the rate of fermentation in the initial phase was increased by increasing the initial biomass ; 190 g of sugar per liter was consumed between 5 and 12 days for amounts of inoculum varying between 9 . 10^4 and 3 . 10^7 viable cells per ml. In grape must of higher sugar concentration, however, it was necessary that the yeasts also ferment in the decline phase, and here the viable populations rapidly became more or less the same, regardless of the amount of inoculum. Generally, the total amount of sugar fermented and the length of the fermentation were not significantly different.

This result shows that for a range of initial populations encountered in practice and varying from 10^4 to 10^6 cells per ml, the concentration of inoculum affect the speed of the fermentation but not its completion (16).

Some yeasts species are particularly good ethanol producers. *S. bayanus* is such a well-known species (figure 6), but certain strains of *S. cerevisiae* also perform well. In most cases the main interest in yeast, management is the inoculation of selected yeast cells with high fermentation capacity, apart from the special case where a low indigenous population requires an increase in the number of cells.

This particular property can be increased by the method of growth of the yeast starter. The following example is given. A must which is aerated before inoculation is seeded with the same *S. cerevisiae* yeast grown either aerobically or anaerobically (traditional yeasting) (table 6). Aerobic pre-culture results in a more rapid and more complete fermentation (20). This property can be related to a cellular structure richer in sterols (6) which constitute a yeast survival factor.

The qualities are combined in industrial preparations of active dry yeasts. Several products are available on the market based on *S. cerevisiae* or on *S. bayanus*. The latter species produces more alcohol, but has the disadvantage of producing a little more volatile acidity, particularly in musts with high sugar concentrations where the application would otherwise be especially useful.

Table 6

Fermentation activity of yeasts as a function of the method of seed preparation (20)

	Yeast for Seed Grown Aerobically	Yeast for Seed Grown Anaerobically
4th day of must fermentation		
viable cells (10^6/ml)	32	20
sugar fermented (g/l)	171	100
11th day of must fermentation		
viable cells (10^6/ml)	23	11
sugar fermented (g/l)	251	199
23th day of must fermentation		
viable cells (10^6/ml)	2,8	0,7
sugar fermented (g/l)	275	229

Must aerated before fermentation and fermented under strict anaerobiosis ; initial must sugar concentration : 298 g/l ; yeast : *Saccharomyces cerevisiae* ; initial population : 10^6 cells/ml

The activity of inoculated exogenous yeasts depends on their specific characteristics but also on the composition of the must and its natural microflora. Sometimes one observes a more difficult fermentation after inoculation with active, dry yeast (16). This phenomenon, which is quite rare, may be due to the presence of indigenous Killer yeasts (6) which produce toxins acting on the exogenous yeasts.

The addition of dry yeasts must be made before the start of the fermentation in medium still free from ethanol. They provide a useful means of control of the fermentation, because of their convenience in use and because of their enological properties. They do not significantly affect to the organoleptic quality of wine.

2.4. Concept of survival factors

Sterols (ergosterol, lanosterol, cholesterol) (25) and oleanolic acid (30), which is an essential constituent of grape cuticle wax (3), were shown as anaerobic growth factors of yeasts. When must was initially deprived of oxygen by heating and then inoculated with anaerobically cultivated yeast, the maximum yeast population doubled resulting in more rapid and complete fermentation (30). However, these same substances, acted differently when they were added to an aerated grape must, inoculated with the indigenous yeasts (table 7). Under these conditions, no differences were noted in the maximum populations nor in the rate of the fermentation during the first stage. However, in the second stage, which corresponds to the decline phase of the growth cycle, a more important population was observed (2 to 10 times larger at 25°C) in the presence of additives ; as a result, a greater quantity of sugar was fermented (2).

It was possible,on the same fermentation to demonstrate both activities : growth activator (diammonium phosphate) and survival factor (ergosterol) (figure 7). The growth activators increased the fermentation in the beginning, but did not intervene at the end ; the steroids acted at the end of the fermentation. The association of the two allowed a more abundant population with a greater viability, resulting in a maximum of fermentation at all stages.

The principal conditions which favor the efficacity of survival factors (sterols and oleanolic acid) are the following (20) : (a) a fermentation temperature which is too low (less than 18°C) or too high (35°C) ; (b) grape musts with high sugar concentrations particularly those infected with *Botrytis cinerea* ; (c) highly clarified musts.

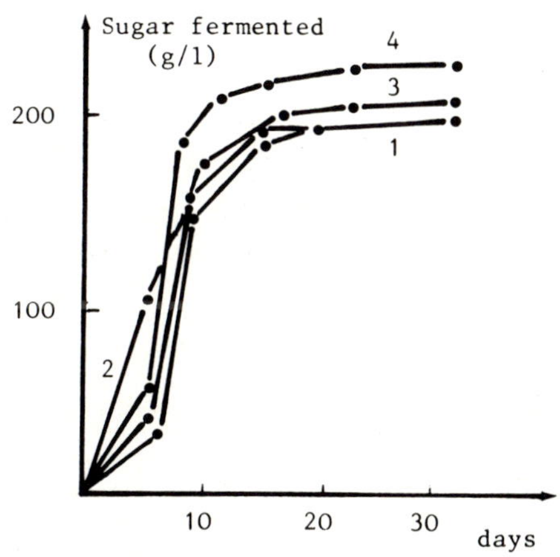

Fig. 7 - Effect of ergosterol and nutritional growth activators on the anaerobic fermentation of fresh grape must. Symbols : (1) control ; (2) plus nutritional growth activators (ammonium monophosphate, thiamin) ; (3) plus ergosterol ; (4) plus ergosterol and nutritional growth activators.

(S. LAFON-LAFOURCADE and al, 1979, Appl. Env. Microb. 38, 6, 1069)

Table 7

Variations in sterols concentration of yeasts during the alcoholic fermentation of grape musts (20)

Conditions	Permanent Aeration C	+E	+OA	Strict Anaerobiosis C	+E	+OA
2nd day						
Sugar fermented (g/l)	30	27	24	37	36	23
Sterols (% of dry weight)	2,7	2,8	2,3	1,6	1,4	1,7
Viable cells (10^6/ml)	12	16	21	22	20	17
5th day						
Sugar fermented (g/l)	116	101	95	113	111	105
Sterols (% of dry weight)	1,9	1,9	1,5	0,6	1,1	0,4
Viable cells (10^6/ml)	21	16	20	13	10	12
9th day						
Sugar fermented (g/l)	187	175	155	164	169	154
Sterols (% of dry weight)	1,2	1,1	0,7	0,4	1,0	0,3
Viable cells (10^6/ml)	20	16	20	5	7	5
End of fermentation						
Sugar fermented (g/l)	256	234	211	170	199	185
Sterols (% of dry weight)	1,0	0,8	0,4	0,3	0,6	0,2
Viable cells (10^6/ml)	18	12	16	0,05	0,5	0,1

C Control must ; +E Ergosterol (25 mg/l) ; + OA Oleanic Acid (50 mg/l)
Must sugar after enrichment : 250 g/l ; yeast : active dry *Saccharomyces cerevisiae* ; initial concentration : 1,5 %; initial cell population : 2,2 . 10^6/ml.

In contrast, under conditions of constant aeration, the addition of sterols acts as a fermentation inhibitor (table 7). Therefore, the sterols can function as : (a) growth factors, (b) survival factors, (c) inhibitors.

The principal technological interest in survival factors is a concern with their ability to activate non-proliferating yeast populations so that a complete fermentation can be assured, in a medium rich in sugar, such as grape must. Practically it is not permitted to supplement the must with sterols. But a knowledge of the factors which control their biosynthesis permits modifications of the fermentation process which favor their presence. Furthermore, oleanolic acid is naturally present on grapes. In white wine fermentations it is removed with the pulp. Its lack may be a factor in the early begin of the declining phase of the yeast population and of the deceleration of the fermentation. This role of survival factor makes this substance technologically most interesting.

The concept of survival factors complements the notion of growth factors. It opens new perspectives for research on the problem of stuck fermentations.

2.5. Prevention and treatment of stuck fermentations by the removal of fatty acids with the addition of yeast ghosts

2.5.1. Action of yeast ghosts on alcoholic fermentation

The activating action of charcoal already described has been well known in oenology for a long time, without being explained. It was shown that the addition of 2 g/l of charcoal,in a medium containing 10 and 3 mg/l of octanoic and decanoic acids respectively,removes 97,2 % of each of these fatty acids after 1 hour of contact. Probably the action of the charcoal is related to the adsorption of these fatty acids, toxic to the yeasts ; but the usage of charcoal is limited by legislation and the difficulty of its removal.

Preliminary observations showed (24) that the yeasts themselves adsorb the fatty acids and their ethyl esters added in the medium ; for this reason we thought to exploit this property. Yeast ghosts prepared in the laboratory were tried ; this technique was found to be very effective (patent n° 83 09 215). Therefore we tested an industrial preparation of these yeast ghosts (Fould-Springer, 94701 Maisons -Alfort, France).

The addition of 0,2 to 1 g/l of yeast ghosts to the must permits the fermentation of a greater quantity of sugar and helps to avoid stuck fermentations. This technic is mainly effective in the following cases : (a) must with an initial high concentration of sugar (table 8) ; it is much more effective than additional of ammonium salt ; (b) must containing residual chemical substances used to protect the vines such as Euparène or Mikal ; (c) must fermenting at high temperature. For example in a must initially containing 220 g/l of sugar incubated at 30°C, 150, 192 and 213 g/l are fermented due to the presence of 0, 0,2 and 1 g/l respectively of yeast ghosts (24).

In the case of stuck fermentation, the additions of yeast ghosts at a concentration of 0,5 g/l , 24 hours before an inoculation of *S. cerevisiae* (10^6 cells/ml) facilitates the fermentation of residual sugar by this second population (table 9)

Table 8

Stimulating action of alcoholic fermentation by yeast ghosts
(Results at the spontaneous arrest of fermentation) (24)

Moment of addition	Measurements	Control	$(NH4)_2SO_4$ added (0,2 g/l)	Yeast ghosts added (g/l) 0,2	0,5	1
Before Fermentation	Fermented sugar (g/l)	206	212	247	256	257
	Total populations (cells/ml)	9.10^7	10^8	$1,1.10^8$	$1,1.10^8$	$1,4.10^8$
	Viable populations (cells/ml)	$3,5.10^6$	$2,7.10^6$	10^7		$2,6.10^7$
5th day of fermentation	Fermented sugar (g/l)	209		239	254	254
	Total populations (cells/ml)	9.10^7		$9,8.10^7$		$8,2.10^7$
	Viable populations (cells/ml)	$3,5.10^6$		$1,6.10^7$		$2,9.10^7$

(temperature : 19°C ; yeast = *S. cerevisiae* ; inoculum : 10^6 cells/ml ; initial sugar content : 260 g/l)

Table 9

Stimulation of the second fermentation after arrest of alcoholic fermentation, by addition of yeast ghosts (0,5 g/l) to the stuck wine (24)

Conditions of the first fermentation	Second fermentation (residual sugar (g/l) as a function of days of fermentation): Days.......	9	12	16	27	36
Temperature (°C) : 19 Initial sugar content (g/l) : 250 Residual sugar (g/l) : 67,1 Ethanol formed (%vol) : 10,5	Control.....	57,6	49	35,9	15,6	13,4
	Yeast ghosts added.......	53,3	37	23,6	2,8	1,4
Temperature (°C) : 30 Initial sugar content (g/l) : 250 Residual sugar (g/l) : 121 Ethanol formed (%vol) : 7	Control.....	66,4	53	33	9,9	8,6
	Yeast ghosts added.......	40,1	27,6	12,9	0,6	

2.5.2. Mechanism of action of yeast ghosts

The presence of yeast ghosts in the must permits to maintain a higher number of viable cells during the declining phase of yeast population ; they act essentialy as survival factors (2). This is shown by the fact that the addition of yeast ghosts during the stationary phase (5th day of fermentation) is just effective as an addition before fermentation begins (table 9). This property makes this material the most effective survival factor of yeasts.

The yeast ghosts act by taking off the fatty acids and their ethyl esters which are formed during the alcoholic fermentation and are toxic for the yeasts. The mechanism of action can be shown by another experiment : 0,2, 0,5, 1 g/l of yeast ghosts are added to a synthetic medium; after 24 hours, samples are centrifuged at 7000 g and the quantities of each of the residual toxic substances are determined in the remaining solution. The results (table 10) show that the yeast ghosts adsorb mainly ethyl decanoate and ethyl octanoate ; also they depress the amount of decanoic acid, which is the most toxic substances for the yeasts (3). The quantities adsorbed increase with the concentration of yeast ghosts ; prolonging the time of contact has no effect.

Table 10

Removal of fatty acids and their ethyl esters
by yeast ghosts in synthetic medium (24)

(Results are expressed in percentage of the substance adsorbed in 24 hours)

Weight of yeast ghosts (g/l)	Acids Hexanoic	Acids octanoic	Acids decanoic	Ethyl esters of acids Hexanoic	Ethyl esters of acids octanoic	Ethyl esters of acids decanoic
0,2	0	1,2	20,2	25,8	58,2	88,8
0,5	0	4,5	40,7	36,2	73,8	94,6
1	0	7,2	54,5	36,8	77,7	94,6

However when the yeast ghosts are added into the must before fermentation (0,2 to 0,5 g/l), the concentration of fatty acids and their ethyl esters in the wine is not significantly changed (table 11). This suggests that, as the yeast ghosts adsorb ethyl esters, the yeasts in turn excrete these esters to maintain a balance between endo and exocellular concentrations. This highly significant point will be further explored by studying the rate of synthesis of these subtances.

No important change is observed in the sensory evaluation of wine prepared using yeast ghosts.

Table 11

Effect of yeast ghosts on the volatile substances content in wine (24)
(initial sugar content 250 g/l ; temperature : 19°C)

Volatile substances (mg/l)	Control	Yeast ghosts added (g/l) 0,2	0,5
Acetic acid................	0,29	0,23	0,23
Hexanoic acid	8,00	7,20	6,45
Octanoic acid..............	4,96	4,4	4,13
Decanoic acid..............	2,03	1,7	1,25
Isoamyl acetate............	1,84	2,13	2,52
Ethyl hexanoate............	0,71	0,62	0,61
Hexyl acetate..............	0,15	0,11	0,11
Hexanol....................	0,58	0,57	0,51
Ethyl octanoate............	0,70	0,66	0,63
Ethyl decanoate............	0,49	0,48	0,39
Diethyl succinate..........	0,14	0,24	0,20
Phenylethyl acetate........	0,18	0,24	0,26
Phenylethanol..............	23,9	39,4	35

As far as we know, this is the first time that such a technique has been proposed. It is likely that the applications of this process will extend beyond the field of oenology and that they could be employed in a large number of fermentations in closed media, in which stoppages are not due to nutritional exhaustion. In such conditions the difficulties in fermentation are due to the accumulation in the medium of toxic metabolics product by the microorganism which carries out the fermentation. Among these products the most inhibitive could be adsorbed by the "ghosts" of the microorganism themselves or by the "ghosts" of a similar species which possesse the same cell-wall properties.

Furthermore the addition of microorganisms ghosts could be used more generaly as a means of a selective adsorption of certain impurities from a wide range of chemical and organic media.

2.6. Biological stabilization of wine

2.6.1. Use of traditional chemicals

With regard to the development of microorganisms, the wine is protected by its acidity and the presence of alcohol. Nevertheless complete biological stabilization requires the use of chemicals. Only addition of SO_2, sorbic acid and physical processes, such as pasteurization or hot filling of bottles, are allowed according the law. SO_2 is an antiseptic universally employed in wine making pratice ; it inhibits yeast and bacterial metabolisms temporarily or permanently. It is adsorbed by the cell, and combines with intracellular acetaldehyde as it is formed, blocking the alcoholic fermentation. It also denatures proteins by reducing S.S. bridges

The mechanism of action of sorbic acid was reported by Reinhard and Radler (32). It acts on the yeast membranes by altering their permeability. Sorbic acid forms a complex with thiol groups of the cell-wall and this is responsible for its inhibitory activity. At the concentrations used in wine it has no effect on lactic and acetic acids bacteria.

5. nitro furylacrylic acid is an inhibitor of yeasts and lactic acid bacteria (34). Its use is permitted in some countries.

2.6.2. New yeast inhibitors

We have already shown that octanoic and decanoic acids are toxic against yeasts. The population of *S. cerevisiae* are more or less susceptible to the inhibitory action of products of their metabolism depending of their physiological state. The most resistant populations are, in increasing order : yeasts in the proliferation phase, in the stationary phase, dry active yeasts and populations in the decline phase (patent n° 83 09 214).

Given the resistance of yeasts in the decline phase to toxic substances, to inhibit the fermentative activity of a viable population of 10^7 cells/ml of a fermenting must containing 11 % ethanol, it is necessary to use 100 mg/l of octanoic acid or 30 mg/l of decanoic acid. At such concentrations, the addition of these acids to a fermenting must affects the sensory qualities of the wine. Nevertheless one can imagine their utilization as adjuvants of SO_2. Laboratory tests show that the addition of these two acids to fermenting musts at a total level of 9 mg/l allows reduction, by about 100 mg/l, of SO_2 necessary to stop the fermentation in sweet wines.

These substances are equally effective against the yeasts acting for refermentation (table 12) : *S. bayanus*, *S. bailii*, *Saccharomycodes ludwigii* and against *Schizosaccharomyces pombe* and *Pichia fermentans*. Certain formuls are prefered by the tasters (patent n° 83 09 214).

The use of fatty acids is authorized in other food industries. Their use in oenology is perfectively acceptable, taking into account : a) their effect at small amounts ; b) their absence of intervention on the taste ; c) the fact that there is an increase in the concentration of substances naturally present in the wine. It seems much better to use 10 mg/l of fatty acids instead of 200 mg/l of sorbic acid which is allowed by the official regulations.

Table 12

Use of octanoic and decanoic acids, as adjuvants of sulphur dioxide, for the biological stabilization of sweet wines (unpublished results)

(fatty acids = octanoic acid (3 mg/l) plus decanoic acid (6 mg/l))

	SO_2 (mg/l)	Fatty acid (mg/l)	Viable yeasts (cell/ml) 24 h after treatment
S. cerevisiae	0	0	10^7
	250	0	$4 . 10^2$
	150	9	$2 . 10^2$
S. bayanus	0	0	$7,5.10^6$
	250	0	<10
	150	9	<10
S. bailii	0	0	$4,5.10^4$
	250	0	70
	150	9	150
Saccharomycodes Ludwigii	0	0	$1,1.10^6$
	250	0	0
	100	9	<10
Pichia mambranaefaciens	0	0	$1,4.10^4$
	250	0	<10
	100	9	<10

3. MALOLACTIC FERMENTATION OF WINE

3.1. New concepts

The concept of the beneficial role of the malolactic fermentation in red wines is well established. Pasteur described the lactic acid bacteria as spoilage organisms. After the early work in the 1930s in Burgundy (35) and Bordeaux (36), full understanding the usefulness of this second fermentation did not take place until 1960 (36).

The malolactic fermentation is desired for two main reasons : (a) the bacterial transformation of malic acid to lactic acid deacidifies the wine conveniently and confers a mellowness to it ; (b) the disappearance of malic acid provides a biological stability with regard to the action of lactic acid bacteria. It occurs more often in red wines than in white.

Traditionally, the malo-lactic fermentation is based on bacterial growth. But in musts and wines the bacteria also find other substrates such as sugars, amino acids, organic acids, and glycerol. The metabolism of most of these compounds by lactic acid bacteria leads to spoilage. Particularly the degradation of sugars is a serious problem which leads to a "lactic flavor". One tries to prevent this by delaying the onset of the malo-lactic fermentation until all sugars have been fermented by yeasts, that is, to a level at which malic acid becomes the most readily fermentable substrate (1). But wine is not a particularly good growth medium. Consequently, the enologist is confronted by two problems which have not yet been solved well : (a) control of bacterial growth ; (b) control of the nature of the degradable substrate.

Recently, a bacterial biomass was prepared industrially (Microlife Technics, Sarosata, Florida U.S.A.) in a frozen form. After reactivation, the implantation of these selected microorganisms to the wine was facilited. The inoculation with these microorganisms stimulates the malolactic fermentation (37).

In warm regions musts have only small concentrations of malic acid and a relatively higher pH. They are better suited for bacterial growth ; but this would lead to an excessive deacidification of the wine and to spoilage. In some cases one tries to prevent the malo -lactic fermentation (38).

3.2. Lactic acid bacteria growth in must and wine

3.2.1. Development of lactic acid bacteria during vinification

Numerous studies have been conducted on the lactic acid bacteria that occur on grape, grape musts and wines (1). However, little information exists, on the development of these bacteria during the vinification process. Some recent studies suggest that a succession of species actually occures (30) (40) (41) (37).With the development of rapid methods for the identification of wine lactic acid bacteria, it is now practical to quantitatively examine the species that evolve during the different stages of vinification (42) (Api batteries).

It is common practice in wine making to add various levels of sulphur dioxide (SO_2) to grape musts, before fermentation, to control the development of unwanted yeast species and bacteria (1). However, little is known of what effects such additions have on the levels of lactic acid bacteria in the musts or on the ability of these bacteria to grow in the wine later and to conduct a desirable malolactic fermentation.

It is known that the capacity of lactic acid bacteria to grow in wines after the primary alcoholic fermentation is determined by wine pH, alcohol concentration, sulphur dioxide concentration and temperature (1) (17). Such information is largely qualitative, and little attempt has been made to examine these influences on a quantitative basis. A recent study (17) examines the levels and species of lactic acid bacteria that develop in wines, according the conditions, during the different stages of vinification, from the grape must until several months after the completion of the malolactic fermentation. Three typies of wine were considered : a red wine, a white produced at Bordeaux and a white wine produced at Cognac for Brandy distillation.

In both red and white musts with high concentrations of sugar, there was a notable and consistent decrease in the population of lactic acid bacteria during the alcoholic fermentation (figure 8). In some cases it was not possible to detect any lactic acid bacteria in the newly fermented wine. These observations are consistent with previous reports. The reasons for this decrease are not known but may be related to the fact that the strains of lactic acid bacteria found on grapes and in musts, namely *Lactobacillus plantorum*, *Lactobacillus hilgardii* and *Leuconostoc mesenteroides*, may not be tolerant to the alcohol generated during fermentation (table 13).

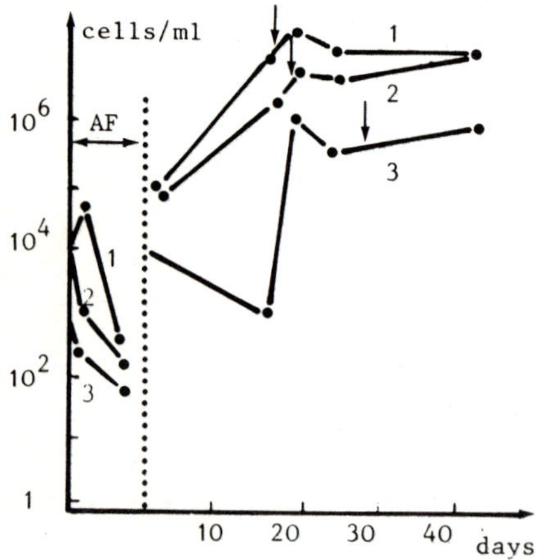

Fig. 8 - Evolution of lactic acid bacteria in red wine during production.

(temperature of conservation : 14°C)

(1) must without SO_2 ; (2) must with 50 mg/l SO_2 ; (3) must with 100 mg/l SO_2

A.F. : alcoholic fermentation
↓ : malolactic fermentation

(LAFON-LAFOURCADE and al, 1983, App. Environ. Microbiol. , 46,874)

Table 13

Enumeration and identification of lactic acid bacteria during the vinification of 1980 red Bordeaux wine (17)

(Draining of wine from fermentation tank occurs at day 14)

Sample	Time of sample (day)	cells/ml	Species isolated and number of strains identified
Grape	1 month before harvest	3	*Lactobacillus higardii* (2)
	harvest time	2	*Lactobacillus plantarium* (2)
Fermenting must (no SO_2)	1	7	*Leuconostoc oenos* (7)
	9	$6\ .10^2$	*L. oenos* (10)
	15	$2,5.10^4$	*L. oenos* (8)
	28	$4,5.10^7$	*L. oenos* (8)
Fermenting must (50 mg/l of SO_2)	1	5	*L. oenos* (2)
	9	$5\ .10^2$	*L. oenos* (8)
	15	$2,5.10^3$	*L. oenos* (8)
	35	$7\ .10^7$	*L. oenos* (3)
Fermenting must (100 mg/l of SO_2)	1	0	
	9	$2\ .10^2$	*L. oenos* (9)
	15	$4,5.10^2$	*L. oenos* (8)
	35	$2,5.10^6$	*L. oenos* (8)

The addition of sulphur dioxide, up to 50 mg/l, to musts did not reduce the initial population of lactic acid bacteria, although 100 mg/l did reduce it by around 10-fold.

The depressed population, however, was reinstated after the various manipulations for subsequent conservation at the end of the alcoholic fermentation. Immediately after these operations, levels of lactic acid bacteria were generally around 10^2 to 10^4 cells/ml. It would seem that these bacteria, largely *Leuconostoc oenos*, for the most part originate from winery equipment and materials. They serve as "natural" inoculum for the subsequent malolactic fermentation in the case of red wines (figure 8).

This study illustrated how the degree of sulphur dioxide addition and the temperature of storage may affect the subsequent development of *Leuconostoc oenos*, the species responsible for the malolactic fermentation (table 14, figure 8).

In the case of the white wine base for distillation, lactic acid bacteria did not decrease during the alcoholic fermentation, but actually increased (figure 9). Furthermore, the commencement of their growth, after alcoholic fermentation, was rapid and strong. This different behavior might be associated with the lower concentration of sugars in this unsulphured and unclarified must ; these conditions make the alcoholic fermentation easier and more rapid.

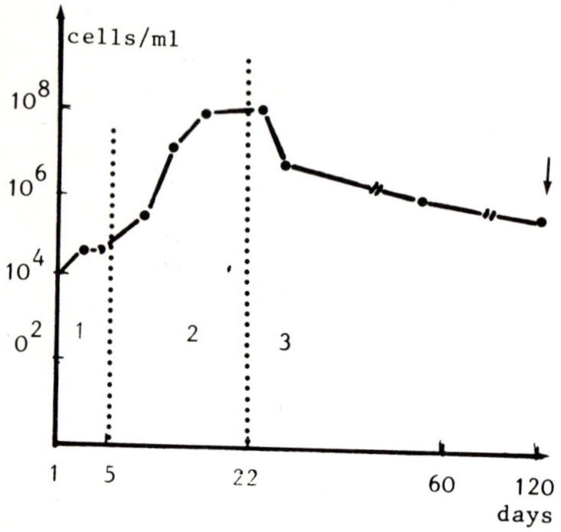

Fig. 9 - Development of lactic acid bacteria during vinification of a base wine for the production of Cognac

(1) alcoholic fermentation ;
(2) malolactic fermentation ;
(3) wine kept on lies containing yeasts and bacteria.
Distillation occurs at day 120

(E. CARRE, Recherches sur la croissance des bactéries lactiques en vinification. Desacidification biologique des vins. Thèse Doctorat 3ème Cycle, Université de Bordeaux II, 1982)

Table 14

Effects of SO_2 addition and temperature of storage on the development of lactic bacteria populations and the time for completion of malolactic fermentation in the same wine (vintage 1979) (17)

SO_2 to must (mg/l)	SO_2 to wine (mg/l)	Temp of storage (°C)	Maximum populations (cells/ml)	Time of completion of malolactic fermentation (days)
0	0	14	$3 . 10^7$	16
50	0	14	$1 . 10^7$	21
100	0	14	$1 . 10^6$	31
0	50	14	$1 . 10^4$	$-^a$
50	50	14	$5 . 10^4$	$-^a$
100	50	14	$1 . 10^4$	$-^a$
0	0	19	$1 . 10^8$	10
50	0	19	$9 . 10^7$	17
100	0	19	$4 . 10^7$	19

$-^a$, no malolactic fermentation after 200 days.

This study has also provided some insight into the factors that affect the growth and survival of lactic acid bacteria in wines during conservation after completion of the malolactic fermentation (figure 10, A, B, C, D,). Under standard conditions, lactic acid bacteria (*Leuconostoc oenos*) remained viable in the wine during storage, exhibiting no tendency for further growth and showing only a slow progressive decline in viability over the 200-days storage period. Temperature, pH, SO_2, alcohol, exert a very important influence on this survival. Althought SO_2 addition to the wines resulted in rapid loss of cell viability, growth recommended at a later stage (figure 10, D). The basis of this regrowth is not understood and is being investigated further.

The various factors act on levels of lactic acid and acetic acid formed in wine during conservation after malolactic fermentation (table 15).

Althought *Lactobacillus plantarum*, *Lactobacillus hilgardii*, *Leuconostoc mesenteroides*, and *Leuconostoc oenos* may be found on grapes and in musts , only *Leuconostoc oenos* survives the alcoholic fermentation ; even this species occurs in very low numbers at the end of this fermentation. In all cases, *Leuconostoc oenos* was the only species found in wines after the alcoholic fermentation and it was the species responsible for the malolactic fermentation (table 15). This specific involvement of *Leuconostoc oenos* appears to be unique to the Bordeaux region, as in other parts of the world *Pediococcus* and *Lactobacillus* species may develop in wines during and after the malolactic fermentation (39) (40).

Table 15

Effects of various factors on levels of lactic and acetic acids in wine during conservation after malolactic fermentation (17)

Factors		Acid concentration (g/l)	
		D(-)- lactic acid	Acetic acid
pH			
	3,3	0,31	0,39
	3,6	0,41	0,56
	3,9	0,46	0,64
Temperature			
	4°C	0,31	0,35
	12°C	0,32	0,36
	18°C	0,41	0,56
	26°C	0,32	0,38
Alcohol			
	10 % vol.	0,41	0,62
	11,25 % vol.	0,41	0,56
	12,5 % vol.	0,34	0,54
SO_2			
	0 mg/l	0,41	0,56
	20 mg/l	0,34	0,55
	40 mg/l	0,31	0,45

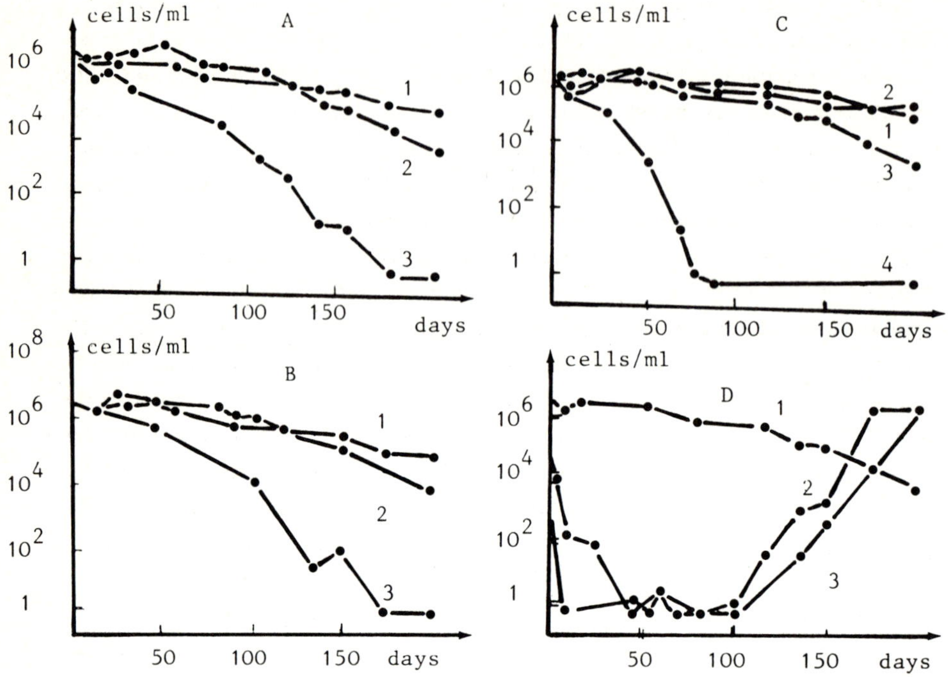

Fig. 10 - Survival of lactic acid bacteria in wines after the malolactic fermentation

A affect of pH : (1) pH 3,9 ; (2) pH 3,6 ; (3) pH 3,3
B affect of alcoholic concentration : (1) 10 % vol. ; (2) 11,25 % vol. ; (3) 12,5 % vol.
C affect of temperature : (1) 4°C ; (2) 12°C ; (3) 18°C ; (4) 26°C
D affect of SO_2 addition (mg/l) : (1) 0 ; (2) 20 ; (3) 40

(LAFON-LAFOURCADE and al, 1983, Appl. Environ. Microbiol, 46, 874)

Variability in the physiological properties of strains of *Leuconostoc oenos* has been observed (17) but its enological significance is not clear. Finally, it was noted that all *Leuconostoc oenos* isolated gave a strong reaction for esculin hydrolysis. This could mean that such strains may have the ability to hydrolyse phenolic heterosides in wine.

3.2.2. Parameters affecting the development of lactic acid bacteria in wines

It is well known (1) that lactic acid bacteria growth and malolactic fermentation are often limited by nutritional and physicochemical factors : temperature of storage, acid pH, alcohol concentration and SO_2 content. The role of aeration is controversial. The energy requirements can be fulfilled by the residual sugars of the wine (1).

In addition some aminoacids are indispensable and can be considered as growth factors (43). It has been shown (44) that the wine contains also some oligopeptides, in very small concentrations, which function as growth factors for lactic acid bacteria. Certain B vitamins are indispensable. The thiamin and riboflavine requirements are mainly shown by the lactobacilli and folic acid requirement by cocci. A specific growth factor has been found by Amachi (40) in tomato juice : the 4'-O-β-D-glucopyranosyl-D-(R)-pantothenic acid.

Sugar concentrations up to 350 g/l do not affect growth in a medium that is otherwise optimal (46). The presence of lactic, succinic, and tartaric acids has an unfavorable effect on the malo-lactic fermentation (47). Tartaric reduces the amount of biomass formed (table 16) and so does malic acid at high concentrations (46). Fumaric acid stimulates bacterial growth at low concentrations but becomes inhibitory at concentrations above 600 mg/l (38). Its use for the acidification of wines, to prevent the malolactic fermentation, has been suggested (38). The concentrations of glycerol normally encountered in wines favor bacterial growth, even if it is not used as substrate (46).

The strain of yeast which carries out the alcoholic fermentation also effects the malo-lactic fermentation. An inhibitory effect of a yeast strain is sometimes attributed to the formation of SO_2 by these yeasts (48). It also affects the malo-lactic fermentation by non-proliferating biomass (49). *S. cerevisiae*, compared to *S. bayanus*, produces wines with malic acid more easily fermented by bacteria. The phenomenon may be explained by a greater richness of amino and peptidic nitrogen in wines from the fermentation by *S. cerevisiae* (44). These results allow interpretation of certain differences in the malolactic fermentability of wines sometimes seen between two neighbouring wineries.

On the other hand, the interruption of the malolactic fermentation of wine due to an attack by bacteriophage which lyses bacteria of the genus *Leuconostoc* has been demonstrated for the first time (50). Two types of bacteriophages with different morphology and lytic capacity have been isolated from wine and cultivated in the laboratory. This discovery provides an explanation for the interruptions of malolactic fermentation sometimes observed in cellars. The possibility of bacteriophage attack should stimulate further research on the malolactic fermentation resulting in a better understanding and giving wine makers the ability to better control it.

Table 16

Effect of some acids on the growth and metabolic activity of *Leuconsotoc oenos* (synthetic medium of pH 3.5) (46)

		Time (in days) 2	5	9	15	22	60	Final pH
Basic medium	cells 10^6/ml	4	10	56	57	50	-	3.35
	acetic acid (g/l)	0,07	0,23	0,23	0,30	0, 40	0,70	
+ Citric acid 3 g/l	cells 10^6/ml	0,5	0,1	15	18	17	-	
	acetic acid (g/l)	0,72	0,95	1,11	1,25	1,52	1,93	3,95
	citric acid (g/l)	1,07	0	-	-	-	-	
+ malic acid 3 g/l	cells 10^6/ml	3	37	70	47	24	-	
	acetic acid (g/l)	0,11	0,17	0,36	0,64	1,00	1,18	3,95
	malic acid (g/l)	1,17	0,55	0	-	-	-	
+ tartaric acid 3 g/l	cells 10^6/ml	4	3	0,3	0,04	0,02	-	3,15
	acetic acid (g/l)	0,02	0,06	0,08	0,15	0,40	0,43	
+ glycerol 10 g/l	cells 10^6/ml	4	5	78	70	36	-	3,60
	acetic acid (g/l)	0,13	0,17	0,36	0,68	0,72	1,00	

3.3. Biochemistry of malolactic fermentation

In the must and wine L-malate is converted exclusively to L-lactate and CO_2. The enzyme active in the fermentation is not the same as the malic enzyme previously described (5). Malic enzyme catalyzes the formation of pyruvate from L-malate with a concomitant reduction of NAD^+. Althought the pyruvate can be converted to L- or D-lactate in the presence of the appropriate lactate dehydrogenase, some bacteria responsible for the conservation of L-malate to L-lactate in wine contain only D-lactate dehydrogenase (52). Thus, the malolactic transformation is catalyzed by a particular activity named "malolactic enzyme" which is active in the presence of NAD an Mn^{2+} and produces only L-lactate from L-malate (53) (64).

The reaction of the malolactic fermentation is not a direct decarboxylation as previously shown (54).

It was not possible to demonstrate any L-lactate dehydrogenase activity (53) . A weak malic enzyme activity measured by the production of NADH was found which was ca. 2 000 times lower than malolactic activity ; it was probably the same activity described as the NADH-producting activity (55). The results obtained strongly indicate that malolactic activity was not a combination of malic enzyme and lactate dehydrogenase activities, but was carried out by a complex that produced the sum of these reactions without catalyzing the partial reactions. Stereochemical studies (56) pointed to the possibility that both oxalacetate and pyruvate might remain attached to the malolactic complex during the course of the reaction and not appear as free intermediates.

NAD was essential for malolactic activity, but no net reduction of NAD occurred. The NADH produced during the first step was probably immediately reoxidized by an L-lactate dehydrogenase contained within the complex. To function, the complex must contain malic enzyme and L-lactate dehydrogenase components or, alternatively, malate dehydrogenase, oxalacetate decarboxylase and L-lactate dehydrogenase components.

The kinetics studies of NAD, Mn^{2+} and malate indicated that the NAD and Mn^{2+} attached first to the protein (57). Malate then was bound by the enzyme-NAD-Mn^{2+} complex. Moreover, the correlation observed between malate molecules at pH values substantially different from the optimum and also in the absence of Mn^{2+}, pointed to the possibility that the protein of the complex which binds malate was oligomeric. It was also observed (57) that the noncompetitive inhibitors fructose -1,6-diphosphate and L-lactate showed allosteric behevior, i.e. there was cooperative binding of the inhibitors to the protein. It should also be noted that the inhibition produced by oxamate was efficient even at the lower concentration tested.

On the other hand, citric, succinic, and tartaric acids produced an effective competitive inhibition due to their structural similarity to malate. The amount of inhibition produced by these acids, at the usual concentrations encountered in wines, were as follows : L-tartrate, 40 % ; succinate, 10 % ; and citrate, 5 %. At the same concentrations, the inhibition was higher for whole cells than for the extracts prepared from them (57). These characteristics may partially account for the difficulties found in practice during the initiation and completion of the malolactic fermentation in wines when pH, temperature and SO_2 are not limiting.

3.4. Stimulation of malolactic fermentation by inoculation of wine with an industrial biomass of *Leuconostoc oenos* after reactivation

In spite of all the work that has been done, the malolactic fermentation, in practice, remains difficult to control. For a long time, stimulation of the malolactic fermentation by bacterial inoculation has been researched for, but with little success. In the hours following the transfer of bacteria to wine, a large regression of the population is observed (1), varying with the composition of the wine, nature of the bacterial strain and the conditions of its preculture. This phenomenon suggests the importance of the physiological state of the inoculum with regard to its cellular resistance to the wine. Nonetheless this question has never been answered. On the other hand, the problem, to date, has been to provide a sufficient quantity of biomass of a suitable quality.

Recently, a bacterial biomass was prepared industrially for experimentation by "Microlife Technics" (Sarasota, Florida, U.S.A.). It was obtained in a frozen form. Viability was good, about 5.10^{10} viable cells/ml. Nevertheless when used directly after thawing, the viability, after inoculation in the wine decreases greatly. It is necessary, therefore, to reactivate the bacteria before use.

Reactivation consists essentially of bringing the bacteria to the end of the proliferation phase, in a medium inducing, in particular, an important synthesis of reserves of the malolactic enzyme. Two reaction techniques have been suggested (37) :

a) The medium of reactivation is sterile grape must, diluted two-fold with water, enriched with 5 g/l of yeast extract. It is deacidified to bring the pH to 4,5. A suspension of thawed biomass in ten times its volume of this medium is made. The initial population is 2 to 7.10^9 cells/ml. After 24 hours of incubation at 25°C, the culture is in proliferation phase containing from 5.10^9 to 1.10^{10} cells/ml. This is used at 1 in 1 000 vol. of wine, which assures an initial viable population of about 10^6 to 10^7 cells/ml in the wine, necessary to induce the growth within a convenient time. In this case, the growth accompanying the reactivation is limited ; it is not the essential factor seeking for.

b) Suspend the thawed biomass in 10000 times its volume of fresh grape must with its indigenous yeasts, after diluting and enriching with yeast extract as in previous process. The initial population is of the order of 10^6 cells/ml. After 4 to 6 days of incubation at 25°C, the sugar is fermented by the yeasts and the medium contains from 2 to 5.10^9 bacteria cells/ml. It can be used as indicated previously. In this case, there are both gowth and reactivation. This second technique is longer but less onerous. It requires less bacterial biomass and the sterilisation of medium is not necessary.

In spite of their apparant simplicity, these procedures require regular quality control, to avoid bacterial deviations and contamination, mainly by acetic acid bacteria. For increased security, it would be advisable that the reactivation is done by specialised laboratories.

The speed of malolactic fermentation is directly linked to the size of inoculum in the wine. Below 10^6 reactived bacteria/ml, the growth is eventual and the duration of malolactic fermentation is prolonged.

Concerning the most favorable moment for bacterial inoculation during vinification, the problem is very complex and it is impossible, with the current state of our knowledge, to give a valuable solution in all cases (37). It would be necessary to study systematically all the factors and their interactions which control the antagonism between yeasts and bacteria in the grape must : nature of the yeast and bacterial populations, ratio of their initial number, composition of the must, eventual levels of anti-yeast and anti-bacterial substances, level of SO_2, temperature, pH etc... Depending on the conditions, inoculated lactic acid bacteria may be inhibited by the yeast metabolism or inhibit the yeast metabolism (37). For safety, it is recommended that the wine is inoculated after total degradation of hexoses.

3.5. Prevention of lactic acid bacterial alteration

To avoid the "piqûre lactique" it is necessary to prevent bacterial growth in presence of sugar. This is obtained by sulphiting the must with an amount of SO_2 high enough to prevent the growth of bacteria, until the end of sugar fermentation by yeasts. As a matter of fact, it is the yeast metabolism which inhibits bacterial growth. But if, for any reason, alcoholic fermentation stops before all sugar has disappeared, SO_2 will prevent the residual sugar to be fermented by bacteria with production of acetic acid (37).

During storage, lactic acid bacteria metabolism can be controled by considering the level of D(-)-lactic acid in the wine. Lactic acid, the principal product of bacterial metabolism exists in two forms, L(+)-lactic acid and D(-)-lactic acid. The degradation of L-malic acid by lactic acid bacteria gives the stoichiometric formation of L(+)-lactic acid. All the other substrates give simultaneously the two isomers, in variable proportions depending on the bacterial species and the substrate.

Yeasts, as well, form lactic acid during the alcoholic fermentation. But the yeasts generally found in vinification form relatively small quantities, with maximum of 200 mg/l of D(-)-lactic acid.

Detection of lactic acid may be interpreted as follows (16) : (a) high levels of D(-)-lactic acid must be considered as a probable indication of bacterial activity depending upon an alternative substrate to malic acid ; (b) all increases of D(-)-lactic acid levels indicate a potential alteration ; (c) large amount of L(+)-lactic acid without a substantial quantity of D(-)-lactic acid mean a pure malolactic fermentation without alteration.

The determination of this substance by enzymatic dosage is rapid and precise. It makes possible the detection of initial lactic acid bacteria activity before an excessive level of acetic acid appears.

In the wine the incidence of sulphiting merits emphasis. It is not sufficient to sulphur the wine to keep it definitively free from bacterial alterations. The survival of some resistant cells suffices to give a new growth. On the other hand, non proliferating bacterial populations, when important, can liberate appreciable quantities of histamine. So, for all these reasons, it is imperative from the termination of the malolactic fermentation, to eliminate the lactic acid bacteria, as rapidly as possible by sulphuring followed by a filtration. In fact the best "natural" protection of wine against lactic alterations resides in its acid pH and in the relatively low temperature of the storage cellars.

4. SPOILAGE OF MUST AND WINE BY ACETIC ACID BACTERIA

4.1. New concepts

In the field of oenology, acetic acid bacteria have been the subject of a fairly small number of publications (58, 59, 60). This lack of interest can be explained by the fact that, as these microorganisms were considered strictly aerobic, it was thought that they could only develop on the wine air-contact surface and it was sufficient to avoid air penetration to prevent an alteration (1). The application of new microbiological analysis techniques has facilited the selective isolation of acetic acid bacteria (42). A recent work (18) studied the evolution of such populations, both in number and nature, from the grape to the wine during storage. The results have shown acetic acid bacteria are always present ; their growth and metabolism requires only on a small amount of oxygen.

4.2. Occurence of acetic acid bacteria during ripening, vinification and storage of the wine

4.2.1. Ripening of grapes

In some cases the acetic acid bacteria were found in concentrations of approximatively 10^2 cells/ml of must. At the beginning of the harvest they were a little more numerous on grapes infected by *Botrytis cinerea* (18). However, at the end of the harvest, the population were between about 10^4 to 10^6 cells/ml of juice, on spoiled grapes. *G. oxydans* was the only species found on sound red and white grapes. Along with this species, mainly *A. aceti* and also *A. pasteurianus* were found when the grapes were a little damaged ; they constitute 75 to 85 % of acetic acid bacteria on entirely rotten grapes. The studies were carried out during three successive years 1978, 1979 and 1980 ; the density of acetic acid bacteria population was found to be linked to the degree to which the grapes were spoiled.

4.2.2. White wine production

Musts extracted from sound grapes contained small concentrations of acetic acid bacteria, exclusively *G. oxydans*, which became almost totally extinct during alcoholic fermentation. However the populations were larger in musts extracted from grapes infected by *Botrytis cinerea* (pourriture noble). For example (table 17) an infected grape must, whether treated with sulphur dioxide (100 mg/l) or not, fermented spontaneously. The acetic acid bacteria were present in the must in the same concentrations as the yeasts (about 10^6 cell/ml).

Table 17

Evolution of acetic acid bacteria during alcoholic fermentation of must from white grape infected by *Botrytis cinerea* (18)

	Acetic and bacteria (cell/ml.10^3)	Identification
Before fermentation	2000	*G. oxydans* : 80 % *A. pasteurianus* : 20 %
5th day of fermentation	80	*G. oxydans* : 70 % *A. pasteurianus* : 30 %
12th day of fermentation	100	*G. oxydans* : 50 % *A. pasteurianus* : 50 %
20th day of fermentation	0,6	*G. oxydans* : 50 % *A. pasteurianus* : 50 %

After 5 days, when the yeasts population was maximum (about 10^7 cells/ml), 8.10^4 cells/ml of acetic acid bacteria were still viable. Afterwards the number of bacteria decreased ; by the end of the alcoholic fermentation, they were present in concentrations of 10^2 to 10^3 cells/ml. During this time, *G. oxydans* progressively disappeared, while the number of *A. pasteurianus* increased. At the end of the fermentation the residual population were composed exclusively of *Acetobacter*.

4.2.3. Red wine production

After crushing Cabernet Sauvignon grapes the must contained 10^2 cells/ml of acetic acid bacteria (table 18) ; this number decreased during fermentation to just a few cells/ml. After the draining the wine from the tank, the population increased from 20 to 30.10^3 cells/ml ; the cause of this increase was an infection due to contact with the equipement used during the operation.

G. oxydans was progressively replaced by *A. pasteurianus* (table 18) ; the two species were present in almost equal quantities by the end of the alcoholic fermentation. After draining, *A. pasteurianus* became predominant.

Table 18

Evolution of acetic acid bacteria during
alcoholic fermentation of crushed red grapes (18)

		Acetic acid bacteria (cell/ml.10^3)	Identification	
Before fermentation		16	100 %	*G. oxydans*
3th day of fermentation		3	75 % 25 %	*G. oxydans* *A. pasteurianus*
7th day of fermentation		0,1	55 % 45 %	*G. oxydans* *A. pasteurianus*
10th day	before draining	0,02	30 %	*G. oxydans*
	after draining	30	70 %	*A. pasteurianus*

The wine from this experiment was next stored in other tanks at 18°C. During the malolactic fermentation which lasts ten weeks, the acetic acid bacteria population stayed constant at about 10^2 cells/ml. After the malolactic fermentation the wine was stabilised by addition of 20 mg/l of SO_2, then filtered and stored in wooden barrels. The filtration removed the yeasts and lactic acid bacteria. The number of acetic acid bacteria was only very slightly decreased, from 640 to 400 cells/ml. Samples were analized every week during the eleven months long conservation of the wine in barrels at 10°C (table 19). Between 10^2 and 10^4 cells/ml of acetic acid bacteria still survived in the same concentrations at the top and at the bottom of the barrels ; these were always A. *aceti sp aceti*.

During this period wine was added twice a week to the barrels in order to compensate evaporation. In one of them, G. *oxydans* was again identified, immediately after this operation. A week later this species has disappeared ; it suggests that its presence was due to an infection brought, either by the wine or by the equipement used in filling barrel.

Table 19

Evolution of acetic acid bacteria during malolactic fermentation and storage in barrels of red wine (61)

	Acetic acid bacteria (cell/ml.10^2)	Identification
Malolactic fermentation		
beginning	4	20 % *G. oxydans* 60 % *A. pasteurianus* 20 % *A. aceti*
at the end	1,2	20 % *A. pasteurianus* 80 % *A. aceti*
Storage in barrels		
4 months	1,6	20 % *A. pasteurianus* 80 % *A. aceti*
6 months	9	10 % *G. oxydans* 30 % *A. pasteurianus* 60 % *A. aceti*
9 months	6	20 % *A. pasteurianus* 80 % *A. aceti*
11 months	6	100 % *A. aceti*

4.3. Effect of the metabolism of acetic acid bacteria

4.3.1. Metabolism of acetic acid bacteria in grape must

The growth cycle and metabolism for one strain of *G. oxydans* and one of *A. aceti*, in the grape must, have been described (61). For the two microorganisms, the proliferation phase lasts 5 days (figure 11). The maximum population is above 10^8 cells/ml. The decline phase is faster for *G. oxydans*.

The degradation of glucose is slower during the proliferation phase of *A. aceti* than of *G. oxydans*, about 70 % of fermented glucose is transformed in gluconic acid. For the two strains, the degradation of fructose starts during the decline phase. Assimilation is limited with a maximum of about 10 mg/l being degraded.

The preferential substrate for acetic acid bacteria in grape must is glucose. Fructose is degraded only after a period of adaptation. The affinity of acetic acid bacteria for glucose explains the important levels of gluconic acid found in must infected by *Botrytis cinerea* ; the mold splits the skin of the berry and makes possible the growth of acetic acid bacteria in contact with the juice. The same work (61) stated that A. *aceti* exclusively degrades partially, malic acid during the proliferation phase and citric acid during the decline phase. Another result is the formation in the must, starting from a small amount of ethanol (1,5 g/l), of acetic acid during the proliferation phase.

It disappears by oxydation only in the culture of A. *aceti* during the decline phase.

As a consequence, the microorganism responsible for the high volatile acidity of musts extracted from spoiled grapes seems to be essentially G. *oxydans* which does not possess this property.

During the growth of acetic acid bacteria, there is also formation of a few mg/l of acetaldehyde, D(-)-lactic acid, L(+)-lactic acid and succinic acid.

The influence of bacterial development on the formation of compounds which combine SO_2 was also studied (61). In the culture medium of A. *aceti* compounds were not produced. With G. *oxydans*, bound SO_2 increased during the decline phase but was limited. At the end of the cultivation keto-2-gluconic acid and keto-5-gluconic acid were found in the medium but these substances do not combine with SO_2 (62).

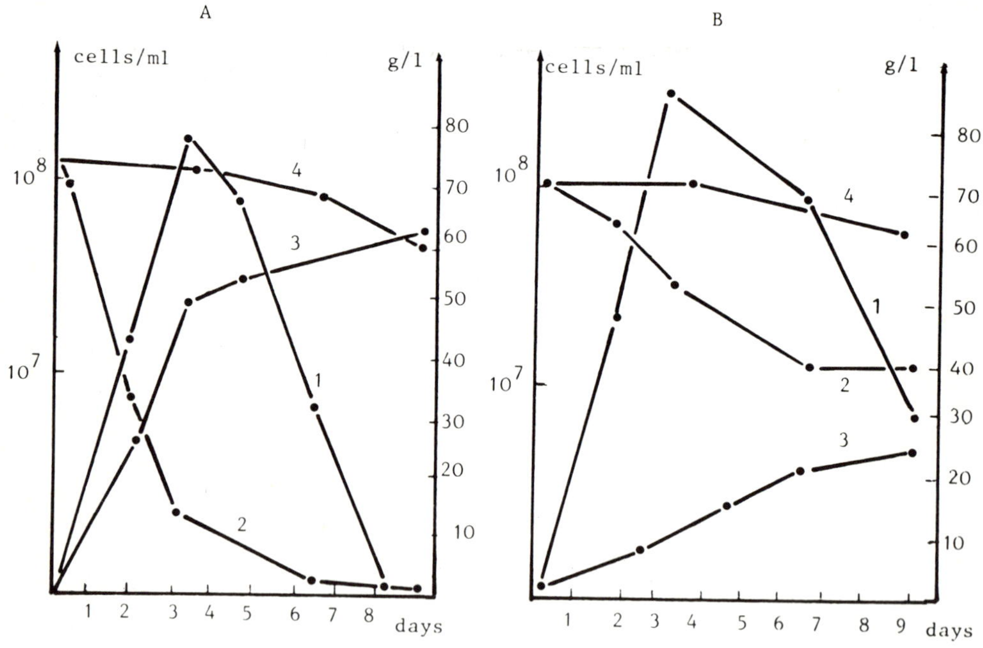

Fig. 11 - Growth and metabolism of G. *oxydans* (A) and A. *aceti* (B) in the grape must : (1) growth curve (cells/ml) ; (2) changes in glucose (g/l) ; (3) changes in gluconic acid (g/l) ; (4) changes in fructose (g/l).
(A. JOYEUX, Analyse microbiologique des raisins, des moûts et des vins. Application à l'étude des bactéries acétiques, (1983). Thèse Doctorat 3ème Cycle, Université de Bordeaux II).

In oenology it is widely believed that the degradation of hexoses by acetic acid bacteria is accompanied by the formation of ketonic substances which are responsible for the excessive levels of combination of SO_2 in musts and wines from damaged grapes (1). The results described above, however, appear to weaken this statement.

4.3.2. The consequence of the metabolism of acetic acid bacteria on the alcoholic and malolactic fermentations

The alcoholic fermentation of the grape must is affected by a preculture during 7 and 14 days with certain strains of *G. oxydans* and *A. aceti* (61) (table 20). A decrease in the total yeast growth and an early arrest of the fermentation were observed.

The wines obtained in this experiment (table 20) were then inoculated with *L. oenos* (table 21). A stimulation of malolactic fermentation was observed (table 21) in all the wines from cultures of *A. aceti*. With *G. oxydans*, the same stimulation was observed with the 3 days culture and to a lesser extent with the 7 day culture. In wines from 14 days cultures of *G. oxydans* the malolactic fermentation was delayed. Generally the development of acetic acid bacteria in must cause a stimulation of the malolactic degradation.

Table 20

Influence of the length of culture of *G. oxydans* and *A. aceti* in regard to the fermentability of musts (61)

Yeasts : *S. cerevisiae*. Fermentation at 25°C under light CO_2 pressure (initial levels of hexoses per l : glucose 85 g ; fructose 85 g)

Acetic acid bacteria			Caracteristics of wines at the arrest of the alcoholic fermentation (18 days)		
Lenght of culture	Species	Collection number	Total yeasts (cell/ml. 10^6)	Residual hexoses (g/l) glucose	fructose
0	0		58	0	0,6
3 days	*G. oxydans*	50	57	0	1,8
	G. oxydans	310	58	0	1,2
	A. aceti	241	56	0	1,7
7 days	*G. oxydans*	50	48	1,32	9,1
	G. oxydans	310	42	0,9	6,6
	A. aceti	241	52	0,29	4,6
14 days	*G. oxydans*	50	48	2,53	13,5
	G. oxydans	310	47	2,93	18,5
	A. aceti	241	48	1,21	9,4

Table 21

Influence of acetic acid bacteria metabolism in the must in regard to the malolactic fermentation of resulting wines (61)

(initial inoculum : 10^7 cell/ml of *Leuconostoc oenos*)

Acetic acid bacteria			Evolution of the malolactic fermentation of wine by *L. oenos*			
			7th day		9th day	
Time of culture prior to alcoholic fermentation	Species	Collection number	cell/ml $.10^7$	malic acid fermented (g/l)	cell/ml $.10^7$	malic acid fermented (g/l)
0	0		65	3,18	10	3,89
3 days	*G. oxydans*	50	82	4,85		
	G. oxydans	310	54	4,20		
	A. aceti	241	92	4,08		
7 days	*G. oxydans*	50	80	3,70	40	4,20
	G. oxydans	310	85	3,42	18	4,10
	A. aceti	241	82	4,51	40	
14 days	*G. oxydans*	50	62	2,27	7	2,58
	G. oxydans	310	13	1,45	3	2,57
	A. aceti	241	86	4,63	2	

4.4. Relationship between the growth of acetic acid bacteria and the production of acetic acid in the wine

Some observations indicate, when air penetration allows the residual population (about 10^2 cell/ml) of acetic acid bacteria existing in wine to grow to about 10^4 cell/ml, acetic acid increases about from 0,3 g/l to 0,7 g/l.

More occurate results come from a laboratory experiment. The wine sample was taken from winery conditions. The growth of lactic acid bacteria was inhibited by the addition of penicillin. The action of pH (3,4-3,8), temperature (10 - 18°C) and momentary aeration (3 mn) were checked (table 22).

After two weeks, in the wines stored under completely anaerobic conditions, at both temperatures, the acetic acid bacteria population decreased faster at the lower pH. The level of acetic acid stayed constant in all conditions.

In the case of momentary aeration, wines were then stored under anaerobic conditions. One week after air introcuction, the acetic acid bacteria population had multiplied 3 or 30 times, at temperatures of 10 and 18°C respectively ; the level of acetic acid in the wine increased from 0,37 g/l to 0,52 g/l, according pH value and temperature.

The presence of oxygen is indispensable for the growth of acetic acid bacteria. However during its storage in barrels, the wine contains a small residual population which progressively decreases, when, as in bottles, the conditions become more strictly anaerobic.

Table 22

Production of acetic acid in wine in relation to the growth of A. *aceti* and A. *pasteurianus*. Action of momentary aeration at the beginning of the experience (61)

(The aeration occurs one week after the beginning)

Time of experimentation	Temperature of storage	pH	Acetic acid bacteria (cell/ml)	Acetic acid (mg/l)
At the beginning			5.10^2	370
After two weeks of incubation	10°C	3,4	15.10^2	380
		3,8	15.10^2	420
	18°C	3,4	10.10^3	500
		3,8	20.10^3	520

Nevertheless a very small population of about 10^2 cells/ml can, if a favorable aerobic vonditions occurs, produce acetic acid in measurable quantities and can have a damaging effect on the quality of the wine. For this to take place, a contact of only 2 or 3 minutes between wine and air is enough ; as it was believed, a long contact between wine and air is not indispensable. In normal wine making procedure, an aeration can occur during the normal racking processes. The temperature and, to a lesser degree, the pH value are the most important factors which command the growth of bacteria and their production of acetic acid after such an aeration. Free SO_2, in concentrations generally used for red wine storage, does not sufficiently protect the wine in barrels from the metabolism of acetic acid bacteria. The practical consequences which arise from these results are : (a) the necessity of a low temperature (15°C) for the storage of the wine in barrels ; (b) the necessity of avoiding any aeration at higher temperatures ; however in new red wine this aeration is indispensable to take off carbon dioxide and for the normal evolution of phenolics compounds involved in the red colour (63).

CONCLUSION

Wine fermentation may be seen as the first biotechnology, even older than that of bread making. All the many prestigious wines produced throughout the world have been created empirically. Therefore one easily attributes the merits of wines quality to the natural specific conditions (soil, climat) ; sometimes the role of scientific oenology is questionned. A precise answer may be supplied to this question. The quality of wines has not ceased to improve and to diversify, attesting knowledge of wines microbiology, biochemistry and chemistry. Modern oenology is required to obtain better quality, regularly and with certainty and by the most efficient means. Moreover oenology has provided a source of a comprehensive program of fundamental research which is of a general interest (1).

REFERENCES

1 J. Ribéreau-Gayon, E. Peynaud, P. Ribéreau-Gayon and P. Sudraud, Sciences et Techniques du Vin, Tome II, Dunod Press, Paris, 1975

2 S. Lafon-Lafourcade, F. Larue and P. Ribéreau-Gayon, Appl. Environ. Microbiol., 38 (1979) 1069-1073.

3 C. Geneix, S. Lafon-Lafourcade and P. Ribéreau-Gayon, C.R. Acad. Sc., 296 (1983) 943-947.

4 K. Nordström, J. Inst. Brew, 70 (1964) 233-242.

5 P. Barre, VIe Symposium Internation on Yeasts, Montpellier, 1978.

6 E.A. Bevan and M. Makover, in S.G. Geets (Ed), Genetics today, XIth Int. Congr.Genet., Pergamon Press, Oxford, 1963.

7 S. Hara, Y. Iimura and K. Otsuka, Am. J. Enol. Vitic., 31, (1980) 28-33

8 K. Ouchi, N. Kawase, S. Nakaso and H. Akiyama, Agric. Biol. Chem., 42 (1978), 1-5.

9 R.G.E. Palfrey and H. Bussey, Eur. J. Biochem 93 (1979); 487.

10 E.J. Middelbeek, J.M.H. Hermans, C. Stum and H.M. Muytjens, Antimicrobiol. Agents and Chemotherapy, 17 (1980) 350-354.

11 H. Bussey and D. Sherman, Biochem. Biophys. Acta, 298 (1973) 868.

12 E.J. Middelbeek, J.M.H. Hermans and C. Stumm, Antonie Van Leeuwenhoek, 45 (1979) 437.

13 P. de la Pena, P. Barros, S. Gascon, S. Ramis and P.S. Lazo, Biochem, Biophys. Res. Comm., 96 (1980) 544.

14 F. Radler, Bull O.I.V., 53 (1980) 568-572.

15 L.V. Tiourina, N.I. Buryan and T.K. Skorikova, Bull O.I.V., 53 (1980) 573.

16 S. Lafon-Lafourcade, in P. Ribéreau-Gayon and P. Sudraud (Eds), Actualités oenologiques et viticoles, Dunod Press, Paris 1983.

17 S. Lafon-Lafourcade, E. Carre and P. Ribéreau-Gayon, Appl. Environ Microbiol., (1983) 46, 874.

18 S. Lafon-Lafourcade and A. Joyeux, Bull O.I.V., 54 (1981), 803-829.

19 P. Ribéreau-Gayon, S. Lafon-Lafourcade, D. Dubourdieu, V. Lucmaret and F. Larue, C.R. Acad Sc; 289 (1979) 441-444.

20 F. Larue, S. Lafon-Lafourcade and P. Ribéreau-Gayon, Appl. Environ. Microbiol. 39 (1980) 808-811 and Ann. Microbiol. (Institut Pasteur), 130A (1979) 231-243.

21 P. Ribéreau-Gayon, S. Lafon-Lafourcade and A. Bertrand, Conn. Vigne Vin, 9 (1975) 117-139.

22 J.M. Navarro and G. Durand, Ann.Microbiol. (Institut Pasteur), 129 (1978) 215-221

23 F. Larue, S. Lafon-Lafourcade, P. Ribéreau-Gayon, C.R. Acad Sci., 294 (1982) 587-590.
24 C. Geneix, S. Lafon-Lafourcade and P. Ribéreau-Gayon, Conn. Vigne Vin, 17 (1983) 205-217.
25 A.A. Andreasen and T.J.B. Stier, J. Cell Comp. Physiol., 41 (1953) 23-36
26 R.A. Demel and B. de Kruyff, Biochem. Biophys. Acta, 457 (1976) 109-132
27 P.P. Slonimski, Proc. Inter. Congr. Biochem. Bruxelles, Acad. Press, New York 1955 242-247.
28 R.H.De Deken, J. Gen. Microbiol. 44 (1966) 149.
29 B.H. Kirsop, J. Inst. Brew.,80 (1974) 252-259.
30 P. Bréchot, J. Chauvet, P. Dupuy, M. Croson and A. Rabatu, C.R. Acad. Sc., 272 (1971) 890-893.
31 F. Radler, Conn. Vigne Vin, 2 (1968) 271-294
32 A.H. Rose and J.S. Harrisson, the Yeasts, Academic Press,London 1968.
33 L. Teinhard und F. Radler, Lebensmittel Untersuchung und Forschung, 172 (1981) 278-283
34 S. Lafon-Lafourcade, Ann. Fals. Exp. Chim. , 67 (1974) 275-282
35 L. Ferre, Traité d'OEnologie Bourguignonne, INAO Press, Paris 1958.
36 J. Ribéreau-Gayon et E. Peynaud, Traité d'OEnologie, Tome I, Beranger Press, Paris 1960.
37 S. Lafon-Lafourcade, E. Carre, A. Lonvaud-Funel and P. Ribéreau-Gayon, Conn. Vigne Vin, 17 (1983) 55-71.
38 G.J. Pilone in J.G. Carr, C.V. Cutting and G.C Whiting (Eds), Lactic acid bacteria in Beverages and Food, Academic Press, London 1975.
39 Y. Chalfan, I. Goldberg and R.I. Mateles, J. Food, Sci., 42 (1977) 939-943.
40 R. Maret and T. Sozzi, Ann. Technol. Agric., 27 (1977) 255-273.
41 R. Maret, T. Sozzi and D. Schellenberg, Ann. Techn. Agric., 28(1979) 41-55.
42 S. Lafon-Lafourcade et A. Joyeux, Conn. Vigne Vin, 13 (1979) 285-310.
43 E. Peynaud, S. Lafon-Lafourcade and S. Domercq, Bull. O.I.V., 38 (1965) 947-958 .
44 M. Feuillat, P. Bidan and Y. Rosier, Ann. Technol. Agric., 26 (1977) 435-44 .
45 T. Amachi, in J.C. Carr, C.V. Cutting and G.C. Whiting , Lactic acid Bacteria in Beverages and Food, Academic Press, London 1975.
46 S. Lafon-Lafourcade, V. Lucmaret and A.Joyeux, Conn. Vigne Vin, 14 (1980) 183-194 .
47 S. Lafon-Lafourcade, Ann. Technol. Agric., 19 (1970) 141-154.
48 M. Feuillat and G. Bureau, C.R. Acad. Agric., 65 (1979) 647-651.
49 S. Lafon-Lafourcade, Conn. Vigne Vin, 7 (1973) 203-207.
50 T. Sozzi, F. Gnaegi, N. d'Amico and H. Hose, Revue Suisse Vitic. Arboric. Hortic.,14 (1982) 17-23.
51 S. Ochoa, A.H. Maheler and A.J. Kornberg, J. Biol. Chem., 174 (1948) 979-1000.
52 E. Peynaud, C.R. Acad. Sci., 257 (1968) 121, 122.
53 M. Schütz and F. Radler, Arch. Mikrobio., 91 (1973) 183-202.
54 M.A. Alizade and H.Simon, Hoppe-Seyler's Z. Physiol. Chem., 354 (1973) 163-168.
55 R.E. Kunkee, in J.G. Carr, C.V. Cutting and G.C. Whiting (Eds), Lactic acid bacteria in Beverages and Foods, Academic Press, London, 1975.
56 A. Krauss, W. Dessau and H. Simon, Hoppe-Seyler's Z Physiol. Chem. , 357 (1976) 1209-1214.
57 A. Lonvaud-Funel and A.M. Strasser de Saad, Appl. Environ,Microbiol.,43 (1982) 357-361.
58 P. Dupuy, Ann. Techn. Agric., 6, (1957) 217-233.
59 P. Dupuy and J. Maugenet, Ann. Techn. Agric., 11 (1962) 219-225.
60 P. Dupuy and J. Maugenet, Ann. Techn. Agric., 12 (1963) 5-14.
61 A. Joyeux, (1983), Thèse d'Université, Université de Bordeaux II.
62 F.N.Beech, Bull O.I.V. 52 (1979) 1002-1022.
63 P. Ribéreau-Gayon, P. Pontallier and Y. Glories, J. Sci. Food. Agric., 34 (1983) 505-516.
64 M. Lonvaud, A. Lonvaud-Funel and P. Ribéreau-Gayon, Conn. Vigne Vin, 11 (1977) 73-91.

CURRENT APPROACHES TO BREWERY FERMENTATIONS

By: J. MacDonald, P.T.V. Reeve, J.D. Ruddlesden and F.H. White
Bass Brewing Ltd., High Street, Burton on Trent, Staffordshire,
and Bass Brewing (Runcorn) Ltd., Aston Lane, Preston Brook, Runcorn, Cheshire

CONTENTS

A. INTRODUCTION

Although the origins of beer brewing are lost in antiquity (most sources suggest sometime between 4,000 and 6,000 BC (1)), it was not until the latter half of the last century that there emerged any real appreciation or rational understanding of the process which lies at the very centre of beer production - the transformation by yeast of brewers' wort into beer. Thus, up to a hundred years ago, although the physical characteristics of the fermentation process were well recognised, in particular the evolution of gas and the formation of foam, yeast was known only as a thick mass of either surface froth or sediment, so that its role as the causative agent of fermentation was only guessed at and surrounded in mystery. The brewer's approach to the fermentation process was therefore one based almost entirely on empirical observation and pragmatism.

It was the revolution in basic knowledge stemming from the classic studies of first Pasteur (2,3) and later Hansen (4) into respectively the nature of fermentation and the biology of yeast, which gave the brewer for the first time a rational basis on which to conduct and control his fermentations. The implications and ramifications of this work were both immediate and far reaching. It resulted in a major emphasis on microbiological control of the brewing process, and established the principles on which modern fermentation practice and yeast management are based. Moreover, it stimulated a scientific interest in beer and brewing which has continued unabated to the present day, and has given the brewer an ever increasing appreciation of the principles on which his operation is based, and the manner in which both raw materials and processing conditions determine the character of finished beer. The end result, therefore, of almost a century of scientific endeavour, has been to give the brewer a much improved ability to control his process, in order to produce consistently beers with the characteristics his customers desire.

It is this increased understanding of the influence and effects of fermentation conditions and yeast performance on beer quality, together with the perennial demand on the brewer to produce beers of good, uniform, and dependable quality, which together are the prime determinants of current approaches towards brewery fermentations. Accordingly, a major section of this chapter reviews in detail the current level of understanding with regard to the interrelationship between the fermentation process and beer quality, and the manner in which increased knowledge of this relationship has modified the brewers approach, both to the fermentation process itself, and to the design of plant and systems.

Fundamental though beer quality is in dictating the brewer's approach to fermentation, it is also important to realise and appreciate the other factors affecting his approach. Thus, the huge technological advances made over the last quarter of a century, in terms of both vessel fabrication and control systems, have increased considerably the options available with regard to vessel design, vessel size, and automation of control and cleaning systems. In addition, advances in brewing technology have not only shown benefits in terms of improved yeast propagation and handling, but have also given the brewer increased opportunities to optimise the resources of his fermentation plant, particularly with regard to throughput.

The impetus for these advances has been in the main financial; over the last two decades brewers have become increasingly cost conscious and are required to manage all facets of the production process in order to maintain competitive unit costs. However, the requirement for a cost effective approach to fermentation has different implications (and is somewhat easier to achieve) when beer sales are increasing, than when as at the time of writing, beer volumes are in decline.

Another major factor influencing the brewers approach to fermentation is the voice of the consumer, and the need to meet his or her changing and changeable requirements - in terms of product range (lagers, ales, and stouts), nutritional requirements (low carbohydrate, alcohol-free), or package type (cask, keg, bottle or can). Fermentation systems which are flexible in their ability to produce beers capable of meeting these various criteria, are therefore an essential feature of any successful modern brewing operation.

In the sections which follow, the influence of these and other factors determining current approaches to brewery fermentations are reviewed with respect to their effect on the development of modern fermentation systems. This is followed by a critical examination of the various alternatives available with regard to modern fermentation plant and systems. For reasons of familiarity, we have of necessity concentrated on the approaches of British brewers, and developments within the United Kingdom brewing industry. Nevertheless, where relevant, reference is made to the systems, plant, and approaches, of brewers throughout the world.

We begin by defining what constitutes the fermentation process in modern brewery practice, together with a brief reference to brewers' wort and its main constituents. This is followed by a short review of each of the component stages in brewery fermentations and their main objectives.

B. THE FERMENTATION PROCESS

B1. Definition of The Fermentation Process

In a modern, integrated brewery the limits of that part of the process comprising the fermentation system, are not easily recognised. This is especially the case in breweries producing traditional British 'draught' beers, since fermentation is still continuing in such beers when they are in their final container, and even when they have left the brewery and are at the point of sale. For the purposes of this review, it is convenient to define the fermentation system as the plant and processes required to convert hot, sterile, fermentable feedstock (brewers' wort) into a wholesome, potable, alcoholic beverage, with the required characteristics of taste (flavour and aroma), and appearance (colour, clarity and foaming potential).

As the main agent for this conversion is brewers' yeast, (strains of either Saccharomyces cerevisiae or S.uvarum (5)), consideration will also be given to the selection, maintenance, and practical handling of yeast in a brewery.

However, before reviewing in outline the component stages of the fermentation process and their objectives, brief reference must be made to the raw material of brewery fermentations: wort, and in particular its composition and origin.

B2. Wort

Wort is primarily an aqueous extract of malted barley, containing complex mixtures of sugars and proteinaceous material which vary widely in molecular weight. Also present are a range of mineral ions, vitamins and growth factors, and various classes of lipid material. In addition, the carbohydrate fraction of wort may be supplemented with refined sugar or starch, derived from a wide range of alternative sources.

In combination, these various materials provide a complete source of nutrients to brewers' yeast, which are thus able to metabolise the majority of wort constituents. Those constituents which are not metabolised survive into the final beer and contribute to beer flavour (eg. sweetness) or appearance (eg. foaming properties).

Wort also contains bittering compounds and flavour-active essential oils, both derived from hops during the wort production process. These have little or no nutrient value to yeast, but are significant in suppressing the growth of other microorganisms and in contributing to the flavour of the final beer. (For a detailed review of wort composition see MacWilliam (6)). The strength of wort is commonly measured by reference to its specific gravity, albeit expressed in various manners around the world. In general terms, an all-malt wort of SG 1.040 (1040° in United Kingdom brewing parlance or 10° Plato in some continental countries (7)) contains approximately 10%(w/w) sugars. The disaccharide maltose predominates (~45%), followed by maltotriose (~15%), glucose (~10%), sucrose (~5%), and fructose (~2%). The remainder (~23%) consists of dextrins, starch breakdown products containing upwards of four glucose units. These are not utilised by strains of brewing yeast, and persist unchanged in the majority of finished beers (cf. Section D2.5.1).

B3. Stages in the Fermentation Process

Despite the great variety of systems currently in use, a common succession can be identified in all, as follows:-

i Wort Cooling,

ii Wort Collection,

iii Primary Fermentation, and

iv Secondary Fermentation (warm conditioning or maturation).

At the end of the fermentation process, beers which are to be filtered are normally given a period of cold storage (or cold conditioning) prior to filtration.

In addition to the stages listed above, all systems employ some means of yeast separation and handling.

B3.1 Wort Cooling

Although arguably not directly a part of the fermentation system, the wort cooling process plays an essential part in fermentation control.

The hot wort produced by the brewhouse should be an essentially complete feedstock for the yeast cell. One of the objectives of wort boiling is to sterilise this feedstock, and sterility is maintained by storage and handling at high temperatures up to the point of wort cooling.

The principle objectives of the wort cooling process are:

i to reduce the wort temperature to that desired for the start of fermentation;

ii to introduce a controlled level of oxygen into the wort;

iii to maintain sterility;

iv to precipitate from the wort excess proteinaceous material (cold-break) insoluble at fermentation temperatures.

In addition, rising energy costs have forced a fifth requirement on the cooling process: to recover the maximum amount of utilisable energy from the hot wort.

B3.2 Wort Collection

Wort collection describes that phase of the fermentation process during which the volume of cooled wort is measured and the yeast added. This may or may not be carried out in a separate vessel to that in which the fermentation is to be conducted, depending on practice and circumstance.

Many brewers, particularly those producing traditional lagers, also use this stage to remove the cold-break formed during wort cooling. Additional roles for collection vessels are fulfilled in breweries where the main fermentation is carried out in a number of small vessels (eg. the Burton Union System - see Section E5.2), when the collecting vessel serves as a convenient stage in the process in which to standardise the strength of the wort ex-the brewhouse, by admixture with water, and in which to dispense the yeast uniformly, prior to sub-dividing into the smaller vessels.

B3.3 Primary Fermentation

During primary fermentation yeast is allowed to grow and ferment in a controlled manner, in order to produce the majority of the alcohol, carbon dioxide (CO_2), and flavour compounds required in the final beer. This is normally achieved by controlling the temperature of fermentation, and by permitting a controlled ingress of air or oxygen, the extent of which is determined by the requirements of the particular yeast strain (see Section C3.3.2). In addition, access of spoilage microorganisms is restricted to a minimum to avoid unwanted (and uncontrolled) flavour development.

Primary fermentation is normally considered to be complete once the majority of the wort fermentable sugars have been utilised. At this stage most strains of brewers yeast flocculate (8) and come out of suspension, either by rising to the surface and forming a thick yeast-head (top-fermenting yeast) or by sedimenting to the base of the fermenting vessel (bottom-fermenting yeast). In either case, because of the reduction in numbers of yeast cells in suspension, the rate of utilisation of the residual sugars slows. However, in some systems primary fermentation is deliberately curtailed at a point when up to approximately 10% of the fermentable sugars are still remaining. This is in order to promote CO_2 gas formation and yeast activity during secondary fermentation. Nevertheless, at whatever point is designated the end of primary fermentation, the bulk of the yeast crop is removed, leaving only sufficient yeast in suspension to carry out the required degree of secondary fermentation.

B3.4 Secondary Fermentation

In order to produce a finished beer from the raw or 'green' beer remaining at the end of primary fermentation, it is necessary to:

i Allow flavour maturation by continued yeast metabolism.

ii Develop the amount of CO_2 in solution to the required level, by fermentation of residual or added sugars.

iii Allow unwanted volatiles to be removed in the gas stream formed by the generation and release of excess CO_2.

iv Separate excess yeast from the beer.

In addition, at this stage, some pale ales have their hop aroma enhanced by the further addition of hops to the 'green' beer.

B3.5 Cold Conditioning

At the end of the fermentation process, beers which are to be filtered prior to sale normally receive a period of cold storage, known as cold conditioning. This may occur within the same vessel as secondary fermentation, or (in the case of traditional lager brewing) at the same time.

The objectives of cold conditioning are as follows:

i To reduce the haze-forming potential of the filtered beer by precipitating protein/polyphenol complexes insoluble at low temperatures;

ii To sediment excess yeast, possibly with the aid of collagen (isinglass or bovine) finings (9).

These are achieved by storage under an inert atmosphere at $0^{o}C$ or below, for a varying period of time.

B3.6 Yeast Handling and Storage

Yeast is not only the causative agent but also a major by-product of fermentation. Its recovery in good condition and subsequent handling, are essential factors in ensuring the health of future fermentations, and also exert a major influence on the economics of the brewing process.

The main objectives of any yeast removal and handling system are as follows:

i the removal of excess yeast at the end of each fermentation;

ii minimum entrainment of beer with yeast in order to minimise beer losses;

iii the maintaining of yeast in a clean and microbiologically pure state, particularly when it is to be used in subsequent fermentations (pitching yeast);

iv the maintaining of yeast in a viable and healthy state during storage;

v the optimisation of treatment and handling of surplus yeast in order to maximise the gain to the brewery resulting from its sale, and to simplify its handling for transport.

C. THE EFFECTS OF FERMENTATION CONDITIONS AND YEAST PERFORMANCE ON BEER QUALITY

The increases in understanding and basic knowledge of yeast metabolism and the nature of fermentation have made their greatest impact in more closely defining the relationship between fermentation conditions and beer quality. Accordingly, in this section we review in some detail the current understanding of the various aspects of this relationship, and also attempt to relate the manner in which increased knowledge has modified the brewers approach to fermentation.

The term 'beer quality' is used in general terms to describe the overall flavour and appearance of a beer. In addition to being major determinants of the flavour of finished beer, fermentation conditions and yeast performance can also affect the visual appearance of a beer, in particular its potential to produce a good 'head' of foam when poured into a glass, and its clarity or brilliance. We therefore start by reviewing briefly the influence of yeast and fermentation conditions on these physical characteristics of beer, before turning to the relationship between yeast, fermentation, and flavour.

C1. Beer Foaming Potential

The ability of a beer to form a good 'head' of foam when dispensed into a glass is determined by the ratio of 'foam - positive' material to 'foam-negative' material in the finished beer (10). Both these types of material are derived and extracted from malt during wort production, and attempts at their control centre on both maximising the removal of 'foam-negative' material (mostly in the brewhouse by ensuring good 'break' formation and removal), and on minimising the loss of 'foam- positive' material throughout beer production.

Fermentation conditions can have a significant influence on the level of 'foam-positive' material in beer, because the surface-active properties of this material result in its concentration and removal from solution whenever foaming occurs. Thus losses occur during fermentation, in the foam produced at the surface of actively fermenting worts, and also during the subsequent transfer of beer between vessels where, depending on the degree of turbulence involved, foam break-out can be significant. This latter can be a particular problem in large modern breweries, in part because of the distances involved in the transfer of beer in such plants, and the rates at which beer is moved, but also because of the generally increased degree of pumping and pressure fluctuations to which beer is subjected in modern plants, compared with more traditional operations. Hence, there is a preference in modern fermentation systems for fermenting vessels with reduced surface/volume ratios, and for minimising the numbers of vessels and/or transfers involved in beer production. Also, in the case of chilled and filtered beer, process foam suppressants or 'antifoams' are widely used, to minimise foam formation in beer during fermentation and later processing (see Section D2.5.2).

C2. Beer Clarity

Whilst not directly determined by fermentation conditions, the degree of clarity, or brilliance, of a beer is dependent on it being free from yeast. The ability of brewers' yeast to separate out of suspension at the end of fermentation, and the manner in which this is achieved, have therefore always been important characteristics of particular yeast strains. (For reviews of yeast flocculation and the possible mechanisms involved see the articles by Rainbow (8) and Stewart (11)). Nowadays the widespread use of centrifuges and filters in order to achieve consistent and convenient control over both yeast concentrations and yeast removal, has somewhat lessened the dependency of the brewer on the flocculation characteristics of his yeast. Nevertheless, the ability of yeast to flocculate is still an important feature of the rough separation, or 'cleansing', of yeast from beer, required at the end of primary fermentation.

Another property of brewers' yeast important in ensuring beer clarification is its response to the action of finings agents, in particular collagen finings (9). Use of this material to accelerate yeast sedimentation at the end of secondary fermentation is still common practice in the United Kingdom, either as a preliminary to filtration, or as the sole means of achieving beer clarification when beers are conditioned in cask. (For a review of the action of finings and their use in present day brewing operations see Ballard (12)).

C3. Beer Flavour

By far the most important single influence of the fermentation process on beer quality lies in its effect on beer flavour. The flavour of a beer is perceived by the consumer as the twin sensations of aroma and taste, and it is in the formation of the flavour-active compounds comprising the aroma of beer that fermentation conditions play an especially dominant role. This is because the majority of the volatile flavour-active components of beer are formed during fermentation, as by-products of yeast metabolism. These components can be classified as follows:

i higher alcohols,
ii esters,
iii aldehydes and ketones,
iv inorganic and organic sulphur-containing compounds,
v organic acids, and
vi fatty acids.

Before reviewing in detail the formation and control of these quantitatively minor products of the fermentation process, brief mention should also be made at this point of the main volatile component of beer - carbon dioxide.

C3.1 Carbon Dioxide

As well as exerting a taste sensation in its own right (variously described as 'CO_2-tingle' or 'carbonation'), the level of carbon dioxide in a beer has an important influence on the degree of perception of most of the other flavour-active constituents. Moreover, in addition to its role as a flavour enhancer, the CO_2 content of a beer also contributes to the perceived 'fullness' or 'body' characteristics, and enhances the foaming potential (13). Thus, the effect of CO_2 on beer quality - both flavour and appearance - is such that its required level in a beer is very tightly controlled.

C3.2 Higher Alcohols

Higher alcohols contribute to the flavour and aroma of beer, not only by virtue of their concentration in beer, but also because they provide the alcohol moieties for certain esters. The most important higher alcohols are n-propanol, iso-butanol, 2-methyl-1-butanol (optically-active amyl alcohol) and 3-methyl-1-butanol (iso-amyl alcohol). These compounds have flavour threshold values in beer of 600, 100, 50 and 50mg/litre respectively (14). Higher alcohols impart an 'alcoholic' or 'solvent-like' aroma to beer, and produce a warming effect on the palate. The aromatic higher alcohols -tryptophol, tyrosol and phenylethanol - have flavour threshold values of 10, 20 and 40mg/litre respectively (14,15). Phenylethanol has a fragrant aroma and may contribute positively to beer flavour, whereas tryptophol and tyrosol impart a 'bitter' or 'yeast-bitten' character which is generally regarded as undesirable (16).

C3.2.1 Formation of Higher Alcohols

The higher alcohols in beer are formed during fermentation by both catabolic and anabolic routes. In the Ehrlich, or catabolic, pathway wort amino acids are taken up by yeast and transaminated to give α-keto acids, which in turn are decarboxylated and reduced to alcohols (see Fig. 1).

The anabolic pathways originate with pyruvate and acetyl CoA, formed by the metabolism of wort carbohydrates, and acting as precursors in amino acid biosynthesis. The penultimate stage in the biosynthesis of all amino acids is the production of the appropriate α-keto acid, which is transaminated to form the respective amino acid. As with the Ehrlich pathway, it is the decarboxylation and reduction of these α-keto acids which are responsible for the formation of higher alcohols. In the case of those higher alcohols, such as n-propanol, for which there is no equivalent naturally occurring amino acid, the anabolic pathway is the only means of synthesis.

The biochemical pathways leading to the formation of higher alcohols, and their interelationship, are shown in Figs. 2 and 3. (For a detailed review of the supporting evidence and

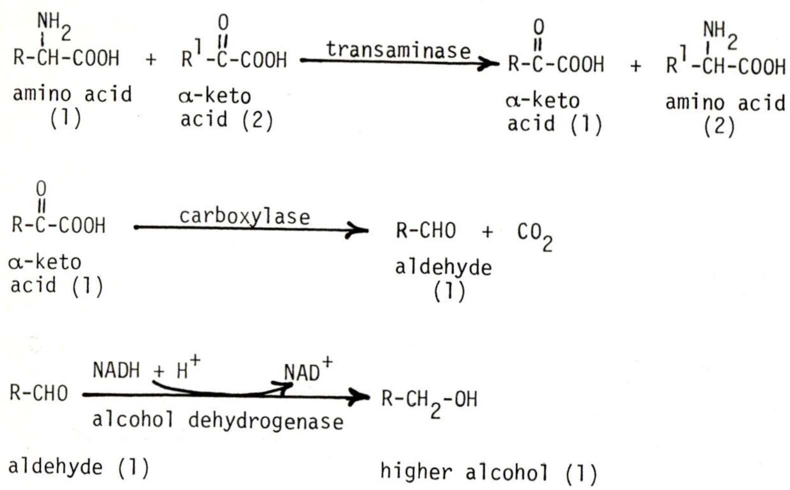

Fig. 1 Generalised Scheme For The Formation Of Higher Alcohols In Brewers' Yeast By The Ehrlich (Catabolic) Pathway (17,18,19)

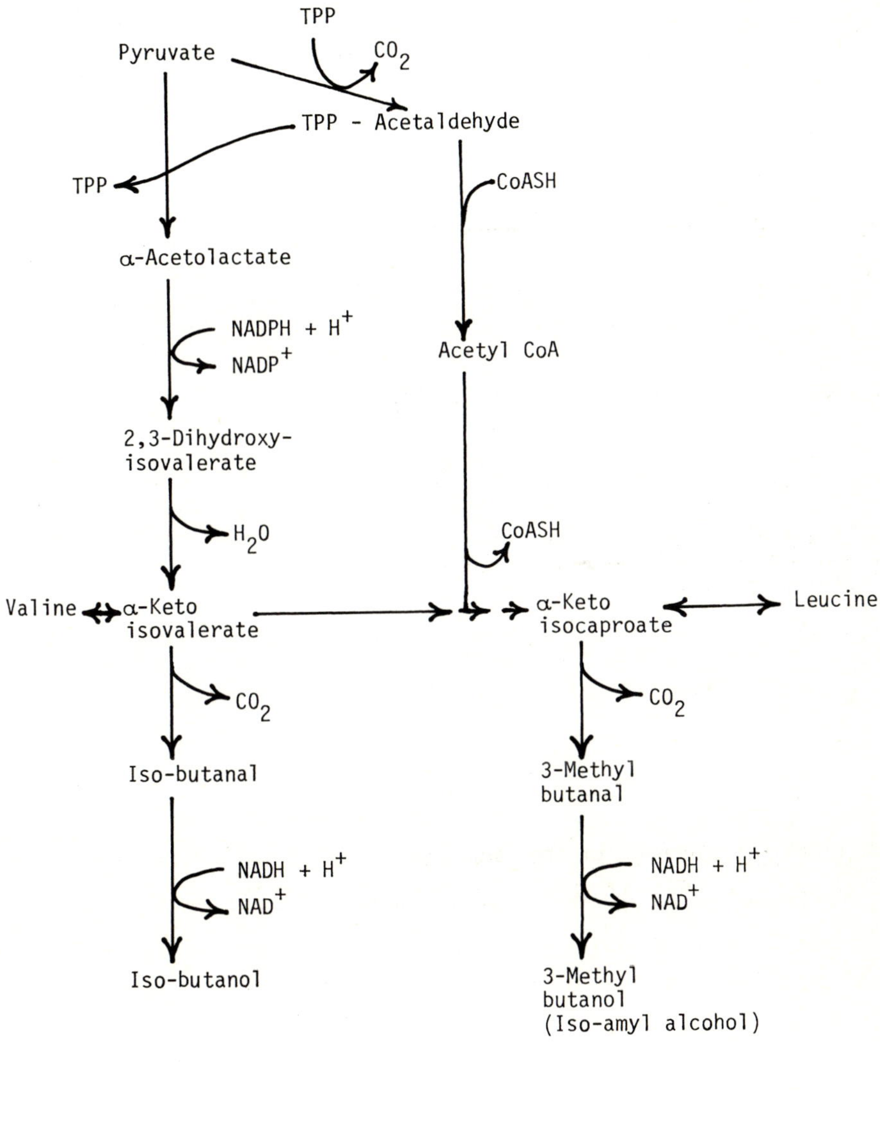

Fig. 2 The Formation Of Iso-Butanol and 3-Methyl Butanol By Strains Of Brewers' Yeast

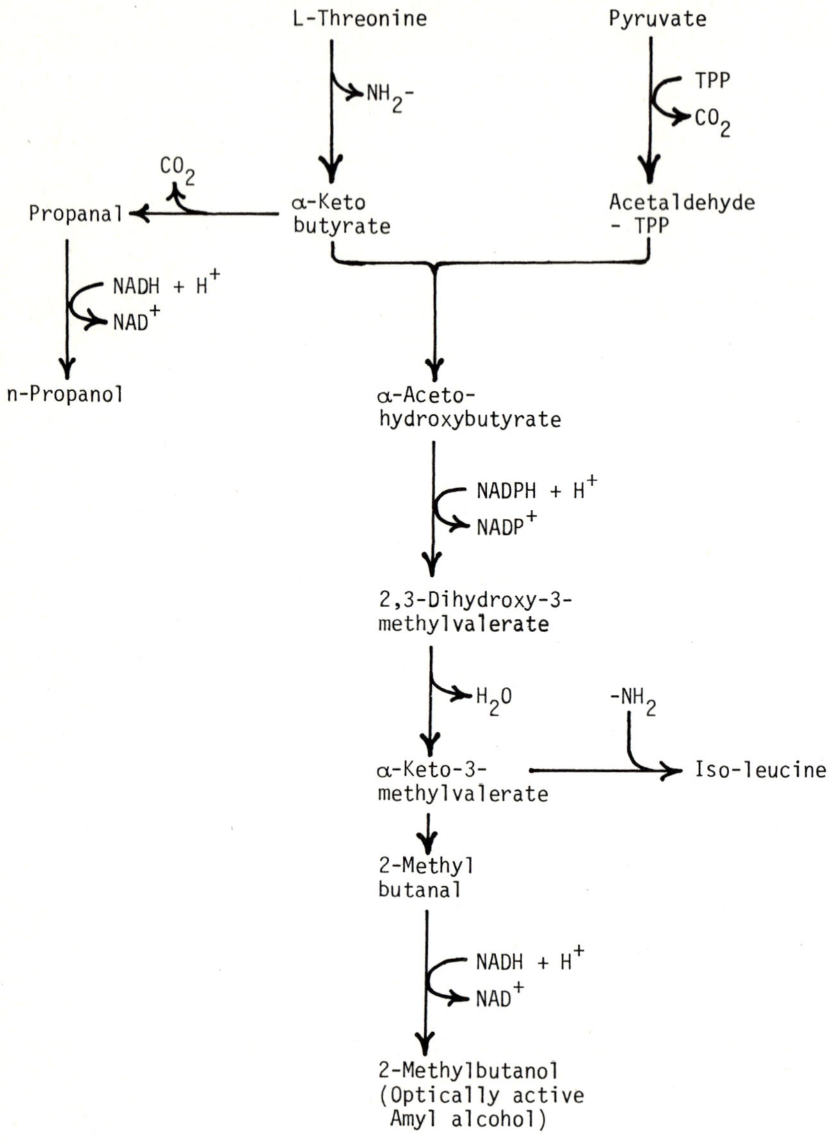

Fig. 3 The Formation Of n-Propanol And 2-Methyl Butanol By Strains Of Brewers' Yeast

mechanisms for both the anabolic and catabolic pathways of higher alcohol formation see Rainbow (20)).

The relative contribution of the Ehrlich and biosynthetic pathways to the formation of higher alcohols by yeast has been the subject of a number of studies, using either radioactive labelling techniques, or careful monitoring of the uptake of amino acids and excretion of higher alcohols (21,22,23,24,25,26).

Because higher alcohols are by-products of amino acid metabolism, the level and composition of the assimilable nitrogen in wort have a major influence of their formation. At low levels of assimilable nitrogen, yeast growth is limited and there is a high yield of higher alcohols by the synthetic pathway (22). As the level of assimilable nitrogen in the medium increases, the total amount of higher alcohols formed reaches a peak and then decreases. Two different effects are in operation at this stage. Firstly, the presence of high levels of amino acids reduces the anabolic formation of their respective higher alcohols by feedback inhibition of key enzymes in the biosynthetic pathway (27). Secondly, as the nitrogen level increases, the formation of higher alcohols by the Ehrlich pathway becomes more pronounced (22). Fig. 4 shows the general pattern for total, catabolic, and anabolic formation of higher alcohols as a function of the nitrogen level in the medium (22).

Recently Chen (26) has shown that the relative contributions of the two pathways vary with different individual higher alcohols and that, in general, the contribution of the anabolic pathway decreases as the number of carbon atoms in the higher alcohol increases.

In a simple medium containing ammonium sulphate as the sole nitrogen source, all higher alcohols are formed via the anabolic pathway (22). In complex media containing mixtures of amino acids, where higher alcohols can be formed by either catabolic or anabolic routes, the pattern of higher alcohol formation is also complex and varies between individual yeast strains (28). However, addition of specific amino acids to such media results in increases in total and catabolic formation of the corresponding higher alcohols (29,30).

C3.2.2 Control Of Higher Alcohol Formation

The principal factors influencing the formation of higher

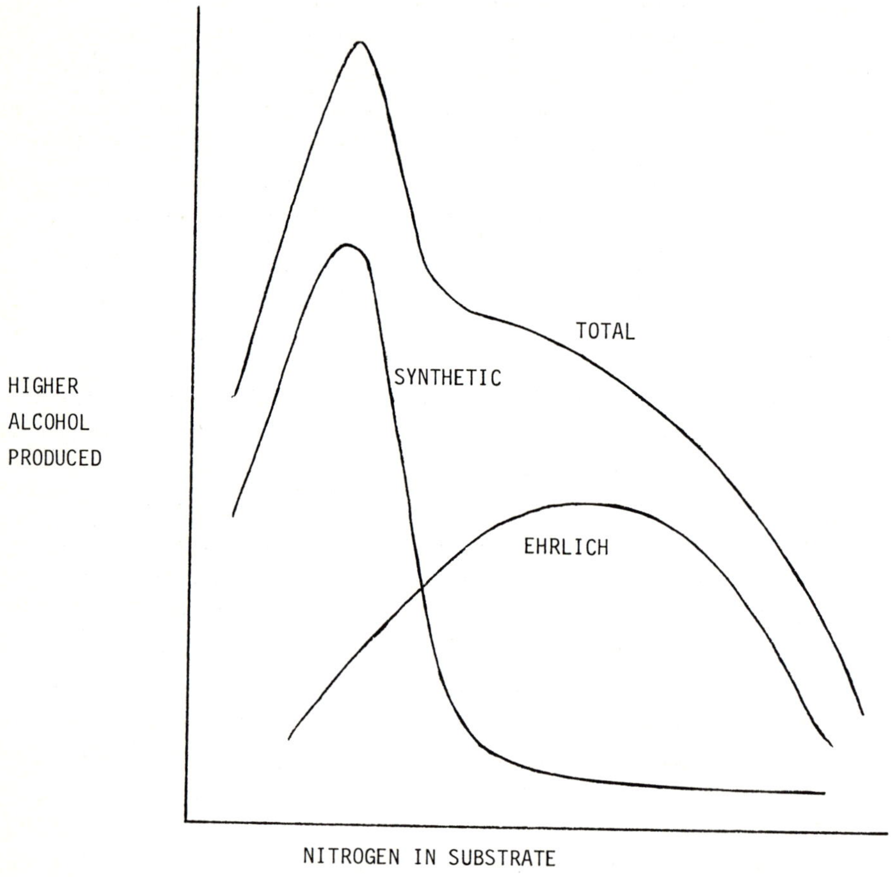

Fig. 4 An Idealised Scheme For Higher Alcohol Formation In Strains Of Brewers' Yeast, Showing The Relative Contributions Of The Synthetic And Ehrlich Pathways.

alcohols during fermentation are yeast strain, fermentation conditions, and wort composition.

The particular yeast strain used in a fermentation appears to be the most significant factor (16,27,28,31,32). Thus, in a series of trials conducted at three different breweries using three different strains of lager yeast, Engan (32) found that yeast strain effects dominated over the different environmental conditions presented by the three breweries. In general, strains of ale yeast tend to produce greater levels of higher alcohols than strains of lager yeast (31).

Fermentation conditions such as temperature (33,34,35,36,37), pitching rate (33,38,39), pressure (34,35,36), and aeration (33,38), modify the levels of higher alcohols in the final beer. Increases in fermentation temperature markedly increase both yeast growth and the rate of fermentation, and at high temperatures, excessive levels of higher alcohols can be formed (33,34,35,37).

However, higher alcohol formation is reduced when fermentations are carried out under pressure (34,35,36). Thus, the application of pressure during fermentation can be used to avoid the excessive formation of both higher alcohols and esters occurring at increased fermentation temperatures, allowing advantage to be taken of the accelerated fermentation rate (27,35,36). Control of both temperature and pressure during fermentation therefore make it possible to produce a satisfactory beer in a much reduced fermentation time (see Section D2.2.1).

Changes in pitching rate have only a slight effect on higher alcohol formation (33,38,39). However, over-aeration leads to excessive yeast growth and over-production of higher alcohols (33,38).

As has already been stated, both the level and composition of assimilable wort nitrogen affect the levels of higher alcohols in the final beer and alter the relative contributions of the Ehrlich and biosynthetic pathways. The nitrogen content and composition of the wort is determined by the choice of raw materials (grist composition) and mashing (wort extraction) conditions (40,41,42). At both low and excessive levels of assimilable nitrogen, the formation of higher alcohols is enhanced (16,22). Care must be taken therefore with the proportion of nitrogen diluents, such as wheat flour or glucose syrup, in the grist, in order to ensure that the level of free amino nitrogen in the wort does not limit yeast growth. A free amino nitrogen content of 140-150mg/litre for a 1040° gravity wort is recommended (43).

When the availability of free amino nitrogen is limiting, yeast growth and yeast crops are reduced and, as a consequence, the fermentation rate decreases and the fermentation tends to 'stick' at a higher-than-normal racking gravity. Under these conditions of restricted yeast growth, the formation of higher alcohols is enhanced, and the flavour of the final beer is atypical. For product consistency, it is therefore important to regulate free amino nitrogen levels in the wort to take account of alterations in grist composition and seasonal variations in raw material composition.

C3.3 Esters

Esters make up numerically the largest family of beer aroma compounds produced by yeast, and in general impart a 'fruity' character to beer (44). Quantitatively and qualitatively the two most important esters produced during fermentation are ethyl acetate and iso-amyl acetate, which have approximate flavour threshold values in beer of 25 and 2.0mg/litre respectively (14). Part of the reason for the importance attached to these two esters in particular is that, in high gravity fermentations ie. fermentations carried out using stronger worts than is normal (see Section D2.4), there is a disproportionate increase in the levels of ethyl acetate and iso-amyl acetate compared with the increase in wort gravity (45). The overall flavour balance of the finished beer is therefore distorted, such that it has an intense 'fruity' aroma or 'strong ale' character. The trend over the last two decades towards high gravity brewing, has thus necessitated the development of satisfactory procedures to control ester formation during such fermentations.

C3.3.1 Formation of Esters

The first detailed studies into the biosynthesis of esters by yeast were carried out by Nordström (46,47,48,49). He concluded that the only significant means by which ethyl acetate was formed in brewery fermentations was by intracellular synthesis, followed by diffusion of the ester into the medium. He further proposed that esters are formed by alcoholysis of acyl-coenzyme A (-CoA) compounds within the yeast cell:

$$R.CO \sim S.CoA + R^1OH \rightleftharpoons R.COOR^1 + CoASH$$

According to this scheme ester formation is therefore dependent upon a supply of alcohols and acyl CoA compounds, the alcohol moieties being provided either by ethanol or higher alcohols. The most common acyl CoA compound is acetyl CoA, which is formed by oxidative decarboxylation of pyruvate and has a central role in yeast metabolism. The higher acyl CoA compounds are intermediates of fatty acid metabolism.

The first stage in the biosynthesis of fatty acids, as elucidated by Lynen and his coworkers (50,51,52) is the ATP-dependent synthesis of malonyl CoA from CO_2 and acetyl CoA:

$$CH_3.CO{\sim}S.CoA \;+\; CO_2 \xrightarrow{ATP \;\rightarrow\; ADP + Pi} HO_2C.CH_2.CO.{\sim}S.CoA$$

In the presence of the multi - enzyme complex 'fatty acid synthetase' or 'acyl - carrier protein' (ACP), malonyl CoA is then rapidly converted into fatty acids by condensation with further acetyl CoA moieties (52), each addition of acetyl CoA extending the chain length by two carbon atoms. (The biosynthesis of fatty acids containing an odd number of carbon atoms is accounted for by replacing acetyl CoA with propionyl CoA in the initial condensation reaction with malonyl CoA). Leakage of acyl CoA compounds from the ACP enzyme complex can thus give rise to small quantities of either volatile esters or the corresponding fatty acids ((53) and see Fig. 5).

The biosynthesis of both lipids and esters is therefore closely linked, and utilises the key metabolite acetyl CoA. This relationship between lipid and ester synthesis is inverse. Thus, during the earlier stages of fermentation, glycogen is dissimilated into acetyl CoA, which fuels sterol and fatty acid synthesis (54). At this period in the fermentation, the specific rate of formation of ethyl acetate and iso-amyl acetate is low (55,56). As yeast growth becomes restricted by the availability of sterols and unsaturated fatty acids (57,58), there is a related decline in fatty acid synthesis and a marked increase in the specific rate of ester formation (55,56). This increase is short-lived and can be suppressed by linoleic acid addition (55,56,59). Wort aeration or oxygenation also delays the induction of ester synthesis, by stimulating yeast growth and creating a demand for acetyl CoA in lipid biosynthesis (55,56,60,61).

The short lived peak in ester formation at the end of lipid synthesis has been described as a mechanism for generating

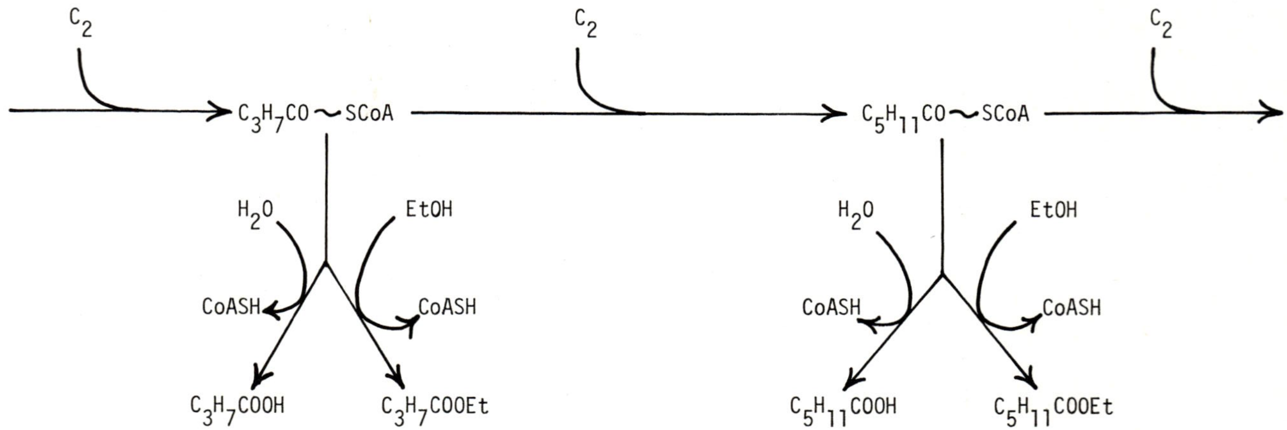

Fig. 5 A Simplified Scheme For The Formation Of Esters And Medium Chain Length Fatty Acids By Strains Of Brewers' Yeast

CoASH and controlling the balance of acetyl CoA and CoASH within the yeast cell (55,56). Alternatively, it has been proposed that the enzymes involved in esterification of glycerol-3-phosphate with acyl CoA intermediates to give phosphatidic acid could utilse acetyl CoA and ethanol as alternative substrates.

Therefore, as lipid biosynthesis declines, the availability of acyl CoA compounds reduces and the esterification enzymes in phospholipid synthesis switch to acetate ester synthesis (56). This reaction would also reduce the ratio of acetyl CoA: CoASH.

Methods of controlling ester formation during fermentations are therefore best directed towards eliminating or reducing the peak of ester synthesis.

C3.3.2 Control of Ester Formation

As already mentioned, the impetus for research into the control of acetate esters has been the need for methods of controlling the disproportionate production of esters occurring in high gravity fermentations, thereby facilitating the flavour matching of 'high' gravity - and 'sales' gravity - brewed beers. A number of different control methods have been evaluated as follows:

i Avoidance of wort starting gravities above 1060°. The relationship between wort starting gravity and beer ester levels is exponential (45,62). Thus, up to starting gravities of approx 1060°, beers fermented at high gravity, when diluted, have only a marginally higher ester content than their sales gravity counterparts. However, above 1060°, the distortion in flavour increases rapidly, and the final diluted beers have a pronounced 'strong ale' or 'fruity' aroma (45). Thus, a doubling of wort gravity from 1040° to 1080° can result in a 4-to 8-fold increase in the ester content of the finished beer (45). Avoidance of wort gravities in excess of 1060° therefore makes flavour matching simpler, but restricts the potential gains from high gravity brewing (see Section D2.4).

ii Choice of Yeast Strain

Yeast strains vary in their ability to produce esters (63,64). However, the influence of wort gravity on ester and higher alcohol formation is essentially similar for all strains (45).

The intention in most high gravity brewing operations is to obtain a satisfactory flavour match with an already existing beer, brewed and fermented at sales gravity (65). Therefore, the use of yeast strains producing low levels of esters, specifically for high gravity fermentations, is not feasible, because each yeast strain produces its own characteristic pattern of flavour components (32).
Thus, a change of yeast strain results in a change in the levels of all yeast-derived flavour components, and the end result is an unacceptable flavour shift in the final beer.

iii Oxygenation or Aeration of Fermentations

A number of workers have found that formation of volatile esters in high gravity fermentations can be adjusted by stimulating yeast growth through increased oxygenation or aeration of wort, prior to, or during, primary fermentation (45,60,61,62,65,66,67,68,69). Thus, the control of wort oxygen levels at pitching and/or controlled access of oxygen (or air) during the active phase of fermentation, allows flavour matching of high gravity and sales gravity beers. However, there is a cost penalty from this approach to ester control, because it results in both increased yeast growth and attendant increased beer wetting losses (65).

The solubility of oxygen in wort decreases with increasing wort gravity (see Section E3.1 and Table II). A change from aeration to oxygenation may therefore be necessary in order to achieve the wort oxygen levels (and thus yeast growth) required.

In some instances, the particular yeast strain and pitching rate limit the amount of oxygen that can be utilised for growth at any one time (70,71 and see Section D2.2.2), so that increasing the dissolved oxygen level in the wort at pitching has little effect on beer ester levels (69). In such cases, either the incremental supply of oxygen throughout fermentation, or the incremental filling of fermenters with batches of air-(or oxygen-) saturated wort, may be necessary to achieve the required ester levels. A continuous low rate of aeration during fermentation gives the maximum suppression of ester synthesis (72,73).
However, the effectiveness of aeration or oxygenation during fermentation is dependent upon the time of

application (66): oxygen must be supplied prior to the cessation of lipid synthesis and the onset of the rapid increase in specific rate of ester synthesis(55,56).

The role of oxygen in brewery fermentations has been the subject of much detailed study over the last two decades (70,74). In normal brewery fermentations, yeast requires molecular oxygen for key reactions in the biosynthsis of sterols and unsaturated fatty acids (75,76,77,78,79). Without a supply of sterols and unsaturated fatty acids the yeast cell cannot grow or reproduce (57,58). Therefore oxygen, by virtue of its requirement in sterol and unsaturated fatty acid biosynthesis, is growth-limiting for yeast in brewery fermentations.

Although the relative requirement for oxygen varies between yeast strains, it is sufficiently characteristic for any one strain that brewers' yeast can be categorised into four classes on the basis of their oxygen requirements (70,74,80).

The availability of oxygen is therefore a key parameter in the control of yeast growth in brewery fermentations, the extent of which affects both the rate and extent of fermentation, and the level of yeast -derived flavour compounds in the resultant beer (70,74).

The role of oxygen in reducing beer ester levels can be explained as follows: the biosynthesis of lipids and esters utilises the common precursor, acetyl CoA. Oxygenation or aeration prior to the cessation of lipid synthesis stimulates lipid biosynthesis (and yeast growth), and creates a demand for additional acetyl CoA (55,56). Any prolonging of lipid biosynthesis delays the onset of ester synthesis, and reduces the availability of acetyl CoA for ester formation.

iv Wort Lipids

The requirement of molecular oxygen for yeast growth in brewery fermentations can be eliminated by the presence of unsaturated fatty acids or sterols (57,58,67). The presence of unsaturated fatty acids in the wort stimulates yeast growth, and as a consequence, reduces the synthesis of both acetate esters and the esters of medium chain fatty acids (81,82). (In contrast, saturated fatty acids may slightly increase the levels of acetate esters (81)).

The supplementing of worts with unsaturated fatty acids has therefore been proposed by a number of workers as a possible means of reducing ester formation in high gravity fermentations (45,62,65,69,83,84).
Spent-grain 'pressings' are a rich source of both long chain unsaturated fatty acids and sterols (83,85), the main sterol components being β-sitosterol and campesterol, whilst linoleic acid is the most abundant free fatty acid (83). Addition of spent grain pressings to fermentations has been found to both increase yeast growth and dramatically reduce acetate ester levels (62,69,83). Similar effects have been observed following the addition of pure oleic acid (45,81,82). The effect of spent-grain pressings on fermentations is due to assimilation by yeast of both the sitosterol and free unsaturated fatty acids, thus supplementing the effects of wort oxygenation.
It has been proposed that the incorporation of unsaturated fatty acids into the yeast cell stimulates triglyceride and phospholipid synthesis, thereby utilising additional acetyl CoA and delaying ester synthesis (55,56,83).
Linoleic acid has been found to be more effective in reducing acetate ester levels than oleic acid (84). Moreover, unlike oleic acid, linoleic acid achieves a reduction in ester formation with little or no increase in yeast growth (84). Linoleic acid is rapidly incorporated into yeast lipids, and alters the pattern of fatty acid and sterol synthesis, such that the consumption of acetyl CoA in lipid synthesis is reduced by up to 40%.
It is known that linoleic acid has to be activated with CoASH prior to its incorporation into yeast lipids. However, competition for CoASH between linoleic acid activation and acetyl CoA synthesis cannot account for the observed repression of lipid and ester synthesis (84).
It is believed that the enzymes involved in ester synthesis are located in the yeast cell membrane (86). Thus, incorporation of linoleic acid into membrane lipids may alter the physical properties of the membrane, and modify or inhibit the activity of the membrane-bound ester-synthesising enzymes (55,56,84). This mode of action would explain the inhibition of ethyl acetate formation in the presence of high intracellular levels of acetyl CoA.

A reduction in beer ester levels by manipulating the levels of unsaturated fatty acids carried into the wort from the raw materials is feasible, but suffers from certain drawbacks. Thus, the presence of lipid material adversely affects beer foaming properties, thereby resulting in beer with a poor visual appearance (87 and Section Cl). Also, unsaturated fatty acids are the precursors of certain compounds responsible for staling flavours in beer, so that the addition of lipids to wort can reduce the flavour stability of the final beer (88).

On balance therefore, because of the generally undesirable effects of lipids on both the appearance and stability of beer, control of beer ester levels in high gravity fermentations by addition of lipids to the wort, is neither a popular nor practical proposition.

v Wort Composition

Adjusting the ratio of total fermentable sugars to total soluble nitrogen is a successful and convenient method of controlling ester formation (65,68,69,89). Pfisterer & Stewart (89) noted that the levels of ethyl acetate in beers fermented at high gravity, were less when the increased gravity was obtained by addition of glucose syrup, than when additional malt was used. They suggested that, in the presence of high concentrations of glucose, yeast metabolism was subject to the 'Crabtree Effect' or 'Contre Effect Pasteur'. (In the Crabtree Effect, the mitochondrial respiratory enzymes are repressed even under aerobic conditions, so that utilisation of glucose occurs solely by alcoholic fermentation). They further speculated that under the influence of the Crabtree Effect, more acetyl CoA is channeled into lipid synthesis, thus reducing its availability for acetate ester formation. The basis for this argument was the observation that with increasing proportions of fermentable sugar in the wort, lipid accumulation in yeast is increased (89).

The results of Pfisterer and Stewart have since been confirmed (68,69). Additions of either sucrose, maltose, or glucose, were observed to give similar reductions in both yeast growth and acetate ester levels. However, when casein hydrolysate was added together with glucose to maintain the balance of fermentable sugars and total soluble nitrogen, no reduction in acetate ester levels or yeast growth occurred (68,69). It was argued therefore, that the observed reduction in acetate esters was simply a result of the reduction in yeast growth occurring in fermentations of worts containing sugar adjuncts, compared with fermentations of all-malt worts. White and Portno (68,69) further proposed that the reason why acetate ester levels were increased in high gravity fermentations at the same time as yeast growth was restricted, was because growth was being limited in the presence of excess assimilable nitrogen and high concentrations of alcohols. Thus, in such fermentations they observed that dilution of wort nitrogen with fermentable sugar syrups, in addition to reducing the ratio of assimilable nitrogen to fermentable sugars in the wort, caused a reduction in ester levels. Similarly, they argued that stimulating yeast growth (by oxygenation) created an extra demand for assimilable nitrogen, (thereby in effect again reducing the ratio of assimilable nitrogen to fermentable sugars), and therefore reduced acetate ester levels (68,69).

vi Application of Pressure

The reported effects of pressure on fermentation performance and the formation of volatile flavour components are contradictory (34,35,36,61,90,91). Increasing pressure inhibits, and reduced pressure stimulates, the formation of acetate esters. In contrast, the formation of the ethyl esters of the C_8, C_{10}, and C_{12} fatty acids reaches a maximum at around atmospheric pressure, and only slightly declines with further increases in pressure (61). Ethyl hexanoate formation is relatively unaffected by pressure (61). However, during fermentations in spheroconical vessels, it was found that, with increasing pressure, ethyl acetate levels were reduced but levels of iso-amyl acetate and phenyl acetate were increased (36,90). Some of these conflicting observations may be due to differences in fermentation temperature rather than pressure (36,90), but usually the combination of high fermentation temperatures and high pressures results in a net reduction in fermentation time, combined with only minor flavour variations (34,35,92) (and see Section D2.2.1).

In general, application of pressure during fermentation reduces yeast growth (61,93), and inhibits the production of acetate esters (34,61), diacetyl (34), and higher alcohols (34,35,92). However, the formation of free medium chain fatty acids and their esters (35,61), acetaldehyde (93), and pyruvate (92), is promoted by increased pressure.

There is strong evidence that the effects on yeast growth are not due to pressure 'per se'. Thus, increased pressure during fermentation also produces increased concentrations of CO_2, and removal of CO_2 by nitrogen purging enhances yeast growth and stimulates acetate ester production (61,94).

The mode of action of CO_2 is not known. However, it has been proposed that carbon dioxide, as bicarbonate ions, can alter the physical properties (such as fluidity and permeability) of the yeast cell membrane, in addition to affecting the equilibrium of decarboxylation and carboxylation reactions (95). (A key decarboxylation reaction is the production of acetyl CoA from pyruvate by the multi-enzyme complex pyruvate dehydrogenase). The cause of the reduction in yeast growth and ester formation observed in the presence of high concentrations of CO_2, may therefore be a combination of both membrane and metabolic effects.

Factors mitigating against the adoption of pressure fermentations solely to control ester levels are:

(a) the high capital cost of fermenting vessels capable of operating at pressures in excess of atmospheric pressure;

(b) in addition to affecting ester synthesis, increased pressure affects both the fermentation rate and the synthesis of other important flavour constituents;

(c) high levels of CO_2 in the resultant beer.

The net result of an increase in fermentation time together with a fundamental change in beer flavour is unacceptable. (However, see Section D2.2.1 and refs. 34,35 and 92 for the combined use of high pressures and high temperatures in order to achieve accelerated fermentation rates.)

C3.4 Acetaldehyde

There are many flavour-active aldehydes present in beer, formed at various stages in the brewing process by both chemical and enzyme-mediated reactions. Together they constitute a most important group of beer volatiles, in particular because of their involvement in many of the flavour changes which occur during beer aging.

Quantitatively the most important aldehyde in beer is acetaldehyde, commonly present at levels ranging from 2-20mg/litre (44), and with a flavour threshold ranging from 5-50mg/litre depending on the type of beer (96). Excess levels of acetaldehyde in beer impart a flavour variously described as 'grassy', 'bruised/raw apples', or 'cidery'.

C3.4.1 Formation of Acetaldehyde

Acetaldehyde is formed in the penultimate stage of the process by which strains of brewers' yeast catabolise wort sugars to ethanol and CO_2. Thus, in the Embden-Meyerhof-Parnas, or glycolytic, pathway (97), acetaldehyde is formed by the enzymic decarboxylation of pyruvate, and is then reduced to ethanol by $NADH^+$ in a reaction catalysed by alcohol dehydrogenase. This reaction sequence is analogous to those previously described in Section C3.2, in which aldehydes are formed by the decarboxylation of their respective α-keto acids during the formation of higher alcohols (Fig. 1).

Levels of acetaldehyde in beer reach a maximum during primary fermentation, normally declining to low levels towards the end of primary fermentation and during secondary fermentation (44). In addition to being reduced to ethanol, a minor portion of the acetaldehyde produced is oxidised to acetic acid, a process influenced by the concentration of ethanol present (98).

C3.4.2 Control of Acetaldehyde Levels

Excess levels of acetaldehyde in beer can be the result of bacterial spoilage, especially by strains of *Zymomonas anaerobia* (99). High levels of acetaldehyde can also be caused by high air levels during fermentation, and acetaldehyde levels in bottled beer have been observed to increase during pasteurisation and storage, especially if there is a high air content in the bottle head-space (96). The addition of sulphite or sulphur dioxide as antioxidants also influences acetaldehyde levels, since these compounds form complexes with acetaldehyde and other carbonyls (100,101).

Many workers have stressed the significance of yeast strain in determining acetaldehyde levels (96), although in plant trials comparing three different strains of *S.uvarum* in three breweries, the effect of processing conditions in the breweries was shown to be paramount (102). The reported effects on acetaldehyde levels of changes in pitching rate, fermentation temperature, and pressure are varied (44,96).

However, since acetaldehyde levels commonly decline towards the end of primary fermentation and during secondary fermentation, as a result of normal yeast activity, high levels of acetaldehyde in beer usually reflect either premature separation of yeast from the beer or insufficient yeast activity. Corrective action is therefore normally based on ensuring that adequate levels of metabolically-active yeast are maintained in the beer during secondary fermentation, in order to achieve the required degree of flavour maturation (96).

C3.5 Diacetyl and 2,3- Pentanedione

Both diacetyl and 2,3- pentanedione are normal by-products of yeast metabolism, and are found in all brewery fermentations. They have a similar aroma and taste, and when present in excess impart a characteristic 'sickly-sweet/butterscotch' or 'diacetyl' flavour to beer. The flavour threshold of diacetyl is lower in lagers (0.1 - 0.2mg/litre) than in ales (approximately 0.4mg/litre); whereas the threshold value of 2,3- pentanedione is approximately 1mg/litre in both types of beer.

In most instances diacetyl and 2,3- pentanedione are present in beers at levels below their taste threshold, but their presence still makes a significant contribution to the overall flavour balance of a beer. Diacetyl is considered to be undesirable in lagers at levels above its taste threshold, whereas it is a characteristic and desirable flavour in certain British ales.

C3.5.1 Formation of Diacetyl and 2,3-Pentanedione

The many different theories concerning the formation of vicinal diketones in brewery fermentations have been the subject of an extensive review (103). It is now accepted that diacetyl and 2,3- pentanedione are formed outside the yeast cell, by oxidative decarboxylation of α-acetolactate and α-acetohydroxybutyrate, respectively (103,104,105,106). These α-acetohydroxy acids are intermediates in the biosynthesis of certain amino acids, and are leaked into the wort by yeast during fermentation. Once diacetyl and 2,3- pentanedione have been formed in the beer, they are normally converted respectively to acetoin and 2-3-pentanediol by the action of yeast reductases (107).

Thus, the final concentration of diacetyl in beer is the net result of three separate steps:-

i synthesis and excretion of α-acetohydroxy acids by yeast,

ii oxidative decarboxylation of α-acetohydroxy acids to their respective diketones,

iii reduction of diacetyl and 2,3- pentanedione by yeast.

Acetohydroxy acid biosynthesis is both linked to yeast growth and dependent upon wort nitrogen composition (108,109).

Acetolactate is formed by the condensation of pyruvate and 'active' acetaldehyde, and is an intermediate in the pathway leading to the production of valine as well as leucine and pantothenate ((103) and see Fig. 6).

Similarly α-acetohydroxybutyrate is the condensation product of α-ketobutyrate and 'active' acetaldehyde, and is an intermediate in isoleucine biosynthesis ((103) and see Fig.7).

The same enzyme, acetohydroxy acid synthetase, is responsible for both condensation reactions (110), and the control of this enzyme is complex. Thus, valine inhibits α-acetolactate production, but stimulates 2,3-pentanedione formation (111,112,113,114). Moreover, isoleucine inhibits α-acetohydroxybutyrate synthesis but increases diacetyl formation (115).

Acetolactate can decompose to either acetoin, or, in the presence of oxygen, to diacetyl ((103) and see Fig 8), and it is the spontaneous conversion of α-acetolactate to diacetyl which is the rate-limiting step in the balance between the formation and removal of diacetyl (107).

This is evident from observations that, at the peak of fermentation, beers can contain up to 1.1mg/litre α-acetolactate and 0.6mg/litre α-acetohydroxybutyrate, but very low levels of diacetyl (0.05 - 0.10mg/litre) and 2,3-pentanedione (0.05mg/litre). The levels of α-acetohyroxy acids and diketones during a pilot brewery ale fermentation are shown in Fig 9.

Diacetyl formation from α-acetolactate has been shown to be dependent upon pH, the concentration of α-acetolactate, temperature, the presence of oxygen, and certain metal ions (107).

The role of oxygen in diacetyl formation is not fully understood (103,116,117,118), although it is generally recognised that ingress of oxygen towards the end of

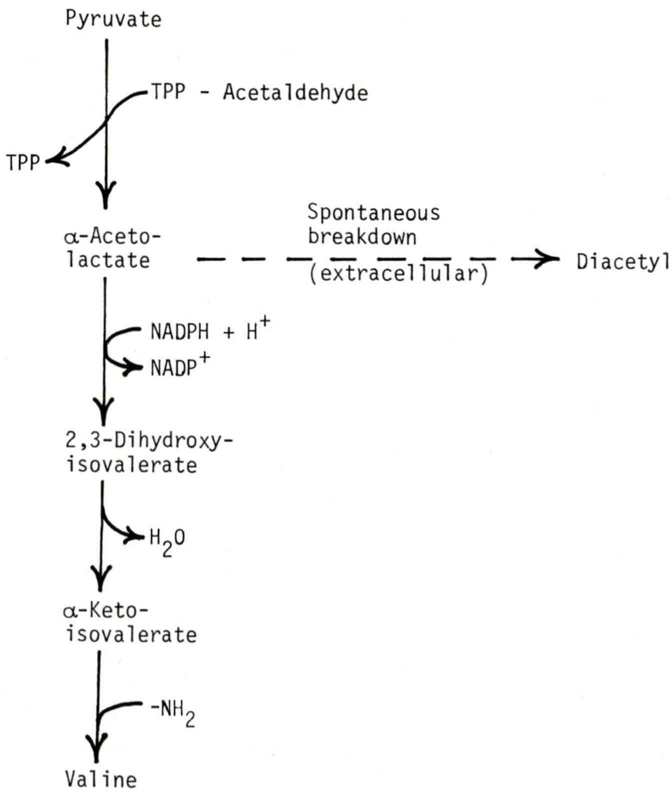

Fig. 6 The Formation Of Diacetyl By Strains Of Brewers' Yeast

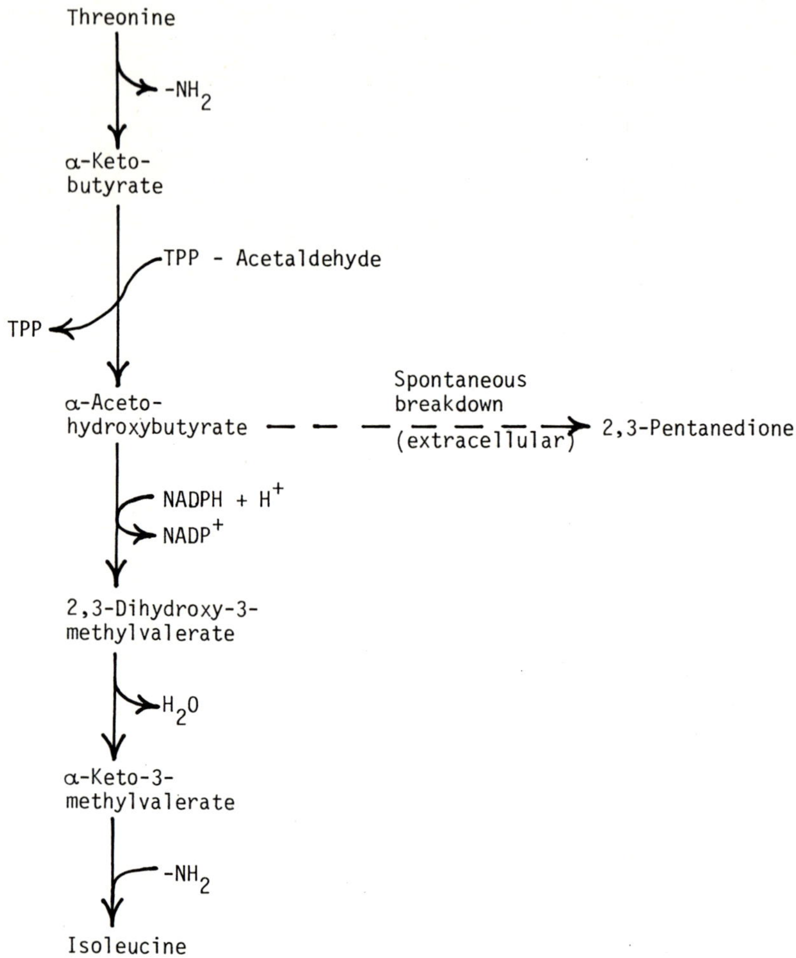

Fig. 7 The Formation Of 2,3-Pentanedione By Strains Of Brewers' Yeast

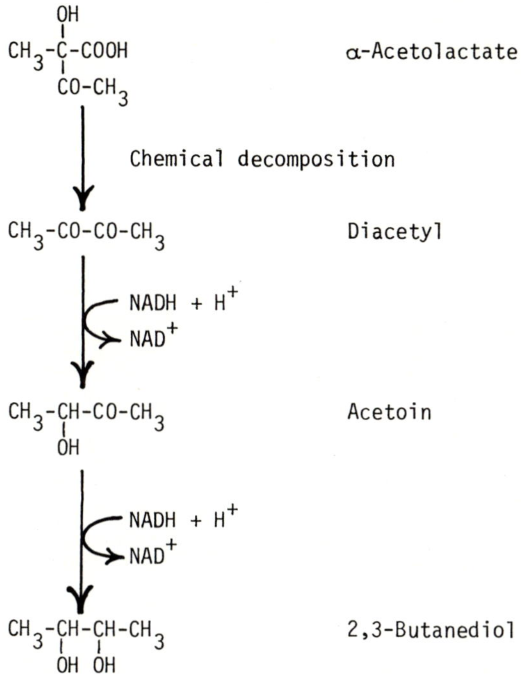

Fig. 8 The Formation And Removal Of Diacetyl In Brewery Fermentations

Fig. 9 Typical Diketone And α-Acetohydroxyacid Levels In Beer During An Ale Fermentation (106)

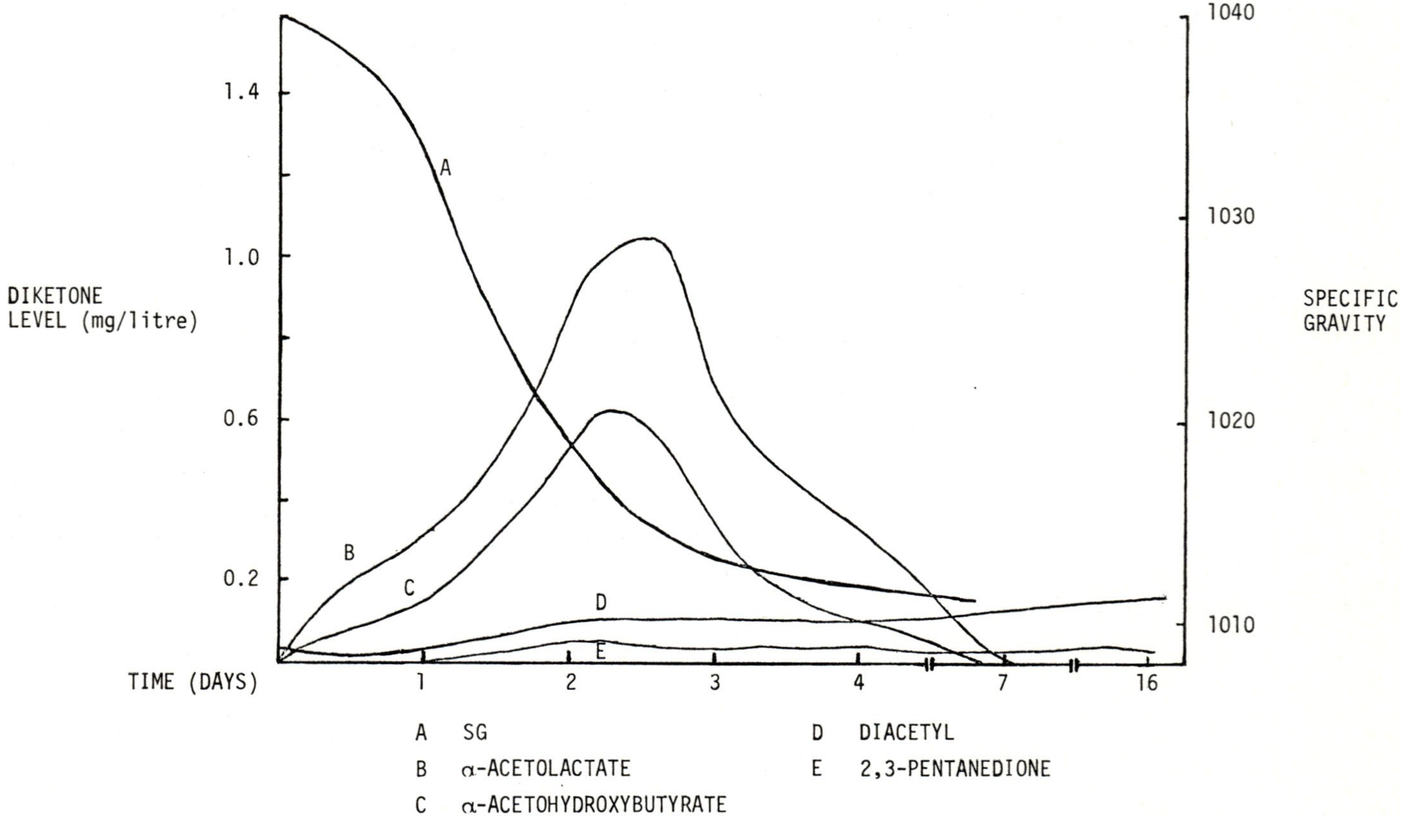

fermentation can result in high diacetyl levels in the final beer. The effects of oxygen at this point are two-fold. Firstly, it results in an increase in the rate of conversion of α-acetolactate to diacetyl, and secondly, the yeast is stimulated to produce α-acetolactate at a stage in the fermentation when there is unlikely to be sufficient residence time for it to be reduced to diacetyl (103,118).
Actively metabolising yeast are able to reduce diacetyl to acetoin and butanediol, and 2,3- pentanedione to 2,3-pentanediol (Fig.8) (103). Although the ability to reduce diketones is strain-dependent (107), the diacetyl-reducing activity of all yeast is at its peak during the active phase of fermentation. Thereafter it declines (103). This explains the lack of diacetyl in beer at the point in fermentation when α-acetolactate concentrations are at their highest (Fig.9). Since yeast contains no specific diacetyl reductase, it is thought that alcohol dehydrogenase is the enzyme responsible for diacetyl reduction (119,120).

C3.5.2 Control of Diacetyl and 2,3- Pentanedione

Excessive levels of diacetyl can be the result of beer spoilage by certain strains of *Pediococcus* and *Lactobacillus* (121,122). Also, certain wort spoilage bacteria, although unable to synthesise diacetyl, can be the cause of abnormal diacetyl concentrations in beer (123). However, the main reason for high diacetyl levels in beer is the conversion of α-acetolactate to diacetyl, either at the stage in fermentation when there is a lack of actively metabolising yeast to reduce diacetyl, or after the yeast has been removed. Diacetyl levels can thus be controlled by ensuring that there is sufficient active yeast in contact with the beer at the end of fermentation to reduce diacetyl to acetoin and 2,3-butanediol. Yeast strain and fermentation conditions play an important role in determining the level of residual diacetyl in beer at the end of fermentation (103). Vigorous fermentations produce more acetohydroxy acids, but the decomposition of acetohyroxy acids to vicinal diketones is also more rapid (103). In addition, since diacetyl is formed earlier in the fermentation, there is more time for diacetyl removal by the yeast. Higher fermentation temperatures (124) and pitching rates (118) accelerate fermentation rates, and providing the overall fermentation time remains unchanged, produce beers with reduced diacetyl levels.

The importance of an actively metabolising yeast for diacetyl reduction is generally accepted. Thus, the yeast in a 'sticking' fermentation has a low vigour and can reduce diacetyl only slowly. The final beer therefore tends to have an abnormally high level of diacetyl. Two of the main causes of 'sticking' fermentations are lack of oxygen at pitching, and insufficient α-amino nitrogen in the wort (114,117,125,126). Lack of aeration (oxygen) at pitching can be overcome by changing from aeration to oxygenation if worts are already air-saturated, or by increasing the level of oxygen in the wort. The effect of wort α-amino nitrogen levels is more complex. Dilution of wort nitrogen with nitrogen-free adjuncts, such as glucose syrups, can lead to a deficiency in assimilable nitrogen, and result in excessive diacetyl concentrations. It is important to stress that high diacetyl levels in such fermentations are not due solely to a lack of valine, and no correlation has been found between wort valine levels and diacetyl production (127). Rather, it appears that a wort deficient in assimilable nitrogen simply results in yeast which is less active towards the end of fermentation, and less able to reduce diacetyl. It is therefore recommended that worts should have a minimum free amino nitrogen level of 140-150mg/litre at 1040° gravity (43).

The application of cooling to fermentations towards their end-point reduces the metabolic activity of yeast and encourages it to sediment. Thus, if cooling is applied too early, excessive diacetyl levels can result (128). The introduction of a 'warm' rest, prior to cooling, allows breakdown of acetolactate to diacetyl, and reduction of diacetyl by the active yeast (35,129,130).

At the end of classical lager fermentations, beer contains about 3.5% (w/w) residual extract (35,129,130). The beer is then cooled from 9°C to 4°C for lagering. During lagering, about 1.3% (w/w) of the remaining extract is fermented, so that the beer becomes carbonated and the diacetyl is reduced (35,129,130). An alternative method is to "Krausen" the lager after primary fermentation. Krausen is actively fermenting wort containing 8 - 9% (w/w) fermentable extract, and a yeast count of approximately 40×10^6 cells/ml (129,130). It is added to cooled 'green' beer, containing approx 2% (w/w) residual extract to promote secondary fermentation and remove diacetyl (129,130). The volume of Krausen added is 10-12% of the volume of the 'green' beer (129,130).

Certain continuous fermentation systems produce beers with high α-acetolactate and diacetyl levels. This is a function of shortened residence time in such systems, allowing only partial breakdown of α-acetolactate and reduction of diacetyl (131,132). Progressive conversion of α-acetolactate to diacetyl occurs during subsequent storage and pasteurisation, to give beers with unacceptable diacetyl levels (133). A novel conditioning process has been described to overcome this particular problem (134,135). It involves heating the beer from either a tower (131) or plug fermenter (132) to 80°C for 4 mins (or 60°C for 15 mins) to convert all the α-acetolactate to diacetyl. The beer is then passed through a plug of active yeast and kieselguhr to remove diketones and produce an acceptable beer (134,135).

Enzymic procedures have also been reported, either for the rapid conversion of acetohydroxy acids to their respective diols (136) or for the reduction of diacetyl (137,138,139). Because the rate-limiting step in the removal of diacetyl from fermentations is the conversion of α-acetolactate to diacetyl (107), the enzyme acetolactate decarboxylase has been isolated from species of Enterobacter aerogenes, and shown to be capable of decarboxylating α-acetolactate to acetoin (140). Addition of this enzyme to freshly fermented beer at 10°C results in a rapid reduction in acetohydroxy acids to a level below the taste threshold of their respective diketones (136). In addition, some bacteria contain a specific diacetyl reductase, which might be used to accelerate beer flavour maturation by rapidly reducing diacetyl to acetoin (137,138,139).

C3.6 Organic and Inorganic Sulphur Volatiles

Volatile sulphur compounds such as hydrogen sulphide, sulphur dioxide, and thiols, have very low flavour threshold values and make significant contributions to beer flavour. When present in trace amounts only, sulphur compounds may be acceptable or even desirable, but in excess they give rise to unpleasant off-flavours.

The development of the flame photometric detector in gas chromatographic analysis has allowed the detection of very low levels of sulphur compounds, and led in recent years to a rapid programme of research into their origin and influence in the brewing process (141,142). Three main sources of sulphur compounds in beer have been recognised:-

i Raw materials - in particular malt (143,144,145,146) and hops (147,148,149,150).

ii Yeast metabolism (143,151,152,153,154).

iii Spoilage organisms - in particular Zymomonas anaerobia (143,151,154), Enterobacter aerogenes (143,151) and Hafnia protea (143,151).

The first two sources are to a certain extent interrelated, and their relative contributions to the levels of sulphur compounds in beer are dependent upon the brewing conditions employed. The only way of controlling the levels of sulphur compounds produced by spoilage organisms however, is to trace and eliminate the source of infection. The formation of volatile sulphur compounds by spoilage organisms is therefore not discussed in this review.

C3.6.1 Formation of Volatile Sulphur Compounds

Yeast can utilise inorganic sulphur, in the form of sulphate, as the sole source of sulphur for incorporation into cellular constituents. However, in wort yeast is provided with a complex mixture of inorganic sulphate, sulphur-containing amino acids (methionine and cysteine), present in either their free forms or in peptides, and sulphur - containing vitamins. In general, organic sources of sulphur are preferred by yeast to inorganic sources. However, in the absence of suitable sources of organic sulphur, or when these have been utilised, yeast can assimilate sulphate. Minor variations in these important wort constituents can dramatically alter the formation of volatile sulphur compounds, and thus markedly affect beer flavour.

The production of both sulphur dioxide and hydrogen sulphide has been shown to correspond to yeast growth and the vigour of fermentation, formation of both compounds normally reaching a maximum rate on the second or third day of fementation, and subsequently declining to low levels at the end of fermentation (152). Recent work on synchronous cultures has shown a relationship between hydrogen sulphide production and the budding index of the yeast culture (155,156). In particular, hydrogen sulphide levels were found to be at a maximum in the period between consecutive budding. Thus, in synchronous cultures hydrogen sulphide production undergoes a series of peaks and troughs.

Both sulphur dioxide and hydrogen sulphide are intermediates in the biosynthetic pathways leading from inorganic sulphate to the sulphur-containing amino acids cysteine and methionine (See Fig.10).

Sulphate is assimilated by yeast and biologically reduced to sulphite by two ATP-dependent reactions (157,158). Sulphite is further reduced to sulphide which, in turn, is utilised in the biosynthesis of cysteine and methionine (157,158). The production of sulphur dioxide and hydrogen sulphide occurs when the reduction pathway to sulphite and sulphide exceeds the rate of incorporation of sulphide into cysteine and methionine (159).

Hydrogen sulphide can also be formed by desulphydration of cysteine, cystine, or glutathione. This pathway is important under conditions of yeast autolysis, and is responsible for increases in the levels of hydrogen sulphide in beers during excessive storage in the presence of yeast (160).

The control of the biosynthesis of cysteine and methionine from sulphate is complex. The enzymes involved in the reduction of sulphate to sulphide are regulated by repression and derepression of enzyme synthesis. In addition, the enzyme ATP-sulphurylase, involved in the activation of sulphate, is subject to feedback inhibition by sulphide, adenosine-5′-phosphosulphate (APS) and 3′-phosphoadenosine-5′- phosphosulphate (PAPS) (159).

Certain yeast strains have a specific growth requirement for the vitamin pantothenate (161,162,163), which is necessary for the activity of cysteine synthetase. A shortage of pantothenate in the medium therefore limits yeast growth, and leads to excessive production of hydrogen sulphide (161,162,163).

The sulphate-reducing pathway is inhibited by methionine (163). However, in its absence, other non-sulphur containing amino acids, such as threonine and glycine, stimulate hydrogen sulphide production (162). In such an event, the over-production of hydrogen sulphide by these amino acids is prevented by addition of methionine (162).

S-methyl thioacetate (MeSAO) has an objectionable 'rotten-vegetable' aroma, and has been detected in beer (164). It is possible that S-methyl thioacetate can be produced by metabolism of methionine, via the intermediate methyl mercaptan

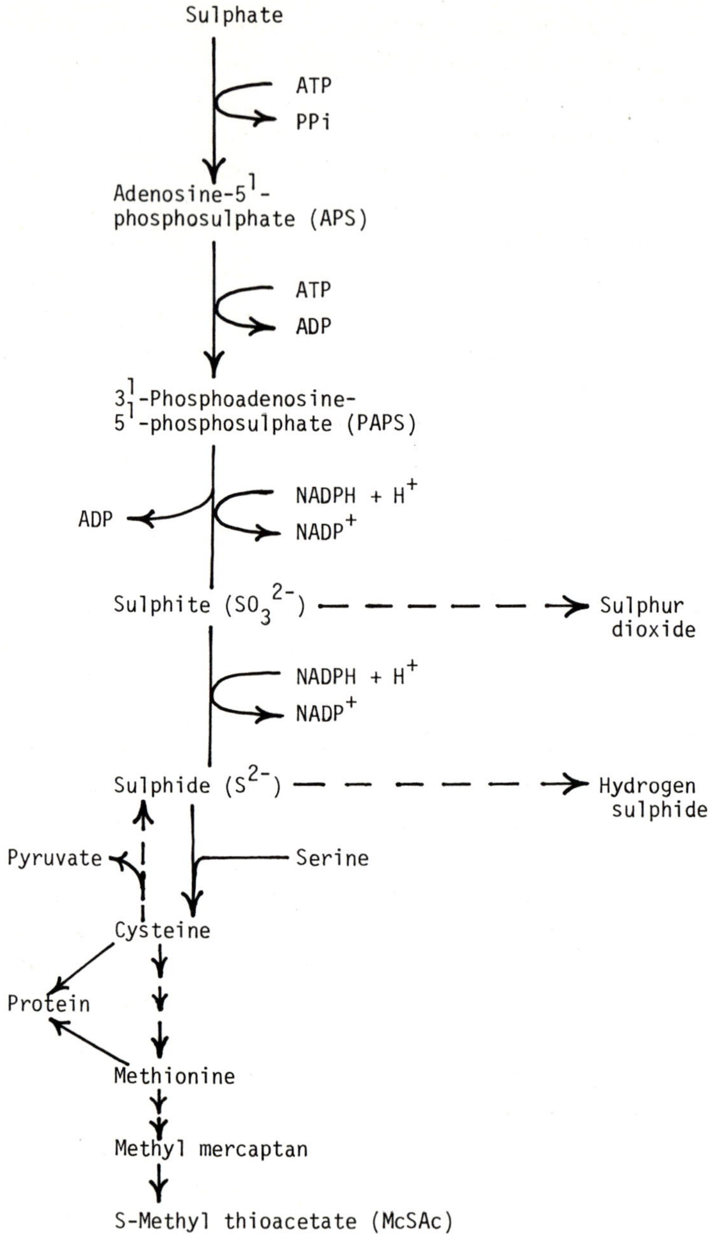

Fig. 10 Pathway For The Biosynthesis Of Sulphur-Containing Amino Acids In Strains Of Brewer's Yeast

(Fig.10) (165). Thus, enrichment of the medium with methyl mercaptan is more effective than methionine in producing S-methyl thioacetate (165).

C3.6.2 Control of Volatile Sulphur Compounds

Yeast strain selection is the most important factor influencing the production of hydrogen sulphide, since the amount formed under defined conditions is characteristic for a particular strain. However, the advent of methods to genetically manipulate yeast strains (166,167,168) may allow the pathways involved in sulphur metabolism to be modified, and give an altered yeast strain producing acceptable levels of both hydrogen sulphide and sulphur dioxide.

There are usually two causes of excessive hydrogen sulphide levels in fermented beer - poor yeast health or vigour, and wort composition.

Under normal conditions, much of the hydrogen sulphide formed during fermentation is purged from the beer by the evolution of CO_2. If a yeast is lacking in vigour then the fermentation rate is reduced, and the fermentation may 'stick'. The rate of evolution of CO_2 is thus reduced, and this can lead to less hydrogen sulphide being purged from the beer. Yeast that has a poor viability and vigour is also more prone to autolysis, and as has already been stated, yeast autolysis can produce 'yeasty' or 'sulphury' aromas by desulphydration of cysteine (160).

Wort contains four components which can affect the formation of hydrogen sulphide:-a) sulphate and sulphite, b) metal ions eg. zinc, c) pantothenate or vitamin B_6, and d) amino acids. Studies with pantothenate - requiring and methionine-requiring mutants of yeast, and with wild-type strains of S. cerevisiae and S. uvarum, have shown that large amounts of hydrogen sulphide are formed from inorganic sulphur when yeast is deficient in pantothenate or vitamin B_6 (162,163). This excess production of hydrogen sulphide can be inhibited by the addition of methionine (162,163). A shortage of trace elements, such as vitamins or metal ions, may also be responsible for poor vigour in yeast. The problems caused by a lack of trace elements can usually be rectified by supplementing worts with commercially available yeast 'foods', which contain a mixture of vitamins, minerals, and amino acids (see Section D2.5.3).

The presence of large amounts of trub in the fermenting wort, and rapid cooling at the end of fermentation have also been associated with 'sulphury' aromas in beer (128). However, the mechanism by which they can cause high levels of hydrogen sulphide is not known.

Excessive concentrations of hydrogen sulphide can normally be removed from beer by gas washing with either carbon dioxide or nitrogen. In addition, in some cases, sulphury aromas can be removed by small additions of copper sulphate to precipitate the volatile sulphur compounds as copper sulphides or mercaptides (169). However, this procedure must be carefully monitored, since an excess of cupric ions can adversely affect both the non-biological stability and the flavour of the treated beer. A maximum addition of 0.1mg/litre cupric ions as copper sulphate is safe (169). Equipment for electrolytically dosing copper into beer, so that the copper reacts almost quantitatively with the volatile sulphur present, thus removing the possibility of excess copper, has been designed (169), but is not used commercially.

C3.7 Dimethyl Sulphide

The nature of the differences in aroma and taste between ales and lagers was first investigated (170,171) using the flavour profile analysis technique developed by Clapperton (172). These studies identified dimethyl sulphide (DMS) as a characteristic and desirable flavour component of both British and Continental lagers.

DMS has a flavour threshold value in lagers of approximately 30ug/litre (173) and imparts a 'cooked-sweetcorn', 'cooked-vegetable' or 'blackcurrant-like' aroma. Table I shows the typical DMS levels found in British ales, British lagers, and Continental lagers.

The DMS levels in British Ales are well below its threshold value, and DMS contributes little to the overall character of ales. However, in both British and Continental lagers DMS is present at concentrations in excess of its threshold value, and contributes to the distinctive flavour of lager-type beers.

Beer Type	Dimethyl Sulphide Levels (µg/litre)	
	Range	Average
British Ale	0-13	11
British Lager	18-27	24
Continental Lager	44-114	74

Table I <u>Typical Dimethyl Sulphide Levels In Ales And Lagers (174)</u>

C3.7.1 <u>Formation of DMS</u>

The formation of DMS in the malting and brewing process and its survival into the final beer has been studied in detail, and is the subject of recent reviews (175,176). Three potential sources of DMS in beer have been identified:-

i Thermal decomposition of S-methylmethionine (SMM) to DMS. This occurs during both malt kilning (144,170,177,178) and wort boiling (144,179,180,181), the DMS so-formed being volatilised and lost. However, decomposition can also occur during the period of wort clarification which takes place after wort boiling and before wort cooling. DMS formed during this later stage of wort production is retained in solution and contributes to the DMS found in beer.

ii Reduction of dimethyl sulphoxide (DMSO) to DMS by yeast during fermentation (144,182,183,184,185).

iii Bacterial contamination of worts by <u>Hafnia protea</u> and <u>Enterobacter aerogenes</u> (143,186,187).

DMS can be released from malt by heating with alkali (177,188). This heat-labile DMS precursor has been identified as SMM (177,178,181,188), and is present in green malt, but absent from raw barley (177,178). SMM is therefore formed during barley germination, and its final level in green malt is dependent upon a number of factors such as barley variety (189,190), barley nitrogen content (191), barley storage time (175) and malting conditions (145,192,193). In the latter respect, application of potassium bromate to restrict rootlet growth (and thus reduce malting losses), lowers the level of SMM in green malt (145,177). This is because SMM is concentrated in the rootlet and acrospire regions of barley grain (145).

Malt type and kilning conditions exert a considerable influence on the DMS content of finished beer (171,177,178). During kilning SMM decomposes to DMS, which is volatilised and lost in the kiln exhaust gases. However, at temperatures in excess of 60°C, some DMS is oxidised to DMSO, which is non-volatile and remains in the malt.

It is the concentrations of SMM and DMSO in malt that determine beer DMS levels, and not the free DMS in the malt. The choice of kilning procedure can be used to control the SMM and DMSO contents of malt. Thus, kilning temperatures of 60°C or less, tend to preserve SMM and produce a malt relatively rich in SMM. Temperatures in excess of 60°C destroy SMM, but the final malt has a high content of DMSO (177,178,193).

During mashing, SMM and DMSO are readily extracted from the malt, but are not converted to free DMS. Any free DMS from the malt is lost (189). With the possible exception of decoction mashing, the mashing procedure is therefore unlikely to exert any significant effect on DMS levels in beer (170).

Further thermal decomposition of SMM to DMS occurs during wort boiling; however, the DMS so-formed is lost by evaporation (179,180,181). The breakdown of SMM is a first order reaction (179,180). Therefore, with a knowledge of the wort processing variables, it is possible to construct an equation that predicts the free DMS in the pitching wort from the malt SMM level (180). The concentration of SMM remaining in wort at the end of the boiling stage is very important, and can be controlled by varying the wort boiling time (179). During wort clarification in the whirlpool, SMM continues to break down to DMS, but the DMS is now retained in the wort (180). Therefore, the free DMS level in the pitching wort is dependent upon the amount of SMM remaining in the wort at the 'copper-cast' stage, and the time period between the end of wort boiling and wort collection (180).

The DMS content of most worts changes little between pitching and the end of fermentation (180) but there is biosynthesis and loss of DMS (182,183,184,185). SMM is utilised by yeast in methionine biosynthesis, but is not converted to DMS (189). Beer does contain a heat and alkali-labile DMS precursor (182), identified as S-adenosylmethionine (SAM), a yeast excretion product (183).

DMSO is converted to DMS by the enzyme DMSO reductase in yeast (144,184,185,194). The yeast DMSO reductase is NADPH - specific, and may be similar to yeast methionine sulphoxide reductase (195). In lager fermentations carried out in the United Kingdom the amount of DMSO reduction is low (144,184), and is very dependent upon yeast strain (177,185) and fermentation temperature (184). Ale yeast are more efficient at reducing DMSO than lager yeast (184). In contrast, the lower temperatures of lager fermentations favour DMS production from DMSO (184).

Methionine sulphoxide has been identified as an inhibitor of DMSO reductase present in brewery worts (195a), and may account for the observed failure of DMSO to be significantly reduced in identified instances in United Kingdom breweries (144). This may be related to the use of more highly modified lager malts in the United Kingdom, compared with continental Europe, since the level of methionine sulphoxide in worts is dependent 'inter alia' on the methionine content of malt (195a).

In addition to yeast strain and fermentation temperature, wort gravity (184), wort pH (175), and fermentation systems (170,182,196) have been reported to influence beer DMS levels. Disproportionate increases in beer DMS levels with increasing wort gravity have been observed in pilot plant brews (182,184), but these have not been confirmed on a commercial scale. Increases in wort pH lead to increases in beer DMS levels (175).

DMS is purged from solution by the fermentation gases, and is lost from open vessels (170). Thus, fermentations carried out in closed conical vessels tend to produce beers with higher levels of DMS (182,196).

Contamination of pitching yeast with Hafnia protea or Enterobacter aerogenes can result in the production of undesirably high levels of DMS, hydrogen sulphide, and other sulphur - containing compounds (143,186,187). Beer infected with H. protea has a characteristic 'parsnip' flavour, due to excessive levels of DMS (186). These wort enterobacteria are able to reduce the DMSO in wort to DMS (184), and the possession of a DMSO reductase allows them to regenerate NADP under anaerobic conditions (197).

The routes by which DMS is formed in beer from malt - derived DMS precursors, and their inter-relationships, are summarised in Fig.11.

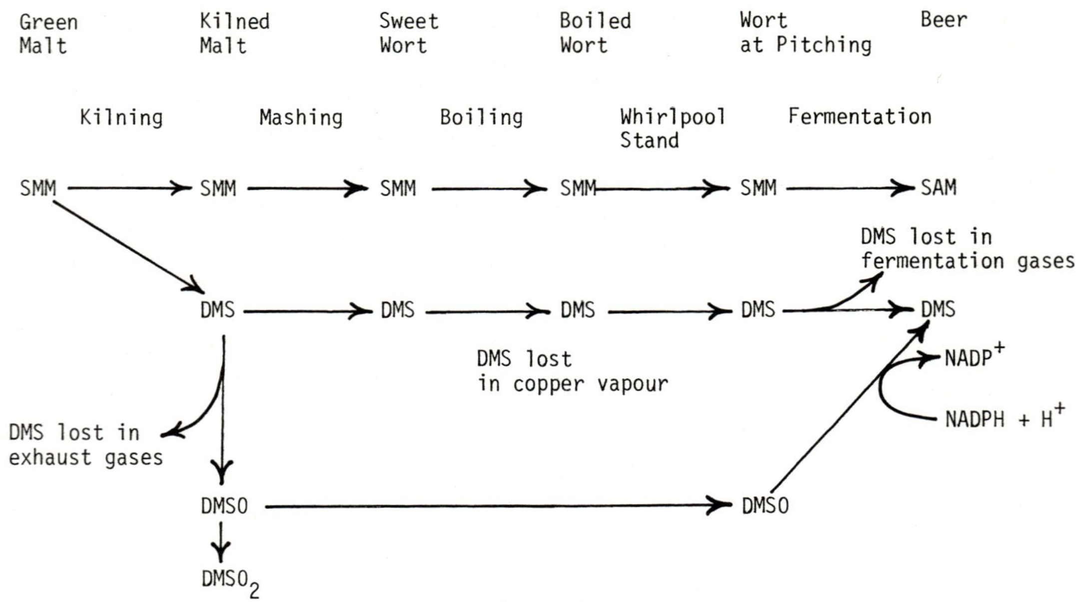

Fig. 11 The Inter-Relationships Between The Routes Leading To DMS During Malting And Brewing

C3.7.2 Control of DMS Levels in Beer

Since the levels of DMS in British lagers are close to the flavour threshold value, it is important for the brewer to be able to control the DMS levels in such beers, in order to produce reliably and consistently beers with the desired flavour.

The levels of SMM, DMSO, and DMS can be carefully monitored from mashing through to the completion of fermentation, to determine the relative contributions of SMM breakdown and DMSO reduction to the final beer DMS concentration. Alternatively, laboratory-scale fermentations, using wort previously purged to remove free DMS, can be used to assess the extent of DMSO reduction.

However, control of wort DMSO levels is not practical, because malts rich in DMSO have a high colour, and are therefore unsuitable for (low colour) lager production. The alternative is to control the thermal breakdown of SMM to DMS.

If there is sufficient SMM remaining in the wort at the end of wort-boiling, then the DMS level in the pitching wort is dependent on the time lapse between 'copper-cast' and wort cooling (144,179). Lengthening this time period increases the DMS level and, conversely, decreasing it reduces the DMS level.

When there is insufficient SMM remaining at copper-cast to give the desired DMS level, it is necessary to increase the level of SMM, either by using a well-modified, lightly-kilned malt (144,177,178), or by altering the wort boiling regime (144,179). The control of kilning schedules to produce malts containing high levels of SMM is well documented (171,177,178,193). Altering the brewhouse regime to lessen the thermal destruction of SMM is a viable alternative, and can be achieved satisfactorily without necessarily adversely affecting the physical stability of the finished beer. It may be necessary to reduce the boiling time to preserve SMM (179). However, it has been found that significant destruction of SMM takes place when worts are simmered during the time they are being collected in the wort boiler. Thus, in order to preserve SMM, worts can be collected in the wort-boiler at lauter-tun run-off temperatures (less than 75°C), and then boiled for the minimum acceptable time.

In summary, the DMS levels in finished beer are best controlled in practice by regulating the content of SMM in the malt, and controlling the wort boiling procedure in the brewhouse.

C3.8 Organic Acids

The most important organic acids found in beer are acetate, pyruvate, lactate, succinate, pyroglutamate, malate and citrate (198). In addition, α-ketoglutarate is usually detectable in beer, and α-hydroxyglutarate is present in some beers (198). Some of these organic acids are derived from malt, and are present at low levels in wort, their concentrations increasing during fermentation. Other acids are produced solely as a result of yeast metabolism.

Organic acids have a two-fold effect on beer flavour. Certain acids - pyruvate, succinate and lactate - may be present at concentrations above their flavour threshold, and modify beer flavour directly (199,200,201). Thus, at levels of 50-400mg/litre, pyruvate can alter beer flavour by affecting 'mouthfeel' (199). In addition, all acids contribute to beer pH. The accumulation of organic acids during fermentation leads to a reduction in pH which, in turn, affects the colloidal stability of beer, perceived bitterness, and the susceptibility of beer to microbiological spoilage (202). However, it has been found that the addition of organic acids to wort gives a smaller pH change than that which occurs during fermentation (202). Since the buffering capacities of wort and beer are similar, changes in the concentrations of basic amino acids and organic acids cannot explain the observed reduction in pH during fermentation. It has thus been concluded that 60% of the change in pH must be due to excretion of hydrogen ions into the medium (202).

C3.8.1 Formation of Organic Acids

The biosynthesis of organic acids during the fermentation of several alcoholic beverages has been extensively reviewed by Whiting (203). Organic acids formed early in fermentation are thought to be a result of amino acid metabolism, which produces α-keto acids. Thus α-ketoglutarate is formed by transamination of glutamate, and its concentration is dependent upon the levels of glutamate in the medium (198). Later in the fermentation, carbohydrate metabolism is responsible for the formation of acids, via the Embden-Meyerhof-Parnas pathway and certain reactions in the Kreb's cycle.

The patterns of accumulation of pyruvate and acetate are complex (198,199). Acetate accumulates during both the lag phase and early stages of fermentation, and is then re-utilised so that its concentration falls to a low level. At later stages during fermentaion, acetate begins to accumulate again, due to conversion of pyruvate to acetate by yeast (198). In the case of pyruvate, this acid accumulates during the growth phase, levels subsequently declining due to its conversion to acetate (198,199). The final concentration of pyruvate is therefore dependent upon the contact time between yeast and beer at the end of fermentation (199).

Under aerobic conditions, operation of the Kreb's or tricarboxylic acid cycle could explain the biosynthesis of many of the organic acids. However, during fermentation, the levels of Kreb's cycle enzymes are greatly lowered or even, in some cases, absent altogether (204). Under these strictly anaerobic conditions, the Kreb's cycle becomes non-cyclic and forms two branched pathways (204).

A key metabolite in both pathways is oxaloacetate, which is normally regenerated by the Kreb's cycle. Oxaloacetate can be formed by fixation of CO_2 by the action of pyruvate carboxylase (205), according to the equation:

$$\text{Pyruvate} + CO_2 + \text{ATP} \longrightarrow \text{Oxalacetate} + \text{ADP} + \text{Pi}$$

The oxidative part of the Kreb's cycle operates from the condensation of oxaloacetate and acetyl CoA to form citrate, through to α-ketoglutarate. However, as already stated, α-ketoglutarate may also be formed by transamination of glutamate (206). In the reductive branch of the Krebs cycle, reversed reactions occur from oxaloacetate through malate and fumarate, to succinate (203,204).

Fig. 12 summarises the main pathways involved in the biosynthesis of organic acids during fermentation.

C3.8.2 Control of Organic Acid Formation

Organic acids are important not only because they contribute directly to beer flavour, but also because they influence the pH of beer. The decrease in pH during fermentation affects beer properties such as non-biological haze stability, flavour stability, susceptibility to spoilage organisms, and perceived bitterness, which increases with increased pH. Beer pH is therefore an important characteristic, requiring careful control.

The final pH of beer is determined by the following factors:

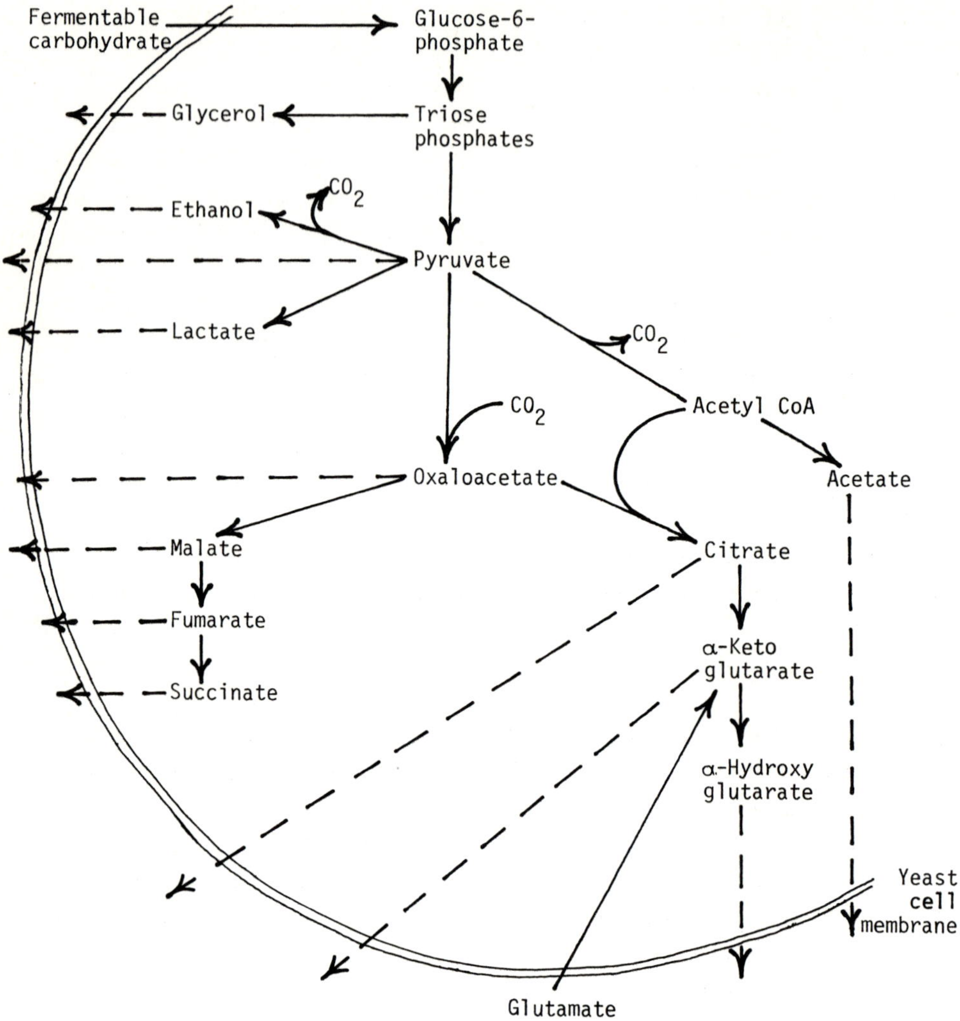

Fig. 12 Scheme For The Formation Of Organic Acids By Strains Of Brewers' Yeast

i the accumulation of organic acids during fermentation;
ii the absorption of basic substances from the wort during fermentation; and
iii the buffering capacity of the wort or beer (202).

Although the buffering capacities of a wort and its derived beer are similar, they can be influenced by the choice of raw materials and wort gravity. Thus, dilution of the nitrogenous constituents of wort with brewing syrups or low nitrogen adjuncts reduces the buffering capacity of wort.

The accumulation of organic acids is dependent upon a rapid, vigorous, fermentation; the fall in pH is greatly reduced in sluggish fermentations. Worts should therefore be pitched with viable healthy yeast to ensure a short lag phase and rapid fermentation (202). In some lager fermentations the pH is reduced by addition of lactic acid to the wort (202).

C3.9 Medium Chain Length Fatty Acids

Medium chain length fatty acids (C_6-C_{10}) are minor constituents of wort, increasing in concentration during fermentation and maturation. These acids can account for 85-90% of the total beer fatty acids. The production of medium chain length fatty acids during fermentation is associated with 'caprylic' flavour in some beers, whereas the medium chain length fatty acids formed during maturation, as a result of yeast autolysis, have been correlated with the development of adverse 'yeasty' flavours.

'Caprylic', 'goaty' or 'soapy/fatty' flavours have been recognised as common flavour characteristics in both lagers and ales, but they are more prevalent in lagers (207,208,209). Caprylic flavours have been associated with the aromas of hexanoic acid (C_6), octanoic acid (C_8), and decanoic acid (C_{10}) (210). It is interesting to note that the trivial names of these three acids, caproic, caprylic, and capric, respectively, are derived from the Latin word, 'caper', meaning goat.

The intensity of caprylic flavour in beer correlates with the concentration of octanoic and decanoic acids (208). At the low levels found in beer, hexanoic acid makes only a small contribution to caprylic flavour. The observation that caprylic flavour is more common in lagers is related to the tendency of certain strains of lager yeast to produce greater quantities of medium chain length fatty acids during fermentation than strains of ale yeast (208).

Marked variations in the flavour threshold values of the various medium chain length fatty acids have been reported (200,209,211,212).

This is due in large measure to the difficulties inherent in dissolving and dispersing these compounds quantitatively in beer (209,213). However, Clapperton (209) has recorded difference threshold values for octanoic, decanoic, and dodecanoic acids (C_{12}) of 4.5,1.5, and 0.6mg/litre, respectively. The flavour threshold values of these fatty acids are within the range of concentrations of these compounds in beer, and suggest that they make a significant contribution to the flavour of certain beers. As the effects of these acids on caprylic flavour are additive, small changes in their sub-threshold levels can still significantly affect beer flavour (209). 'Yeasty' off-flavours may develop during maturation due to the release of yeast autolysis products. The important factors governing the development of 'yeasty' flavours are the yeast concentration in contact with the beer, the maturation temperature, and the length of the maturation time (214,215,216). High yeast concentrations, long contact times, and high temperatures should be avoided (216). Recent studies have shown that hexanoic, octanoic and decanoic acids are released into beer from the yeast intra-cellular pool during maturation (215,216), the rate of excretion being temperature-dependent (216). Moreover, in the case of decanoic acid, there is strong evidence to suggest that this acid is still synthesised during maturation (215). These results therefore implicate the release of short and medium chain fatty acids from yeast cells in the development of 'yeasty' off-flavours during beer maturation.

C3.9.1 Formation of Medium Chain Length Fatty Acids

It is known that medium chain length fatty acids are only minor wort constituents, and that their levels increase during fermentation (85,217,218,219,220,221) and maturation (214,215,216). Their final concentration in beer is therefore associated with yeast activity. Initially there were two possible explanations for the accumulation of medium chain length fatty acids in beer during fermentation. Firstly, it was proposed that catabolic processes within the yeast cell degraded wort or yeast-derived long chain fatty acids to medium chain length fatty acid intermediates by β-oxidation (222,223). Alternatively, because medium chain length fatty acids are intermediates in fatty acid biosynthesis, it was suggested that they accumulated in the cell as a result of the anabolic processes involved in lipid biosynthesis (224).

In a series of experiments, Taylor and Kirsop (220) investigated the relative contributions of these alternative proposals to the final concentration of medium chain length fatty acids in beer. When worts were supplemented with heptadecanoic acid (C_{17}), no nonanoic acid (C_9) was found in the medium, despite incorporation of heptadecanoic acid into the yeast cells. In addition, yeast rich in heptadecanoic acid and pentadecanoic acid (C_{15}) was obtained by growth in wort containing high concentrations of propionic acid (C_3). When this yeast was pitched into normal wort, no nonanoic acid was detected in the beer. These results therefore indicated that the medium chain length fatty acids formed during fermentation are not a result of β-oxidation of wort or yeast long chain fatty acids (220).

When worts were supplemented with propionic acid, the excretion of nonanoic acid was dependent on the concentration of propionic acid (220). At high concentrations of propionic acid, the fatty acid composition of the yeast was altered, and the excretion of medium chain length fatty acids was suppressed. No nonanoic acid was detected in the resulting beer.

At low propionic acid concentrations, the fatty acid composition of the yeast was normal and nonanoic acid was detectable in the beer. Similarly, addition of valeric acid (C_5) to wort led to accumulation of nonanoic acid in the beer (220). These results therefore suggested that medium chain length fatty acids are formed by fatty acid biosynthesis during fermentation. Moreover, it was observed that the concentration of medium chain length fatty acids in beer was directly related to their level in yeast.

The findings of Taylor and Kirsop have been confirmed by Chen (225) using a different approach. Chen found that uniformly labelled long chain fatty acids in wort were taken up by the yeast and incorporated into yeast cell lipids. No labelled medium chain length fatty acids were found in the beer. However, on addition of uniformly labelled ^{14}C acetate to wort, labelled medium chain length fatty acids were found in the beer, thereby confirming the synthetic pathway as the main source of these acids in beer.

The biosynthesis of fatty acids and its inter-relationship with ester formation has already been described in Section C3.3.1. Fig. 13 shows diagrammatically the inter-relationship between fatty acid and ester synthesis in yeast.

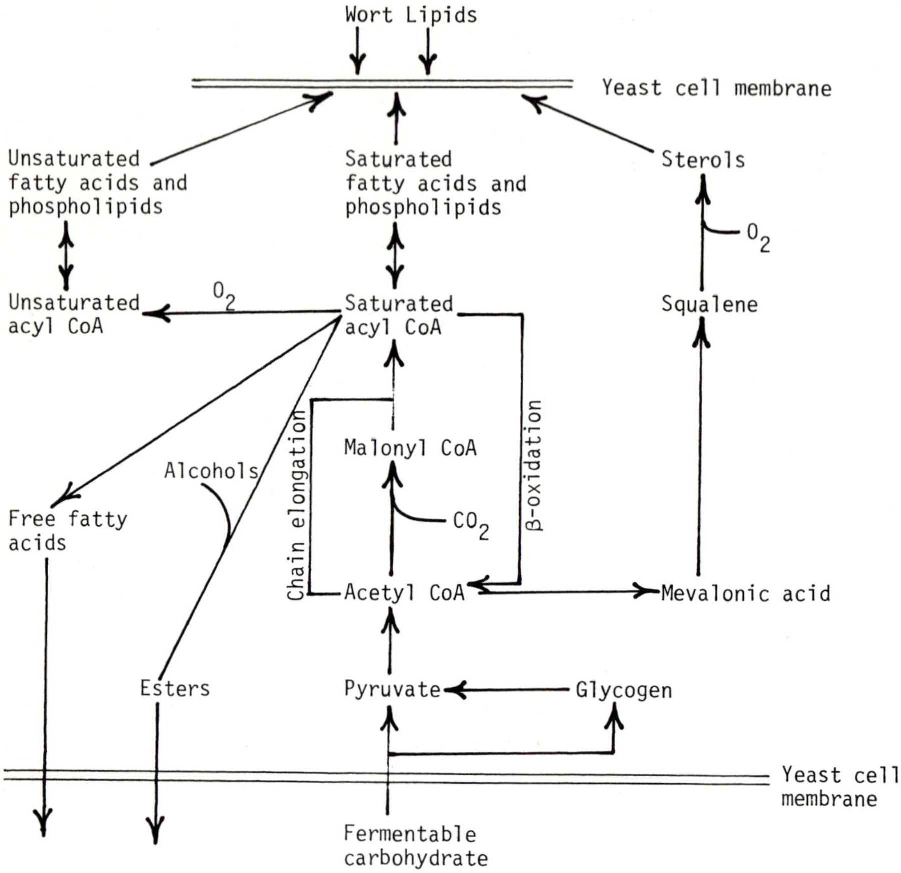

Fig. 13 <u>The Inter-Relationship Between The Biosynthesis Of Esters And Medium Chain Length Fatty Acids In Strains Of Brewers' Yeast</u>

C3.9.2 Control of Medium Chain Length Fatty Acid Formation

Because of the close relationship between ester formation and fatty acid biosynthesis, it is likely that the mechanisms operating to control ester levels are also applicable to the control of medium chain length fatty acids in beer. Thus, as with esters, the major factors influencing the concentrations of these acids in beer are yeast strain, the degree of wort aeration or oxygenation, the presence of wort lipids, wort composition, and pressure. Factors such as fermentation temperature (61,217,226), pitching rate (61,227), and fermentation system (227), produce variable results, and are of secondary importance in determining the concentration of medium chain length fatty acids in the final beer.

The importance of yeast strain to the final fatty acid levels in beer has been reported by several workers (102,208,227,228,229). In general, lager yeast strains produce more medium chain length fatty acids than ale yeast strains (208). In a study carried out at three different breweries with three different yeast strains, it was found that yeast strain was the most important factor determining the level of fatty acids in beer (102).

The effects of oxygen, or aeration, are also dependent on yeast strain (70). A number of workers have observed that increased aeration, or oxygenation, of the wort results in a decrease in the levels of medium chain length fatty acids (61,219,220,224,227). If aeration is continued after pitching, very large reductions in the levels of these acids are achieved (61,219,220).

The role of oxygen in fermentations has been reviewed at length (see Section C3.3.2). When yeast is transferred from anaerobic to aerobic conditions there is a shift in the fatty acid spectrum of yeast towards long chain unsaturated fatty acids (219,230). Under fermentation conditions, long chain saturated fatty acids inhibit the activity of acetyl CoA carboxylase, and medium chain length fatty acids tend to be released (231). In the presence of oxygen, the biosynthesis of unsaturated fatty acids is stimulated and the inhibition of acetyl CoA carboxylase is removed. Under these conditions, the medium chain length intermediates are more effectively elongated to C_{16} and C_{18} fatty acids, and as a result, the intracellular pool of medium chain length fatty acids is reduced (231).

Additions of unsaturated fatty acids, such as oleic acid, linoleic acid, or complex mixtures in spent-grains pressings, substitute for the effect of oxygen on yeast metabolism (59,82,220,232). At low concentrations, oleic acid inhibits the synthesis of medium chain length fatty acids, but at higher concentrations the formation of decanoic and dodecanoic acid is stimulated (82).

Unsaturated fatty acids are incorporated into cell membranes, and alter the pattern, rather than the total amount, of fatty acid biosynthesis. It has been postulated that the incorporation of unsaturated fatty acids alters the membrane properties, thus affecting the activity of acetyl CoA carboxylase, and therefore the chain lengths of the fatty acids which are produced by fatty acid synthetase (231).

The influence of variations in wort composition on the formation of medium chain length fatty acids has also been studied by a number of workers (61,217,227,229,233). The most important factor appears to be the ratio of nitrogen to carbohydrate. Thus, the addition of sugar solutions to wort dilutes the nitrogen content, and diminishes the levels of medium chain length fatty acids in the final beer (61,229,233). However, when the fermentability of wort is increased with glucose syrup containing yeast extract to maintain the α-amino nitrogen content constant, there is a slight increase in the level of medium chain length fatty acids. This was considered to be a result of dilution of the unsaturated fatty acid concentration in the wort (229). Unlike ester formation, the production of medium chain length fatty acids is proportional to the original gravity of the wort (229).

Finally, the effect of hydrostatic pressure, or more correctly carbon dioxide concentration, is different for esters and for medium chain length fatty acids (61). Purging with nitrogen, or a reduction in pressure, increased both the levels of acetate esters and yeast growth, but decreased the formation of octanoic, decanoic, and dodecanoic acids, and their esters (61). Pressure increases above atmospheric pressure, raise the levels of decanoic and dodecanoic acids, but have little effect on the concentration of hexanoic and octanoic acids (61).

C3.10 Non-Volatile Flavour Components Influenced By Fermentation Conditions

Of the four basic taste sensations: sweetness, bitterness, acid, and salt, the influence of the fermentation process on acidity has already been covered (Section C3.8), and the influence on sweetness and bitterness can be summarised as follows:

i Sweetness - this is determined in beer either by the level of fermentable sugars remaining at the end of fermentation, the so-called 'fermentable residue', or by the post-fermentation addition of 'priming sugars'. This latter practice is the preferred manner in which to achieve a controlled level of sweetness in a beer, since sugars vary in their ability to be perceived as sweet. Thus maltose, quantitatively the major wort fermentable sugar (6) and, together with maltotriose, the last to be utilised by yeast (20,234), does relatively little to increase sweetness. It is common practice therefore, for brewers to add controlled amounts of the sweet sugar sucrose (or less commonly glucose) to beers in which sweetness is a characteristic of the perceived flavour.
The concentration of alcohol in a beer can also contribute to the level of perceived sweetness.

ii Bitterness - the bitter characteristics of a beer are determined primarily by the hop resins extracted during wort boiling, in particular the so-called 'α-acids' (235), which are isomerised during wort boiling to 'iso α-acids'(235). These compounds are extremely surface-active, and are removed from solution to a greater or lesser extent during the subsequent stages of beer production and transfer - on vessel walls, in the foam produced during fermentation, and on yeast (236). Surface/volume ratios in fermentation vessels, and the amount of foam produced during fermentation can therefore significantly affect the loss of hop bitters from a beer, and thus the rate at which hops must be added to worts in the copper.
The presence of high levels of SO_4^{--} ions in beer can also contribute to the perceived bitterness (237).
There are two remaining perceived flavour characteristics of beer, the levels of which are influenced by the fermentation process: beer 'fullness' or 'body', and beer 'smoothness'. These are determined principally by the level of alcohol in beer (238), less significant contributions being made by maltose and glucose, and by carbon dioxide levels.

D. OTHER FACTORS AFFECTING THE DEVELOPMENT OF MODERN FERMENTATION SYSTEMS

In addition to the requirements imposed by the nature of the fermentation process itself, and the restrictions resulting from the effects of the fermentation process on beer quality (see Section C), five other factors can be listed as influencing and/or restricting the development and selection of fermentation systems over the last two decades. These are:-

i technological developments - both in basic engineering and in the brewing process itself;

ii economic requirements;

iii market requirements;

iv legislation;

v the requirements of other plant and equipment within the brewery.

Each of these factors is in itself complex, as is evident in the following sections, where their influence on the development of modern fermentation systems is reviewed in turn.

D1. Developments In Engineering Technology

Advances in engineering technology, particularly over the past thirty years, have given the brewer many opportunities to change and improve his fermenting systems. The main developments lie in four areas:

i Materials of Construction

ii Cleaning-in-Place

iii Automation and Control Equipment

iv Carbon Dioxide Recovery and Recycling

D1.1 Materials of Construction

The properties required of materials depend on detailed applications and local conditions, eg. cleaning techniques and available technology. However, several basic requirements can be identified:

i Rigidity - materials for applications such as vessel walls must be self-supporting (or capable of being economically supported), and capable of bearing pressure and weightloads.

ii Ability to be shaped and joined - both in manufacture and in subsequent modification.

iii Good thermal conductivity for heat exchange surfaces.

iv Chemically inert to beer, wort, and chemical cleaning solutions.

v A smooth working surface to enable dirt and microorganisms to be easily removed, and to minimise the deposit of insoluble calcium salts or scale, known as 'beer stone'.

vi Economic to use.

vii Impervious to liquids and gases.

With the increased tendency to install vessels outside, and the need for improved temperature control, lagging materials have assumed a greater importance.

A wide range of materials have been, and continue to be, used in the fabrication of fermentation equipment:

a. Wood

Wood is probably amongst the earliest materials used, particularly for vessel construction. Nowadays, a bitumastic lining is normally applied to give a smooth waterproof interior coating.

In modern breweries wood is too costly to fabricate and maintain, and offers an inferior interior surface to modern materials. However, the use of wooden beer casks is still favoured by traditionalists, despite its many disadvantages.

b. Stone and Slate

Although widely used in the last century, because its high thermal capacity offered consistent temperatures for both fermentation and settling vessels, stone is too costly and inflexible to use in modern breweries.

c. Copper and Brass

Copper has been widely used in the brewing industry for vessels and equipment, often in conjunction with brass or gun metal fittings, where greater mechanical strength is required. Copper has the advantages of being easily worked, having excellent heat exchange properties, and is generally non-reactive with beer and cleaning materials.

However, the high cost of copper in modern terms has lead to its replacement by stainless steel as the material of choice. Copper is also too soft to fabricate pressure or vacuum vessels without extensive support, and is prone to slow erosion due to its solution in beer and cleaning solutions. Copper is normally jointed in vessel manufacture either by riveting (which is prone to leaks and gives an unhygienic inner surface) or soldering. However, modern food regulations require expensive lead-free solders to be used to minimise lead pick-up in beer.

d. Concrete

Concrete has been widely used in vessel manufacture (normally lined with metal, plastics, or butumastic coating), particularly for open vessels. Earlier vessels suffered from many of the disadvantages of stone in terms of weight and difficulties in modifying. However, modern concrete technology, and the developments in improved linings, make the use of large enclosed concrete vessels a possibility for the future.

e. Mild Steel

Mild steel is a cheap and versatile material for the manufacture of most equipment, but is attacked chemically by most aqueous solutions. Early mild steel vessels were lined with glass, but suffered from the disadvantages of chipping, resulting in exposure of the base metal, and difficulties of modifying. More recently a wide range of epoxy resins, which can be repaired 'in situ', have been used as lining material (239).

Although early materials had problems due to chipping and flavour taints, modern resins are totally acceptable (240), and the use of lined mild steel vs. stainless steel is a balance of local, capital, and repair costs.

f. Stainless Steel

Stainless steel has been the material of choice for most brewers since the 1950's, although relative costs have changed rapidly with world market prices of the nickel and chromium required for its manufacture. The chemical inertness of stainless steel (with the notable exception of high chloride ion concentrations (241)), enables equipment manufactured in stainless steel to have a very long, maintenance-free, life.

Cold-rolled stainless steel sheets provide an excellent smooth surface (239,242), and can be pressed or rolled into a wide variety of shapes. Pipework can be formed either by rolling sheet metal, or by spinning the molten metal to give a seam-free material.

Welding techniques require the use of sophisticated inert gas systems and, although a flowed weld gives a satisfactory surface for pipework, time-consuming grinding and polishing of welds is required for vessel manufacture.

Stainless steel is sufficiently rigid to form a wide range of brewing vessels unsupported. However, the gauges of metal required for very large pressure vessels are prohibitively expensive (240).

To overcome this problem, thin sheets of stainless steel can be cold-rolled onto mild steel, which acts as a mechanical support. This method, although offering substantial material savings, requires more sophisticated welding techniques, and so fabrication costs may be increased.

Although the thermal conductivity of stainless steel is considerably less than that of copper, its greater rigidity enables thinner gauges to be used, making it a satisfactory material for heat exchange. Modern plate heat exchangers are built almost exclusively from stainless steel, and most designs of vessels can be attemporated by heat exchange panels in the vessel wall, which replace less hygienic coils within the vessel.

g. Aluminium

Aluminium has been used extensively in the past as a lining for vessels, and is still popular in some third world countries, due to its light weight for transport even with its associated frames. The metal can be fabricated into most shapes and smooth joints can be obtained, although welding techniques are costly. Notable disadvantages of aluminium are its softness, which results in it being easily damaged by dents and scoring, and its rapid corrosion by the alkaline chemicals used in many cleaning applications.

h. Plastics and Epoxy Resins

Although polypropylene has been widely used for the manufacture of small vessels and lining traditional vessels (243), and epoxy resins have been used for many years to line mild steel vessels, a widespread resistance to their use remains. This is possibly due to the early reputation of plastics for chipping and imparting flavour taints, and ignorance of modern plastic technology (240,243).

The low modulus of rigidity of plastics ($3 - 8 \times 10^5$psi) and reinforced plastics (approximately 30×10^5psi), compared with metals ($100 - 450 \times 10^5$psi), limits the maximum size of vessels when dimensional stability is important (eg. in collection vessels for volume measurement). However, vessels of upto 30,000brls (50,000Hl) have been constructed from bisphenolic polyester resins, strengthened with glass-fibre applied by filament (240).

Other disadvantages of plastics include poor heat transfer characteristics, a tendency to age by oxidation, and a susceptibility to damage by some chemicals and excess heat. However, the economics of construction, a wider acceptance of plastics as beer containers, and improved technology, make plastic equipment a likely prospect for the future.

i. Lagging Materials

The major requirements of lagging materials are that they should be;

i non-toxic and non-irritant,

ii resistant to fire,

iii effective thermal insulators,

iv chemically inert (non-corrosive),

v capable of being sealed against ingress of moisture.

Most lagging materials are covered with a sealed metal or plastic layer both for decorative effect and in order to prevent wetting, either by direct wetting or by condensation. The most common insulating materials are mineral wool, urethane insulants, and phenolic foams. Urethane materials suffer from the disadvantage that they can release toxic fumes in the event of fire, and their use is either restricted or banned by some authorities (244). Some materials also contain high levels of soluble chlorides, which can corrode stainless steel when water and heat are present (245).

D1.2 Cleaning-in-Place (CIP)

'In site' cleaning systems were probably developed primarily because of their labour-and time-saving potentials (246). However, the effect of the development of highly efficient cleaning systems has been to:

i improve microbiological standards due to greater cleaning efficiencies, and make sterilisation practical, if required;

ii permit the installation of enclosed vessels to contain the cleaning spray;

iii encourage the installation of simple vessel shapes and pipework systems to facilitate effective cleaning (although some modern complex systems would be totally impractical to clean manually, and can only be cleaned with modern automated CIP equipment);

iv enable very large vessels to be cleaned;

v facilitate CO_2 recovery, particularly when acid- based detergents (which do not react with CO_2) are used;

vi greatly reduce cleaning costs;

vii improve operator safety by automating the handling of hazardous cleaning solutions, and eliminating the necessity for men to enter vessels potentially containing CO_2.

Most CIP systems are fundamentally simple, and the design of the fermentation equipment itself determines the efficiency of cleaning. Good design requires that all working surfaces are exposed to vigorous cleaning flows, thus the need to avoid 'dead-legs' in pipework systems, and 'shadows' or stagnant liquid traps in spray-cleaned vessels.

The basic cleaning cycle is common to all systems:

i removal of loose soil by water rinsing - pre rinse;

ii removal of soil by chemical action and heat, if required - detergent wash;

iii removal of chemical by clean water rinsing - post rinse. Post rinse water can be recovered for use as pre-rinse in the next cycle, thus facilitating detergent and water economies.

iv sterilisation by chemical circulation or steaming.

Pipework systems are cleaned by routing flows through the whole system at sufficient velocity to ensure that the systems are full and that highly turbulent flow exists.

Vessels may be cleaned either:

i by high pressure rotating sprays - these give a good mechanical cleaning action, but are expensive to install, operate, and maintain, and may also cause excessive foaming, particularly where yeast residues exist; or

ii by low pressure sprays which are directed at the roof of the vessel, and clean the sides and base by a cascade action. These systems are more dependent on the effect of chemical cleaning, but can be cheaper to use.

Detergents have traditionally been based on caustic soda, but may include added sequesterants and chelators. Such detergents are normally used at approximately 1% active caustic, and at temperatures of 40-60°C to lower surface tension and promote wetting. Low foaming chemical wetting agents can be used to reduce cleaning temperatures and save energy, where this is economic (247,248). Detergent solutions are normally recovered for re-use, because of their value in terms of both the chemicals and the heat they contain. However, some systems use lower concentrations and/or lower volumes of detergent and are designed for single useage. Hypochlorite solution (to give up to 80ppm free chlorine) can be added to reinforce the low detergent strength. These systems can be not only efficient but also cost effective (249), especially in situations where high soil or CO_2 lead to heavy contamination of bulk tanks, or where cleaning loads are infrequent, making storage of hot detergent expensive in energy terms. Such 'single use' systems often incorporate a dilute solution of phosphoric and nitric acids as a final rinse to act as a mild sterilant, remove 'beer-stone' not removed by the action of dilute caustic detergents, and restore the corrosion-resistant oxidation coating of stainless steel removed by the action of chlorine (241).

Acid detergents (based on phosphoric and/or nitric acid, normally with low foam wetting agents) have been introduced into routine use in some breweries for cleaning vessels (250,251). Although these are more expensive per unit volume than caustic, they have a number of advantages which may make them more cost effective:

i they produce a low pH effluent which may be required by local effluent disposal authorities;

ii they are unaffected by CO_2 (although sodium carbonate, produced by the reaction of CO_2 with alkaline solutions, has some detergent power);

iii they can be effective cold and do not absorb CO_2, thus saving energy and eliminating the two major causes of vacuum collapse of vessels (thereby reducing the cost of vacuum protection);

iv they enable a CO_2 environment to be maintained during CIP, thus reducing the ingress of air, and preventing the oxidation of beer during subsequent vessel filling;

v they are more effective at removing 'beer-stone' and hop resins than caustic detergents;

vi they are more effective against gram-positive bacteria (251);

vii sterilants can be encorporated in the detergent formulation.

D1.3 Automation and Control Equipment

Development of reliable, automatic, control of the fermentation process, to a higher degree of accuracy and consistency than could be achieved manually, has led directly to the development of larger batch sizes and improved beer quality. The ability to measure and control process parameters and fermentation equipment remotely has also enabled larger and more complex systems to be designed, which can be installed in less hospitable environments than would be required for purely manual control.

Electronic and electro-mechanical instruments, measuring fermentation control parameters, are capable of feeding signals directly to modern, reliable, and sophisticated electronic control systems (252). In many countries, breweries have been built where the entire brewing process is now under computer control, with a single operator monitoring the plant. However, with a few notable exceptions (253,254,255,256), British brewers have been relatively slow to apply sophisticated sequence control equipment to what is fundamentally a simple process. Arguments against computerisation have included:

i the high capital cost;

ii the loss of flexibility, eg, in response to changes in raw materials quality or yeast performance;

iii the increased number of highly skilled staff required to maintain and operate sophisticated equipment;

iv the complexity of changing the process to meet changing needs;

v historically, the lack of reliability of control equipment, particularly field sensors.

Nevertheless, the increasing accuracy and reliability of electronic sensing instruments, coupled with the relatively low cost and simplicity of the latest generations of micro-processors and computer control systems, can be expected to result in their wider application in the UK brewing industry.

D1.4 Carbon Dioxide Recovery and Recycling

The rise in cost of the fossil fuels used to produce industrial CO_2 has made the economics of CO_2 recovery more favourable. Other arguments in favour of CO_2 recovery from brewery fermentations include the independence to be gained from external suppliers, and national laws prohibiting the use of industrial CO_2, such as exist in West Germany (257).

Theoretically, approximately 14.71b per brl (4.1Kg per Hl) of CO_2 is evolved from a typical fermentation of a wort of SG 1040° (10^oPlato). System losses, due to solution in product, vessel and system purging losses, leaks, etc., may account for over 50% of the total gas generated, but the remainder is more than is required to make a brewery self-sufficient in CO_2 (258).

The cost of liquefying CO_2 for storage is high in plant and energy terms (257). However, this cost can be significantly reduced by CO_2 conservation measures within the brewery, and the use of low pressure recycling systems, which conserve CO_2 without incurring the costs of liquefaction (259). When CO_2 must be liquefied - for long term storage - the liquid gas evaporator can be coupled with the primary refrigeration systems within the brewery to recover the latent energy within the gas (260).

D2. Developments in Brewing Technology

Advances in understanding of the fundamental processes taking place during fermentation have enabled the modern brewer to make major modifications, and to control his process to a far greater extent than ever before, resulting in both improved beer quality and reduced costs.

D2.1 Control of Yeast

D2.1.1 Use of Pure Cultures

The pioneering work of Hansen (4) resulted in the introduction of pure-culture methods to lager brewing in Europe at the end of the 19th Century. However, acceptance of the new techniques by British ale brewers was slower, due in part to the complex mixtures of yeast then used in many British breweries, and the difficulties in isolating a single strain capable of performing in a comparable manner to a mixture. Thus, problems with lack of secondary fermentation, poor fermentation performance, and failure to fine were not uncommon results of early attempts at introducing pure cultures in United Kingdom breweries (261). Neverthless, new fermentation systems and an increase in lager brewing, have inevitably lead to the use of pure culture yeast in most British breweries. It has been suggested that the widespread adoption of pure cultures in the brewing industry worldwide, and the concomitant reduction in the potential of spoilage microorganisms to influence the quality of finished beer, have resulted in a general increase in the degree of uniformity in beer characteristics (262). The extent to which this is true is difficult to assess. However, on balance, any loss in the individuality attached to some beers, because of the (uncontrolled) action of spoilage organisms, is more than offset by the obvious benefits to be derived from the use of pure cultures, in terms of the improved ability to predict and control beer quality and consistency.

D2.1.2 Yeast Propagation

In essence, pure culture techniques provide a reliable source of a standard yeast with known characteristics, and maintain that yeast in a pure form, free from competition or infection by other strains of yeast or bacteria.

The technique involves the selection of yeast from a brewery fermentation, its subculturing under laboratory conditions to obtain a pure single cell isolate, and the reintroduction of this yeast into the brewery after growth to a suitable inoculum size in a yeast propagation plant. Because yeast is normally selected only from those fermentations which have given totally satisfactory results, in terms of yeast performance and beer quality, the procedure introduces a selective pressure on the yeast. It also provides an effective means of controlling bacterial infection in pitching yeast, and therefore subsequently in the beer.

Single cell isolation can be achieved by micromanipulation, but is more commonly assumed by isolation of a single colony from a streak plate. The risk of a single colony being developed from more than one cell can be reduced to practically insignificant levels by repeated reisolation (261).

Having isolated a culture stock, the yeast is normally checked for type, usually by an examination of its giant colony morphology (263) and flocculation characteristics (264,265). Further tests for specific characteristics of a particular yeast strain might include growth factor requirements, the formation of specific by-products, and performance in laboratory-scale fermentations (266).

Using the latter technique, parameters such as fermentation rate and extent, the extent of yeast growth, and yeast flocculation characteristics can be assessed. In addition, the resulting beer can be analysed, particularly for yeast-derived flavour components (267). As with the other tests, results can be compared with those from a 'standard' yeast, taken direct from a brewery fermentation.

Once a yeast strain has been isolated, tested, and judged acceptable for brewery use, it is essential for it to be maintained in a pure form, often for prolonged periods. A popular method of storing yeast cultures is in liquid or on solid MYGP medium at 4°C (268), subculturing at 2-3 monthly intervals. Yeast can also be maintained in the freeze-dried state (269), although this technique can result in excessive loss of viability (270). An alternative method, which appears more promising, involves storage of yeast cultures in liquid nitrogen (271).

Propagation schemes vary slightly from brewery to brewery, but all owe much to the basic techniques developed by Hansen. A typical procedure is described in the figure below:

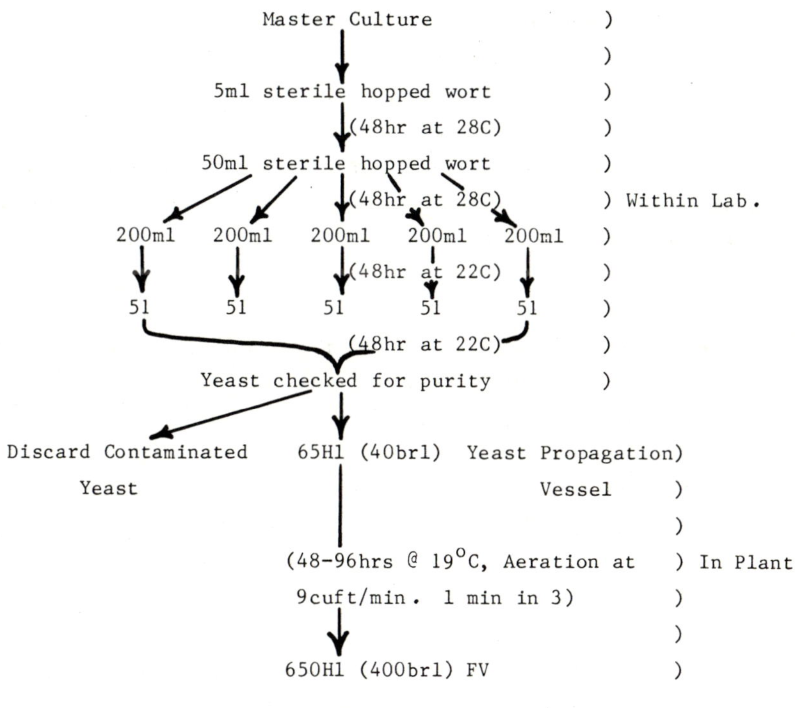

Fig. 14 Scheme Showing The Stages Between a Laboratory Master Culture And The Final Inoculum For a Brewery Fermenting Vessel (261).

Although a ten-fold dilution is generally considered the ideal maximum in such a culturing system, substantially higher dilutions can be used when control of sterility is adequate. Development of yeast during a propagation cycle is normally monitored by estimation of growth (cell count/ml), microscopic examination, and differential plating procedures to detect infection or atypical yeast morphology(268).
Aeration in the culture vessel may be altered to suit the growth requirements of individual strains. It is also possible to combine yeast strains in a propagator when mixed strains are required. Fig. 15 shows a typical modern yeast propagation plant.

D2.1.3 Genetic Manipulation of Brewing Yeast

It is almost fifty years since it was first shown that Saccharomyces species exhibit both haploid and diploid phases in their life cycle (272). However, despite the efforts of a number of workers over the intervening years, in both pure and applied aspects of yeast genetics, the results, in terms of application within the brewing industry, have to date proved disappointing. There are four main reasons for this lack of progress:

i Most strains of brewing yeast are polyploids, with a very poor ability to sporulate. Moreover the ascospores, once formed, have only a low viability (273,274,275).

ii Up to relatively recent times, there has been great difficulty in clearly defining yeast-dependent aspects of the brewing process or of beer quality amenable to genetic manipulation.

iii Attendant with (ii), the difficulties in defining the properties desireable in a yeast for a given brewery situation (276,277,278).

iv The degree of difficulty involved in modifying any one property of a yeast without changing any of the other (desireable) properties (279,280).

The problems with regards to definition of the interrelationships between yeast behaviour and the brewer's requirements have diminished in recent years, mostly as a result of the increases in detailed biochemical understanding of yeast metabolism and its influence on beer characteristics. Thus today, a number of potential applications and targets for the yeast geneticist can be identified, in terms of either

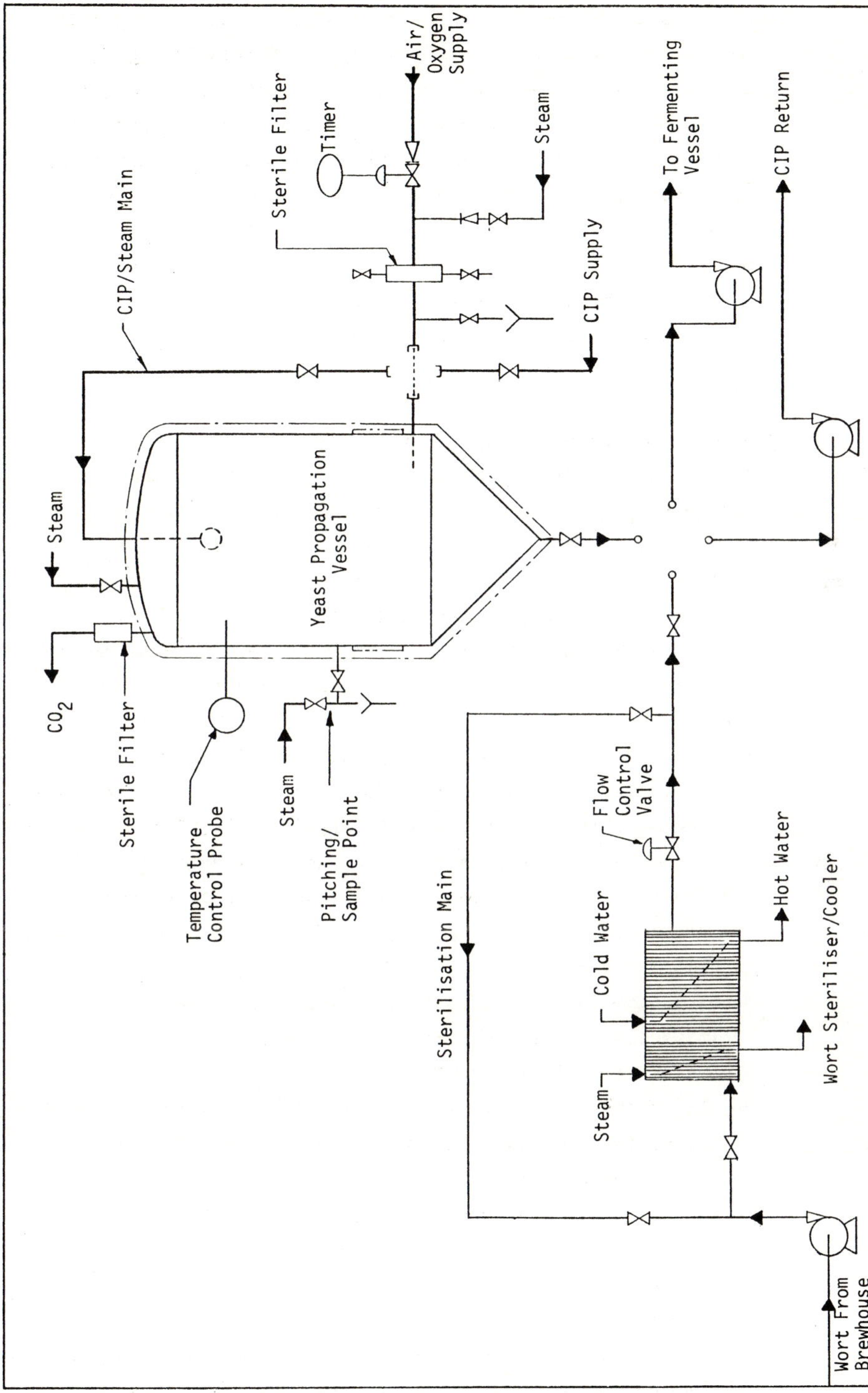

Fig. 15 Schematic Diagram Of A Typical Modern Yeast Propagation Plant

deleting certain yeast characteristics, amplifying others, or introducing totally new characters (168,278,280).
Moreover, the advances in molecular genetics and genetic engineering made over the last decade have largely removed the dependency on sporulation to achieve recombination of genetic material from different yeast strains (167). Another significant advantage of the techniques now available, is that they enable specific characteristics of yeast to be altered, without incurring any adverse effects on other characteristics.
Thus, in addition to the classical techniques of sexual recombination (hybridisation) (273,274,276,281), have been added 'rare-mating' (282,283,284), spheroplast (protoplast) fusion (167,285,286), and transformation using either crude, whole - DNA preparations (167,287,288), or discrete sequences of DNA inserted into plasmid vectors(284,289,290).
It is these latter techniques especially which enable highly specific changes in the yeast genome to be achieved. (For comprehensive reviews of the techniques currently available in yeast genetics and the potential for their application to brewing yeast see refs. 167,168,278,279,280).
The above methods, allied to the more traditional approaches of mutation (either spontaneous or induced (166)), and selection (which brewers have always (unwittingly) imposed by the manner in which they select yeast for repitching (see Section E7)), give the present day geneticist considerable scope. Perhaps at last therefore the genetic manipulation of brewers' yeast is becoming a practical proposition.
Thus, in the last three years reports have appeared of the construction of strains of brewing yeast with the ability to assimilate and utilise maltotriose (287), and even wort dextrins (284,286). In addition, an induced mutant of a strain of brewing yeast with a reduced ability to produce acetate esters from glucose, has been used to combat the flavour shift resulting from fermentation of worts treated with amyloglucosidase (238).

D2.2 Accelerated Fermentation

In response to the economic pressures stemming from the capital cost of additional plant, and the cost of financing stock in process, a major part of the developmental work carried out on fermentation systems over the past thirty years has been aimed at reducing both fermentation and maturation times (35).

In addition to giving the brewer increased insight into the various factors affecting the rate of fermentation, and their interrelationships, the net result of much of the work on accelerated fermentations has been to show the brewer how he can best aim to optimise his process. Thus, in many instances, he has been able to significantly reduce fermentation times without impairing the quality of the finished beer - and without incurring any hidden on-costs. The two major determinants of fermentation rate-and the two most readily under the brewers control-are the yeast pitching rate and the fermentation temperature. However, other variables capable of exerting a significant effect are yeast strain, wort original gravity, wort composition, and the concentration of the end products of fermentation - especially carbon dioxide and ethanol.

D2.2.1 Temperature and Pressure Effects

In general, increases in temperature (up to approx 30°C) result in accelerated fermentations because there is both a reduction in the initial lag period of fermentation and an increase in fermentation rate (see Fig. 16).

The magnitude of the increase in rate varies, depending on the other factors mentioned above, especially yeast strain, but typical figures show a reduction in primary fermentation time from 10 days at 8.5°C to less than 2 days at 20°C (35). However, in addition to the acceleration in fermentation rate, both the extent of yeast growth and the level of yeast-derived flavour volatiles (especially higher alcohols and acetate esters) are significantly increased at high fermentation temperatures (33,34,35,37). Thus, over the same range of fermentation temperatures mentioned above, iso-butanol concentrations at the end of primary fermentation increased from approximately 7mg/litre at 8.5°C to approximately 16.5mg/litre at 20°C (35).

The mode of formation of these and other classes of yeast-derived beer flavour components has already been reviewed (see Section C3), together with the various options available for their control. However, it is worth mentioning in the context of accelerated fermentations, that the application of CO_2 pressure during fermentation has been shown by a number of workers to be effective in combating the excess production of higher alcohols found at elevated fermentation temperatures (34,35,36).

Pressure fermentations also exhibit lower levels of some of the

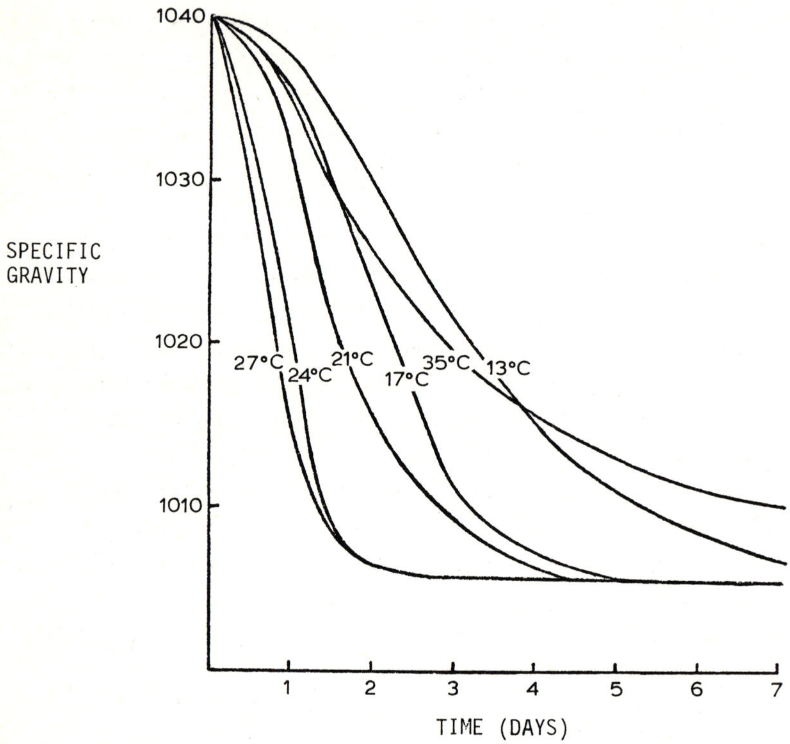

Fig. 16 <u>The Effect Of Temperature On Fermentation Rate</u>

other flavour-active beer components, notably esters (34,61), diacetyl (34) and acetaldehyde (93). However, the magnitude of the effect varies, indicating the influence of other factors - notably yeast strain. The action of increased CO_2 pressure during fermentation, in retarding the production of flavour compounds, is undoubtedly linked to the inhibitory effect of increased CO_2 concentrations in solution both on yeast growth and fementation rate (61,93,94). However, the exact mode of action of CO_2 is as yet unknown (see Section C3.3.2 and ref. 95).

The application of elevated CO_2 pressures to fermentations, in order to counteract the unacceptable flavour shift caused by increased temperatures, thus negates to a certain extent the acceleration in fermentation rate. Nevertheless, by careful adjustment of both parameters, it has been shown that significant increases in rate can be achieved whilst maintaining an acceptable flavour balance (34,35,92). However, the cost of flavour control, in terms of both increased beer losses (due to the increased yeast growth incurred), and the higher capital cost of pressure fermenters, must be balanced against any savings made due to accelerated rate. Nowadays, typical fermentation temperatures can be up to 14°C for lagers and 22°C for ales. These compare with the more traditional figures of 8-10°C for lager fermentations and 15-17°C for ales. At such temperatures the time for primary fermentation of lagers is reduced typically from approximately 10 days to about 3 days, and for ales from 5 days to 2-3 days, depending on pitching rates.

D2.2.2 Pitching Rate

In simple terms, the more yeast that is added at the start of fermentation, the shorter the time taken for the fermentation to reach its end-point (racking gravity). This relationship holds up to a certain maximum pitching rate, which varies depending on individual yeast strains (see Fig. 17).

The reason for the shorter fermentation time is that, at high pitching rates, there is both an increase in the rate of fermentation and a decrease in the lag period at the start of fermentation. However, the amount of yeast growth occurring also increases as pitching rates are raised, up to a maximum after which it declines (see Fig. 18).

Again therefore, as with increases in fermentation temperature,

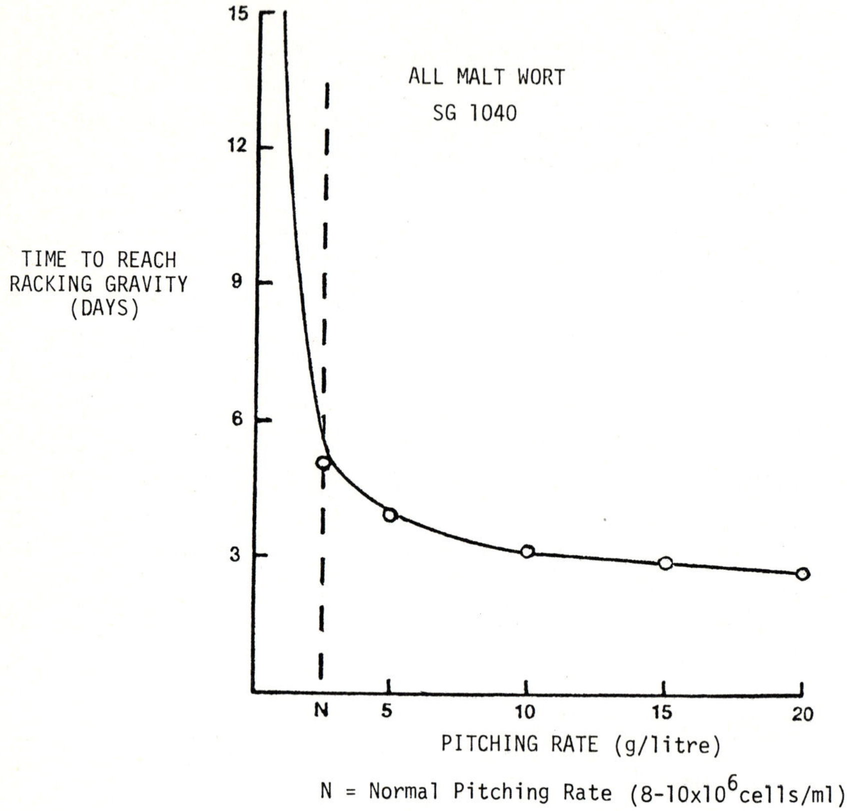

Fig. 17 <u>The Effect Of Pitching Rate On Fermentation Time</u>

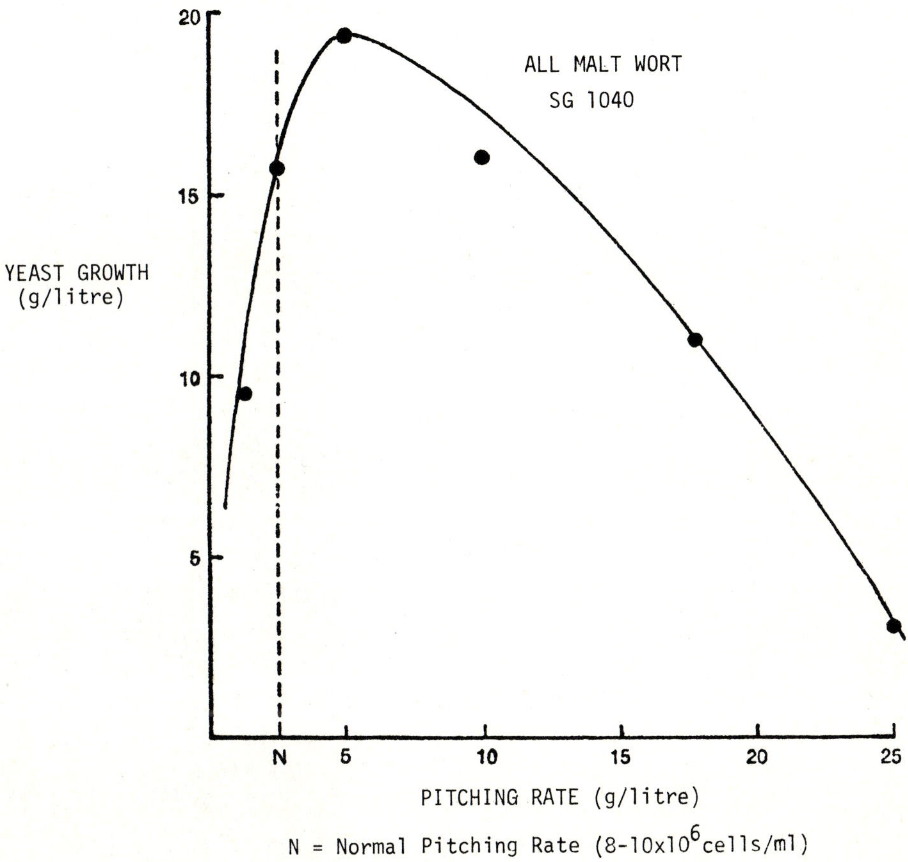

Fig. 18 The Effect Of Pitching Rate On Yeast Growth

increased pitching rates result in accelerated fermentations, but at the expense of increased yeast growth and associated beer wetting losses. Moreover, since these changes are combined with increases in the levels of yeast-derived flavour compounds (particularly higher alcohols and esters), there can also be an unacceptable flavour shift in the finished beer (33,38,39).

The magnitude of this flavour shift is very much dependent on the particular yeast strain, and can be compensated for to varying extents by a number of the techniques discussed in Section C3 for the control of higher alcohols and esters. Nevertheless, excessive yeast growth represents a reduction in fermentation efficiency, since the primary objective of any brewery fermentation is to produce beer, and not yeast cell mass. The cost of achieving accelerated fermentations by permitting excessive yeast growth is therefore unacceptable, especially in countries where tax is paid on the wort. Thus, pitching rates are normally limited to between 8 - 12 x 10^6 cells/ml for worts of OG 1040, and yeast growth to a maximum increase of four-five times the pitching rate.

The explanation for the initially somewhat surprising finding, that the extent of yeast growth increases as the initial inoculum level is raised, is that, as explained in Section C3.3.1, the extent of yeast growth in brewery fermentations is limited by the amount of dissolved oxygen present (70,74). Exogenous oxygen is required by the yeast for the synthesis of essential sterols and unsaturated fatty acids (79), and it is the intracellular levels of these compounds and the yeasts ability to synthesise them, which ultimately determine the amount of growth achieved. Therefore, because in the majority of cases oxygen is present in excess in worts at pitching, the greater the number of yeast cells added at pitching the greater the number that grow and are capable of division. However, over the normal range of brewery pitching rates, no oxygen can be detected in worts approximately six hours after pitching (291). Hence the significant effect on yeast growth of any ingress of oxygen to a fermentation after this point - such as may occur by the addition of a further batch of air-saturated wort, by rousing, or by transfer of an active fermentation from vessel to vessel.

D2.2.3 Effective Mixing

The effective concentration of active yeast cells can be increased by agitation during fermentation (292,293). Agitation both prevents premature sedimentation of yeast, and ensures faster transfer of metabolites to and from the yeast cell wall (292). Moreover, by virtue of its effect on the levels of dissolved CO_2 present during fermentation, agitation increases both the extent and rate of yeast growth (293). The overall effect of agitation therefore is to accelerate the fermentation rate.

Agitation of brewery fermentations can be achieved either by mechanical rousing (293) or simply by virtue of the stream of CO_2 gas bubbles released during fermentation (292). Thus, the reason fermentations conducted in very tall vessels tend to be more rapid than those within traditional vessels, is almost certainly due to the agitation and mixing imparted by CO_2 gas bubbles rising through the vessel. The power of agitation generated by CO_2 evolution can be deduced from mathematical studies (294), and has been confirmed in practice (295).

Early studies on agitated fermentations suggested that both diacetyl and ester levels are raised because of the increased yeast growth (296). Neverthless, in practice ester levels in beers produced from tall vessels are lower than in those produced from conventional vessels (292), although it is possible that this is due, at least in part, to the increased CO_2 concentrations resulting from the increased hydrostatic pressure (61,94).

D2.3 Accelerated Maturation

The objectives of secondary fermentation, as described in Section B3.4, are:

i flavour maturation

ii carbonation

iii removal of excess yeast

These objectives can be combined with those of cold conditioning (Section B3.5) when discussing accelerated maturation techniques, since, although not strictly a part of the secondary fermentation process, chill-haze formation has become the major reason for the cold storage of beers in fermentation systems employing accelerated maturation techniques.

The main impetus for accelerating the maturation process is the same as that given for accelerated fermentations: any reduction in the time taken to achieve satisfactory maturation results in substantial cost savings.

D2.3.1 Flavour Maturation

In general terms flavour maturation describes the utilisation by yeast of excess levels of metabolites either excreted or not utilised during primary fermentation. The net result is to change the perceived flavour of the beer from its raw or 'green' state at the end of primary fermentation, intó the more fully balanced flavour of a finished beer.

The flavour components primarily responsible for raw or 'green' beer flavours are diacetyl, acetaldehyde and hydrogen sulphide (297,298). The formation of these compounds during primary fermentation has already been reviewed in Section C, together with the factors determining their levels in finished beers. Suffice it to reiterate at this stage, that both their formation and removal are dependent upon maintaining the presence of a vigorous and active yeast population.

Thus, the maturation phase has always been a particularly important feature in the production of lager-type beers, since not only is the primary fermentation of such beers conducted traditionally at temperatures of approx 8-10°C, but maturation is classically achieved by progressive further cooling of the beer to approx 0°C. Such practices have the effect of both maintaining the yeast at a very low level of metabolic activity, and encouraging yeast to sediment - thus reducing the numbers of yeast in contact with the beer. The end result is that storage times of weeks or even months are required in order to achieve the required degree of maturation (129,299).

Since maturation is dependent on yeast activity, the practical route to reduce maturation times is to raise the temperature (35,130,300). Modern practice is therefore to remove the yeast sediment at the end of fermentation, leaving active yeast still in suspension at normal fermentation temperatures. Thus fermentation and maturation are achieved in the same vessel.

In fermentations conducted at lower temperatures, one way of accelerating the maturation process is by applying a 'warm rest'. This is achieved by removing the external cooling in the final stages of primary fermentation, and allowing metabolic heat to raise the temperature of fermentation (130,214). Thus, a fermentation carried out at 8°C may be allowed to rise to 12°C (the exact temperature being limited by the amount of heat generated). Alternatively, heat can be applied to the fermentation by circulation through a heat exchanger (35,130). Under such conditions it is possible to achieve the required degree of maturation within approximately 72hrs, depending on yeast strain, wort composition, and primary fermentation conditions (35,130).

D2.3.2 Carbonation

Traditional methods of increasing the concentration of carbon dioxide in beer involved the generation of CO_2 by yeast under controlled conditions of temperature and pressure, which effectively established the CO_2 concentration. Storage at lower temperatures and higher pressures resulted in increased CO_2 levels. Hence lagers, which ordinarily have a high CO_2 content, have traditionally been conditioned at low temperatures (129).

Most modern breweries have liquid CO_2 storage facilities, and any loss of CO_2 from the maturation vessel can be made good by injection of CO_2 during transfer to the cold conditioning tank, or later in the process if required (301). Even in Germany, where purity laws prohibit the use of industrial CO_2, the gas recovered from primary fermentation can be employed in this manner (129).

D2.3.3 Excess Yeast Removal

This is achieved by 'skimming' (see Section E5.2) in traditional ale fermentations using 'top fermenting' strains of S.cerevisiae, and where flavour maturation in fermenting vessel has long been the norm (302). However, in lager fermentations, using 'bottom fermenting' strains of S.uvarum, the sedimentary yeast is not easily removed (especially when the fermentations are conducted in square vessels), other than by decanting the beer into another vessel. The introduction of fermentation vessels from which it is practical to remove sedimented yeast (notably vertical cylindroconical vessels) has helped remove this limitation.

The removal of sedimented yeast is particularly important when maturation is carried out at elevated temperatures. This is in order to prevent the development of off-flavours due to yeast metabolism and autolysis within the sediment, where temperatures can rise well above that of the main body of the beer, particularly in the deep sediments which can form in cylindroconical vessels (303). Thus, many of the perceived off-flavours in beer sometimes described as 'yeasty' (214,215,216), are due to the excretion of medium chain fatty acids by yeast, a process which is accelerated at high temperatures (see Section C3.9).

Another reason why it is preferable to remove yeast from beer at the end of primary fermentation is because yeast vigour and storage potential are reduced at high temperatures, as discussed in Section E7.4. This factor is of particular relevance when the yeast is to be used for repitching purposes.

D2.3.4 Precipitation of Haze-Forming Material

Beer which is required to remain bright ie. free from haze, for long periods of time (in particular beer for bottling and canning) is filtered prior to packaging, and also subjected to a period of cold storage before filtration. This latter process accelerates the precipitation of protein/polyphenol complexes from the beer (304), which if not removed by filtration, can cause the formation of 'chill-hazes' in packaged beer, when it is stored at low temperatures.

Complexes of protein and polyphenols can also give rise to 'permanent-hazes' in packaged beer, formed as the beer ages, especially in the presence of oxygen (305).

The precipitation of haze-forming material from beer is maximised by rapid chilling and subsequent storage of the beer at as low a temperature as is practical without danger of it freezing (306).

Thus, storage for seven days at $0^{o}C$ has been advocated for some lagers (35,130), although it has been shown that periods of 3-5 days can be sufficient, provided the beer is held at temperatures of $-1^{o}C$ or below, and not allowed to rise before or during filtration (307). The total time required for the conditioning of lager can therefore be reduced from approximately 6 weeks to 5-7 days.

D2.4 High Gravity Brewing

By increasing the strength (specific gravity) of worts supplied to fermentation and hence to conditioning vessels, it is possible to produce beers of greater strength than normal. In simple terms, this means more alcohol from a given plant capacity. If such beers are then diluted at the end of processing back to normal strength (with deaerated water), provided that processing times can be held constant, the increased alcohol production is translated into an increase in output from the plant. This, in essence, is the theory behind high gravity brewing, as practiced over the last two decades, mainly in the USA and United Kingdom (308).

However, as wort starting gravities are increased, although the extent of fermentation, and thus the amount of alcohol produced, increase proportionately, the production of some of the other, quantitatively minor, products of fermentation does not. In particular, yeast produces disproportionately high levels of acetate esters as the wort gravity is increased, causing unacceptable flavour shifts in the resulting beer (45).

Section C3.3 reviews in detail the factors affecting the biosynthesis of esters in fermentations, and in particular the manner and extent to which their production in high gravity fermentations can be controlled. In general terms, the level of control required to maintain normal beer flavours increases disproportionately with increasing gravity, as do the costs of the operation due to increased yeast growth and the attendant heavy losses (65).

Cost savings in high gravity fermentations accrue due to:

i reduced capital plant requirements;

ii reduced beer pumping and refrigeration costs (during cold storage);

iii reduced labour input;

iv a solution factor 'gain' (of about 0.4% if wort is diluted from 80°-40°), when duty is paid on collected worts.

However, on-costs arise due to:

i the cost of flavour control - in terms of both the provision of wort oxygenation equipment and the increased yeast growth required;

ii the (potential) need for pressure fermentation vessels;

iii a reduction in the 'duty-free' element of wort when duty is levied on full degrees. (For example, a wort of SG 40.9^{o} attracts the same duty payment as a wort of 40.0^{o}. The relative percentage of 'duty-free' extract therefore reduces as gravity increases).

iv the higher unit value of high gravity beer losses in process;

v the plant and energy costs of deaerated liquor production.

As the revenue on-costs exceed the purely revenue savings across the brewery as a whole, brewing gravities should be adjusted in line with production demand, to ensure plant is fully utilised. In practice this leads to control problems, and most breweries operate at one or two brewing gravities as a maximum.

In addition to purely commercial considerations, high gravity brewing offers other attractions and problems throughout the brewery. Thus, beer flavour is generally thought to be smoother and better balanced from high gravity fermentations, when diluted to 'sale' gravity, than from comparable 'sale' gravity fermentations(65,309). Against this, it is normally desirable to use process foam supressants (see Section D2.5.2) in high gravity fermentations with top-fermenting yeast, to control foam formation during fermentation. This is unacceptable in traditional cask beer production, since the foaming potential of the finished beer may be affected.

For a useful summary of the advantages and disadvantages of high gravity brewing, together with the control techniques required and the potential costs and savings, see the review by Whitworth (65).

D2.5 Additives

Many aids to brewing have been developed over the last two decades, offering opportunities for increased control of beer flavour and appearance, reduced production costs, and improvements in both fermentation performance and beer quality.

D2.5.1 Brewing Enzymes

Probably the most widely used enzyme in brewing fermentations is amyloglucosidase (310). Food-grade preparations of this enzyme are normally added at the start of fermentation to break down wort dextrins to glucose, thus increasing the amount of fermentable sugars available to the yeast. Beers produced in this way are generally known as 'Lite' beers (311,312), and have higher alcohol contents and lower carbohydrate levels (although only marginally lower total calorific values) than beers produced conventionally.

Commercial preparations of pullulanase (310) and barley β-amylase have also been used in the production of 'Lite' beers (313), and low dextrin worts can be produced by adding unboiled worts to cold 'finished' worts, breakdown of dextrins then being due to the action of malt enzymes(311,312). Although the latter process has serious disadvantages in terms of the risk of microorganisms being brought forward from the malt (notably species of lactobacillus) and can cause haze instability in the finished beer, it is used to produce 'Lite' beers in countries where the addition of prepared enzymes is prohibited or impractical.

Potential disadvantages reported with the use of amyloglucosidase are:

i an increase in acetate ester and higher alcohol levels in the finished beer, associated with the higher than normal glucose levels present during fermentation (238); and

ii residual enzyme activity in the finished beer, which can result in a slow release of glucose, thus increasing the perceived sweetness of the beer and the potential for microbiological spoilage.

D2.5.2 Process Foam Suppressants (Antifoams)

Addition of food grade antifoams to the fermentation vessel can:

i increase the effective vessel capacity (since there is then no need to leave a head space to accomodate the foam produced during fermentation) or, alternatively, reduce fabrication costs (314);

ii maintain top-fermenting yeast in suspension when used in cylindroconical vessels;

iii reduce wetting losses on vessel sides;

iv reduce cleaning costs (foam tends to dry onto the sides of vessels and is difficult to remove);

v allow more vigorous fermentations, particularly when higher temperatures and high gravity brewing are used;

vi improve the utilisation of hop bittering compounds, since these tend to concentrate in the foam produced during fermentation, resulting in their loss (236);

vii improve the foaming performance of finished beers by reducing the losses of foam positive material during fermentation.

Because they are surface-active, antifoams become associated with the yeast in beer, and are thus removed quantitatively on filtration (315,316). It is because there is no filtration stage, that their use in cask conditoned beer is not normally feasible.

Silicone-based antifoams were the first to be introduced into the brewing industry (317,318), with typical useage rates in fermentations of 20-40mg/litre.

However, alternative antifoam preparations based on sorbitan monolaurate are also available (65,319).

D2.5.3 Yeast Nutrient Supplements

Yeast nutrient supplements have long been added by some brewers to worts, in order to achieve enhanced fermentation performance. Commonly, these are simply dried yeast extracts, and as an alternative some breweries routinely add surplus yeast to the hot wort. However, detailed examination of the constituents and composition of malt worts (6) and the (relatively simple) nutritional requirements of most strains of brewers' yeast (20), makes it difficult to justify the use of such supplements, with the possible exception of zinc.

It has been recognised for a number of years that zinc is one of several trace elements essential for the satisfactory fermentation of wort by strains of brewers' yeast (320,321). Thus, worts deficient in zinc can give rise to problems due to incomplete fermentations, and additions of zinc to wort, over the range 0.2-2.0mg/litre, generally result in increases both in fermentation rate and in the extent of yeast growth (321,322,323). The most important site of action of zinc is considered to be the prosthetic group of alcohol dehydrogenase (320,322).

D2.5.4 Flavour Additives

The hop flavour and aroma which is required in some beers - particularly bottled pale ales and some draught ales - is due to the presence in the beer of the essential oil of hops (324). Traditionally this 'hop character' is imparted into beer by the introduction of a small quantity of prime hops into the warm conditioning vessel (tank or cask) - a process known as 'dry hopping' (325). However, this practice is microbiologically unsound and gives variable results, in particular because the aroma quality of hops deteriorates during storage (326), and because of variations between individual batches of hops (327). Moreover, dry hopping is costly and wasteful, since long contact times are required, and only approximately 25% of the available essential oil is transferred to the beer after conditioning for 21 days (324). Alternative processes for imparting hop character to beer are available, but all suffer from certain disadvantages (324).

However, the recent developments in liquid CO_2 extracts of hops (328), have made available preparations containing a high proportion of the essential oil of hops (329), thus offering the potential for a viable commercial alternative to dry hopping (330). Although commercial quantities of these products have only recently become available, initial reactions are very favourable (331).

Overall, the improved efficiency of utilisation of the extracted hops is expected to compensate for the high cost of the extraction process (328,332). However, before it becomes economic to use the aroma fraction of hops extracted with liquid CO_2, similar extracts of the hop bittering substances must also be used, either as a direct replacement for whole hops in the brewhouse, or as a partial replacement as post-fermentation bittering agents. Trials to date with these extracts have given encouraging results (332,333).

D2.6 Carbon Dioxide Control

The use of tall fermenting vessels has resulted in greatly increased CO_2 concentrations in beer at the end of fermentation. The actual concentration is dependent on several factors: fermentation temperature, pressure, wort gravity, and vessel height. Nevertheless, typical figures for CO_2 levels in an ale fermentation conducted in a vessel approximately 60' high are in the region of 1.5 - 1.8 vols. CO_2 (128,334).

However, many draught ales only require CO_2 concentrations of approximately 1.0 vols., therefore necessitating the removal of excess CO_2 from the fermented beer. This can be difficult, the most common method being to purge the vessel with gas from the base, after removal of the sedimented yeast. The purge gas in supersaturated conditions may be CO_2 itself, but a reduction in CO_2 content to (or below) equilibrium saturation requires the use of an inert gas - normally nitrogen (128,239).

Introduction of purge gas requires great care, because gas break-out from supersaturated beer at the base of a tall fermenter can set up a 'chain reaction' - leading to a violent release of gas from the top of the fermenter (128). Very small gas flows are used, and antifoam may be applied to the vessel surface where permitted, to reduce loss of foam - positive material, and beer itself, as a result of overflowing.

A novel method of removing excess CO_2 has been described (335), which employs surface active materials acting as nuclei for bubble formation during fermentation. However, practical trials with these materials have mostly proved disappointing due to:

i the difficulty of cleaning the (rough) surfaces;

ii fouling of the surface by yeast and wort solids;

iii their overall inefficiency.

D2.7 Novel Methods of Fermentation-Adjunct Fermentations

As described in Section C3, the formation by yeast of the flavour-active components of beer is determined fundamentally by the yeasts' ability to grow, and in so doing to assimilate and metabolise the nitrogenous constituents of brewers' wort. However, the conversion of wort fermentable sugars to alcohol during fermentation is far less growth dependent, and is even inhibited by oxygen, a growth promotor. It has therefore been suggested that the two processes - the conversion of fermentable sugars to alcohol, and the formation of growth-related flavour volatiles - could be optimised if they were carried out separately (262). Most brewers derive their worts from malt supplemented with sugar adjuncts (the latter acting as both an extender of brewhouse capacities and as a means of controlling wort nitrogen levels). A system of fermentation has therefore been proposed in which solutions of sugar adjuncts are fermented separately from all-malt worts, using pitching rates upto four times the normal, high temperatures, and agitation, all to maximise the fermentation rate (336). The end product is essentially an aqueous solution of alcohol which, since it contains no proteinaceous or polyphenolic materials, does not require cold conditioning, and which can be added as a diluent to all-malt beer prior to filtration. Substantial savings are therefore possible in cold conditioning plant capacity, refrigeration costs, and the value of beer stocks held in cold conditioning. It is claimed that adoption of such a system makes it possible to increase a brewery's capacity by between 40 and 100%, without a major change in beer quality (337). Moreover, if the adjunct fermentation is carried out at low gravity, the (oxygen-free) alcoholic liquor produced can be used as a means of diluting beers brewed and fermented at high gravity, without incurring deaeration costs (262,337).

D3. Economics

All commercial enterprises must make a profit to survive, and therefore must be at least as cost effective as their competitors.
Thus economics has been, and continues to be, one of the most powerful influences on the development of fermentation systems.

D3.1 Scale

For many years, the economic advantages of increasing the scale of unit production have been recognised, in terms of cost and utilisation of plant, labour, and materials (239,338).
However, increasing scale usually results in reduced flexibility, since large production units normally require large throughputs of

single products to be efficient. Thus, changes in product requirements, varying demand levels, or the introduction of small volume products, can greatly reduce, or even reverse, the economic advantages of scale.

Brewing has always been a highly seasonal industry, in which peak demands during the summer periods and at holiday times can approach double the lowest demand periods. The market can also change dramatically in response to weather - a hot spell in the summer months can increase trade dramatically in the United Kingdom.

Moreover, in recent years market trends have been towards a diversity of products (see Section D4). In addition, the introduction of a number of new products to take the market of an established brand, inevitably reduces the volume throughput of each product.

A combination of these factors has caused the maximum size of fermenting vessels to have 'peaked', and the most modern breweries in the United Kingdom often have a range of vessel sizes to accommodate varying requirements. Thus, vessels of 2,400brls (4,000Hl) capacity, installed in the late 1960's and early 1970's in the larger breweries, have mostly been superceded by vessels of 600-1,200brls (1-2,000Hl) capacity (339,340).

In addition, the energy crisis and subsequent rise in transport costs have dramatically affected the balance between production and distribution costs of a product that is primarily water. The 2-5×10^6 brl. p.a. breweries planned in the United Kingdom in the 1960's have never been fully developed, and maximum capacities of around 1-1.5×10^6 brl. p.a. are generally considered the most economic.

Even in the USA, where production volumes of individual products are very high, and long distance transport is highly efficient, there has been a trend towards the development of regional breweries.

D3.2 Duty (Taxation)

For a variety of reasons, alcoholic drinks have attracted high levels of taxation, and the system of tax collection has had significant effects on the development of fermentation equipment.

In the United Kingdom duty is levied according to both the volume and strength of the 'collected' wort at the start of fermentation (341). In the past this has been one reason for the installation of separate 'collection' vessels, accessible for both the sampling and gauging of wort (see Section E4). In more recent times the cost of installing access and gauging very large outdoor vessels has been an argument against their use as dual-purpose collection/fermentation vessels. However, because on balance the advantages of such vessels outweigh the disadvantages (see Section E4), the latter are now commonplace. In addition, the processing of beer as 'duty-paid' greatly increases the amount of capital tied up in stock, and has therefore acted as a strong stimulus to the development of accelerated production techniques (see Section D2.2).

A less obvious form of taxation affecting the development of fermentation systems has been the system of local taxation of plant and buildings, or 'rating' system. Within the United Kingdom, outdoor vessels can be classed as 'plant', and attract a lower charge than buildings containing large vessels. However, vessels in excess of 1,200brls (2,000Hl) may be classed as buildings, thus affecting the balance of costs in increasing scale.

D3.3 Space

Many present-day breweries were built in the developing towns and cities in the 19th Century, and subsequent development around them together with soaring land values have limited physical expansion. In recent years a rising market and the rationalisation of production companies have resulted in tremendous pressure to increase output from existing sites, and promoted the use of large space-efficient vessels, notably vertical cylindroconical and/or stacked horizontal cylindrical units (see Section E5.3).

D3.4 Labour Costs

In general terms, the relative rise of labour costs has promoted the use of increased labour-saving mechanical systems and automation. Moreover, large batch volumes tend to be more efficient in labour cost terms.

However, in a relatively simple process such as brewing, the additional risks, capital costs, and complications, brought about by utilising large and sophisticated equipment can outweigh the advantages gained. Many very old, simple, plants have survived and are cost effective when compared with larger and newer breweries, despite higher unit labour costs. Nevertheless, the latest generations of microprocessors can be expected to give a fresh impetus to reducing manning levels.

D3.5 Inflation Costs

In brewing, as in a number of other production industries, rapidly rising construction and engineering costs have discouraged investment in new plant and techniques where existing plant can cope, albeit at higher operating costs.

D3.6 Energy Costs

The recent rises in primary energy costs and the costs of materials which themselves have a high energy cost in their production, eg. liquid carbon dioxide, have had a significant effect on brewery production costs. Thus, fermentation systems which are inefficient in the heat transfer operations of wort and beer cooling are not acceptable. In addition, the cost of pumping between, and top-pressurising, vessels argues against multi-vessel systems, particularly where these are built on a 'bungalow' design principle rather than as a tower (259).

D3.7 By-Products

In recent years the value of the by-products of fermentation ie. yeast and CO_2, have increased significantly.

Yeast is a high value food for both human and animal consumption (342,343,344), and can also attract a premium price if sold to distillaries as viable yeast in good physiological condition (342). However, the legislative requirements proposed by the EEC for handling yeast when it is intended for human consumption, are becoming more stringent. Thus regulations may require storage at temperatures $\not>$ $4^{o}C$, and specify minimum standards of hygiene (343).

Carbon dioxide is used universally in brewing to displace air from beer vessels and containers, as well as for beer carbonation. There is also a growing market for the gas in other food and industrial applications. Moreover, the fine balance between cost of recovery and conservation methods vs. cost of purchase inevitably changes as energy costs rise (257). The rising cost of treating brewery effluent also affects fermentation practice. Thus, any loss of yeast or beer incurs a double cost penalty, and modern systems must seek to exclude such losses (345).

D4. Market Requirements

Although it has been said that brewers are able to manipulate their market to a certain extent to suit their own requirements (eg. the promotion of keg beers and lagers (346)), the changing requirements of the public have a profound effect on the production process (347,348,349).

D4.1 Fashion

In response to commercial pressures and advertising, beer brands have become more susceptible to changes in fashion (346). New brands, particularly 'novel' products, are being introduced at an increasing frequency, with the expectation that they will occupy a large segment of the market for a few years and then probably decline (347). Some of these products may require specialised equipment and procedures for their production.

Relative volumes of products may vary dramatically in response to both weather conditions and the economic climate (348). Thus, consumers of large quantities of premium lager during a hot summer may 'trade down' to mild ale in a cold winter, especially when it is coupled with an economic recession.

Because of these variations in demand, and because the capital cost of fermentation equipment is high and its life long in industrial terms (in excess of 20 years), complex and inflexible fermentation systems have failed to become successful - even when they can show an economic advantage in producing any one given product.

D4.2 Quality

Although brewers have always prided themselves on the quality of their products, the increase in mobility and affluence of the population, combined with the advantage to the brewer of advertising and promoting single 'National' brands, has greatly increased the public demand for beer of good, dependable and consistent quality. Chilled and filtered (and normally pasteurised) beers meet the demand for high standards of consistency and visual clarity, although they are not acceptable to some sections of the market (349).

Key parameters in the production of consistent beers are large batches, blending, and precise control over raw materials, production conditions, and microbiological contamination.

D4.3 Traditions

Brewers are by nature conservative, and have a high regard for tradition. Even though commercial judgements may play an important part in the selection of equipment and process systems, the views of traditional brewers in senior management positions undoubtedly temper the more radical advances proposed by modern technologists.

In addition, the success of smaller breweries (which have not invested as heavily as the major groups in modern equipment and technology) in promoting their products as high quality 'traditional' beers, combined with the effect of consumer pressure groups such as Camra (Campaign for Real Ale) have restricted the expansion of revolutionary products and fermentation systems.

D5. Legislation

Because of its effect on social behaviour, and because of its potential for raising government revenues, the brewing industry is subject to a wide range of legislative controls. These controls differ widely in their extent and effect between countries, but in general those controls with either real or potential effects on the development of fermentation systems can be identified as follows:

i Taxation,
ii Food additive control,
iii (Alcohol) strength control,
iv Health and Safety regulations,
v Consumer protection.

D5.1 Taxation

Taxation systems differ widely, and have changed over the years (341), but in an industry where tax costs can often equal or exceed other material and production costs, any procedure which reduces the liability to taxation, or reduces losses of tax-paid product, must be considered.

As already mentioned (Section D3.2) the point of tax collection can also have a very significant influence on production procedures. Thus the system of tax collection in the United Kingdom, in addition to providing the impetus for accelerated processing techniques, has also resulted in great emphasis being placed on minimising beer losses during processing. Hence the installation of beer recovery systems (eg. from yeast), which otherwise could not be justified in terms of return on capital and operating costs (see Section E7).

D5.2 Food Additive Control

As a food, beer is subject to regulations as to the materials and additives which may be used in its production (344). At one extreme, the German 'Reinheitsgebot', or Purity Laws, prohibit the use of any materials other than malt, hops, sugar, yeast, and water in brewing (351). In more liberal countries the use of food grade additives (see Section D2.5) has allowed many advances in fermentation technology.

D5.3 (Alcohol) Strength Control

Many countries attempt to regulate the strength of beers on sale, either by raising the duty levels on stronger beers, or by direct alcohol strength limits-as in Sweden. Drinking and driving legislation also tends to have a depressive effect on beer strength, and has prompted the introduction of alcohol-free beers (352). For the same reason, the production of lower strength beers with full-bodied, good quality, flavour characteristics has become a key objective in many present day brewing operations (352).

D5.4 Health and Safety Regulations

Safety regulations in the production environment have been strengthened over the years (353,354,355). The major safety hazards in fermentation systems are the presence of CO_2 gas, and chemical cleaning solutions, which may also be hot (355). The need to improve working environments has added to the pressures resulting in enclosed fermentation systems, cleaned-in-place, and operated remotely, thus removing the operator from any hazardous areas.

D5.5 Consumer Protection

Many countries now actively practice comsumer legislation, limiting the choice of materials used in brewing, beer strength (alcohol content), and even the production procedures which can be used. In addition, many countries require consumer information to be published at the point of sale, commonly detailing the ingredients, strength, and shelf-life ('best before' date) of a beer (356). Each of these requirements places additional technical and commercial pressures on the brewer - the most recent being the need to specify the shelf-life of a beer in terms of its flavour and physical stability (357).

D5.6 Others

There are also may indirect influences of legislation on fermentation systems: for example the normal limitation on maximum vehicle width in the United Kingdom (14'6"). This has resulted in most cylindrical vessels being fabricated to this approximate diameter, giving a maximum practical capacity of approximately 1,200brl (2,000Hl), (which also equates with the maximum external vessel size not taxed (rated) as a building).

D6. Requirements of Other Plant in the Brewing Process

Although the fermentation process is a key factor in determining beer quality, any system employed must obviously be integrated with the requirements for the efficient operation of other plant in the brewery - ie. the brewhouse, in which the wort is produced, the conditioning and filtration plant, and the packaging machinery. Although the range of these requirements is too great to discuss in detail, relevant points include:

i Brewhouse:

Large batch operations are normally preferred, since these give a relatively consistent product from variable raw materials, and are economic.

ii Cold Conditioning:

Removal of yeast prior to, or during, cold conditioning is necessary, in order to prevent the development of flavours due to yeast autolysis, and to avoid the over-loading of filters. This normally requires a yeast which is sedimentary, or capable of being sedimented with collagen finings (9,12). The introduction of 'green' or 'raw-beer' centrifuges to remove excess yeast from beer ex-fermenter, reduces this constraint on yeast performance (see Section E7).

iii Stock-Holding:

As already mentioned, brewing is a very seasonal trade with sharp peaks in demand, eg. at holiday times. In order to obtain optimum utilisation of plant, it is essential that an element of stock-building occurs prior to these peaks, and fermentation vessels must be able to play their part in holding beer when the downstream plant is congested.

E. CURRENT BATCH FERMENTATION SYSTEMS

As outlined in Section B, the fermentation process can be divided broadly into four processes:

i Wort Cooling
ii Wort Collection ← Yeast Handling → Excess
iii Primary Fermentation
iv Conditioning

Yeast handling acts in parallel, as a distinct but integral part of these processes.

E1. Traditional Batch Systems

Traditionally each of the above processes was carried out in a separate vessel or group of vessels:

	Vessel	Major Function(s)
Fermentation System	Wort Cooling Vessel (Coolship)	Temperature adjustment. Precipitation and first removal of cold break. Wort oxygenation (aeration).
	Collecting Vessel	Yeast addition control. Collection of duty (where required by law). Additional oxygenation (where required). Removal of cold-break (either by precipitation or on first yeast head as fermentation starts).
	Fermenting Vessel	Fermentation of wort sugars to alcohol by yeast. Temperature control - removal of excess metabolic heat. Controlled ingress (or exclusion) of oxygen. Primary removal of excess yeast.
	Conditioning (Maturation) Vessel	Secondary fermentation for flavour maturation and CO_2 development. Secondary removal of excess yeast.
	Cold Conditioning or Sale (Cask Conditioned beers)	

Many traditional systems of fermentation are still widely used throughout the world, albeit enlarged and slightly modified from their original forms, to take advantage of improved construction techniques, efficient cooling systems, and economic pumps.

Schematic outlines of traditional systems for ale and lager fermentations (as adopted for current use) are shown in Figs. 19 and 20.

Each vessel in the fermentation system represents an additional cost in terms of:

i capital cost

ii wetting losses

iii pumping costs

iv cleaning costs

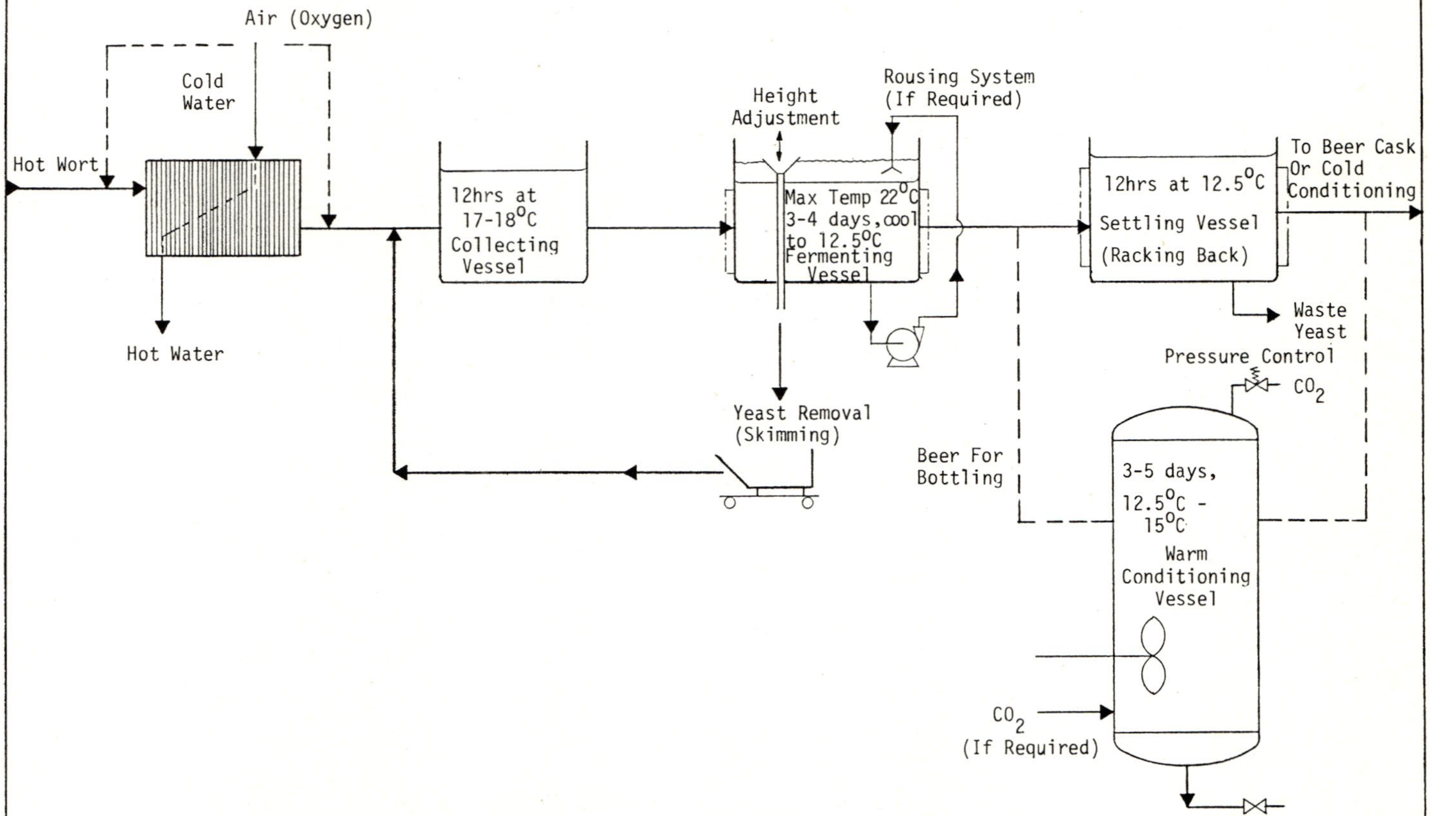

Fig. 19 Schematic Diagram Of A Typical Traditional Ale Fermentation System

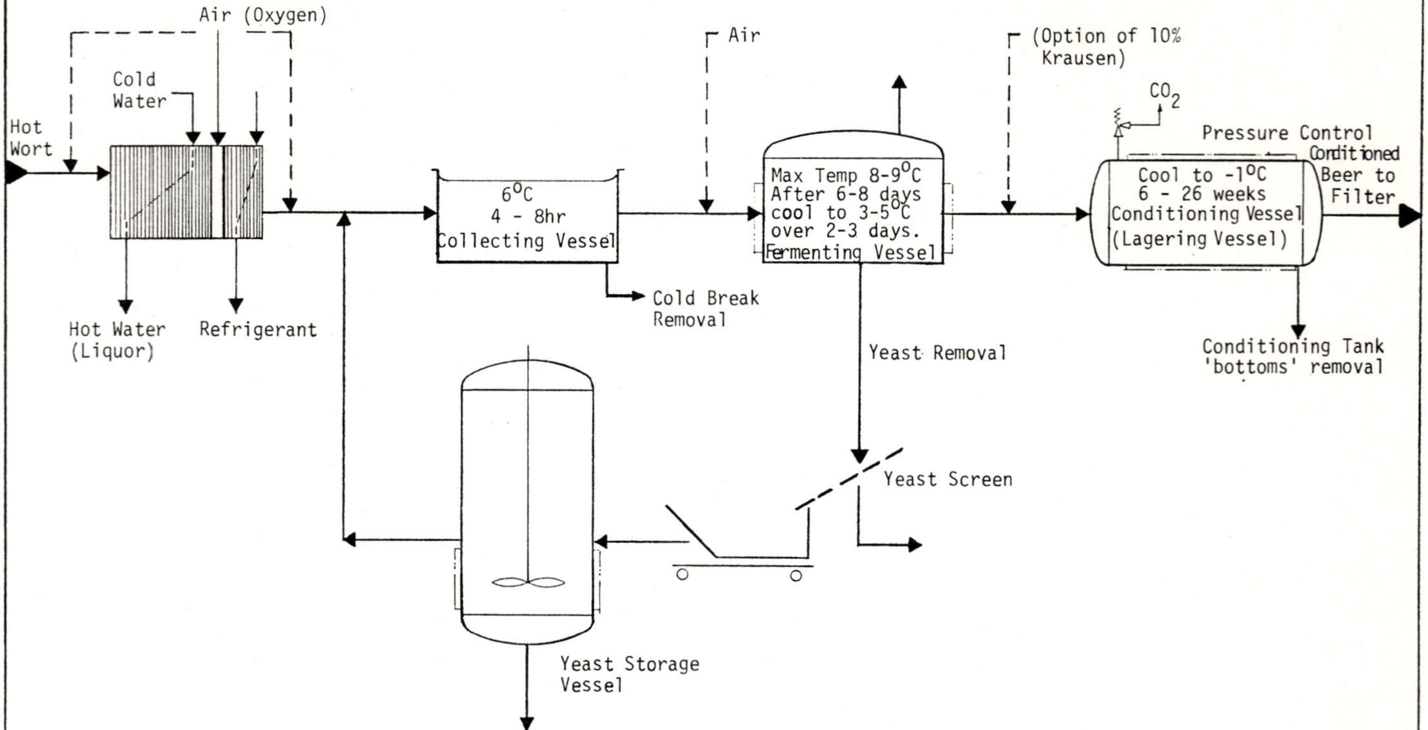

Filters may be used instead of (or as well as) collection vessels for cold break removal

Fig. 20 Schematic Diagram Of A Typical Traditional Lager Fermentation System

v maintenance costs

vi space requirement

vii increased volume of beer in process

viii increase in overall fermentation time

Thus, wort coolships have been replaced almost universally by plate heat exchangers, due largely to the improved thermal efficiency, but also because of the advantages gained by removing one vessel from the process. For the same reason many breweries, particularly those producing predominantly ales, have also discarded the collection vessel, worts being cooled directly into the fermenting vessel.

E2. Modern Batch Systems

The major trend over the last two decades has been towards the use of larger, single vessel, systems, which can be effectively cleaned-in-place and automated, giving optimum operating costs. In addition, vessels nowadays are commonly situated outside, in order to minimise building costs, and are arranged to give high output from a minimum space. A typical system is shown schematically in Fig. 21.

Such systems are flexible, in that they are capable of producing all qualities of beer, and able to utilise a wide range of top-and bottom-fermenting yeasts. Moreover, collection, primary fermentation, and maturation, all take place within a single vessel. Yeast is recovered directly from the base of the vessel, either during, or at the end of, the maturation stage, and the final yeast concentration is controlled by centrifugation into the cask racking or cold conditioning tank.

Traditional lager brewers are able to remove sedimented cold-break from the base of most modern fermenters (129), although some continue to use two vessels or tanks for this purpose.

E3. Wort Cooling

Early industrial forms of wort cooling included the coolship - a large shallow tank in a well-ventilated room in which the hot wort was simply stood and allowed to cool, and open, or 'Baudelot' coolers, in which the wort was allowed to run over pipes supplied with a flow of cooling water (358). Although these systems achieved their main objective, they suffered from being slow, inefficient in energy terms, and unhygienic. They have therefore been superceded almost universally by the plate heat exchanger or 'paraflow' (359). In plate heat exchangers water is the most common primary coolant, and the resulting hot 'liquor' is normally fed to the brewhouse for subsequent wort production. However, when lower start or fermentation temperatures are required than can be easily obtained using the local water supply as the sole coolant, refrigeration can be applied as a second stage.

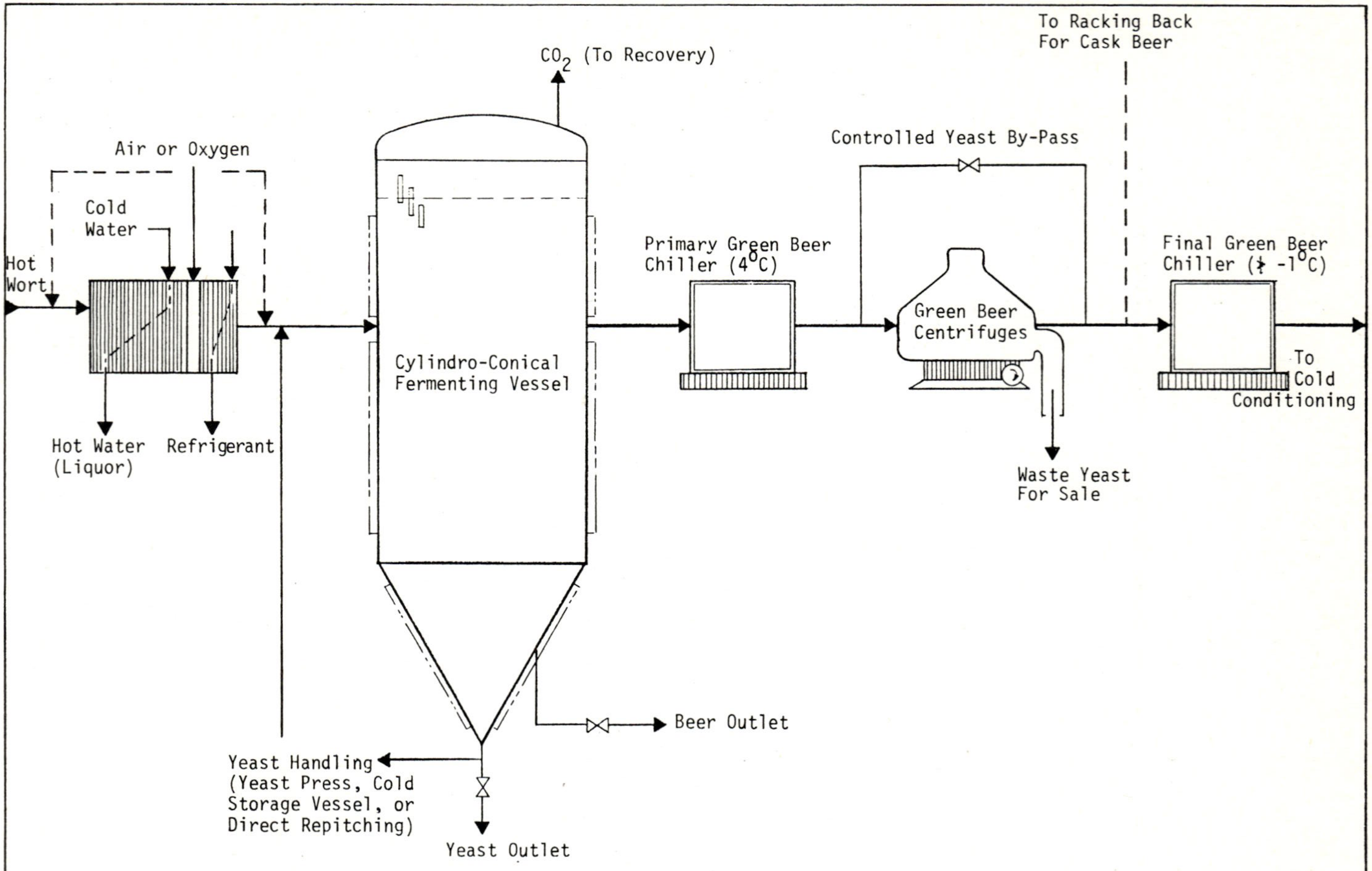

Fig. 21 Schematic Diagram Of A Typical Modern Fermentation System (Ale And/Or Lager)

Modern wort cooling units are almost invariably fabricated in stainless steel, and can therefore be cleaned cheaply and effectively using chemical-based detergents (usually caustic soda). They are normally sized to be able to cool a brew within 1 - 2 hours (359). This gives the brewer control over the time period during which wort is held hot, and thus control over excessive colour development due to caramelisation of wort components, and over the development of important flavour compounds - most notably DMS (see Section C3.7). The very quick cooling which occurs within the heat exchanger, and the highly turbulent conditions, also combine to give a very good cold-break, ie. precipitation of proteinaceous material from the wort brought out of solution by the sudden temperature change. This can have important implications with regard to both yeast performance and the colloidal stability of the finished beer (see Section E6).

E3.1 Wort Oxygenation

Wort requires to be aerated (oxygenated) prior to the addition of yeast, in order to provide the yeast with sufficient oxygen for growth during fermentation (70,74) (and see Section C3.3.2).

Control of wort oxygenation is most commonly achieved by direct injection of either air or oxygen at the wort cooling stage. When dissolved oxygen levels of 8-10 ppm are sufficient, depending on the wort starting gravity and temperature, these are usually readily achieved by direct injection of an excess of air. However, the use of gaseous oxygen is required in order to achieve higher levels. Table II shows the levels of dissolved oxygen in air - and oxygen - saturated worts of varying specific gravities over a range of temperatures:

Wort SG	Dissolved Oxygen (mg/litre) At Air-Saturation			Dissolved Oxygen (mg/litre) At Oxygen-Saturation		
	$15^{o}C$	$20^{o}C$	$30^{o}C$	$15^{o}C$	$20^{o}C$	$30^{o}C$
1080	5.4	4.9	3.9	25.7	23.3	18.6
1060	6.3	5.5	4.7	30.0	26.2	22.4
1040	7.1	6.2	5.4	33.8	29.5	25.7
1030	7.5	6.5	5.8	35.7	31.0	27.6

Table II The Levels Of Dissolved Oxygen In Air-And Oxygen-Saturated Worts Of Varying Gravities (70)

Oxygenation control equipment commonly consists of either a gas volume or mass flow regulator (the wort flow being regulated through the cooler). Alternatively, a dissolved oxygen electrode can be placed downstream of the injection point, the oxygen injection rate being controlled via a feedback control loop.

The point of injection of oxygen has been the subject of much debate. Dissolution of gas injected into the hot wort prior to the cooler is aided by the turbulence generated within the cooler. Proponents of gas injection at this point claim that, not only does the temperature of the wort effectively sterilise the gas, but that the action of oxygen on the hot wort improves cold-break formation (possibly by promoting oxidation of polyphenols).

Using microbiologically pure gas (sterile-filtered), and an alternative dissolving system, such as an inverted U-tube or scintered disc, it is practical to add air or oxygen to the cooled wort. Although there is a greater microbiological risk to this system, studies indicate that improved control is possible, with less oxidation of wort reductones (important in final beer stability), less oxidative colour development, and less interference with the efficiency of the heat exchanger due to gas bubbles.

While many brewers claim a significant flavour effect in moving the point of oxygen addition, it is probable that the final choice is largely a matter of tradition. Many modern plants compromise, by adding oxygen at a point within the cooler, eg. between the water cooling and refrigeration sections, where the wort is at an intermediate temperature.

A novel method of adding oxygen to wort in a controlled manner, by the injection of a supersaturated solution of oxygen in water, has been described (360). This sytem has the advantages of offering accurate control and making possible the measured addition of oxygen to active fermentations, where this is required. However, the complexity and cost of the equipment required has prevented its general adoption within the brewing industry.

E4. Wort Collection

Many traditional European brewing systems, especially in Germany and Scandanavia, involve the use of a separate collection or 'starting' vessel, in which yeast is added to the cooled wort, and the contents then allowed to stand for a period of approximately 24 hours, before transferring ('dropping') to the fermentation vessel proper. The principal reason for this is in order to allow cold-break either to be precipitated within the vessel, or to be lifted onto the foam which forms at the surface when fermentation starts - the 'dirty' or 'rocky' head. This material is then left behind when the vessel contents are transferred and subsequently discarded (35,129,130,361).

Traditionally, British brewing practice has placed far less emphasis on the removal of cold-break, and the most common use for collection vessels within the United Kingdom is as a means of gauging wort volumes for Customs and Excise duty declaration (see Section D3.2).

When the collection period is complete (as dictated either by the time required for cold-break sedimentation or by Customs and Excise regulations), the wort volume is transferred into the fermentation vessel proper. In practice, this process also involves a secondary entrainment of air (oxygen). In those breweries using yeast strains which require more oxygen than can be introduced simply by air-saturation of wort at the paraflow, this stage is therefore of major significance (see Section D2.2.2).

Most modern breweries collect wort directly in the fermentation vessel to avoid the costs incurred in operating a separate vessel stage. Sometimes more than one wort production volume, 'brew', may be collected in a single vessel. Such a system limits the opportunity for removal of cold break, but some brewers remove sediment from the base of cylindroconical fermentation vessels 12-24 hours after filling (129).

E4.1 Cold-Break Removal

Cold-break is a complex mixture of debris carried forward from the wort production process, and protein and polyphenol material precipitated from solution by the temperature change during wort cooling. Associated with this material is a proportion of the hop resins, which are responsible for the bitter characteristics in beer. The principle reasons given for removing this material prior to fermentation are:

i to prevent partial resolubilisation of the protein/polyphenol material during later processing, which can result in a greater tendency for haze formation in the finished beer;

ii because incomplete removal of wort proteinaceous material is often associated with the development of excessive sulphurous and sulphidic flavours in beer (see Section C3.6.2);

iii because the particular bittering substances adsorbed onto the surface of 'break' particles are said to impart a harsh unpleasant bitterness to beer.

It is traditional lager brewers in particular who advocate the removal of cold break prior to fermentation, and some even filter cold worts prior to pitching (35,129,130,361).

However, there are a number of arguments against the removal of cold-break, the implications of which are perhaps of more relevance to ale brewers. Thus, there is increasing evidence that excessively bright worts can result in slow, or 'sticking', fermentations, due to the removal of essential yeast nutrients, probably lipids. In addition, the physical performance (especially flocculation characteristics) of some top - fermenting yeast can be impaired in very bright worts. Moreover, the costs of the operation in terms of equipment, labour, and material losses are substantial.

E4.2 Yeast Addition

The amount of yeast used to seed fermentations (the 'pitching rate') is a principal determinant of the extent of yeast growth during fermentation. In turn, yeast growth determines the rate of fermentation, the viability of the resulting yeast crop, the flavour of the finished beer, and its colloidal and microbiological stability (the latter partially by virtue of the level of α-amino nitrogen remaining in the beer at the end of fermentation). Control of pitching rate is therefore essential to ensure consistent beer quality.

The level of yeast addition commonly used to achieve a rapid vigorous fermentation, the required beer flavour, and minimal yeast 'wetting' losses is in the region of $8\text{-}10 \times 10^6$ cells/ml for a wort of OG 1040°.

Although the requirement for accurate control over pitching rates is universally accepted, in practice methods of pitching vary widely in their precision. Yeast for pitching is usually recovered from previous fermentations either as a 'barm' (sedimented yeast in a suspension in beer), or as a pressed (filtered) 'cake' - sometimes resuspended in chilled water. Addition rates are normally determined empirically, and measured in terms of weight or volume of yeast. A correction factor may then be applied to allow for the solids composition of the source yeast, or for reduced viability.

Traditional pitching methods involve weighing the required amount of yeast (this overcomes specific gravity variations due to entrained gas) into containers, the contents of which are then either pumped or manually transferred into the cool wort stream or collecting vessel. Metering pumps (normally lobe or Mono pumps) are now commonly employed, but these can suffer from a high error factor if seals are worn, or if there is entrained gas present in the yeast. Automatic batch weighing systems are also available (362).

Most modern breweries check yeast counts in the pitched wort, either by microscopic evaluation using a haemocytometer, or by a Coulter Counter (363). Correction for a low yeast count can then be made if required. Other methods of measuring yeast concentration include: direct measurement of solids, centrifugal estimation, or turbidometric methods (363).

Some breweries still infer yeast pitching rate by measuring the yeast crop at the end of fermentation - a low crop requiring an increase in the pitching rate for subsequent brews. This empirical method ignores the effects of wort composition and oxygen supply on the growth performance of any given yeast, and can thus be grossly inaccurate.

E5. Primary Fermentation

In an ancient and worldwide industry such as brewing, it is not surprising to find a very broad spectrum of equipment and techniques employed to suit local materials, technology, and taste. This diversity is well illustrated in the wide range of fermenting vessels, with associated equipment and techniques, which are employed. (For a review of some recent developments see Lindsay and Lansen (364).)

E5.1 Requirements Of Batch Fermenting Vessels

In its basic form, a fermenting vessel is an open tank of any shape convenient to manufacture. Fundamental requirements include:-

i clean, hygienic, interior surfaces;

ii cooling panels or an included cooling pipework system for temperature control;

iii an outlet for emptying at the base;

iv a sampling system for fermentation control;

v a system for removing yeast during and/or at the end of fermentation;

vi sufficient capacity for holding the wort volume, and the foam generated during fermentation.

Individual brewers may also make additional specifications such as:-

i an ability to withstand pressure (for flavour control - see Section D2.2.1 and for CO_2 collection - see Section D1.4);

ii an ability to be cleaned-in-place using sprayball equipment (see Section D1.2);

iii an ability to rouse (agitate) the fermentation (see Section D2.2.3);

iv an ability to make secondary oxygen additions.

E5.2 Traditional Batch Fermenting Vessels

Early industrial vessels were usually fabricated in wood, and were often round and coopered. Developing technologies in materials and methods of construction allowed greater scope for variation in size and shape.

The use of an enclosed vessel gave improved microbiological protection, reduced the need for sophisticated CO_2 extraction and 'clean' ventilation systems in the fermenting room, and also made possible the recovery of the CO_2 generated during fermentation.

Further developments in construction techniques and automatic cleaning systems, gave brewers the option of utilising vessels of almost any size and shape.

Traditionally, both the shape and size of the fermentation vessel were fixed by each individual brewery, and the ratio of dimensions was claimed to affect the flavour of the beer. Modern experience would suggest that these effects were due to the control of temperature within the vessel, and possibly to variations in the degree of ingress of oxygen to the wort surface at the start of fermentation.

Many traditional types of fermenting vessel, particularly in the United Kingdom, are designed to facilitate the separation and removal of top-fermenting strains of yeast. Common systems include Burton Unions (Fig. 22) and Yorkshire Stone Squares (which may be fabricated in many materials) (Fig. 23).

In both of these systems (detailed descriptions of which are given in Lloyd-Hind (302) and by Lasman et al. (365)) the 'excess' yeast remains either in the yeast trough, or on the yeast deck, from which it can be conveniently run-off for recovery. This process is commonly termed 'skimming'. Skimming of top-fermenting yeast from simple open fermenters can also be achieved either by vacuum or by gravity (see Fig. 24).

All such yeast collection systems tend to be complex, potentially unhygienic, and very dependent on the physical behaviour of the yeast. Moreover, many breweries utilising such systems use more than one strain of yeast, each with differing flocculation characteristics. Since the skimming system, by its very design, is selective for top-fermenting yeast, its use in such situations can result in unstable fermentation performances. Another disadvantage is that both the nature of the skimming system and the flat bottom of the vessels, result in sedimentary yeast being left behind in the vessel. This can cause significant volume losses each time the vessel is emptied (366).

Many brewers continue to utilise traditional open, or closed, fermenters, in order to conduct fermentations in a single tank system. Where suitable equipment exists, the capital savings may outweigh revenue costs, particularly when the design is modern, including CIP and automatic temperature control, and where capacities are adequate. However, although traditional fermenting vessels are still widely used (and are still being built), there is a general trend towards cylindrical vessels, because of their many advantages.

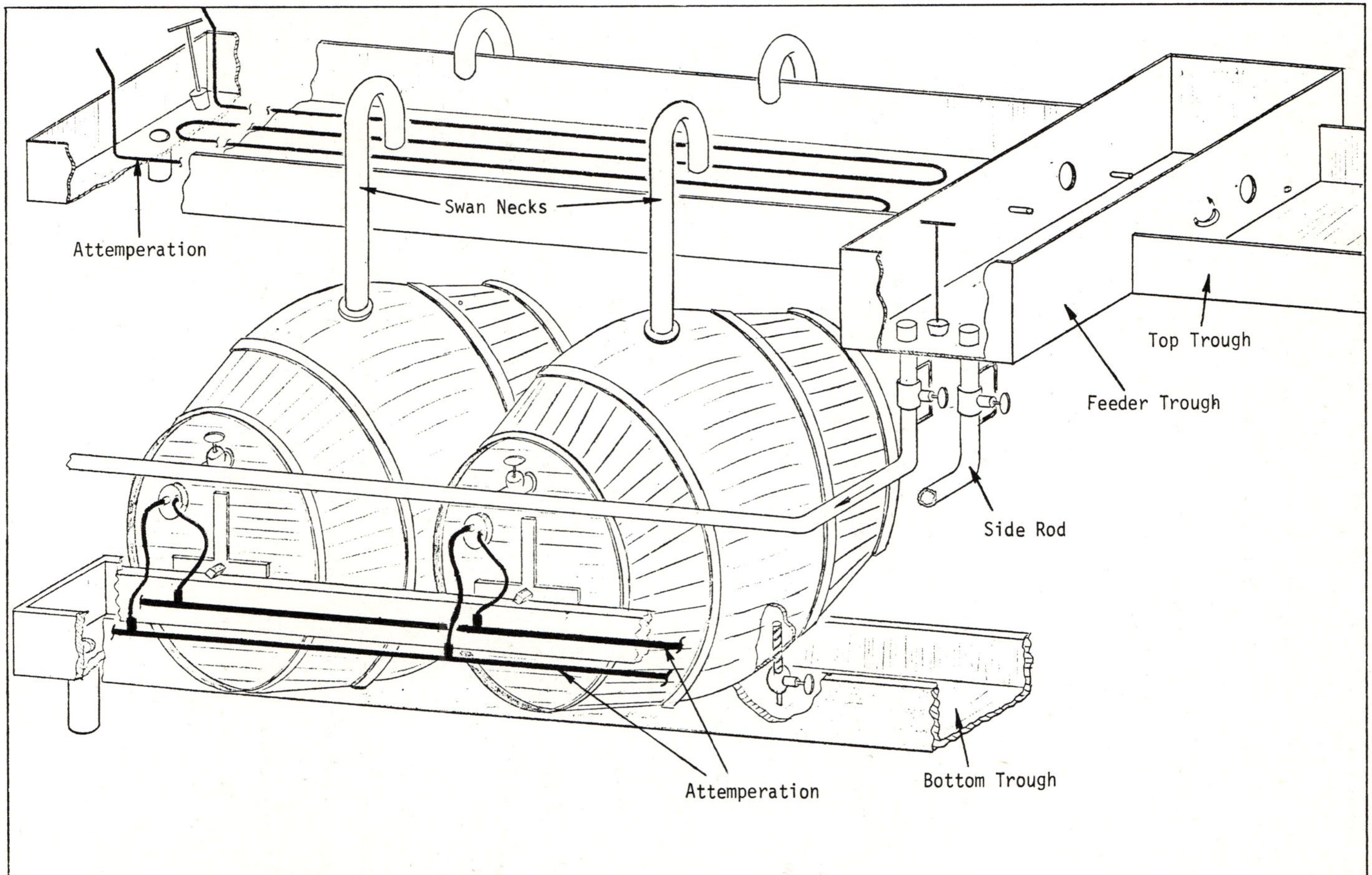

Fig. 22 Outline Diagram Of The Burton Union Fermentation System

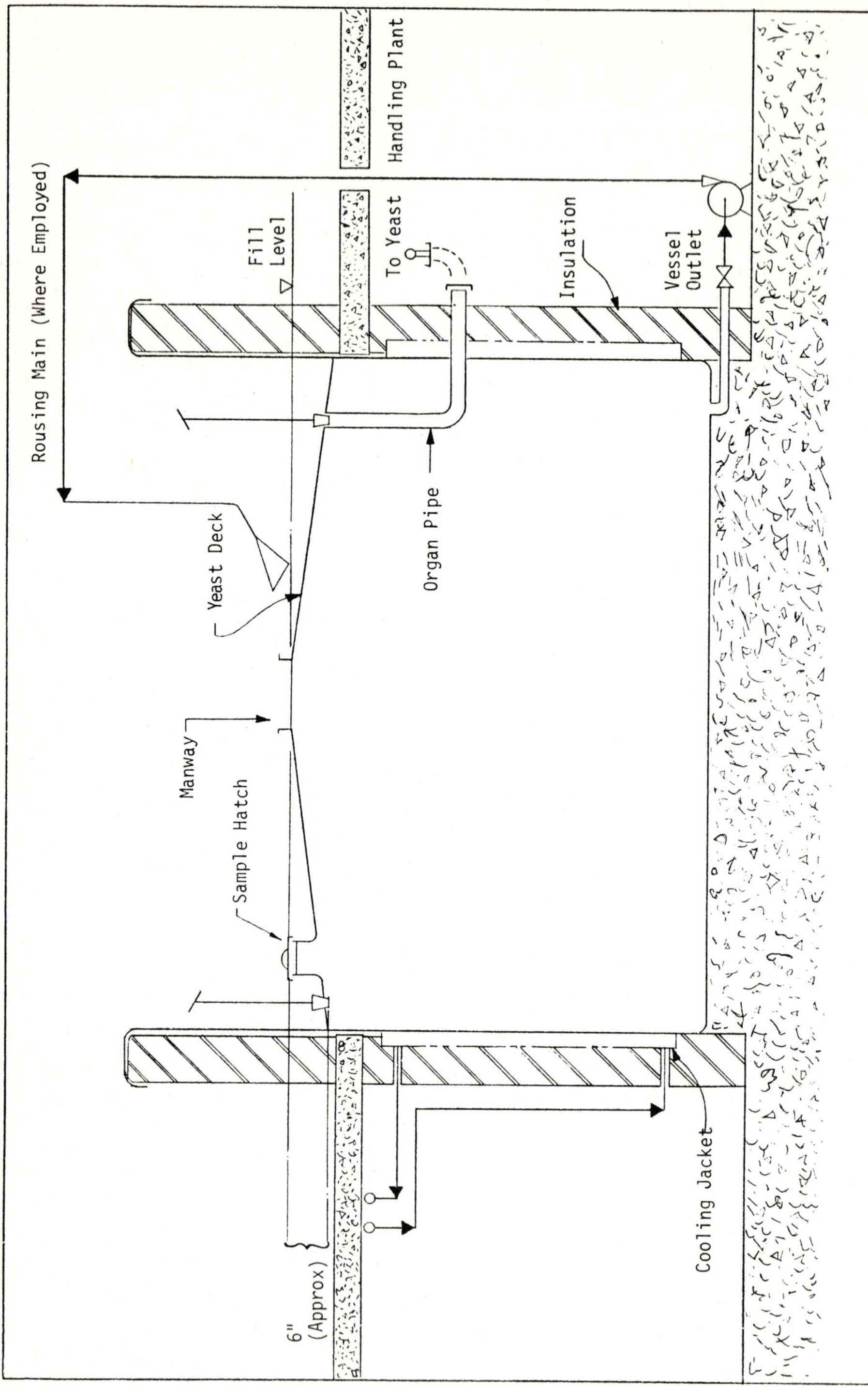

Fig. 23 Schematic Diagram Of A Typical Yorkshire Stone Square Fermenting Vessel

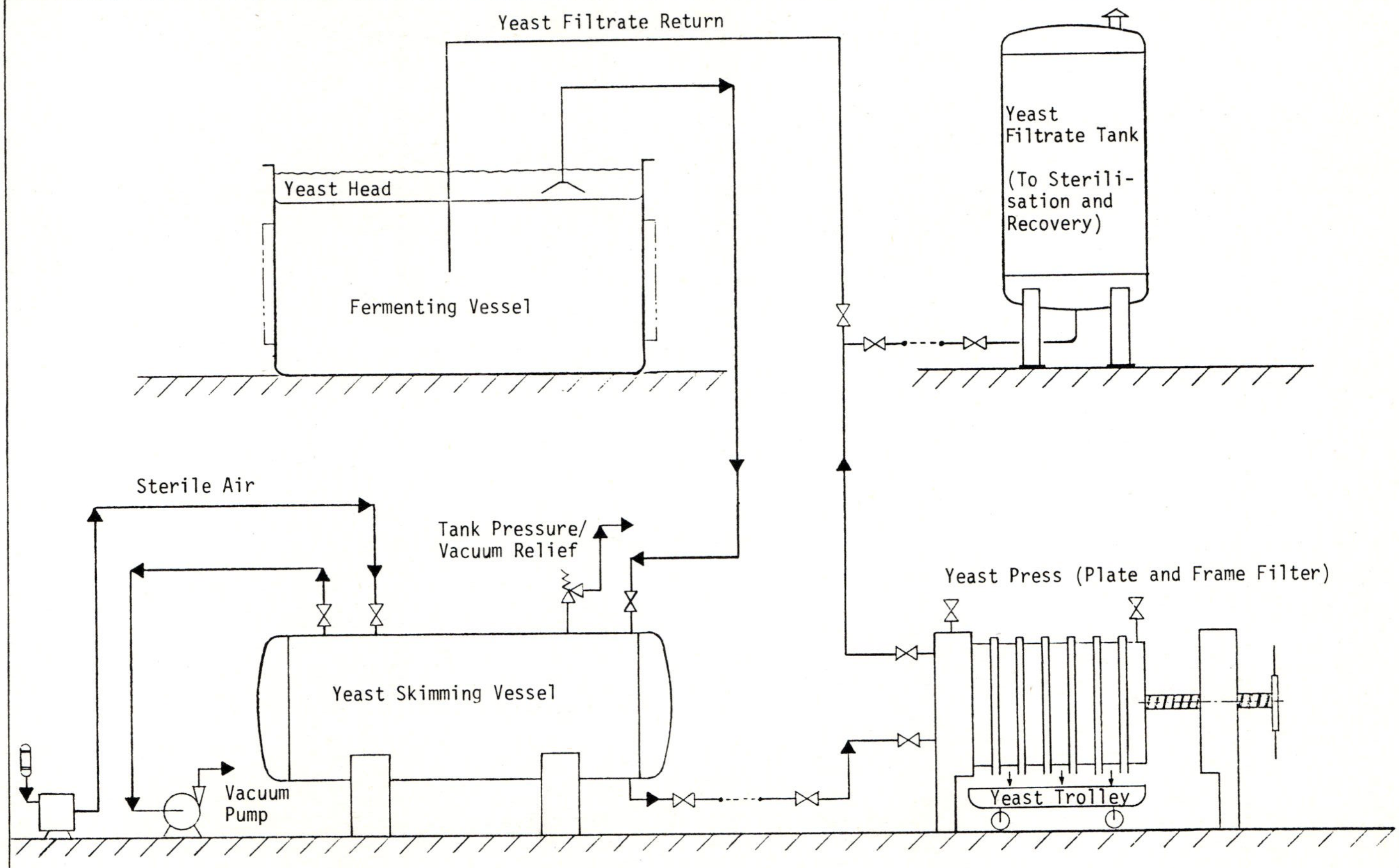

Fig. 24 Schematic Diagram Of A Typical Scott's Skimming (Vacuum Yeast Removal/Filtration) System (302, 365)

E5.3 Modern Batch Fermenting Vessels

E5.3.1 Cylindroconical Fermenters

The vertical cylindroconical fermenter (Fig. 25), first described by Nathan (367), is probably the most common fermenting vessel currently being installed in the United Kingdom. The vessel has a great number of advantages over traditional designs (340,368,369):

i Construction Costs: The vessels are relatively cheap to build to a range of size, pressure, cooling, and overall shape specifications. They can be installed either outside or over a bungalow construction, which provides protected access for operators.

ii Compact Installation: The vessels, by nature of their height and shape, provide a high capacity in a confined land area.

iii Flexibility: Most yeast strains operate well in cylindroconical fermenters, although antifoam may be necessary to inhibit excessive fobbing (and/or reduce vessel 'freeboard' capacity), particularly where top-fermenting strains are used (see Section D2.5.2).
Rousing can also be carried out, either via an agitator, pumped recirculation, or by inert gas rousing, to maintain yeast in suspension, thus accelerating the maturation process. When an individual yeast has a particular requirement for oxygen during primary fermentation, this can be provided by using the rousing system with appropriate oxygen injection.
This flexibility makes the cylindroconical fermenter compatible with most traditional equipment. Excellent beer quality can be achieved, with good flavour matching to beer produced in traditional plant, if required.

iv Rate of Fermentation: agitation imparted by CO_2 evolution inside tall vessels accelerates the rate of primary fermentation (see Section D2.2.3).

v Low 'Wetting' Losses: the shape of the vessel ensures very efficient emptying and draining, thus reducing beer/yeast losses and effluent loads. Some breweries use a flush of water via the CIP sprayball to transfer the last traces of beer and yeast clinging to the vessel wall.

vi Ease of CO_2 Collection from the enclosed fermentation (see Section D1.4).

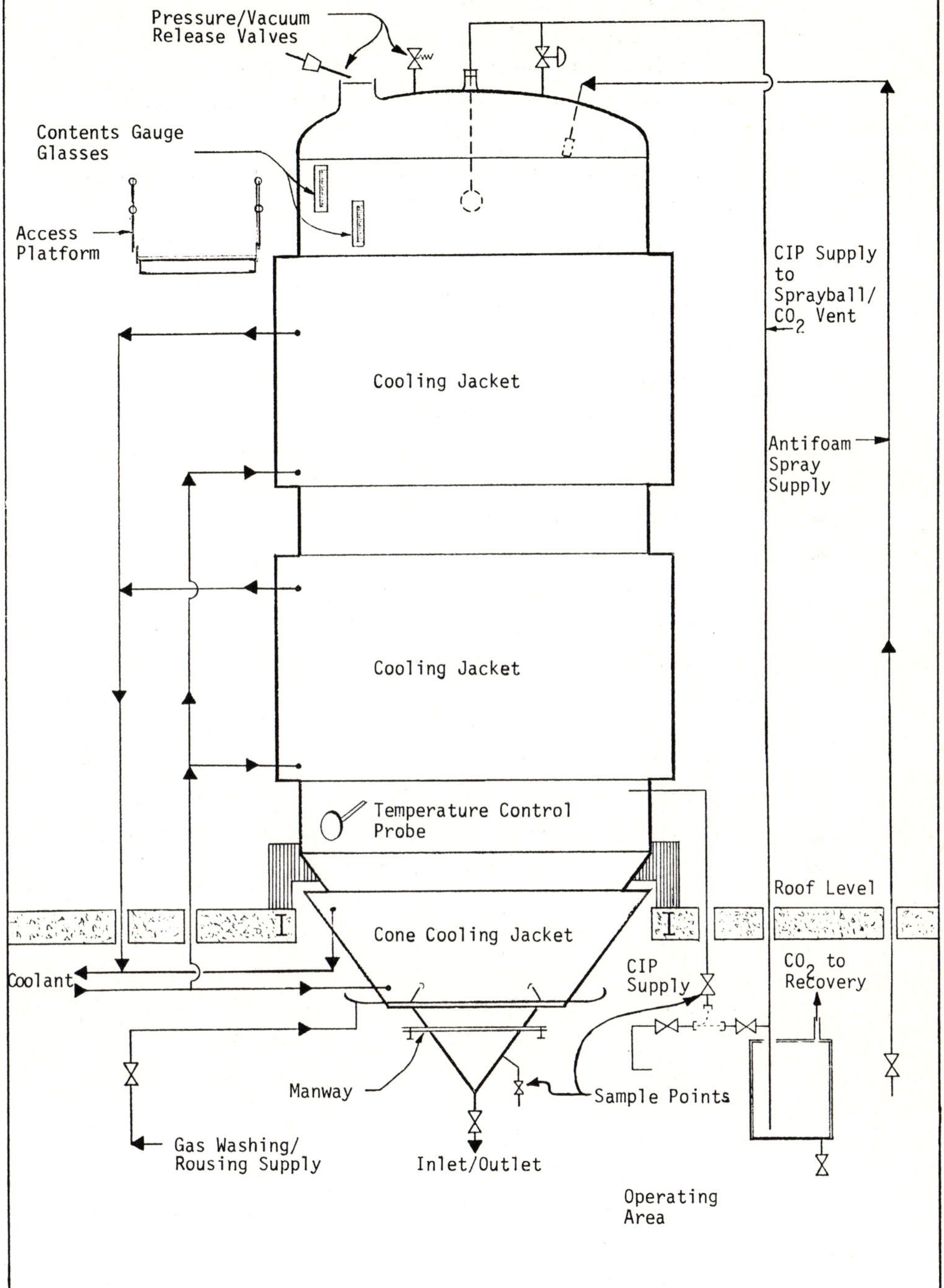

Fig. 25 Schematic Diagram Of A Typical Modern Cylindroconical Fermenting Vessel

vii Ease of Yeast (and Cold-Break) Removal: Most yeast strains sediment rapidly into the cone at the end of fermentation. Cooling may be applied to accelerate this process, although the height of the vessel vs. the settling rate may limit the efficiency of recovery for some strains.
Sedimented yeast solids containing approximately 20% (w/w) dry matter can be achieved with some strains, and yeast may be transfered directly from fermenter to fermenter for repitching. Alternatively, yeast may be further concentrated by traditional presses prior to repitching or sale (see Section E7.2).

viii Minimum Operator Costs: The vessels are fundamentally simple and lend themselves easily to automation. Manual operations can be limited to making and breaking mains connections, or eliminated altogether by fully automated valve systems.
CIP is simply and efficiently achieved via fixed sprayballs. A single top-mounted spray unit effectively cleans the sides of the vessel by a 'waterfall' effect. Temperature control and sampling are efficient due to the agitation effect of CO_2 evolution, and can be carried out remotely if required.

ix Duty Collection: Although the height of the vessels makes access difficult, their tall slender shape makes accurate gauging by liquid depth measurement practical.

x Reduced Losses of Bittering Compounds: These substances are surface-active and a high proportion is lost in the foam and yeast in traditional vessels (236). Practical experience indicates that 15-20% savings can be made in vertical vessels, where foam formation is limited (particularly when antifoams are used), and where foam is allowed to redissolve in beer (368).

The vessels also have some disadvantages:

i High CO_2 Concentrations In The Beer: due to the hydrostatic head of beer in the fermenter, resulting in some instances in the need for additional processing (see Section D2.6).

ii Incompatability Of Some Yeast Strains: and thus the difficulty of achieving flavour matches with some traditional systems.

iii Temperature Stratification Within The Vessel: particularly towards the end of primary fermentation and during maturation, may result in shock excretion by some yeast at the top of the fermenter (where temperatures may be 2°C higher than in the main body), and generation of sulphur aromas and taints by high concentrations of warm yeast in the cone (216,303,370).

Alternatives to the cylindroconical vessel are used, but are less common.

E5.3.2 Asahi Fermenters

These tanks are basically similar to cylindroconical vessels (see Fig. 26), and share many of their advantages. Moreover, the vessels are cheaper to build, since the flat base profile is simple to manufacture. This profile also makes the vessel support cheaper to fabricate, since a direct placement on to ground foundations is possible, rather than the vessel having to be suspended via a strut or building frame (as is necessary with cylindroconical tanks) (371).

Disadvantages however, include less efficient emptying and removal of yeast and cold-break than from cylindroconical fermenters. In addition, some installations have suffered problems arising from 'dishing' of the base plate. The cooling of the base of the vessel (yeast cone) is also difficult or impractical.

E5.3.3 Unitanks

These vessels originated in the USA during the 1960's (372) and are now in use worldwide (373). The basic design is similar to the cylindroconical vessel, but normally includes an agitator and increased cooling capacity (including base coolers). The included angle of the cone is also usually greater than for conventional cylindroconical vessels.

In operation, the Unitank aims to achieve the cold conditioning process within the same vessel as primary and secondary fermentation. Thus, at the end of the maturation period, the yeast in the cone is removed, and the remaining beer cooled to -1°C. This requires great care to avoid freezing, and therefore agitation is usually applied to give a more even temperature distribution. As the tank cools to near-freezing, the beer temperature must be measured from the upper region of the fermenter, and cooling applied at the base, because beer has a maximum density at approximately 3°C (303,334).

Advantages claimed for the Unitank system are (374,375):

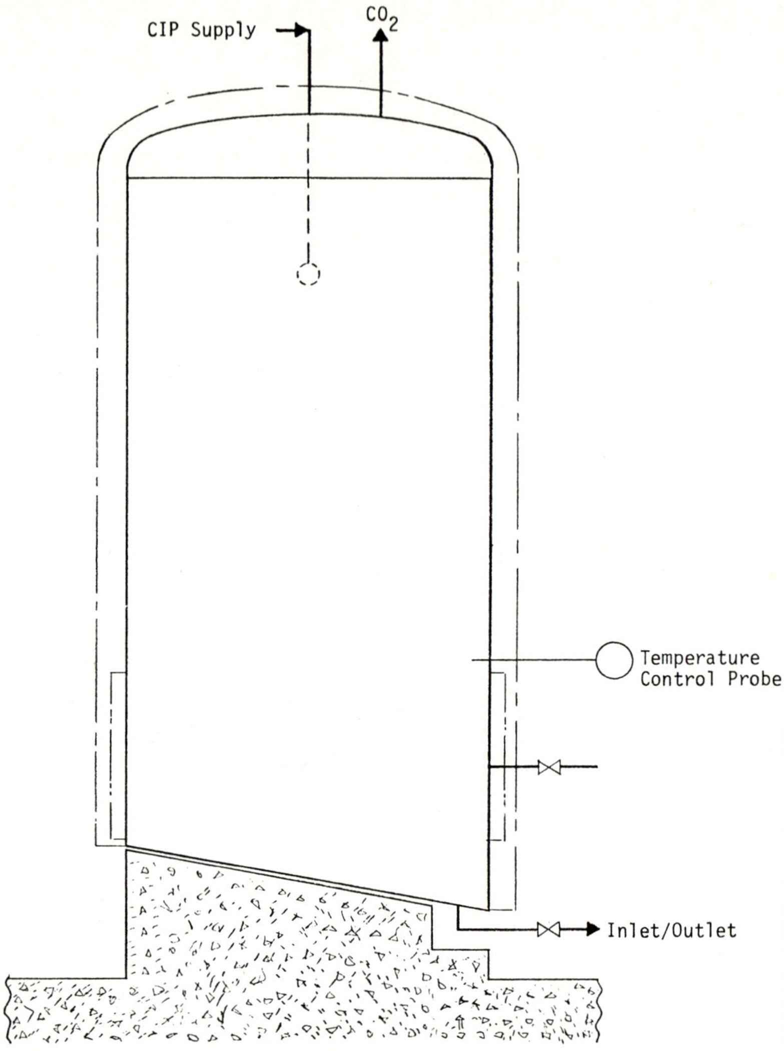

Fig. 26 Outline Diagram Of An Asahi Fermenting Vessel

i The single tank operation eliminates the beer losses, oxygen pick-up, cleaning costs, and pumping costs, involved in beer transfer operations.

ii It is a simple system which is easy to operate and automate.

iii It gives improved product quality and consistency.

iv It gives reduced operating costs - overall costs only 65% of those for a comparable multi-vessel system have been claimed (376).

However, these advantages must be weighed against many other factors:

i Capital Costs: Unitanks are more complex and expensive vessels to build and operate, compared with a simple cold conditioning tank. Moreover, breweries hold buffer stocks of finished beer in cold tank to accomodate peak production periods and unexpected changes (up or down) in demand. Holding this stock in Unitanks requires an increased number of these expensive vessels to be installed.

ii Cooling Costs and Conditioning Period: Cooling in tank is less efficient than by plate heat exchangers. Thus, cooling the contents of a large vessel from fermentation temperature to $-1^{o}C$ can take over 24hrs when carried out in tank, compared with 2-3hrs using a plate heat exchanger. Unitanks offer a potential reduction in the peak demand for refrigeration capacity caused by the loading from 'green' beer plate heat exchangers; but in a well managed brewery such peaks can be avoided.

Because, as discussed in Section D2.3.4, the precipitation of chill-haze material from beer is dependent on the rate of cooling as well as the temperature of cold-storage, the slow rate of cooling in Unitanks requires beer to be conditioned longer to achieve the same degree of haze-precipitation obtained by flash-cooling. This increases both the capital requirement for vessels, and refrigeration costs.

Moreover, the temperature control of very large vessels is normally only sufficiently accurate to cool to about $0^{o}C$ or $-1^{o}C$. Plate heat exchangers are commonly operated to $-2^{o}C$ or less. The turbulence within the transfer system of plate heat exchangers can also accelerate precipitation in supersaturated solutions.

iii Loss of Flexibility: Most brewers operate beer blending programmes on transfer to cold conditioning, to ensure consistency of quality. Additions can also be conveniently blended into the beer stream during vessel to vessel transfer, in order to adjust various quality parameters and/or promote yeast sedimentation, eg. isinglass finings, sugar primings, caramel solutions, CO_2, and beer recovered from yeast presses. Beer quality can then be checked before it is filtered, when a second opportunity to make minor corrections is available.

Addition of these materials into Unitanks is difficult and requires the vessels to be operated below maximum capacity to accomodate the volume of additions (which may exceed 10% in some ale breweries). Uniform distribution is also difficult to achieve in tank, and any layering will result in uneven beer quality if the beer is transferred into a number of filtered-beer tanks.

iv Difficulty in Yeast Collection: The included angle of the conical base in many Unitanks is larger ($\sim 155^{o}$) than in cylindroconical fermenters($\sim 70^{o}$ - 90°), in order to achieve improved temperature control. This can necessitate the introduction of CO_2 gas to the centre of the cone to encourage the yeast to slip and thus aid its removal (373,377).

The vessel requirements in order to achieve an effective and economical single tank operation have been summarised as follows (375):

i A sharp included cone angle ($\ngtr 70^{o}$) to facilitate yeast removal (see (iv) above).

ii A take-off point for beer higher than the base to allow direct removal of bright beer to filters.

iii Sufficient cold-jacketing to prevent yeast autolysis.

iv Agitation to achieve adequate cooling through the temperature inversion point.

E5.3.4 Horizontal Cylindrical Fermenters

These vessels can be used either as dedicated fermenting vessels, or in Unitank systems (see Fig. 27). They share many of the capital advantages of vertical tanks, and can be stacked to give a high capacity per unit land area, albeit at extra cost for support and access buildings.

The vessels are particularly suited to those strains of yeast

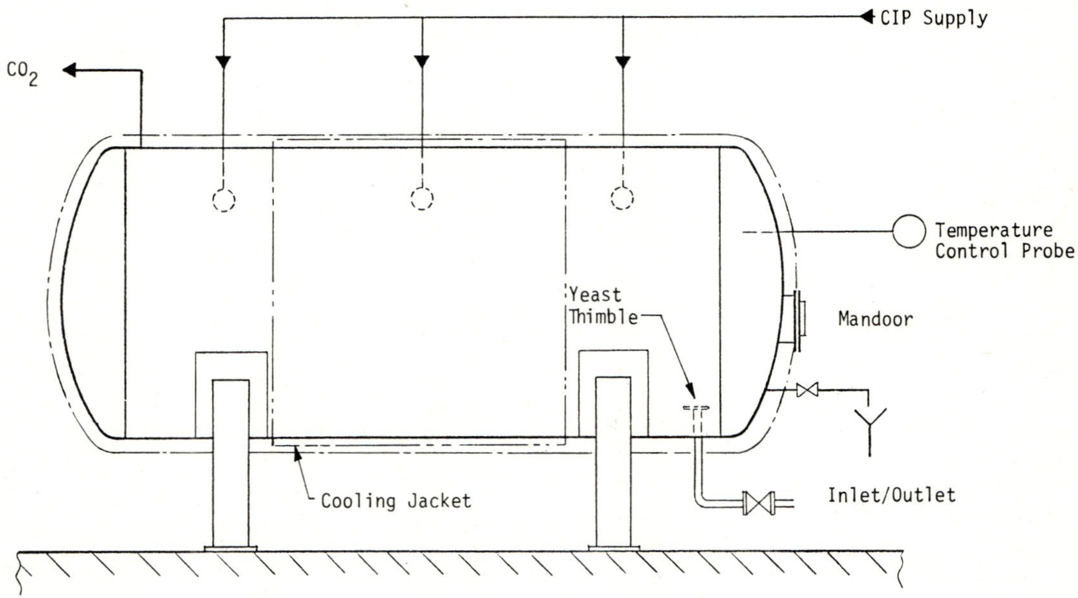

Fig. 27 Outline Diagram Of A Horizontal Cylindrical Fermenting Vessel

which do not sediment well in a vertical tank, since the yeast settling distance is low. Beer can be run off from the yeast, which must then be drawn to the front of the vessel for removal, normally by a mechanical scraper. However, this operation has significant microbiological and operator - safety hazards, the latter due to CO_2 loss from the manway door when using scrapers. (This can be overcome by the use of hydraulic jet installations which drive yeast forward and out of the vessel (377)). Removal of poorly sedimenting yeast can also be achieved nowadays by centrifugation at transfer to cold conditioning tank.

Apart from higher capital installation costs compared with vertical tanks, other disadvantages include:

i Lower vessel capacity utilization, because of the need to provide adequate headspace for foam.

ii Higher wetting losses due to less efficient drainage.

iii More complex means of removing yeast and trub (see above).

iv More complex CIP-multiple spray heads or high pressure systems are required to achieve full coverage of interior surfaces.

v Slower fermentation due to reduced agitation (see Section D2.2.3).

However, the relatively low CO_2 content in the finished beer, due to the lower hydrostatic head, is a possible advantage, depending on the brewer's requirements.

E5.3.5 Spherical (or Spheroconical) Fermenters

It has been claimed that spherical tanks offer the most economic construction in materials and labour of any vessel shape (378). In addition, the shape offers maximum volume to surface ratios, minimising unwanted heat transfer with the environment. However, practical construction costs make the potential savings unachievable (372). Moreover, yeast removal from a completly spherical vessel is difficult, because of the shallow slope at the base. A cone is therefore normally installed at the base of the fermenter to overcome this problem - albeit at some compromise to the 'ideal' shape (see Fig. 28).

Several advantages have been claimed for these vessels (90,378):

i Lower CO_2 levels in the beer: due to the reduced hydrostatic head compared with the equivalent volume vertical tank.

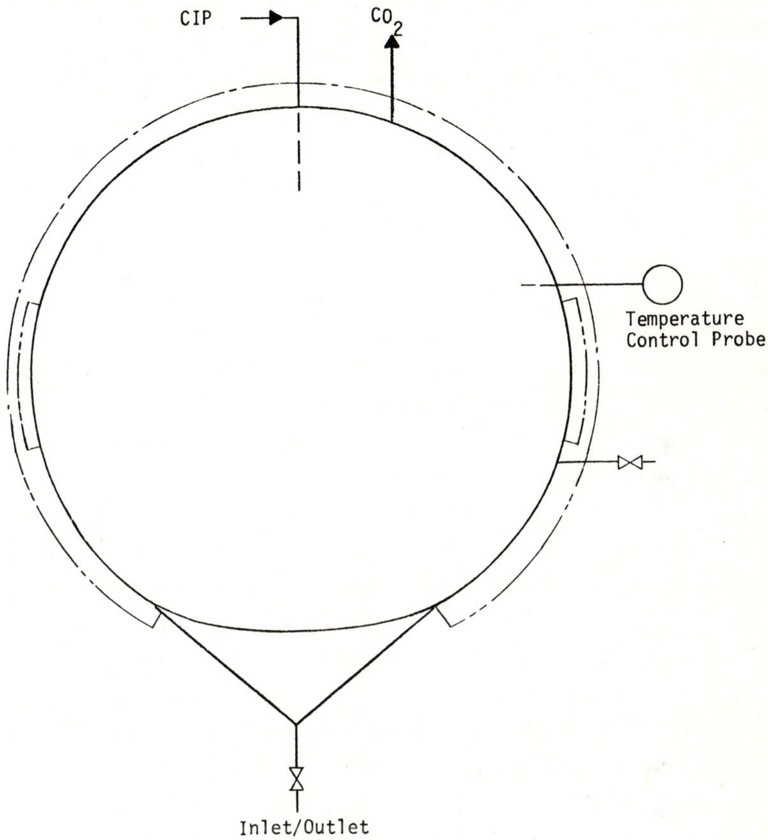

Fig. 28 Outline Diagram Of A Spheroconical Fermenting Vessel

ii An ability to withstand increased pressure: an advantage for accelerated fermentation and maturation (see Section D2.2.1), enabling direct control of carbonation level.

iii Reduced foam area (at top of vessel): enabling effective cleaning by a single, central, sprayball.

iv The ability to be used as Unitanks.

On balance, spherical vessels are probably best suited to breweries where very large fermenter capacities (in excess of 1,800brls. (3,000Hl)) are required, and where pressure fermentations are practiced.

E6. Secondary Fermentation

Primary fermentation is normally considered to be complete when the majority of the wort fermentable sugars have been utilised. At this stage, most strains of brewers' yeast flocculate (8,11) and come out of suspension, either by rising to the surface of the fermentation or by sedimenting to the base of the fermenting vessel. In both cases, because of the reduction in numbers of yeast in suspension, the rate of utilisation of the residual sugars, slows. However, yeast metabolism continues, and is responsible for the resulting 'maturation' of the beer flavour ie. the development of the characteristic flavour of the finished beer. This process is variously known as 'secondary fermentation', 'warm conditioning' or 'maturation'.

Traditional methods of achieving secondary fermentation differ markedly between ale and lager brewing. Ales are traditionally 'conditioned' by warm storage, ie. holding the beer at temperatures between 12-20°C (302). This can be achieved either in the cask from which the beer is to be eventually served (as with traditional draught beer), or in an enclosed tank, for keg or bottled beers. Many ale brewers also add premium hops to the cask or tank prior to filling with beer, in order to enhance hop aroma - a practice known as 'dry-hopping' (325). Additional fermentable sugar (primings) may also be added.

As the temperature is relatively high, 'conditioning' occurs quickly. Residual and priming sugars are metabolised to produce CO_2, excess of which may be vented, either via a porous wooden peg, or via a pressure vent valve. After removal of unwanted volatiles and excess CO_2, the final CO_2 level is set by controlling the temperature of the beer, and sealing the container. When conditioning in tank, the tank pressure can be raised towards the end of conditioning, in order to increase the CO_2 content to that required for packaging.

Typical times required for the secondary fermentation of ales are 2-5 days for draught beers, 1-2 weeks for high-conditioned bottle beers, and longer for strong ales and premium stouts.

Lager beers are traditionally 'conditioned' at low temperatures - usually less than $3^{o}C$ (129). Sugars may either be carried forward from primary fermentation (which is terminated when approximately 10% of the original fermentable sugars still remain), or added in the form of actively fermenting wort from a subsequent brew - the technique known as Krausening (35,129,130 and see Section C3.5.2). The processes involved in the conditioning of both ales and lagers are essentially similar, but the time required is much longer with lagers because of the low temperatures employed. Thus, typical conditioning periods for lagers can be between 6-26 weeks, during which time the temperature is gradually lowered to around $0^{o}C$ (129,299).

The low conditioning temperatures used in traditional lagering also serve to promote precipitation of protein-polyphenol complexes from the beer, which must be removed prior to filtration in order to produce beer with a stable clarity (304,305). The formation of this chill-haze material in ales is achieved by a period of 'cold conditioning' after secondary fermentation, and can normally be completed in 3-5 days at a temperature of $-1^{o}C$ (307 and see Section D2.3.4).

The yeast count during warm conditioning is critical, both to achieve satisfactory conditioning and the required flavour. Too little yeast results in failure to condition, whilst an excess may lead to unpleasant yeasty flavours due to autolysis, particularly if temperatures are allowed to rise (see Section C3.9).

E7. Yeast Handling and Storage

E7.1 Yeast Removal and Collection

As described in Section D2.3.3, the removal of excess yeast from beer at the end of primary fermentation is necessary, both to provide a source of yeast for pitching subsequent brews, and to prevent yeast autolysis products from adversly affecting the flavour of the beer. However, depending on the yeast strain and fermentation system employed, it is normally necessary to leave a controlled level of yeast in suspension, in order to achieve the required degree of secondary fermentation (as described in Section E6). Accordingly, the design of fermenting vessels has been greatly influenced by the need to remove yeast in a controlled manner.

E7.1.1 Traditional Systems

Top-fermenting yeast characteristically rise to the surface of beer at the end of primary fermentation. The removal of excess yeast can therefore be achieved simply by lowering a board in the side of an open vessel to just above the liquid level, and allowing the yeast to flow out. However, many other systems have been designed which take advantage of this property of top-fermenting yeast, including the 'Parachute' Skimming System (see Fig. 19), the Burton Union System (see Fig. 22) and the Yorkshire Stone Square System (see Fig. 23). Alternatively, a vacuum removal (Scott) system can be employed (see Fig. 24) to transport yeast over greater distances.

Secondary fermentation of ales is commonly carried out either in fermenting or warm conditioning vessels (for chilled and filtered beers), or in cask (for traditional draught ales). However, some multi-vessel systems employ secondary sedimentation tanks - 'settling vessels' - to control yeast concentration in suspension prior to conditioning (or racking into cask to condition). Controlled sedimentation of yeast in such systems is normally brought about by chilling to approximately 12-13°C (55°F), and varying the residence time in the settling vessel.

In contrast to top-fermenting yeast, the excess yeast of bottom-fermenting strains sediments to the base of the vessel at the end of primary fermentation. Removal from traditional systems can therefore only be achieved by decanting the 'green' (unconditioned) beer from the yeast, at transfer into conditioning vessel. The sedimented yeast is then held within the vessel by a thimble, or upstand, placed in the vessel outlet.

Control of the level of yeast remaining in suspension can be achieved by chilling towards the end of primary fermentation if required.

Subsequent removal of yeast from the base of the fermentation vessel is achieved by removing the thimble, and drawing the yeast through the vessel outlet. (Some brewers stress the importance of collecting the 'middle layer' of yeast for repitching, to avoid the proteinaceous cold-break at the base of the vessel, and the bitter components associated with the beer foam on the surface layer of yeast).

E7.1.2 <u>Modern Systems</u>

Most modern batch fermenting systems employ fermenting vessels such as cylindroconical vessels (see Fig. 25), which are designed to collect excess yeast at the base of the vessel.

Surprisingly, many yeast, including strains of S.cerevisiae, can be successfully collected from the base of such vessels, particularly if cooling is applied towards the end of fermentation and antifoam is used to minimise foam formation. In many such systems, secondary fermentation is also carried out within the fermenting vessel, thus facilitating accelerated maturation (see Section D2.3), which requires a higher level of yeast in suspension within the vessel than would be normal at the end of traditional primary fermentation.

Although excess yeast can be readily removed from such vessels, the degree of control over the yeast count during conditioning (in either cask or cold tank) is less precise. Thus, in order to increase the degree of control without involving additional vessels or long settling times, continuous self-cleaning centrifuges are commonly used to remove yeast from the 'green' beer (see Fig. 29).

Centrifuges achieve separation by exploiting the difference in specific gravity between yeast and beer. Rotation of beer within the centrifuge causes a very high centripetal force, upto 14,000 times that of gravity, to accelerate the sedimentation of yeast (379). Also, because of the arrangement of the disc stack (see Fig. 29), the settling distance for the yeast is reduced, so that very rapid separation is achieved, even for yeast which would normally separate only very slowly in fermenter.

The separated yeast accumulates in the outer bowl of the centrifuge, and is compacted. As the bowl fills, a sensing device is triggered which causes the bowl to open momentarily, discharging the yeast. Alternatively, where a very constant yeast load is fed to the centrifuge a nozzle system can be used, which allows a continuous small discharge from the outer edge of the bowl (379).

Optimal yeast separation is achieved at temperatures of around 3-5°C. Thus, when beer is transferred from fermenter at higher temperatures, a primary chiller is normally used. The low temperature also serves to reduce any physiological damage to the yeast, caused by the mechanical forces and heating effects within the centrifuge.

Yeast discharged from centrifuges can have a dry solids composition of up to 25% (w/w), although in practice only 20-22% (w/w) is normally achieved. The beer entrained in this

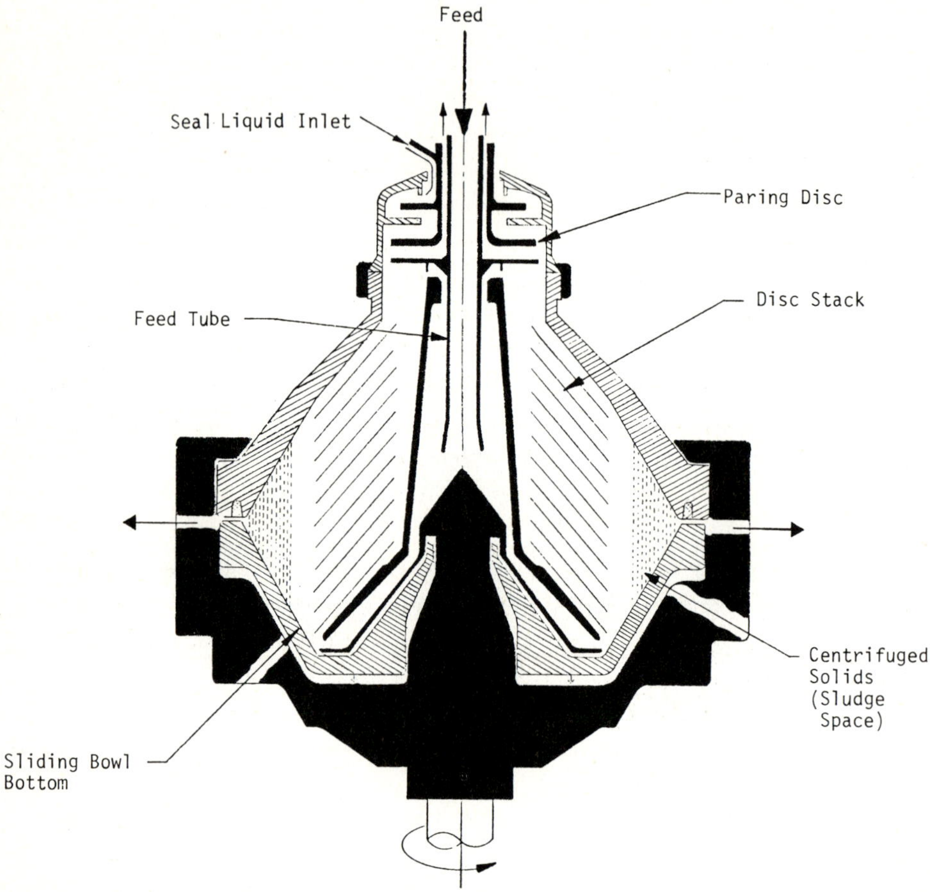

Fig. 29 Vertical Cross Section Through A Modern Self-Cleaning Disc Bowl Centrifuge

yeast can have a high commercial value, particularly when duty has been paid, so that installation of secondary concentration units, such as yeast presses, may be justified to enable its recovery (see Section E7.2 below).

Most of the yeast collected from centrifuges is sold as a by-product for food or pharmaceutical products (see Section D3.7), although some breweries reuse yeast from centrifuges for pitching. This latter practice requires the utmost care in the cleaning and operation of centrifuges, to avoid infection and/or physical damage, and is only normally practical in breweries where a single yeast type is processed.

The yeast count in beer ex-centrifuge can be controlled to a level of 0.05×10^6 cells/ml. Many brewers practicing accelerated cold conditioning, pass beer from fermenter via centrifuges directly into cold conditioning tanks, to minimise the subsequent load on the filters. When a higher yeast count is required, eg. for cask conditioned beer, this is achieved either by a controlled by-pass of the centrifuge, or by direct reinjection of yeast after the centrifuge.

A disadvantage of early models of centrifuges was the high degree of oxygen pick-up in beer within the centrifuge, resulting in the development of oxidised flavours and premature staling of the final beer (380). More modern machines have a modification to the paring disc (see Fig. 29) which acts as a hermetic seal, thus preventing ingress of air into the bowl during discharge. Oxygen pick-up in such centrifuges should not exceed 0.2ppm (379).

E7.2 Yeast Concentration

Yeast removed directly from a fermentation vessel usually has the consistency of thin cream, and has a dry weight solids content ranging from 10% - 20% (w/w), depending on yeast strain and separation technique. The major proportion of the yeast slurry or 'barm' is therefore beer. There are a number of disadvantages to the direct reuse of barm yeast for repitching subsequent brews:

i The yeast concentration can be variable within the slurry, making control of subsequent pitching rates difficult.

ii Any infection present is carried forward.

iii The value of the beer in the slurry is effectively reduced by reprocessing in fermentation. (This is particularly significant in the United Kingdom where duty has been paid on the barm).

These factors therefore all argue in favour of concentrating yeast ex-fermenter, prior to its reuse as pitching yeast. Moreover, the major proportion (up to 80% depending on crop rate) of the yeast produced in brewery fermentations is in excess of repitching requirements, and is subsequently sold as a by-product, either as foodstuff or as distillers' yeast. The concentration of yeast disposed of in this manner is usually a commercial necessity (342,343,381).

When the physiological state of the yeast is not critical, the use of centrifuges to remove and concentrate yeast can be considered. The most efficient units, as discussed in Section E7.1.2 above, are the self-cleaning or nozzle-discharge disc-bowl units, which require a continuous feed of relatively low solids content. Sedimented yeast must therefore be dispersed into the main beer volume prior to centrifugation. Dry weight solids of 25% (w/w) can be achieved by the use of such centrifuges.

Rotary vacuum (Lundin) filters (382) and simple mesh strainers (377) can also be used directly with separated yeast. However, the most common system of yeast filter is the plate and frame filter, or yeast press (383). These units are generally operated at relatively low pressures (up to 4 bar), and separation occurs across a cotton or polypropylene 'cloth'. With careful operation, yeast concentrations of approximately 27% dry weight of solids can be achieved, with little added infection or loss in viability (383).

A variation on the standard yeast press incorporates a hydraulic 'balloon' in the yeast chamber, which can be inflated when the press is full (383). This 'squeezing' action can increase the dry weight of solids to 35% (384), but the yeast viability can be reduced, making it unfit for repitching. In addition, the filtrate beer (or 'yeast pressings'), from such a system contains high concentrations of autolysed yeast, which may adversely affect flavour and haze stability when added back to beer.

Temperature increases during yeast handling, either as a result of metabolic heat or of mechanised handling, can significantly reduce yeast viability, and damage the quality of the recovered beer (due to the presence of yeast autolysis products). The maintenance of low temperatures is therefore desirable, and to this end the use of cold rooms and cooling jackets on vessels used to handle yeast is common.

Vibrating screens are commonly used on the continent to separate and wash yeast (377). These units can be effective, but lack control.

Air introduced into the beer recovered from yeast can promote staling and poor haze stability. All systems are subject to this risk, but especially the Lundin filter, and the Scott System if air is allowed to 'blow through' the yeast press and into the recovered beer tank (384).

E7.3 Yeast Purification

As previously discussed, most breweries nowadays use pure culture techniques and equipment to replace their pitching yeast every 8-12 'generations' (fermentation cycles) (see Section D2.1). When continuous repitching is still practiced, bacterial infection levels can be controlled by 'acid washing' the yeast (385). This is usually achieved by suspending the yeast in a dilute solution of mineral acid, or ammonium persulphate and phosphoric acid, at a pH of 2.0-2.8. Contact times of 2-8 hours kill the majority of spoilage bacteria with minimum loss of yeast viability and vigour (377,385).

E7.4 Yeast Storage

The objectives of yeast storage are to maintain yeast in a pure and viable condition, and to ensure that it retains sufficient endogenous reserves to commence multiplication and fermentation rapidly, once it is repitched.

Because high storage temperatures and exposure to oxygen promote metabolic activity, and thus deplete reserves of both ATP and glycogen (54,386), yeast is normally stored at temperatures of 0°-$4^{o}C$ (387,388,389).

Yeast can either be stored as a barm (ie. as a slurry in beer), as a pressed 'cake', or resuspended after concentration (see above) in water, buffer, or wort. A number of studies have been carried out on the optimum storage conditions for brewing yeast (387,388,389), and it is likely that each strain has its own specific optima. However, most yeast remain viable for at least 7 days when stored in any of these states, provided good temperature control is practiced. Whatever reslurry medium is used, it is necessary to maintain a low pH value, in order to inhibit infection. To this end, 2% phosphate buffer solutions at pH 5.05 (approximately isotonic with the yeast cell) have been suggested (387). Table III compares the maximum times at which optimum viabilities are maintained under four different conditions of storage:

Storage Condition	Temp °C	Max. Storage Time
Pressed	4	10 days
Water Slurry	4	3-4 weeks
Beer Slurry	4	4 weeks
Slurry in 2% KH_2PO_4	0	5 weeks

Table III The Effect of Storage Conditions On The Maximum Storage Time Of Strains Of Brewers' Yeast (387,388)

Exposure of yeast to air, for example when it is laid on shallow trays after pressing, encourages metabolic activity, leading to debilitation and autolysis (387). Thus, in order to maintain anaerobic conditions, yeast is normally stored in bulk. Pressed yeast is stored either in sacks or bins, which are usually packed under pressure to exclude air. Slurried yeast is normally stored in agitated vessels, with attemperation panels to remove any metabolic heat generated.

F. CONTINUOUS FERMENTATION SYSTEMS

F1. Continuous vs. Batch Systems

Continuous brewing (wort production) and fermentation systems have been the ideal of a number of brewers for many years, and early patents date back to the last century (for a review see Portno (390)). Moreover, continuous fermentation systems have been widely developed in other industries, particularly during the past fifty years, as standards of microbiological control have improved.

Development of continuous beer fermenters on a commercial scale dates from the late 1950's, and several large installations were made during the next decade (391,392,393,394). However, with the notable exception of the Dominion System in New Zealand (392), most continuous systems have now either been moth-balled or abandoned. A consideration of the anticipated advantages versus practical experience explains the causes of this.

F1.1 Flexibility

Brewers require plant which is flexible, in terms of its ability to produce a variable output volume and a range of beer brands, but which can also produce a standard product in a reliable manner. The anticipated flexibility of continuous systems, in terms of rapid changes in throughput, could not be realised on a commercial scale. Combined with the inherent loss of production reliability, this lack of flexibility required larger stocks of finished beer to be held, in order to meet planned or unexpected increases in output.

Practical experience also showed that beer quality could vary significantly with throughput, due to the changes in yeast contact time and nutrient supply within a fixed volume system. This type of variability is unacceptable to market requirements.

As has been emphasised throughout this review, the brewer achieves variations in beer flavour by changing either his yeast strain, brewing materials, or process conditions. However, continuous systems require very specific combinations of these factors to achieve optimum output. Thus, relatively few yeast operate well in continuous systems, due to the physical and physiological conditions imposed. Moreover, alteration of a particular operating parameter in order to achieve a desired flavour, may have unsettling effects on a number of other factors. Continuous systems are therefore best suited to markets where a single beer type is dominant, and changes in that market slow.

F1.2 Capital Costs

Although the complexity of continuous systems makes them expensive to construct, if operated at design throughput capacity, the actual capital cost per unit volume of beer fermented, is lower than comparable batch systems (395).

However, for a combination of reasons, continuous fermentation systems failed to reach their designed output, thus increasing the real capital cost per unit of production.

One major reason for this is because beer production volumes vary according to seasonal demand, and in order to meet maximum demand with a continuous system, there must be sufficient capacity on-stream during peak periods, since the volume of beer held in the system is relatively small. This factor, combined with the lack of reliability found with such systems, required large stocks of finished beer to be held elsewhere in the brewery, and a low utilisation of fermentation capacity in off-peak periods.

In addition, start up times for most continuous systems are relatively long, until yeast concentrations are built up, and steady-state conditions established. Thus, if a system is shut-down for any reason (eg. because of spoilage problems, or labour disputes), that part of the brewery's capacity is lost for a considerable period.

Moreover, many of the systems designed and build during the sixties simply failed to achieve their designed output reliably and consistently, particularly when changes in flow rate were required to meet demand (396).

F1.3 Operating Costs

Highly automated continuous fermentation plant was expected to be cost effective in terms of:

i Reduced beer volume in process.

ii Reduced beer and hop losses.

iii Reduced labour costs.

iv Ease of CO_2 recovery.

v Improved energy utilisation.

vi Reduced cleaning costs.

In practice few of these savings were made:

i Reduced beer volume in process - as previously discussed, lack of capacity at peak periods, and doubtful reliability caused higher stocks of finished beer to be held, negating this saving.

ii Reduced beer and hop losses - largely related to an anticipated reduction in yeast growth in continuous systems. However, beer flavour considerations (see Section C3) effectively precluded this saving. Reduced 'wetting' losses due to less frequent filling and emptying of the fermentation system itself were offset by losses from associated buffer tanks and pipework systems. Also, any microbiological infection occurring in a continuous system can develop rapidly, and cause spoilage of a significant volume of beer.

iii Reduced labour costs - were expected to accrue mainly from the use of automation, when the system was operating at steady-state. In practice, continuous systems required very close monitoring, not only by plant operators but also by skilled technicians screening for infection. The sensitivity of the plant to infection also required frequent and precise cleaning regimes for the associated equipment.

Moreover, the sensitivity of the plant to machine failure required a rapid response from engineering technicians, since all downtime represents lost production, and could also seriously affect beer quality (eg. due to hold-up of hot worts). Prolonged shut-down periods also required a slow start-up procedure to allow steady-state conditions to be re-established. Continuous systems require around the clock attendance by all personnel involved in feeding wort to, operating, and receiving beer from, the unit. This involves the payment of continuous shift premiums and continuous manning-even in periods of low production.

iv CO_2 recovery - is potentially easier from units with a predictable, relatively small, continuous output of gas, than from batch units, which have high output peaks. In practice however, a number of batch units operating out of phase can give a similar spread of CO_2 output in a more reliable fashion, albeit at some cost in monitoring purity.

Moreover, the CO_2 from some continuous systems, notably towers, operated with air or oxygen feed at the base, may be contaminated with oxygen.

v Reduced energy consumption - depends largely on more even services loads. However, the main energy loads in a brewery are in the brewhouse, and few successful continuous brewhouses have been built. Moreover, batch brewhouses feeding continuous fermentation systems can be less energy efficient in combination, due to extended production throughout the week, heat losses in hot wort storage, and energy costs involved in the stirring, pumping, and sterilising operations of the fermenter.

vi Reduced cleaning costs - were envisaged due to reduced frequency of emptying vessels. However, this was more than offset in practice by the cleaning of associated tanks and mains, particularly the wort steriliser.

In summary, the reason why continuous fermentation systems have failed in commercial practice is because of their complexity, lack of flexibility, and doubtful reliability, combining to make them more expensive to operate than modern batch systems.

F2. Types of Continuous System

A great number of systems have been described, but very few have reached large-scale commercial production. Many work on broadly similar principles, differing only in points of mechanical design.

F2.1 Cascade Systems

The British System, shown in Fig. 30, can be classified as an 'open' system, where the yeast concentration in each of the fermenting vessels is allowed to find its own level - the net result of growth from the nutrient supply vs. wash-out in the effluent beer. At least one example of this system is still operating successfully in a United Kingdom brewery.

In the Dominion Breweries System (Fig. 31), the yeast count can be regulated by returning sedimented yeast from succeeding vessels into the agitated fermenters. Other examples of such a 'closed' (or partially-closed) multi-vessel system use centrifuges to remove yeast carried forward, and recycle the requisite quantity. This increases the range of yeast strains which can be used, but has the disadvantage of employing a complex and costly item of plant, which is difficult to maintain under sterile conditions.

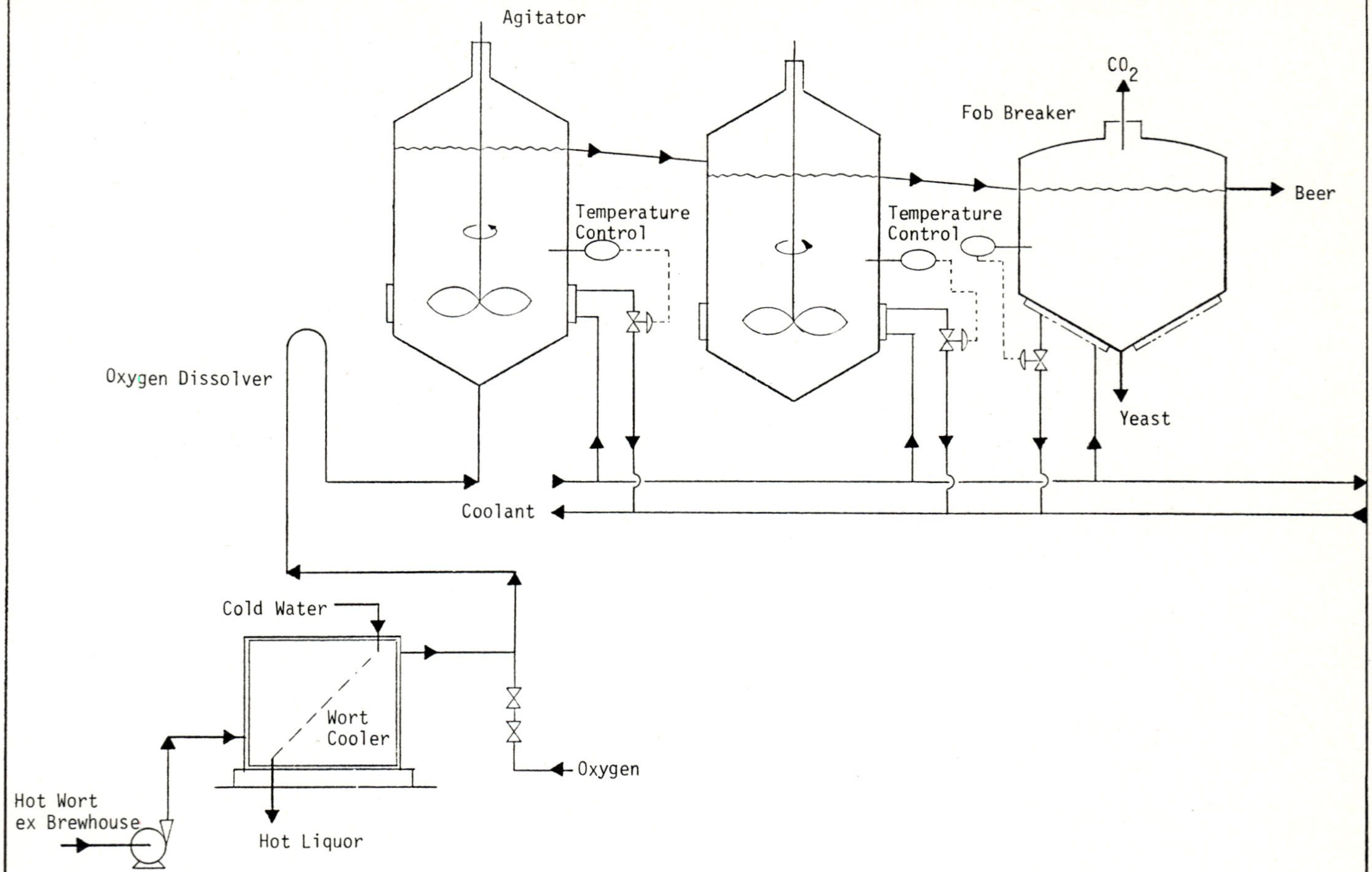

Fig. 30 Schematic Diagram Of Bishop's Continuous Fermentation System (393)

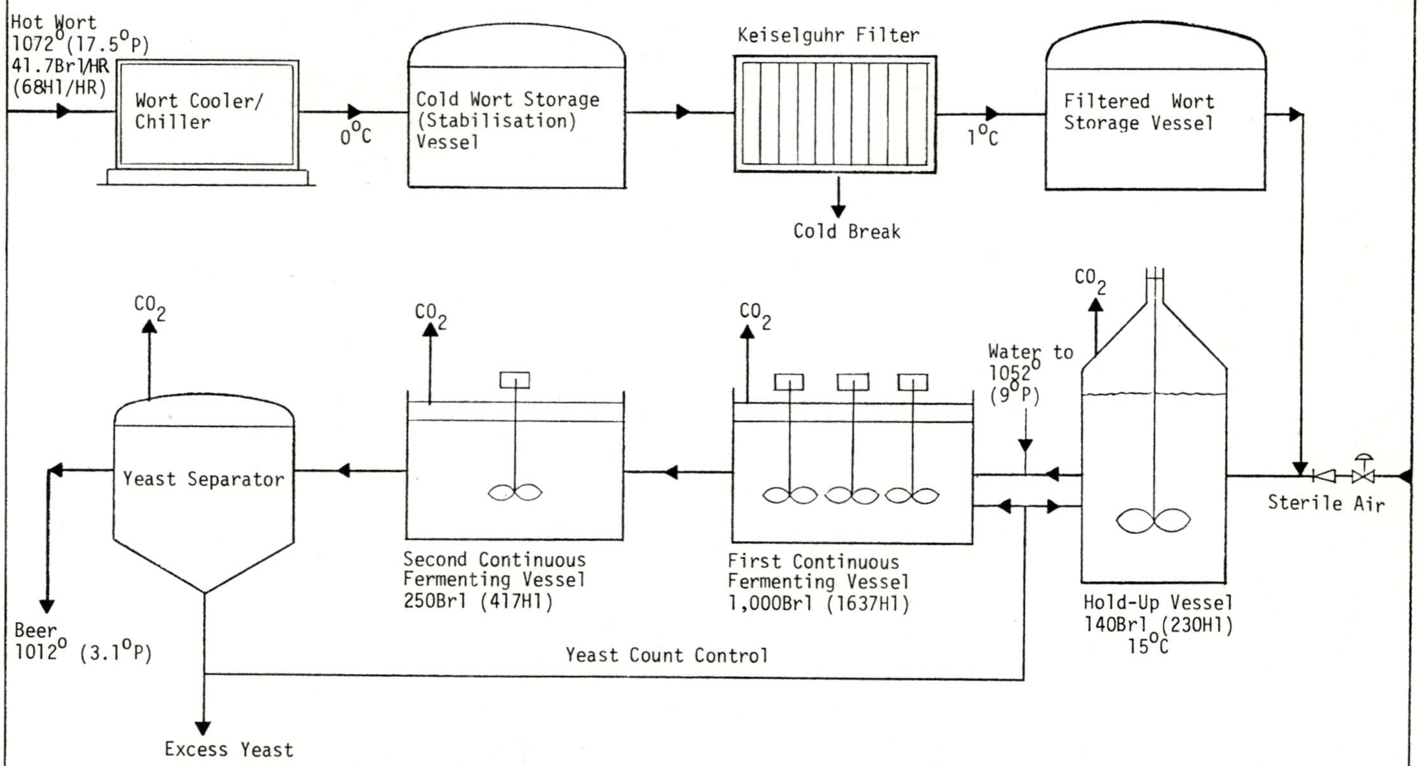

Fig. 31 Schematic Diagram Of Dominion Breweries Continuous Fermentation System (392)

A Canadian system has been described which effectively produces a cascade system within a single vessel, by utilising a system of internal partitions.

The objective in all such systems is similar - to maintain steady-state conditions in each vessel. This is only achieved by careful control over all the variable parameters: wort composition, aeration, and yeast concentration. Given a fixed feed rate of wort, steady-state conditions can then exist in each vessel, giving a graduation to fully fermented and conditioned beer in the final vessel. Total residence times vary between systems, but range typically from as low as 15hrs for ale fermentations to 30hrs for lagers (390). Faster throughputs can be achieved, but require both higher yeast concentrations and high temperatures.

While steady-state conditions can be maintained in such systems in the short term, changes in the physical performance and viability of the yeast can occur which affect the potential throughput. This may be due to the requirement of yeast to be exposed occasionally to high substrate concentrations, rather than continuously to partially depleted media (397). Yeast in continuous fermenters vary in their ability to utilise fermentable sugars (notably maltose) with time. This has been ascribed to the slow growth of individual cells in such systems (397), and to the effect of glucose introduced continuously into the fermenter, in suppressing the metabolism of maltose (390,397).

F2.2 Tubular Fermenters

Tubular fermenters rely on the highly sedimentary nature of some yeast to produce very high yeast concentrations at the base of the fermenter. One development of this type was the inclined tube fermenter (see Fig. 32).

However, the best known of such systems, and the one most widely adopted and used during the 1960's and early 1970's, is the APV tower (131,398,399) (see Fig. 33).

The system operates by providing a fixed flow of sterilised wort to the base of the fermenter. In Fig. 33 the oxygen supply is regulated by injection at the wort cooler (c.f. earlier designs where oxygen/air was introduced at the base of the tower). CO_2 gas may also be introduced in order to break up the yeast sediment and thus avoid over-compaction, which can result in the yeast 'plug' being lifted like a piston ahead of the wort flow (394).

The tower itself consists of a tall narrow tube, up to six feet in diameter, containing the yeast 'plug'. This is headed by a larger

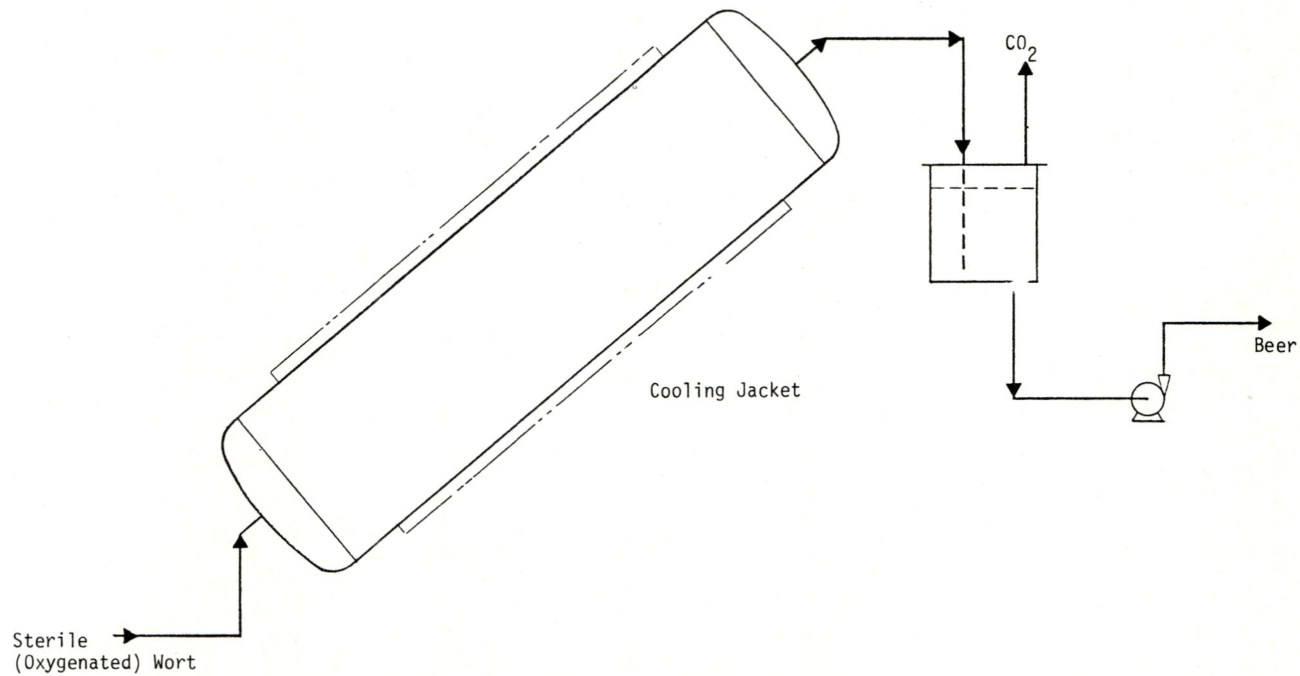

Fig. 32 Schematic Diagram Of An Inclined Tube Fermenter

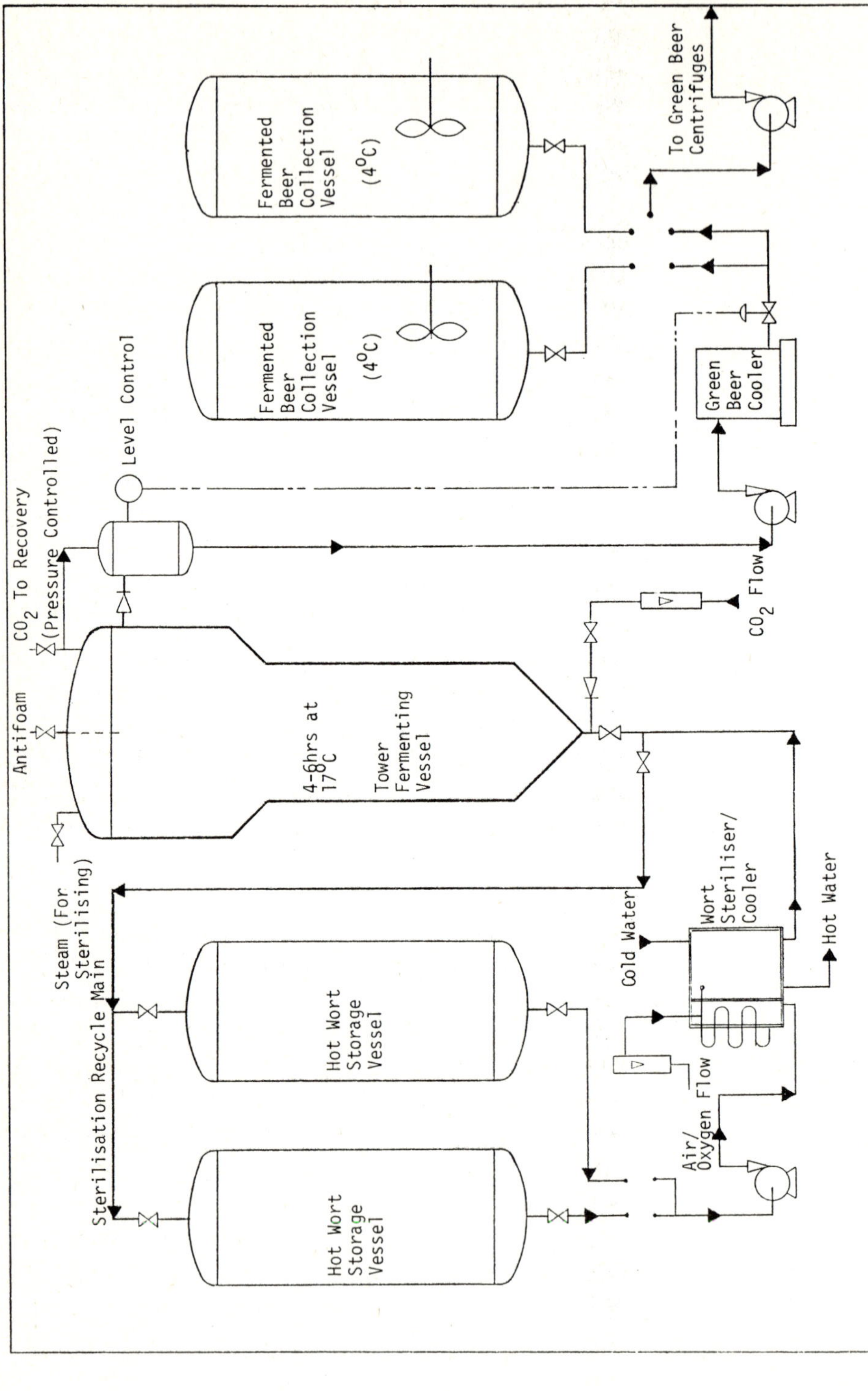

Fig. 33 Schematic Diagram Of Modern Tower Fermentation System

diameter tank which acts as a maturation and settling zone, allowing yeast to sediment back into the lower section. Baffles may also be included to reduce turbulence and prevent chanelling. Effluent CO_2 is released at 0.5-1.0 Bar pressure, and overflow beer is chilled into finished beer tanks.

The action of towers can be compared to a three tank system (shown schematically in Fig. 34 (396)).

A fermentation gradient is maintained in the yeast plug, where the high yeast concentration rapidly utilises nutrients and yeast growth occurs. In the central regions of the tower agitation takes place due to CO_2 evolution, resulting in rapid maturation with relatively high yeast concentrations. The yeast settling zone allows some of the yeast carried upward to recycle back into the main tower, and maturation to be completed. Overflow beer should be fully fermented and matured, and any excess yeast is removed by centrifuges or settling tanks.

Residence times of less than 4hrs have been claimed for tower fermenters (394), although these efficiencies could not be achieved regularly on commercial plant, particularly when the vessel diameter was increased. Six feet diameter towers with a nominal capacity of 120brl (200Hl) had typical minimum residence times of 5hrs at one major brewery. However, actual total system residence times can be much greater, particularly when a batch brewhouse is used to feed the system, resulting in buffer storage requirements.

Tubular fermenters offer greater efficiency than most cascade systems, due to the higher yeast concentrations they contain. They have one considerable disadvantage in that the choice of yeast which can operate in them is very limited, reducing their flexibility with regard to the number of beer qualities that can be produced.

In practice, the yeast in tower fermenters is particularly vulnerable to deflocculation (and loss of sedimentary properties); this is related to the cyclical growth pattern and loss of fermentative capacity already described (390,397).

Tower fermenters are also particularly susceptible to microbiological infections: these can arise in an unpredictable manner and develop rapidly. Because conditions can vary within a tower, the flow rate must be varied also, in order to maintain full attenuation. The resulting difference in residence time and physiological state of the yeast, gives rise to flavour and quality variations in the final beer, which cannot easily be controlled.

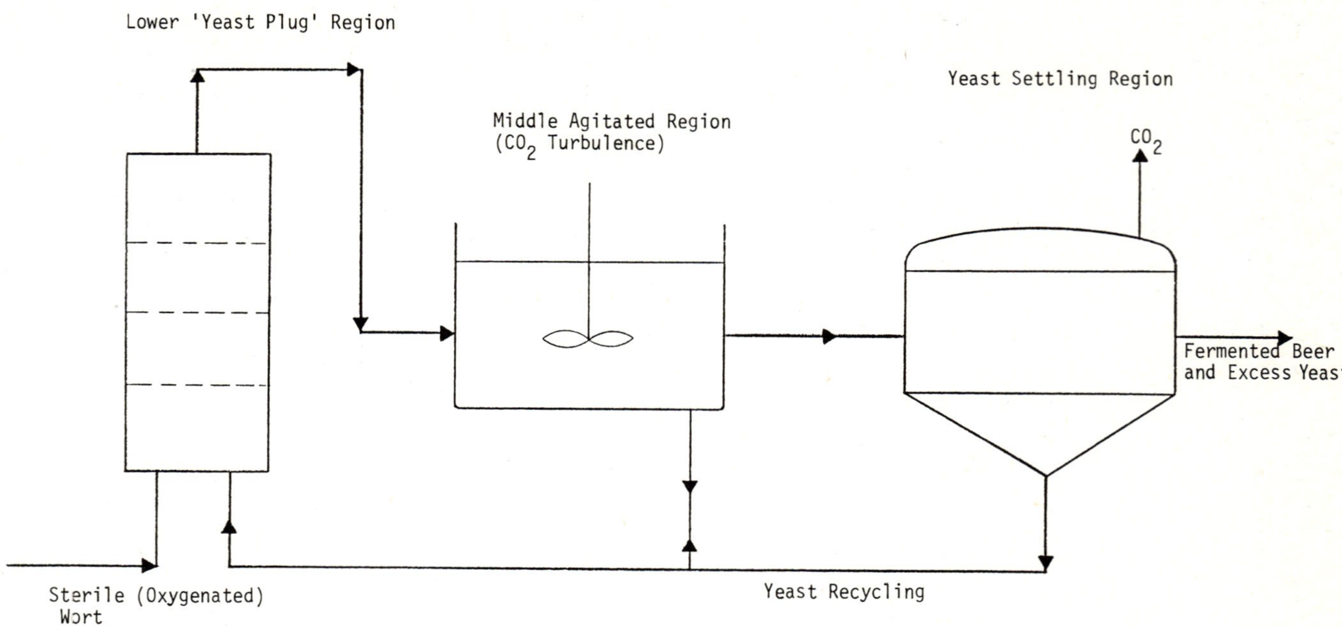

Fig. 34 Schematic Model: Operation Of Tower Continuous Fermentation System (396)

Much development work has been carried out to overcome the problems with tower fermentations. These include the addition of support media to stabilise the yeast plug, and the development of the multi-stage tower fermenter (396). This latter system includes a discrete, cooled, yeast sedimentation zone at the top of the vessel, from which yeast can be recycled to the base at a controlled rate. Its operation is likened to a long series of stirred tanks, culminating with a sedimentation tank with a controlled recycling of yeast. Advantages claimed for this system include:

i Attenuation from 1040° to 1008° (10° - 2^oP) in 1hr.

ii A requirement for yeast which are only slightly flocculent, thus greatly increasing the number of strains which can be employed.

iii It allows control of all parameters, including yeast concentration, to be varied according to required throughput.

Despite the success of these improvements, there are no plans to develop the system commercially, due to the overiding practical difficulties described in Section F1.

G. FUTURE DEVELOPMENTS

The predictions for the future of the brewing industry made over the last twenty years have generally proved incorrect. Thus, the enthusiasm for the high technology brewing systems and novel raw materials developed in the 1960's waned, at least in the industrialised countries, as the real costs and difficulties associated with them became apparent. Similarly, the massive increases in scale of production units, projected before the energy crises and control technology developments of the 1970's, has been reduced, particularly in the United Kingdom. Indeed, within the United Kingdom there is currently a reverse trend to disperse production into smaller, independent, and cost effective regional units with efficient distribution networks. This trend can be expected to continue.

Within the United Kingdom, the initiative for many of the changes which have taken place over the last two decades stemmed mainly from the amalgamation of many of the smaller brewing groups, as a result of mergers during the 1960's. The ensuing rationalisation lead to a sudden demand for new production plant, which could not be sustained once the reorganisation of the new groups was complete.

The economic recession which began early in 1980 has demonstrated, at least in the United Kingdom, that the beer market is now, possibly more than ever before, sensitive to price and consumer incomes. Current beer production forecasts show that it is unlikely that the 15% fall in output experienced over the last three years will be recovered before the end of the decade. This leaves the brewing industry with an excess of relatively modern production plant capacity, and little incentive to invest in its replacement.

Undoubtedly economic pressures will continue to require brewers to improve the control of their process in order to maximise efficiency. In plant terms, this may require the utilisation of microprocessor control technology, to improve accuracy and reliability, whilst reducing manning levels and operating costs. Efficiency improvements within the process itself may also become available, particularly through the use of more precisely defined yeast strains (possibly modified by genetic engineering) and processing conditions.

Arguably, many of the real advances made in the industry over the last twenty years owe as much to the introduction of engineering technology as to the application of brewing science. Certainly up until the last two decades, the main benefit derived by the brewer from the increases in fundamental knowledge of the process was simply a better understanding and rationale for many of the techniques and practices already in common use in breweries, and known by experience to be beneficial to the production of beer of good quality. It is only in recent times that the brewer has sought to apply the increased knowledge and understanding of the science of fermentation, to take advantage of the advances in engineering technology.

In order to exploit fully the increases in knowledge of the nature of fermentation and its effects on beer quality, it will be necessary to establish fine control of wort composition, yeast type, and fermentation conditions. Thus, there is little point in knowing that a fine change, for example in wort composition, gives a specific result if there is no practical means of controlling that parameter within the limits required. Computer technology and new control equipment give some hope for improvements, but process control on a commercial plant is likely to remain relatively coarse because of the scale involved.

Given the failure overall of many of the revolutionary ideas of the 1960's, and the massive investment in relatively long-life plant made in the United Kingdom during the early 1970's, it is unlikely there will be any significant new investment in novel or additional fermentation plants. Cost considerations are now requiring that some of the 'high technology' innovations of the past two decades are re-evaluated in the light of lower production volumes, higher energy costs, and excess plant capacity. Where traditional techniques can be applied, these may offer significant costs savings (albeit at the expense of maximum plant capacities). Obvious examples include the use of high gravity brewing, beer centrifuges, and pressure fermentations.

Another factor resulting from the excess plant capacities available throughout the world may well be a re-evaluation of the possible alternative uses of brewing plant, either during the recession, or during seasonal low production periods. The development of demand for other products produced by fermentation, such as enzymes, single cell protein animal feeds, and fuel alcohol, could be met, in part, by utilising excess brewing capacity.

In direct contrast to the situation in the United Kingdom and many other parts of the industrialised world, Third World countries are now rapidly developing their own brewing industries. These countries have large potential markets, with a growing expenditure on consumer products. Additionally, there are relatively few constraints in matching traditional products in these markets, and this gives free reign to both engineers and brewers to introduce novel equipment, materials, and products, with little fear of adverse reaction from consumers. It is probable therefore, that the more revolutionary ideas and processes will be adopted and developed in these countries rather than in the industrialised world.

Where new plant is installed within the developed countries, it is likely that new materials of construction will be employed. Beer for sale in plastic (PET) bottles, and the use of plastics and other novel materials in the manufacture of various items of brewing plant, can be expected to reduce the resistance of brewers to their use in fermentation equipment. However, in line with the reduction in brewery capacities, a likely increase in the number of beer qualities produced, and the higher plant efficiencies required, it is probable that the scale of new plant will be reduced.

Changes in fashion for beer types may continue, in response to both consumer and marketing pressures. Modern single tank fermentation systems have shown themselves to be flexible enough to produce a wide range of products. By taking advantage of this flexibility, and by applying current (and future) knowledge of the nature of yeast and fermentation, it may well be possible to produce 'novel' beers to predefined flavour specifications, without fundamental changes to the equipment and materials used.

In summary, it is unlikely that there will be any dramatic changes in beer fermentation techniques in the forseeable future. The next few years are more likely to represent a period of consolidation on the advances made over the past two decades, resulting in improved plant efficiencies and better control over the quality of the end product, possibly allowing a further technological explosion in the 1990's as current production plants become due for replacement.

H. SUMMARY

This chapter reviews current approaches to brewery fermentations within the context of the various factors which determine and limit the approaches. Paramount amongst these are an increased understanding of the scientific principles underlying the fermentation process and yeast metabolism, and the technological advances - both in basic engineering and in brewing technology - made over the last two decades.

The influence of fermentation conditions and yeast performance on beer quality, especially the formation of the various classes of flavour-active constituents in beer, is reviewed in detail, together with the options available to the brewer to adjust the levels of these compounds in beer.

Under engineering technology, developments in materials of construction, cleaning-in-place, automation and control, and CO_2 recovery are discussed, together with the impact these advances have made on fermentation systems. The influence of advances in brewing technology includes sections on the increased options for yeast control, ranging from the use of pure culture techniques to the potential for genetically modified yeast, the various alternatives for accelerating the fermentation and maturation processes, and the reasons and rationale for high gravity brewing and fermentation. Developments in the use of enzymes in fermentations are discussed, together with the options now available for controlling such parameters as foam formation and CO_2 levels in deep fermenting vessels. The development and potential for novel fermentation systems are reviewed.

In addition to scientific and technological developments, the importance of economic factors in determining current approaches to fermentation is stressed, and their impact reviewed. The effects of changing market forces and marketing strategies, and the increasing influence of various legislative and legal requirements are also discussed.

Against the background of these various influences, current approaches and options (both traditional and modern) to brewery fermentations are reviewed. Methods and alternatives for wort cooling and collection, primary and secondary fermentation and conditioning, and yeast management are discussed. Batch and continuous systems are compared and evaluated, and the reasons for the commercial failure of continuous fermentation discussed.

In a final section on future developments, it is argued that, certainly within the United Kingdom, because of the high level of capital investment over the last two decades and the current excess capacity within the brewing industry, there are unlikely to be any major new developments during the present decade. When new developments do occur, the emphasis is most likely to favour medium sized breweries, operating to maximum cost effectiveness with highly efficient and flexible plant.

I. ACKNOWLEDGEMENTS

The authors wish to thank the directors of Bass PLC for permission to publish this paper and to express their appreciation to the following individuals for their help in its preparation: Miss S. Gloyne and Mr. C. Fairclough.

J. REFERENCES

1. H.S. Corran, A History Of Brewing, David & Charles Ltd., Newton Abbot, 1975.
2. R. Dubos, Pasteur and Modern Science, W.H. Heinemann, London, 1960, 159.
3. J.R. Porter, Science, New York, 178 (1972) 1249-1254.
4. A. Jørgensen, Microorganisms and Fementation, 15th Edn., Trans. by A. Hansen, Charles Griffin & Co. Ltd., London, 1948.
5. J.P. Van der Walt, in J. Lodder (Ed.) The Yeast, A Taxonomic Study, North-Holland Publishing Co., Amsterdam, 1970, 555-718.
6. I.C. MacWilliam, J. Inst. Brew., 74 (1968) 38-54.
7. F.J. Bates, N.L. Bowman, D.H. Brawns, J.F. Brewster, C.S. Cragoe, H.L. Frush, P.E. Golden, L.D. Hammond, M.L. Hubbell, H.S. Isabell, R.F. Jackson, E.J. McDonald, F.P. Phelps, W.W. Pigman, M.J. Proffitt, J.B. Saunders, C.F. Snyder and A.Q. Tool, Polarimetry, Saccharimetry and the Sugars, National Bureau of Standards Circular No. C440, Washington, D.C., 1942, p810.
8. C. Rainbow, Process Biochem., 1 (1966) 489-492.
9. A.A. Leach, J. Inst. Brew., 73 (1967) 8-16.
10. A.H. Cook, Proc. Europ. Brew. Conv., 13th Congress, Estoril, Elsevier, Amsterdam, 1971, 469-478.
11. G.G. Stewart, Brewers Digest, 50 (1975 Mch.) 42-62.
12. G.P.S. Ballard, The Brewer, 68 (1982) 322-328.
13. E. Helm and O.C. Richardt, J. Inst. Brew., 42 (1936) 413-419.
14. S. Engan, J. Inst. Brew., 78 (1972) 33-36.
15. C.M. Szlavko, J. Inst. Brew., 79 (1973) 283-288.
16. C.M. Szlavko, J. Inst. Brew., 80 (1974) 534-539.
17. S. Sentheshanmuganathan, Biochem. J., 74 (1960) 568-576.
18. S. Sentheshanmuganathan, Biochem. J., 77 (1960) 619-625.
19. S. Sentheshanmuganathan and S.R. Elsden, Biochem. J., 69 (1958) 210-218.
20. C. Rainbow in A.H. Rose and J.S. Harrison (Eds.) The Yeasts, Academic Press, London and New York, 1970, Vol. 3, Ch. 4, 147-224.
21. T. Äyräpää, J. Inst. Brew., 71 (1965) 341-347.
22. T. Äyräpää, J. Inst. Brew., 73 (1967) 17-33.
23. T. Inoue, Rep't. Res. Labs. Kirin Brewery Co. Ltd., 18 (1975) 13-16.
24. G. Thaukis, Amer. J. Enol. Viticul., 9 (1958) 161-167.
25. H. Suomalainen and H. Kahanpaa, J. Inst. Brew., 69 (1963) 473-478.
26. E.C.H. Chen, J. Am. Soc. Brew. Chem., 36 (1978) 39-43.
27. D. Schulthess and L. Ettlinger, Proc. Europ. Brew. Conv., 16th Congress, Amsterdam, Elsevier, Amsterdam, 1977, 471-482.
28. T. Äyräpää, J. Inst. Brew., 74 (1968) 169-178.
29. S. Engan, J. Inst. Brew., 76 (1970) 254-261.
30. D. Schulthess and L. Ettlinger, J. Inst. Brew., 84 (1978) 240-243.
31. J.R. Hudson and R. Stevens, J. Inst. Brew., 66 (1960) 471-474.
32. S. Engan, Europ. Brew. Conv., Monograph V, Fermentation and Storage Symposium, Zoeterwoude, Elsevier, Amsterdam, 1978, 28-39.
33. B. Mändl, E. Geiger and A. Piendl, Proc. Europ. Brew. Conv., 15th Congress, Nice, Elsevier, Amsterdam, 1975, 539-563.
34. J. Kumada, S. Nakajima, T. Takahashi and L. Narziss, Proc. Europ. Brew. Conv., 15th Congress, Nice, Elsevier, Amsterdam, 1975, 615-623.
35. H. Miedaner, Europ. Brew. Conv., Monograph V, Fermentation and Storage Symposium, Zoeterwoude, Elsevier, Amsterdam, 1978, 110-134.

36. J. Posada, J. Candela, G. Calero, J. Almenar and S. Martin, Proc. Europ. Brew. Conv., 16th Congress, Amsterdam, Elsevier, Amsterdam, 1977, 533-544.
37. T. Äyräpää, Brauwissenschaft, 23 (1970) 48-55.
38. B. Drews and J. Riemann, Monats. für Brauerei, 20 (1967) 254-268.
39. K. Wackerbauer, Monats. für Brauerei, 22 (1969) 211-216.
40. M. Jones and J.S. Pierce, Proc. Europ. Brew. Conv., 9th Congress, Brussels, Elsevier, Amsterdam, 1963, 101-134.
41. J.S. Pierce, Process Biochem., 1 (1966) 412-416.
42. J. Barrett, C.M. Griffiths and B.H. Kirsop, J. Inst. Brew., 73 (1967) 445-450.
43. M.O. Jones and C. Rainbow, Proc. Am. Soc. Brew. Chem., (1966) 66-70.
44. S. Engan in J.R.A. Pollock (Ed.) Brewing Science, Vol. 2, Academic Press, London, 1981, Ch.3, 93-165.
45. R.G. Anderson and B.H. Kirsop, J. Inst. Brew., 80 (1974) 48-55.
46. K. Nordström, J. Inst. Brew., 67 (1961) 173-181.
47. K. Nordström, J. Inst. Brew., 68 (1962) 398-407.
48. K. Nordström, J. Inst. Brew., 69 (1963) 142-153.
49. K. Nordström, Proc. Europ. Brew. Conv., 10th Congress, Stockholm, Elsevier, Amsterdam, 1965, 195-208.
50. F. Lynen, Fed. Proc., 20 (1961) 941-948.
51. F. Lynen, Harvey Lectures, 48 (1962) 210.
52. F. Lynen, Biochem. J., 102 (1967) 381-385.
53. K. Nordström, Svensk Kemisk Tidshrift, 76 (1964) 510-543.
54. D.E. Quain, P.A. Thurston and R.S. Tubb, J. Inst. Brew., 87 (1981) 108-111.
55. P.A. Thurston, D.E. Quain and R.S. Tubb, Proc. Europ. Brew. Conv., 18th Congress, Copenhagen, Elsevier, Amsterdam, 1981, 197-206.
56. P.A. Thurston, D.E. Quain and R.S. Tubb, J. Inst. Brew., 88 (1982) 90-94.
57. M.H. David and B.H. Kirsop, Proc. Am. Soc. Brew. Chem., 30 (1972) 14-16.
58. M.H. David and B.H. Kirsop, J. Inst. Brew., 79 (1973) 20-25.
59. P.A. Thurston, R. Taylor and J. Ahvenainen, J. Inst. Brew., 87 (1981) 92-95.
60. R.G. Anderson, B.H. Kirsop, H. Rennie and R.J.H. Wilson, Proc. Europ. Brew. Conv., 15th Congress, Nice, Elsevier, Amsterdam, 1975, 243-253.
61. C. Norstedt, A. Bengtsson, P. Bennet, I. Lindström and T. Äyräpää, Proc. Europ. Brew. Conv., 15th Congress, Nice, Elsevier, Amsterdam, 1975, 581-600.
62. A.K. Palmer and H. Rennie, J. Inst. Brew., 80 (1974) 447-454.
63. S. Engan, Brewers Digest, 49 (1974 Nov.) 40-48.
64. K. Nordström, J. Inst. Brew., 70 (1964) 226-233.
65. C. Whitworth, Europ. Brew. Conv., Monograph V, Fermentation and Storage Symposium, Zoeterwoude, Elsevier, Amsterdam, 1978, 155-164.
66. R.G. Anderson and B.H. Kirsop, J. Inst. Brew., 81 (1975) 111-115.
67. R.J.H. Wilson, Europ. Brew. Conv., Monograph V, Fermentation and Storage Symposium, Zoeterwoude, Elsevier, Amsterdam, 1978, 17-27.
68. F.H. White and A.D. Portno, Proc. Europ. Brew. Conv., 17th Congress, Berlin, Elsevier, Amsterdam, 1979, 447-460.
69. F.H. White, Brewers Digest, 54 (1979 Oct.) 26-36.
70. B.H. Kirsop, J. Inst. Brew., 80 (1974) 252-258.
71. B.H. Kirsop, Brewers Digest, 53 (1978 July) 28-32.
72. T.W. Cowland and D.R. Maule, J. Inst. Brew., 72 (1966) 480-488.
73. D.R. Maule, J. Inst. Brew., 73 (1967) 351-361.
74. B.H. Kirsop, Brewers Digest, 50 (1975 Dec.) 34-46.
75. D. Bloomfield and K. Block, Biochim. Biophys. Acta, 30 (1958) 220-221.
76. D. Bloomfield and K. Block, J. Biol. Chem., 235 (1960) 337-345.
77. H.P. Klein, J. Bact., 69 (1955) 620-627.
78. G. Popjak and J.W. Cornforth, Advances in Enzymology, 22 (1960) 281-312.
79. A.H. Rose, Europ. Brew. Conv., Monograph V, Fermentation and Storage Symposium, Zoeterwoude, Elsevier, Amsterdam, 1978, 96-107.

80. M. Jakobsen and R.S.W. Thorne, J. Inst. Brew., 86 (1980) 284-287.
81. T. Äyräpää and I. Lindström, Proc. Europ. Brew. Conv., 14th Congress, Salzburg, Elsevier, Amsterdam, 1973, 271-283.
82. T. Äyräpää and I. Lindström, Proc. Europ. Brew. Conv., 16th Congress, Amsterdam, Elsevier, Amsterdam, 1977, 507-517.
83. G.T. Taylor, P.A. Thurston and B.H. Kirsop, J. Inst. Brew., 85 (1979) 219-227.
84. P.A. Thurston, R. Taylor and J. Ahvenainen, J. Inst. Brew., 87 (1981) 92-95.
85. M.O. Jones, R. Cope and C. Rainbow, Proc. Europ. Brew. Conv., 15th Congress, Nice, Elsevier, Amsterdam, 1975, 669-681.
86. D. Howard and R.G. Anderson, J. Inst. Brew., 82 (1976) 70-71.
87. G. Krauss, C. Zurcher and M. Holstein, Monats. fur Brauerei, 25 (1972) 113-119.
88. N. Hashimoto in J.R.A. Pollock (Ed.) Brewing Science, Vol. 2, Academic Press, London, 1981, Ch.6, 348-405.
89. E. Pfisterer and G.G. Stewart, Proc. Europ. Brew. Conv., 15th Congress, Nice, Elsevier, Amsterdam, 1975, 255-267.
90. J. Posada, Europ. Brew. Conv., Monograph V, Fermentation and Storage Symposium, Zoeterwoude, Elsevier, Amsterdam, 1978, 207-217.
91. B.W. Drost, Proc. Europ. Brew. Conv., 16th Congress, Amsterdam, Elsevier, Amsterdam, 1977, 519-532.
92. H.J. Wellhoener, Schweiz. Brau. Rund., 86 (1975) 34-37.
93. H. Miedaner, L. Narziss and G. Woerner, Brauwissenschaft, 27 (1974) 20-29.
94. J.F. Rice, E. Chicoye, J.R. Helbert and J. Garver, J. Am. Soc. Brew. Chem., 35 (1977) 35-40.
95. R.P. Jones and P.F. Greenfield, Enzyme Microb. Technol., 4 (1982) 210-223.
96. E. Geiger and A. Piendl, Tech. Quart. M.B.A.A., 13 (1976) 51-63.
97. A. Sols, C. Gancedo and G. Delafuente in A.H. Rose and J.S. Harrison (Eds.) The Yeasts, Academic Press, London and New York, 1970, Vol.2, Ch.7, 271-307.
98. K. Nordström, J. Inst. Brew., 74 (1968) 192-195.
99. M.J.S. Dadds, A.L. Macpherson and A. Sinclair, J. Inst. Brew., 77 (1971) 453-456.
100. G.E. Otter and L. Taylor, J. Inst. Brew., 77 (1971) 467-472.
101. E. Pessa, Proc. Europ. Brew. Conv., 13th Congress, Estoril, Elsevier, Amsterdam, 1971, 333-342.
102. S. Engan and O. Aubert, Proc. Europ. Brew. Conv., 13th Congress, Estoril, Elsevier, Amsterdam, 1971, 407-432.
103. T. Wainwright, J. Inst. Brew., 79 (1973) 451-470.
104. A.D. Haukeli and S. Lie, J. Inst. Brew., 77 (1971) 538-543.
105. A.D. Haukeli and S. Lie, J. Inst. Brew., 78 (1972) 229-232.
106. F.H. White and T. Wainwright, J. Inst. Brew., 81 (1975) 46-52.
107. A.D. Haukeli and S. Lie, J. Inst. Brew., 84 (1978) 85-89.
108. S. Lie, Europ. Brew. Conv., Monograph V, Fermentation and Storage Symposium, Zoeterwoude, Elsevier, Amsterdam, 1978, 40-50.
109. S. Lie, M. Skjeldam and T. Jacobsen, Proc. Europ. Brew. Conv., 17th Congress, Berlin, Elsevier, Amsterdam, 1979, 715-724.
110. P.T. Magee and H. De Robichon-Szulmajster, Europ. J. Biochem., 3 (1968) 502-505.
111. L.F. Chuang and E.B. Collins, J. Gen. Microbiol., 72 (1972) 201-210.
112. T. Inoue, K. Masuyama, Y. Yamamoto and K. Okada, Rep't. Res. Labs. Kirin Brewery Co. Ltd., 11 (1968) 1-23.
113. T. Inoue, K. Masuyama, Y. Yamamoto, K. Okada and Y. Kuroiwa, Proc. Am. Soc. Brew. Chem., (1968) 158-165.
114. A.D. Portno, J. Inst. Brew., 72 (1966) 193-196.
115. A. Scherrer, Wallerstein Lab. Comms., 35 (1972) 5-33.
116. T. Inoue and Y. Yamamoto, Proc. Am. Soc. Brew. Chem., (1970) 198-208.
117. A.D. Portno, J. Inst. Brew., 72 (1966) 458-461.

118. M.W. Brenner, Tech. Quart. M.B.A.A., 7 (1970) 43-49.
119. M.W. Brenner and M. Kamimura, Proc. Am. Soc. Brew. Chem., (1969) 160-170.
120. L.F. Chuang and E.B. Collins, J. Bact., 95 (1968) 2083-2086.
121. C. Rainbow in J.R.A. Pollock (Ed.) Brewing Science, Vol. 2, Academic Press, London and New York, 1981, Ch.9, 491-550.
122. R.G. Ault and R. Newton in W.P.K. Findlay (Ed.) Modern Brewing Technology, MacMillan Press, London, 1971, Ch.7, 164-197.
123. M.A. Cowbourne, F.G. Priest and J.S. Hough, Brewers Digest, 47 (1972 Oct) 76-84.
124. T. Ishibashi, T. Sasahara and K. Okada, Rep't. Res. Labs. Kirin Brewery Co. Ltd., 12 (1969) 1-8.
125. F. Huber, P. Zeisz and M. Natter-Hornbostel, Proc. Europ. Brew. Conv., 12th Congress, Interlaken, Elsevier, Amsterdam, 1969, 281-298.
126. R.H. Roberts and C. Rainbow, The Brewer, 57 (1971) 112-116.
127. P.R. Witt and G.S. Robbins, Tech. Quart. M.B.A.A., 7 (1970) 105-110.
128. B.B. Awford, Tech. Quart. M.B.A.A., 14 (1977) 129-133.
129. H. Kieninger, J. Inst. Brew., 83 (1977) 72-77.
130. H. Miedaner, The Brewer, 64 (1978) 33-39.
131. W.J. Klopper, R.H. Roberts, M.G. Royston and R.G. Ault, Proc. Europ. Brew. Conv., 10th Congress, Stockholm, Elsevier, Amsterdam, 1965, 238-259.
132. L. Narziss and P. Hellich, Brewers Digest, 47 (1972 Sept.) 106-118.
133. M. Burger, P.R. Glenister and K. Becker, Proc. Am. Soc. Brew. Chem., (1957) 110-115.
134. D.A. Baker and B.H. Kirsop, J. Inst. Brew., 79 (1973) 43-44.
135. D.A. Baker and B.H. Kirsop, J. Inst. Brew., 79 (1973) 487-494.
136. S.E. Godtfredsen and M. Ottesen, Carlsberg Res. Comms., 47 (1982) 93-102.
137. V.S. Bavisotto, J. Shovers, W.E. Sandine and P.R. Elliker, Proc. Am. Soc. Brew. Chem., (1964) 211-216.
138. W.E. Sandine, P.R. Elliker and M.Y. Pack, Tech. Quart. M.B.A.A., 2 (1965) 155-159.
139. J.W. Thompson, J. Shovers, W.E. Sandine and P.R. Elliker, Appl. Microbiol., 19 (1970) 883-889.
140. J.P. Loken and F.C. Stormer, Europ. J. Biochem., 14 (1970) 133-137.
141. S. Home, Europ. Brew. Conv., Monograph VII, Flavour Symposium, Copenhagen, Elsevier, Amsterdam, 1981, 79-87.
142. H. Garza-Ulloa, Brewers Digest, 56 (1980 Jan.) 20-26.
143. R.J. Anderson and G.A. Howard, J. Inst. Brew., 80 (1974) 357-370.
144. C.J. Dickenson and R.J. Anderson, Proc. Europ. Brew. Conv., 18th Congress, Copenhagen, Elsevier, Amsterdam, 1981, 413-420.
145. D.W. Hysert, R.L. Weaver and N.M. Morrison, Tech. Quart. M.B.A.A., 17 (1980) 34-43.
146. R. Parsons, T. Wainwright and F.H. White, Proc. Europ. Brew. Conv., 16th Congress, Amsterdam, Elsevier, Amsterdam, 1977, 115-127.
147. T.L. Peppard, J. Inst. Brew., 84 (1978) 337-340.
148. T.L. Peppard and D.R.J. Laws, Proc. Europ. Brew. Conv., 17th Congress, Berlin, Elsevier, Amsterdam, 1979, 91-104.
149. A. Suggett, M. Moir and J.C. Seaton, Proc. Europ. Brew. Conv., 17th Congress, Berlin, Elsevier, Amsterdam, 1979, 79-89.
150. J.C. Seaton, A. Suggett and M. Moir, Europ. Brew. Conv., Monograph VII, Flavour Symposium, Copenhagen, Elsevier, Amsterdam, 1981, 143-155.
151. R.J. Anderson, G.A. Howard and J.S. Hough, Proc. Europ. Brew. Conv., 13th Congress, Estoril, Elsevier, Amsterdam, 1971, 253-264.
152. A. Sinclair, R.D. Hall and D. Thorburn Burns, Proc. Europ. Brew. Conv., 12th Congress, Interlaken, Elsevier, Amsterdam, 1969, 427-444.
153. T. Wainwright, Proc. Europ. Brew. Conv., 13th Congress, Estoril, Elsevier, Amsterdam, 1971, 437-449.
154. R.J. Anderson and G.A Howard, J. Inst. Brew., 80 (1974) 245-251.
155. K. Nagami, T. Takahashi, K. Nakatani and J. Kumada, M.B.A.A. Tech. Quart., 17 (1980) 64-71.
156. T. Takahashi, K. Nagami, K. Nakatani and J. Kumada, M.B.A.A. Tech. Quart., 17 (1980) 210-214.

157. A.B. Roy and P.A. Trudinger, The Biochemistry of Inorganic Compounds of Sulphur, University Press, Cambridge, 1970.
158. R. Eschenbruch, Amer. J. Viticult., 25 (1974) 157-161.
159. G.G. Stewart and I. Russell, Europ. Brew. Conv., Monograph VII, Flavour Symposium, Copenhagen, Elsevier, Amsterdam, 1981, 173-187.
160. G.A. Maw, Wallerstein Lab. Comms., 28 (1965) 49-68.
161. T. Wainwright, Proc. Am. Soc. Brew. Chem., (1970) 127-133.
162. T. Wainwright, J. Appl. Bact., 34 (1971) 161-171.
163. T. Wainwright, J. Gen. Microbiol., 61 (1970) 107-119.
164. O.A. Leppenan, J. Denslow and P. Ronkainen, J. Agri. Food Chem., 28 (1980) 359-364.
165. S. Matsui and M. Amaha, Agric. Biol. Chem., 45 (1981) 1341-1349.
166. S.W. Molzahn, J. Am. Soc. Brew. Chem., 35 (1977) 54-59.
167. G.G. Stewart and I. Russell, Proc. Europ. Brew. Conv., 17th Congress, Berlin, Elsevier, Amsterdam, 1979, 475-490.
168. R.S. Tubb, J. Inst. Brew., 85 (1979) 286-289.
169. R.S.W. Thorne, E. Helm and K. Svendsen, J. Inst. Brew, 77 (1971) 148-153.
170. R.J. Anderson, J.F. Clapperton, D. Crabb and J.R. Hudson, J. Inst. Brew., 81 (1975) 208-213.
171. J.A. Pickett, J. Coates, T.L. Peppard and R.F. Sharpe, J. Inst. Brew., 82 (1976) 233-238.
172. J.F. Clapperton, J. Inst. Brew., 79 (1973) 495-508.
173. G.A.F. Harrison and E. Collins, Proc. Am. Soc. Brew. Chem., (1968) 83-87.
174. A. Sinclair, R.D. Hall, D. Thorburn Burns and W.P. Hayes, J. Sci. Food Agric., 21 (1970) 468-470.
175. B.J. Anness and C.W. Bamforth, J. Inst. Brew., 88 (1982) 244-252.
176. C.J. Dickenson, J. Inst. Brew., 89 (1983) 41-46.
177. F.H. White and R. Parsons, Proc. Europ. Brew. Conv., 15th Congress, Nice, Elsevier, Amsterdam, 1975, 721-735.
178. F.H. White and T. Wainwright, J. Inst. Brew., 82 (1976) 292-296.
179. R.J.H. Wilson and C.D. Booer, J. Inst. Brew., 85 (1979) 144-148.
180. C.J. Dickenson, J. Inst. Brew., 85 (1979) 235-239.
181. C.M. Szlavko and R.J. Anderson, J. Am. Soc. Brew. Chem., 37 (1979) 20-25.
182. C.D. Booer and R.J.H. Wilson, J. Inst. Brew., 85 (1979) 35-37.
183. B.J. Anness, C.W. Bamforth and T. Wainwright, J. Inst. Brew., 85 (1979) 346-349.
184. B.J. Anness, J. Inst. Brew., 86 (1980) 134-137.
185. C.W. Bamforth and B.J. Anness, J. Inst. Brew., 87 (1981) 30-34.
186. F.G. Priest, M.A. Cowbourne and J.S. Hough, J. Inst. Brew., 80 (1974) 342-356.
187. F.G. Priest and J.S. Hough, J. Inst. Brew., 80 (1974) 370-376.
188. C.J. Dickenson, J. Inst. Brew., 85 (1979) 329-333.
189. F.H. White and T. Wainwright, J. Inst. Brew., 82 (1976) 46-48.
190. T.E. Kavanagh, R.C. Derbyshire, R.P. Hildebrand, B.J. Clarke and F.J. Meeker, J. Inst. Brew., 82 (1976) 270-272.
191. C.J. Dickenson, Chem. and Ind., (1979) 896-899.
192. F.H. White, Brewers Digest, 52 (1977 May) 38-50.
193. L. Narziss, H. Miedaner and T. Bourjau, Brauwissenschaft, 33 (1979) 62-69.
194. C.W. Bamforth and B.J. Anness, Soc. Gen. Microbiol. Quart., 6 (1979) 160.
195. C.W. Bamforth, F.E.M.S. Micro Letters, 7 (1980) 55-59.
195a R.M. Gibson, P.J. Large, B.J. Anness and C.W. Bamforth, J. Inst. Brew., 89 (1983) 215-218.
196. M.O. Jones, C.V. Lloyd, R. Parsons and R.W. Ricketts, Proc. 14th Conv. Inst. of Brew., Australian and New Zealand Section, 1976, 37-44.
197. S.H. Zinder and T.D. Brock, Arch. Microbiol., 116 (1978) 35-40.
198. N. Coote and B.H. Kirsop, J. Inst. Brew., 80 (1974) 474-483.
199. N. Coote, B.H. Kirsop and G.K. Buckee, J. Inst. Brew., 79 (1973) 298-304.
200. S. Engan, J. Inst. Brew., 80 (1974) 162-163.
201. G. Charalambous in J.R.A. Pollock (Ed.) Brewing Science, Vol.2, Academic Press, London, 1981, Ch.4, 167-254.

202. B.H. Kirsop, The Brewer, 65 (1979) 270-273.
203. G.C. Whiting, J. Inst. Brew., 82 (1976) 84-92.
204. C. Chapman and W. Bartley, Biochem. J., 107 (1968) 455-459.
205. M. Sumper and C. Riepertinger, Europ. J. Biochem., 29 (1972) 237-249.
206. M.J. Lewis and C. Rainbow, J. Inst. Brew., 69 (1963) 39-45.
207. J.F. Clapperton, C.E. Dalgliesh and M.C. Meilgaard, J. Inst. Brew., 82 (1976) 7-13.
208. J.F. Clapperton and D.G.W. Brown, J. Inst. Brew., 84 (1978) 90-92.
209. J.F. Clapperton, J. Inst. Brew., 84 (1978) 107-112.
210. J.F. Clapperton, Proc. Europ. Brew. Conv., 15th Congress, Nice, Elsevier, Amsterdam, 1975, 823-835.
211. G.A.F. Harrison, J. Inst. Brew., 76 (1970) 486-495.
212. M.C. Meilgaard, M.B.A.A. Tech. Quart., 12 (1975) 151-168.
213. R.T. Roberts and J.F. Clapperton, J. Inst. Brew., 84 (1978) 157.
214. J. Van de Meerssche, A. Devreux and C.A. Masschelein, Proc. Europ. Brew. Conv., 16th Congress, Amsterdam, Elsevier, Amsterdam, 1977, 561-575.
215. J. Van de Meerssche, A. Devreux and C.A. Masschelein, Proc. Europ. Brew. Conv., 17th Congress, Berlin, Elsevier, Amsterdam, 1979, 787-800.
216. C.A. Masschelein, Europ. Brew. Conv., Monograph VII, Flavour Symposium, Copenhagen, Elsevier, Amsterdam, 1981, 211-221.
217. G. Krauss and M. Forch, Proc. Am. Soc. Brew. Chem., 33 (1975) 37-41.
218. G. Rosculet, Proc. Am. Soc. Brew. Chem., (1969) 127-138.
219. V. Aries, B.H. Kirsop and G.T. Taylor, Proc. Europ. Brew. Conv., 16th Congress, Amsterdam, Elsevier, Amsterdam, 1977, 255-266.
220. G.T. Taylor and B.H. Kirsop, J. Inst. Brew., 83 (1977) 241-243.
221. R.C. Tripp, B. Timm, M. Iyer, T. Richardson and C.H. Amundson, Proc. Am. Soc. Brew. Chem., (1968) 65-74.
222. J.K. MacPherson and G.K. Buckee, J. Inst. Brew., 80 (1974) 540-544.
223. R. Von Tressl, T. Kossa, R. Renner and H. Koppler, Monats. für Brauerei, 28 (1975) 109-118.
224. M. Forch, G. Krauss and H. Proksch, Proc. Am. Soc. Brew. Chem., 33 (1975) 148-154.
225. E.C.H. Chen, J. Am. Soc. Brew. Chem, 38 (1980) 148-153.
226. S. Engan and O. Aubert, Proc. Europ. Brew. Conv., 16th Congress, Amsterdam, Elsevier, Amsterdam, 1977, 591-607.
227. G. Krauss, M. Forch and H. Holstein, Monats. für Brauerei, 28 (1975) 229-237.
228. V. Arkima, Monats. für Brauerei, 21 (1968) 247-249.
229. B.J. Clarke, D.F. Davine, D.B. Hawthorne, T.E. Kavanagh and P.J. Moulder, M.B.A.A. Tech. Quart., 18 (1981) 188-194.
230. H. Suomalainen and A.J.A. Keranen, Suomen Kemustilehti, B36 (1963) 88-90.
231. M. Sumper, Europ. J. Biochem., 49 (1974) 469-475.
232. G.T. Taylor, P.A. Thurston and B.H. Kirsop, J. Inst. Brew., 85 (1979) 219-227.
233. S. Engan, J. Inst. Brew., 78 (1972) 169-173.
234. S.R. Griffin, J. Inst. Brew., 76 (1970) 41-45.
235. R. Stevens, Chem. Revs., 67 (1967) 19-71.
236. I.J. Dixon, J. Inst. Brew., 73 (1967) 488-493.
237. D. Taylor, Brew. Dist. Int'l., 11 (1981 Apr.) 35-42.
238. R.C. Hockney, A.D. Portno and F.H. White, Europ. Brew. Conv., Monograph VII, Flavour Symposium, Copenhagen, Elsevier, Amsterdam, 1981, 59-65.
239. Anon, Brew. Dist. Int'l., 12 (1982 Feb.) 20-26.
240. Anon, Brew. Dist. Int'l., 7 (1977 Dec.) 22-23.
241. M.P. Donald and N. Harlow, The Brewer, 65 (1979) 12-17.
242. J.H. Bradbury, Process Biochem., 3 (1968 Apr.) 32-35.
243. L. Forbes, Brew. Dist. Int'l., 12 (1982 Feb.) 29.
244. P. Gregory, Brew. Dist. Int'l., 9 (1979 Apr.) 32-33.
245. A.S. Auty, The Brewer, 64 (1978) 226-229.
246. P. Crispin, Brewers' Guardian, (1978 Nov.) 43-47.
247. M. Barrett, Brewers' Guardian, (1979 Aug.) 35-40.
248. V.G. Thurston, The Brewer, 64 (1978) 223-225.

249. N. Sharp, Brew. Dist. Int'l., 9 (1979 Aug.) 27-28.
250. D. Elmore, CIP Of Mains and Vessels By Acids, Diversey Ltd., 1981.
251. J. Schoorens, The Brewer, 68 (1982) 13-20.
252. D.M. Hudson and R.F. Illingworth, The Brewer, 68 (1982) 239-242.
253. J. Hampton, The Brewer, 68 (1982) 151-154.
254. G. Meekings, The Brewer, 68 (1982) 155-159.
255. W.E. Stephen, The Brewer, 68 (1982) 202-204.
256. R. Boughton, The Brewer, 68 (1982) 205-207.
257. W.R. Turvill, The Brewer, 59 (1973) 607-611.
258. J.T. Leech, The Brewer, 64 (1978) 300-306.
259. A. Gibson, Tech. Quart. M.B.A.A., 17 (1980) 89-97.
260. A. Gibson and C. Winstanley, personal communication.
261. S.W. Molzahn, Brewers' Guardian, (1977 July) 21-24.
262. J.R.A. Pollock, Brauwelt, 21 (1979) 721-729.
263. M. Richards, J. Inst. Brew., 73 (1967) 162-166.
264. R.B. Gilliland, Proc. Europ. Brew. Conv., 3rd Congress, Brighton, Elsevier, Amsterdam, 1951, 35-58.
265. J.S. Hough, J. Inst. Brew., 63 (1957) 483-487.
266. B. Ferguson, M. Richards and C. Rainbow, Proc. 12th Conv. Inst. of Brew., Australian and New Zealand Section, 1972, 169-177.
267. D.R. Maule, J. Inst. Brew., 73 (1967) 351-361.
268. A.H. Rose, Methods in Cell Biology, 12 (1975) 1-16.
269. J. Wynants, J. Inst. Brew., 68 (1962) 350-354.
270. F.W. Beech and R.R. Davenport, in J.R. Norris and D.W. Ribbons (Eds.) Methods in Microbiology, Vol.4, Academic Press, New York, 1971, Ch.5, 153-182.
271. A.M. Wellman and G.G. Stewart, Appl. Microbiol., 26 (1973) 577-583.
272. Ø. Winge, Comptes Rendus Des Travaux Du Laboratoire Carlsberg, Serie Physiologique, 21 (1935) 77-111.
273. E. Anderson and P.A. Martin, J. Inst. Brew., 81 (1975) 242-247.
274. R.R. Fowell in A.H. Rose and J.S. Harrison (Eds.) The Yeasts, Academic Press, London and New York, 1969, Vol.1, Ch.7, 303-383.
275. J.R. Johnston, J. Inst. Brew., 71 (1965) 130-135.
276. Ø. Winge and O. Lausten, C.R. Trav. Lab. Carlsberg, 22 (1939) 337-355.
277. C.C. Lindegren, Proc. Am. Soc. Brew. Chem., (1956) 165-168.
278. G.G. Stewart, Can. J. Microbiol., 27 (1981) 973-990.
279. R. Tubb, Brewers' Guardian, (1981 July) 23-31.
280. M.C. Kielland-Brandt, Proc. Europ. Brew. Conv., 18th Congress, Copenhagen, Elsevier, Amsterdam, 1981, 263-276.
281. J.R. Johnston, J. Inst. Brew., 71 (1965) 135-137.
282. N. Gunge and Y. Nakatomi, Genetics, 70 (1972) 41-58.
283. J.F.T. Spencer and D.M. Spencer, J. Inst. Brew., 83 (1977) 287-289.
284. R.S. Tubb, B.A. Searle, A.R. Goodey and A.J.P. Brown, Proc. Europ. Brew. Conv., 18th Congress, Copenhagen, Elsevier, Amsterdam, 1981, 487-496.
285. I. Russell and G.G. Stewart, J. Inst. Brew., 85 (1979) 95-98.
286. R.F. Freeman, Proc. Europ. Brew. Conv., 18th Congress, Copenhagen, Elsevier, Amsterdam, 1981, 497-504.
287. I. Russell and G.G. Stewart, J. Inst. Brew., 86 (1980) 55-59.
288. M.C. Barney, G.P. Jansen and J.R. Helbert, J. Am. Soc. Brew. Chem., 38 (1980) 71-74.
289. R.S. Tubb, J. Inst. Brew., 86 (1980) 78-80.
290. M.C. Kielland-Brandt, T. Nilsson-Tillgren, S. Holmberg, J.G.L. Petersen and B.A. Svenningsen, Carlsberg Res. Comms., 44 (1979) 77-87.
291. D.R. Maule, M.A. Pinnegar, A.D. Portno and A.L. Whitear, J. Inst. Brew., 72 (1966) 488-494.
292. A.M. Vrieling, Europ. Brew. Conv., Monograph V, Fermentation and Storage Symposium, Zoeterwoude, Elsevier, Amsterdam, 1978, 135-144.
293. J.F. Rice, J.R. Helbert and J.C. Garver, Proc. Am. Soc. Brew. Chem., 32 (1974) 94-96.
294. J. Delente, C. Akin, E. Krabbe and K. Ladenburg, Tech. Quart. M.B.A.A., 5 (1968) 228-234.

295. K. Borkmann and H.J. Manger, Die Lebensmittel-Industrie, 22 (1975) 501-508.
296. K. Nordström, J. Inst. Brew., 70 (1964) 328-336.
297. C.A. Masschelein, C. Jeunehomme-Ramos and A. Devreux, Proc. Europ. Brew. Conv., 13th Congress, Estoril, Elsevier, Amsterdam, 1971, 211-225.
298. N. Hashimoto, Y. Kuroiwa and H. Sogawa, Rep't. Res. Labs. Kirin Brewery Co. Ltd., 12 (1969) 57-68.
299. J.S. Hough, D.E. Briggs and R. Stevens in Malting and Brewing Science, Chapman and Hall, London, 1971, Ch.19, 505-536.
300. E. Pajunen and T-M. Enari, Europ. Brew. Conv., Monograph V, Fermentation and Storage Symposium, Zoeterwoude, Elsevier, Amsterdam, 1978, 181-191.
301. J. Posada, J. Candela, J. Almenar and F.L. Palomero, Proc. Europ. Brew. Conv., 12th Congress, Interlaken, Elsevier, Amsterdam, 1969, 253-262.
302. H. Lloyd Hind in Brewing Science and Practice, Chapman and Hall, London, 1940, Vol.2, Ch.33, 808-833.
303. D. Crabb and D.R. Maule, Europ. Brew. Conv., Monograph V, Fermentation and Storage Symposium, Zoeterwoude, Elsevier, Amsterdam, 1978, 165-180.
304. S. Claesson and E. Sandegren, Proc. Europ. Brew. Conv., 9th Congress, Brussels, Elsevier, Amsterdam, 1963, 221-232.
305. N.S. Curtis in W.P.K. Findlay (Ed.) Modern Brewing Technology, Macmillan Press, London, 1971, Ch.11, 254-291.
306. L. Chapon and M. Chemardin, Proc. Europ. Brew. Conv., 11th Congress, Madrid, Elsevier, Amsterdam, 1967, 389-405.
307. F.H. White, unpublished results.
308. O.O. Schaus, Tech. Quart. M.B.A.A., 8 (1971) 7-10.
309. C. Whitworth, The Brewer, 66 (1980) 339-343.
310. G. Harding, The Brewer, 67 (1981) 365-367.
311. M.W. Brenner, Brauwelt, 3 (1979) 42-45.
312. Anon, Brew. Dist. Int'l., 9 (1979 Nov.) 20-21.
313. I.C. Willox, S.R. Rader, J.M. Riolo and H. Stern, Tech. Quart. M.B.A.A., 14 (1977) 105-110.
314. J.I. Evans, Brewers' Guardian, (1973 Apr.) 59-62.
315. P.S. Vernon and A.H. Rose, J. Inst. Brew., 82 (1976) 336-340.
316. P.M. Jayatissa and A.H. Rose, J. Inst. Brew., 83 (1977) 308-312.
317. R.D. Hall, J. Inst. Brew., 78 (1972) 364.
318. R.D. Hall and J.I. Evans, British Patent 1,290,444 (1972).
319. A.H. Button and J.J. Wren, J. Inst. Brew., 78 (1972) 443.
320. L.R. Bishop, J. Inst. Brew., 77 (1971) 12-24.
321. I.S. Maddox and J.S. Hough, J. Inst. Brew., 76 (1970) 262-264.
322. K. Steiner and U. Länzlinger, Brauerei Rundschau, 89 (1978) 3-7.
323. B. Mändl, K.U. Heyse and A. Piendl, Proc. Europ. Brew. Conv., 16th Congress, Amsterdam, Elsevier, Amsterdam, 1977, 483-494.
324. F.R. Sharpe and D.R.J. Laws, J. Inst. Brew., 87 (1981) 96-107.
325. F. Rabak, Wallerstein Lab. Comms., 7 (1944) 173-178.
326. M. De Mets and M. Verzele, J. Inst. Brew., 74 (1968) 74-81.
327. F.R. Sharpe, Chemistry and Technology of The Essential Oil of Hops, Ph.D. Thesis, Council For National Academic Awards, 1978, p40.
328. D.R.J. Laws, J. Inst. Brew., 87 (1981)24-29.
329. F.R. Sharpe, C.M. Grimmett, D.R.J. Laws and G. Bett, J. Inst. Brew., 86 (1980) 234-241.
330. G. Bett, C.M. Grimmett and D.R.J. Laws, J. Inst. Brew., 86 (1980) 175-176.
331. D.R.J. Laws, C.M. Grimmett, G. Bett, R.W. Scott, A.S. Marsh and A.G. Wheldon, J. Inst. Brew., 89 (1983) 28-33.
332. F.V. Harold and B.J. Clarke, Brewers Digest, 54 (1979 Sept.) 45-68.
333. F.R. Sharpe and D. Grabb, J. Inst. Brew., 86 (1980) 60-64.
334. R. Fricker, Brewers' Guardian, (1978 May) 28-37.
335. R.J.H. Wilson, J. Inst. Brew., 82 (1976) 168-169.
336. J.R.A. Pollock and M.J. Weir, J. Inst. Brew., 81 (1975) 375-380.
337. B. Duteurtre, M. Moll, J.R.A. Pollock and M.J. Weir, J. Inst. Brew., 81 (1975) 381-390.

338. C.A. Masschelein, Brew. Dist. Int'l., 11 (1981 May) 37-42.
339. W. Baxter, Brewers' Guardian, (1976 Nov.) 29-33.
340. D.R. Maule, The Brewer, 63 (1977) 204-208.
341. C.J. Packman, The Brewer, 66 (1980) 375-378.
342. J. Hoggan, The Brewer, 65 (1979) 7-11.
343. A.R. Acraman and A.J.T. Smith, The Brewer, 62 (1976) 318-323.
344. R. Witting and W. Ellenberger, Brauwelt, 118 (1978) 200-210.
345. A.W. Seddon, The Brewer, 67 (1981) 392-395.
346. Anon, Brew. Dist. Int'l., 8 (1978 Oct.) 28-36.
347. A.F.C. Leighton, The Brewer, 66 (1980) 308-310.
348. J.Q. Walker, The Brewer, 64 (1978) 389-395.
349. M.G. Delahooke, The Brewer, 64 (1978) 258-260.
350. A.A. Leach, The Brewer, 65 (1979) 129-132.
351. B. Trolle, Brygmesteren, 27 (1970) 255-262.
352. M.W. Brenner, Tech. Quart. M.B.A.A., 17 (1980) 185-195.
353. B.R. Neal, The Brewer, 66 (1980) 240-242.
354. B.R. Neal, The Brewer, 66 (1980) 276-279.
355. B.R. Neal and M.E.N. Hall, The Brewer, 67 (1981) 43-49.
356. A.G.T. Shave, The Brewer, 65 (1979) 201-203.
357. A.L. Whitear, Brewers' Guardian, (1981 Jan.) 25-31.
358. H. Lloyd Hind in Brewing Science and Practice, Chapman and Hall, London, 1940, Vol.2, 725-726.
359. D.F. Clark, The Brewer, 62 (1976) 8-12.
360. H. Rennie and R.J.H. Wilson, J. Inst. Brew., 81 (1975) 105-106.
361. H. Kieninger, Tech. Quart. M.B.A.A., 12 (1975) 256-263.
362. Anon, Brew. Dist. Int'l., 12 (1982 April) 28.
363. C. Rainbow, J. Inst. Brew., 74 (1968) 427-429.
364. J.H. Lindsay and J.W. Larson, Tech. Quart. M.B.A.A., 12 (1975) 264-272.
365. W.C. Lasman, G.T. Peard and E.G. Peet, J. Inst. Brew., 61 (1955) 192-199.
366. R.W. Ricketts in W.P.K. Findlay (Ed.) Modern Brewing Technology, Macmillan Press, London, 1971, Ch.4, 83-107.
367. L. Nathan, J. Inst. Brew., 36 (1930) 538-550.
368. J. Hoggan, J. Inst. Brew., 83 (1977) 133-138.
369. J. Hoggan, Europ. Brew. Conv., Monograph V, Fermentation and Storage Symposium, Zoeterwoude, Elsevier, Amsterdam, 1978, 194-206.
370. C.A. Masschelein and J. Van de Meerssche, Tech. Quart. M.B.A.A., 13 (1976) 240-250.
371. S. Takayanagi and T. Harada, Proc. Europ. Brew. Conv., 11th Congress, Madrid, Elsevier, Amsterdam, 1967, 473-488.
372. F.B. Knudsen and N.L. Vacano, Brewers Digest, 47 (1972 July) 68-76.
373. J.H. Lindsay, Brewers Digest, 51 (1976 July) 36-44.
374. J.O. Harris and J. Irvine, Tech. Quart. M.B.A.A., 15 (1978) 30-37.
375. J.O. Harris, J. Inst. Brew., 86 (1980) 230-233.
376. M. Amaha, S. Nakakoji and Y. Komiya, Proc. Europ. Brew. Conv., 16th Congress, Amsterdam, Elsevier, Amsterdam, 1977, 545-560.
377. L. Pagh-Rasmussen, Europ. Brew. Conv., Monograph V, Fermentation and Storage Symposium, Zoeterwoude, Elsevier, Amsterdam, 1978, 78-95.
378. S. Martin, J. Bosch, J. Almenar and J. Posada, Proc. Europ. Brew. Conv., 15th Congress, Nice, Elsevier, Amsterdam, 1975, 301-310.
379. D.J. Olsen, The Brewer, 65 (1979) 102-106.
380. C.E. Dalgliesh, Proc. Europ. Brew. Conv., 16th Congress, Amsterdam, Elsevier, Amsterdam, 1977, 623-659.
381. J. Ellison, The Brewer, 59 (1973) 601-606.
382. P.M. Carpenter, The Brewer, 60 (1974) 480-484.
383. I.M. Young, The Brewer, 68 (1982) 272-279.
384. Anon, Brew. Dist. Int'l., 8 (1978 Nov.) 24-25.
385. M.W. Brenner, J. Inst. Brew., 71 (1965) 290.
386. C. Rainbow, unpublished results.
387. L.R. Bishop, W. Hickson, A. Alexander and F.H.M. Cory, J. Inst. Brew., 61 (1955) 150-160.
388. A.W. Scopes, J. Inst. Brew., 66 (1960) 2.

389. B.S. Christensen, Proc. Am. Soc. Brew. Chem., (1964) 74-75.
390. A.D. Portno, Wallerstein Lab. Comms., 33 (1970) 149-164.
391. K.H. Geiger, Tech. Proc. M.B.A.A., (1961) 8-14.
392. M.W. Coutts, Proc. 9th Conv. Inst. of Brew., Australian Section, 1966, 1-8.
393. L.R. Bishop, J. Inst. Brew., 76 (1970) 172-181.
394. A.W. Seddon, The Brewer, 62 (1976) 72-78.
395. M. Royston, Process Biochem., 5 (1970 Apr.) 25-39.
396. A.D. Portno, Europ. Brew. Conv., Monograph V, Fermentation and Storage Symposium, Zoeterwoude, Elsevier, Amsterdam, 1978, 145-154.
397. A.D. Portno, J. Inst. Brew., 74 (1968) 448-456.
398. R.G. Ault, A.N. Hampton, R. Newton and R.H. Roberts, J. Inst. Brew., 75 (1969) 260-277.
399. M.G. Royston, Process Biochem., 1 (1966) 215-221.

THE PHYSIOLOGY AND MICROBIOLOGY OF SCOTCH WHISKY PRODUCTION

DAVID R. BERRY
Department of Bioscience and Biotechnology, Yeast Technology Group,
University of Strathclyde, 204 George Street,
Glasgow G1 1XW, SCOTLAND

INTRODUCTION

Alcoholic beverages have been produced by man for several thousand years in all parts of the world. There are records of the production of non distilled beverages dating back to 5,000 to 6,000 B.C. in Egypt and of distilled beverages dating back to 1,000 B.C. in China (ref 1). All the processes involved the fermentation of a locally produced carbohydrate source with yeast to produce a potable beverage. Barley was the carbohydrate used for producing alcoholic beverages in Mediaeval Europe where processes for producing beer were well established in the monasteries. Whisky is the distilled product of the alcoholic wash produced from malted barley.

The history of whisky production is not clear but the earliest reports indicate that it was produced in Ireland in the 12th century, several centuries before the report of whisky production by Friar John Cor in Scotland in 1494. These early stills used air cooling to condense the spirit. However around 1540-50, distillation apparatus was improved by passing the delivery tube through water and increasing its surface area by coiling the delivery tube. The shape of the still was also modified to improve the reflux properties of the still and give it the characteristic pear shape of the pot still (ref 2).

A large part of the history of the development of the Scotch Whisky industry since these early days has been a game of hide and seek between the distiller and the tax collector. However concurrent with these changes in tax law, the increase in demand for whisky led to increases in the scale and the technological sophistication of the industry. These changes have been comprehensively discussed in a recent history of the Scotch Whisky industry by Moss & Hume (ref 2). Some of the most important of these developments were the use of sherry barrels for storage and maturation, the refinement of the pot still, the invention of the Coffey still, the production of blended whisky by Andrew Usher in 1848 and the development of special distilling yeast strains by Distillers Company Limited. In a recent review, Ramsay & Berry (ref 3) have commented that the history of the technical development of the industry has been wholly concerned with the chemical engineering aspect of the industry and that little or no information is available on the microbiological and biochemical aspects of the process. This can be attributed to the fact that the process of production was well established by the late 19th century when the principles of microbiology and biochemistry were being laid down.

This lack of information however, should not be allowed to hide the fact that whisky production, like the production of other alcoholic beverages, is essentially a biological process. It would not be unreasonable to describe it as one of the earliest biotechnological processes since it exhibits many of the characteristics of modern biotechnological processes. The generation of the amylase itself during the malting process and the breakdown from starch to glucose is an early example of enzyme technology, the fermentation stage involves the growth of an organism in a special vessel or fermenter and the production of a microbial product which was originally and still is to some extent used for medical purposes, although it would be fairer to say that the major consumption at the present time was for pleasure. The distillation and maturation stages are early examples of product extraction and downstream processing.

One of the major problems found by the research worker who wishes to investigate the production of complex fermented food products, such as cheese, beer and whisky as a biological process is that the final evaluation of the product must remain a qualitative rather than a quantitative process. The best quality product is that which has been judged to be the best of the results produced by the existing process. This leads to an understandable reluctance to alter the existing, apparently successful process. However, there can be little doubt that many changes have occurred in whisky production over the past two to three hundred years which have made to-day's product different from that produced by John Cor. Some of the changes are obvious and have been welcomed by the Industry, for example the use of pure yeast strains, aerobically grown yeasts, modern disinfectants to reduce bacterial contamination and the use of stainless steel fermenters. Others are less obvious but may have an equal impact on the nature of the new spirit produced, e.g. temperature, length of fermentation, mashing technique and dimensions of fermentation vessel.

A further difficulty in considering processes such as whisky production as biotechnological processes is the complex nature of the process. Whereas superficially the process of whisky production with its batch fermentation stage may appear simple in physiological terms, it is much more complex than many more recent fermentations. The complexity is introduced by the complexity of the substrate used, the complexity of the product and the need for at least two organisms in the fermentation stage. The perfect whisky fermentation has to maintain a balance between the needs for a high yield of ethyl alcohol and the need to maintain the correct level of a whole range of organoleptic compounds produced by the yeast.

A considerable amount of work has been carried out to analyse the aroma of whisky and some 240 compounds have been identified (ref 4, 5). The majority of these are produced by the yeast during the fermentation, however little work has been published on the influence of different fermentation parameters on the production of these compounds. Although more data is available from studies on the beer fermentation, these differ from the whisky fermentation in several important respects.

Similarly although it is generally recognised that the distillation and the maturation stage play an important role in creating the final flavour of the whisky, little information is available on the chemistry of these processes.

Scotch Whisky

The term Scotch Whisky is applied to two types of product distilled in Scotland. Commercially the most important product is blended whisky which is a mixture of grain whisky and malt whisky. However malt whiskies are also sold alone and are normally the product of a single distillery. Recently it was reported that there were 123 malt distilleries and 13 grain distilleries in Scotland, each producing a unique product. The individual brands of blended whisky are produced by blending some twenty or more malt whiskies and three or more grain whiskies. Each one of them must have been matured for four or more years. Although it is considered that most of the character of a blended whisky is derived from the malt whiskies it contains, the grain whisky also contributes to the flavour of the final product. Since it is the malt whisky which holds the key to the character of Scotch Whisky, this review will be primarily concerned with the little studied malt whisky process. Some reference to the grain whisky process will be made where there is a topic of special interest. The process of whisky production has been extensively written about, however most of the accounts are of a popular nature. The process has however been discussed from a technical point of view by Pyke (ref 6), Simpson (ref 7) and Lyons & Rose (ref 8).

WHISKY PROCESSES

Malt Whisky Production

The process of whisky production can be divided into several stages which are often separated both in time and geographical location. The growth of barley should be considered as the first stage in the process, since special barley strains have been developed for the Whisky Industry and the quality of the grain arriving at the gate of the Maltsters is critical to the whole Industry. It is probably true that more scientific work has gone into the evaluation of barley quality in terms of its ability to produce a high level of sugars during the malting and mashing process than any other aspect of the whisky process. The barley is harvested at the end of the growing season, then stored until it can be malted. Malting was traditionally carried out on stone or tiled malting floors and turned manually. However most malting is carried out by specialist Maltsters using drum maltings or saladins in which the germinated barley is turned mechanically. During the malting process water is added to start the steeping process in which the germination of the barley is initiated. The use of gibberellic acid is not permitted in the malting of barley for whisky production. The germination is allowed to continue until the shoot is half the length of the grain, usually about a week, when the malt is then taken to the kilns for drying. Traditionally this was carried out by burning peat to dry the malt. However since modern kilns are oil-fired, peat is burned simply to provide the smoke for the peated malts. During drying the temperature should not exceed 70°C or the diastatic enzymes would be denatured. The dried malt is then transported to the distilleries where it is again stored until it is required.

The first stage of the whisky process at the distillery is the grinding of the malt to a coarse flour; grist, in a grist mill. The grist is mixed with water in a mashing machine and then transferred into a mash tun. The initial water used for mashing is normally the final wash of the previous mashing. Since the malt used in the whisky industry is of a high quality and well modified, a single stage infusion mashing technique, which takes from 6-8h is normally used. The wort is run off through the slotted plate at the base of the mash tun and transferred to the fermenters or wash backs. This first water wort is cooled to 20-24°C during the transfer to the wash back. A second batch of warm water (68°C) is then added to the grist which has formed a filter bed during the filtration of the first waters. The grist is then raked and allowed to settle before the second extract of wort is run off. The third extract is prepared and used as the water for the first extraction of the next batch of wort. Spent grain or draff, which remains after mashing, is either sold directly as animal feed or processed as a by product. The washbacks are inoculated with either a single distillery yeast or a mixture of distillery yeast and a spent brewers' yeast. In traditional distilleries, fermentations are carried out in 8,000 gallon open topped wooden vessels constructed of larch; however many modern distilleries have larger stainless steel vessels which are closed at the top except for small manholes. No additions are made to the fermentation and no temperature control or mixing is carried out during the fermentation. After 40-44h the wash which contains about 7% ethanol is transferred to the wash still. The traditional pot stills used in Scotland are onion or pear shaped copper stills which were traditionally heated from below using coal or peat. However most are now heated with steel coils. The alcohol water vapour passes through the neck of the still into the worm, a copper coil immersed in a bath of cold running water, where it condenses to form low wines. These are collected in a low wines charger before being transferred to the spirit still for the second distillation. Only a fraction of the distillate which comes off the spirit still can be used directly as new spirit, the first distillate over, the "foreshorts", and the last material over, the "feints", are recycled to the low wines charger. The cut of the second distillation is very critical and is made by the distiller based on the gravity of the spirit and the disappearance of a bluish tinge when water is added to the distillate. These observations are made as the spirit passes through the spirit safe which is situated between the still and the holding vessel. This is a glass container which is kept locked such that access to it is restricted to Excise Inspectors.

The effluent from the low wines still is a thick brown liquid with a BOD in the order of 40,000mg/l., which is used to produce either still solubles or distillers dark grains which are sold as by-products. The effluent from the spirit still however has a lower BOD and is normally treated before being discharged. The only other waste produced in the process, carbon dioxide, is normally vented to atmosphere; however it is collected and bottled in some of the larger distilleries.

Grain Whisky Production

Grain whisky production is carried out on a much larger scale than malt whisky production (ref 6, 7, 8). The major carbohydrate source, maize, (*Zea mays*) is ground in a mill then cooked at 120°C for 1-1.5h to gelatinize the starch. Continuous cooking is now frequently used. The maize is then transferred to the mash tun where it is mixed with a small quantity of malted barley, approximately 10%, which provides the diastatic activity to break down the starch. This stage is very critical since cold water must be added to bring the temperature down to 60-65°C. If the mixing is not adequate the modified starch in the malt can solidify. After mashing, the wort may be filtered, however high yields have been obtained when the whole mash was inoculated.

The fermentation is inoculated with the same distillery yeast which is used for malt whisky production. Little attention has been paid to the physiology of the grain whisky fermentation. However it would be interesting to know the effect of the presence of solids in whole mash fermentations on the amount of yeast growth, the formation of ethanol, glycerol and higher alcohols (ref 9).

The other major difference in the production of grain whisky is the use of continuous distillation. The still used is essentially the same as the plate still developed in the last century; however rather than two columns modern distilleries often have five column stills. The operation of the two column patent still has been described by Pyke (ref 6) who presented information on the distribution of organoleptic compounds in the spirit vapour in different plates in a Coffey still rectifier.

Maturation, Blending and Bottling

The spirit produced from both malt distilleries and grain distilleries is stored in barrels which vary in size from 14 gallon octaves to 108 gallon butts. Most whisky is matured in wood for eight years. During this time some 15% of the whisky is lost by evaporation. The new spirit is colourless when it enters the barrel, however changes in colour and flavour occur during the maturation period. The quality of the final blend is controlled by the blenders, who determine which malt and grain whiskies will be included in the blend. When these have been selected they are collected together and emptied into a trough. They are then run off through filters into blending vats, where the different whiskies are mixed by stirring or by blowing air through them. The blended whisky is then stored in barrels for several more months to allow the blend to "marry". The blend is then taken for bottling. The whisky is chilled and filtered to remove precipitated matter before being bottled and labelled in highly automated bottling plants.

DESCRIPTION AND EVALUATION OF WHISKY

Chemical Analysis of Whisky

The use of temperature programmed GLC has led to several detailed analyses of the chemical nature of the diverse minor components which contribute to the flavour of Scotch Whisky (ref 10). The most extensive work which has been published has come from the laboratory of the Finnish State Alcohol Monopoly. They used GLC to compare the level of organoleptic compounds in cognac, whisky, rum, brandy, wine and sherry. Using a Carbowax 20M column, some 38 peaks were separated and identified representing the most common higher alcohols and esters. The important conclusion from this work was that in general the same aroma components appear in all fermented beverages (ref 11). A spirit produced by distilling the fermented products of a sugar solution contained a similar selection of compounds even though many of them were at different concentrations (ref 12). It was deduced that the main aroma compounds are produced by the metabolic activity of the yeast. Several lists of organoleptic compounds present in whisky have been published (Ref 5, 11, 13). The most recent of these lists some 280 components which have been identifed in whiskies. These can be divided into alcohols, acids, esters, carbonyl compounds, phenols, hydrocarbons, bases, sulphur compounds and others (ref 5).

Such lists of compounds occurring in whisky are not of any great value in evaluating the quality of the spirit, or distinguishing between different spirits. However the ultimate objective of analytical studies on whisky must be to evaluate the quality of a whisky by direct measurement of a selection of flavour compounds.

Several empirical techniques have been described to distinguish between different spirits (ref 14). The level of propanol has been used to distinguish rum; the ratio of 2 methyl butanol to 3 methyl butanol, to distinguish between brandy and whisky; and the ratio of isobutanols to pentanols to distinguish between American and Scotch whiskies.

Good discrimination between Scotch and Irish whiskies on the one hand and whiskies from the U.S.A. and Canada on the other was obtained by plotting the sum of pentanols over 2 methyl propanol against propanol g/l ethanol (ref 14). One interesting technique recently described is the analysis of the stable isotopes of carbon and hydrogen in the ethanol to distinguish between spirits produced from maize starch which is produced by the C_4 (Hatch & Slack) photosynthetic process, and from barley starch and sugar which are produced by a C_3 (Calvin) photosynthetic system (ref 15).

Extensive data on the level of some of the more important organoleptic compounds in different types of whisky have been published (ref 16). The data for eight Scotch whiskies shown in Table 1 gives some indication of the variability of the product. In addition to the alcohols and esters whisky contains 90-140mg/l of acids, of which acetate is the most common. Others include propionic, butyric,

TABLE 1. Alcohol and ester content of a selection of blended Scotch whiskies (mg/l). From M. Lehtonen & H. Suomalainen (ref 16).

Whisky alcohols	1	2	3	4	5	6	7	8
n propanol	209	254	217	197	242	230	225	178
isobutyl alcohol	300	326	252	296	357	342	377	354
Opt. active amyl alcohol	64	71	77	93	81	85	119	119
isoamyl alcohol	212	244	288	267	274	254	329	345
phenyl ethanol	8.1	9.2	9.9	10.4	10.5	10.9	14.2	14.4
Esters								
ethyl acetate	125	137	147	121	147	165	144	173
iso amyl acetate	5	5.7	4.3	5.6	5.7	7.3	5.4	5.3
ethyl caproate	0.6	0.6	0.8	0.7	0.6	1.0	0.9	1.0
ethyl caprylate	3.3	4.1	5.0	4.2	4.7	5.4	6.9	6.4
ethyl caprate	8.8	11.3	12.8	11.6	12.8	13.5	16.6	13.8
ethyl laurate	5.9	8.1	10.2	8.9	9.1	9.8	11.0	8.3
ethyl myristate	0.8	0.9	1.4	1.2	1.0	1.3	1.1	0.6
ethyl palmitate	2.1	2.0	3.2	3.9	1.9	3.1	1.8	1.0
ethyl palmitoleate	2.8	3.7	5.0	4.0	4.2	5.5	6.7	2.4

Table 2. Levels of phenols found in Scotch blended and malt whiskies From Lehtonen & H. Suomalainen (ref 16).

	Scotch blended ug/l.	Scottish malt ug/l.
phenol	190	100
o cresol	75	75
m cresol	35	30
p cresol	50	50
guaiacol	120	90
4 ethylphenol	40	70
Unknown	20	30
4 ethyl guaiacol	30	100
Eugenol	100	50

3 methyl butyric, 2 methyl butyric, valeric, isovaleric, hexanoic, octanoic, decanoic, dodecanoic, tetradecanoic, hexadecanoic and 9 hexadecenoic acids. Whisky also contains phenolic compounds. The relative abundance of phenolic compounds in blended and malt whisky is shown in Table 2. These can arise from the peated malt but also from the mashing of the grain, the fermentation and from

the oak of the barrels used in maturation (ref 17, 18). Since peat is not used in the drying of the barley for grain whisky production the higher level of phenol, m cresol, guaiacol and eugenol in blended whisky must indicate that they can be derived from a source other than peated malt.

Sensory Evaluation of Whisky

A quality Scotch blended whisky contains more than 25 different malt whiskies and three or more grain whiskies. It is generally considered that the character of blended whisky is derived from its constituent malt whiskies although the grain whisky is not a neutral spirit since different grain spirits have different organoleptic characteristics and uses in the preparation of blended whiskies. The creation of individual brands and the maintenance of a constant brand character is a very complex and skilled process which is carried out by blenders.

One problem which has recently received considerable attention is the difficulty of describing the organoleptic character of whisky and communicating this information between blenders and from blenders to production and technical staff (ref 19). This problem of flavour description has been attributed to the use of the same term to describe different flavours and the use of imprecise words, e.g. fruity, which cover a range of flavour (ref 20).

This problem is not unique to the whisky industry and special languages have been developed for other beverages such as cider and beer (refs 21, 22), however these were not found suitable for whisky. Recently two different studies have been carried out in an attempt to develop a suitable language for whisky. One developed at Pentland Scotch Whisky Research Ltd., uses two tiers of descriptive terms which can be arranged in a form of a wheel if this facilitates the use of the language. The language consists of 14 principal or first tier terms and 44 secondary tier terms. A third tier of terms is also available which can be used to give a more precise description of the flavour or odour. The principal tier terms, considered as broad categories of flavour, are arranged in an order which largely reflects the stage in the process at which the flavour can be considered to be introduced into the product (Table 3). The second tier includes a variety of terms which the authors considered to be capable of precise definition. The terms are all terms which are used and understood by blenders in the industry (ref 5, 19).

In the second study carried out at Strathclyde University, a primary list of 35 descriptive terms was prepared by reference to current terminology in the literature. The list was refined, not by reference to trained blenders, but by statistical analysis of results obtained with trained and untrained taste panels. The taste panels were trained using standards which were, where possible, specific chemicals, however less precise reference standards were also used, e.g. dried yeast and malt extract (refs 23, 24).

TABLE 3. Principal and second tier terms used in the flavour wheel proposed by Shortreed *et al.*, (ref 19).

Principal or 1st tier term	2nd tier term
1. Nasal effects	Pungent
	Prickle
	Nose-warming
	Nose-drying
2. Phenolic	Medicinal
	Peaty
	Kippery
3. Feints (fusel oil)	Leathery
	Tobacco
	Sweaty
	Stale fish
4. Cereal	Cooked mash
	Cooked vegetable
	Toasted
	Malt extract
	Husky
5. Aldehydic	Hay-like
	Leafy
	Floral
6. Estery	Fragrant
	Fruity
	Solvent
7. Sweet associated	Glycerin-like
	Honey-like
	Vanilla-like
8. Woody	New wood
	Developed extract
	Defective wood
9. Oily associated	Nutty
	Buttery
	Fatty
	Rancid
10. Sour associated	Sickly
	Cheesey
	Vinegary
11. Sulphury	Stagnant
	Coal-gassy
	Rubbery
	Cabbage-water
12. Stale	Metallic
	Blotting paper
	Musty
	Earthy
13. Primary tastes	Sweet
	Sour
	Salty
	Bitter
14. Mouthfeel effects	Mouth coating
	Astringent
	Mouth warming

One important result which has come out of this work is that the technique of descriptive sensory analysis used could be carried out by untrained personnel. When untrained personnel were used the results obtained were less precise but this problem could be overcome by using a larger panel. This system has been used to solve problems in three areas which normally could only have been undertaken with the help of a skilled blender. These areas are, quality control, process and product development and market research.

The odour of rectified spirit is difficult to describe because the odour notes are relatively weak and masked by the strong odour of the ethanol. Spirit samples were assessed by a panel in comparison to ethanol at the same concentration. The results obtained indicated that the panel could discriminate

TABLE 4.Vocabulary for descriptive sensory analysis of Scotch Whisky. The chemical compounds listed against eleven of the descriptors are the standards used for eleven terms selected for a specialised vocabulary. From Piggott & Canaway (ref 24).

Description	Standard	Description	Standard
Pungent	formic acid	Gassy	-
Solvent	-	Phenolic	Phenol
Spicey	eugenol	Hydrocarbon	-
Grainy	-	Oily	Heptanol
Malty	-	Woody	-
Mouldy	-	Metallic	-
Fruity (ester)	isoamylalcohol	Meaty	-
Fruity (other)	Ethyloctanoate	Sulphury	Dimethyl sulphide
Floral	2phenylethanol	Catty	-
Smooth	-	Soapy	-
Vanilla	ethyl vanillin	Sour	-
Nutty	-	Fish	Trimethylamine
Buttery	Diacetyl	Sweet	-

between good and defective spirits. In another series of experiments, a change in spirit quality was detected after a modification of operating procedure at one of the distilleries under test. Extensive studies by 14 assessors on 111 samples indicated that the assessors could be divided into three groups according to similarities in their preference for different samples. The application of descriptive sensory analysis in this manner opens up the possibility of evaluating in a qualitative manner differences in preference for beverages between different groups of people.

One of the most exciting developments in the last few years has been the availability of capillary column GLC. This has increased the resolution of different components dramatically. Recently this has been combined with sniff analysis on the 210 usable peaks separated by this technique (ref 5). The technique was used to evaluate the dry character of Scotch whisky which has traditionally been attributed to steam volatile phenols. This dry characteristic which has also been referred to as peaty, has been described as having both the phenolic aromatic character and a smokey character. No direct correlation between the intensity of the medicinal or smokey descriptors with any phenols or with the total phenol content was obtained. They have also obtained evidence that individuals who appeared to be incapable of detecting steam volatile phenols by sniffing could detect a dry peaty character in certain whisky samples but not the medicinal character. This evidence supports the view originally expressed by Bathgate & Taylor (ref 28) that dryness and steam volatile phenols may not be directly related. However comparison of GLC analysis of four peaty whiskies and four non-peaty whiskies identified 13 peaks which showed a significant difference. Of these, six were found to correspond to phenol acetate, guaiacol, phenol, o-cresol, m-cresol and eugenol. The other seven peaks were not identified, but most were described as having a nutty flavour. Analysis of this data has been used to predict the phenolic sensory score of a range of whiskies with a high degree of success. Cluster analysis was also carried out and three groups were identified based on analytical data which corresponded well with three regional groups, Speyside's, Islay's and the remainder, which had been previously identified by sensory analysis (ref 26). Peatiness appears to play a major role in determining the character of the whisky although it only contributes 7% of the total odour units (ref 27). In contrast an analysis of the principal esters did not give rise to any correlation with geographical groups or other sensory groups.

Results obtained with capillary GLC indicated that esters of almost every combination of acid and alcohol from the fermentation and maturation stages are present in blended whisky. They are believed to make a major contribution to the rounded flavour of the whisky (ref 5), although no individual estery descriptor is conspicuous in many whiskies. The contribution of fatty acids and carbonyl compounds to the character of whisky has not been fully investigated. This is surprising in view of the fact that carbonyls contribute 40% to the total odour composition of the whisky.

MALTING AND MASHING

Special varieties of barley have been developed for the Scotch whisky industry since malting for whisky requires certain unique properties. Since yield is of critical importance in the whisky industry an ideal barley for whisky production has a high starch content relative to the nitrogen level of the grain and a large corn size which gives high fermentable extracts. Since the addition of gibberellic acid is forbidden a variety which gives rapid enzyme synthesis without the addition of exogenous gibberellic acid is required. The varieties Golden Promise, Porthos and Athos satisfy these requirements. Porthos and Athos are newer varieties which have larger corn sizes and produce higher fermentable extracts but tend to shed grain in the high winds which are common in Scotland during the harvest period (ref 28). The processes of malting and mashing although they are carried out at different times and also in different premises miles apart are best viewed as part of the same process. Since the raw materials of whisky production are restricted to water, barley and yeast, the barley must provide all the nutrients required by the yeast for its growth and the production of ethanol and flavour compounds. In the malting process enzymes are generated which are required for the conversion of the stored nutrients in the barley grain into the soluble form which can be metabolised by the distilling yeast. Normally the quality of a barley is measured in terms of the amount of extract which can be obtained from a given weight of barley and the fermentability of this substrate. However such calculations assume that the product of the whisky fermentation is solely ethyl alcohol and that the organoleptic quality of the spirit is not affected by the quality of the barley and the techniques of malting and mashing. If this were so there would be little advantage in adhering to traditional processes since the use of maize and fungal enzyme is much cheaper (ref 28). Since wort is not boiled in the whisky process it perhaps not surprising that measuring hot water extracts has not been found to give good correlation with ethanol production (refs 29, 30). It has been necessary to develop different techniques for analysing the fermentability of worts. Dolan (ref 31) reported that a distiller requires malt which on mashing will release 99% of the available carbohydrate to the wort. The fermentable carbohydrate should represent 86% of the extract. A technique was reported for predicting spirit yield from hot water extracts using a factor which was derived by correlating the spirit yield of extracts with the standard hot water extracts obtained from experimental samples of grist. Subsequently extensive trials have been carried out to produce a standard method of evaluating the fermentability of unboiled worts for use in the whisky industry (refs 32, 33). This can be summed up in the following equations:

$$\%\,FE = \frac{\%\,SE \times \%\,F}{100}$$

where FE = fermentable extract

SE = hot water soluble extract

F = fermentability

$$\%\,SE = G \times 2.63$$

where G = excess degrees of gravity of the filtrate taking water at 20°C as 1,000

$$\%\,F = \frac{OG - FG \times 81.4}{OG - 1{,}000}$$

where OG = original gravity

FG = final gravity

The fermentability was determined in small scale fermentations carried out in 500ml. flasks fitted with fermentation locks and inoculated with the "M" strain of distilling yeast. Using this technique the following data was obtained: (Table 5) .

Most of what we know about malting and mashing is derived from studies on these procedures in breweries (refs 34, 35). Descriptions of the processes as they apply to the whisky industry have been made by Simpson (ref 7) Lyons & Rose (ref 8), and Rutherford (ref 36). Bathgate (ref 28) discussed the requirements of malting and mashing by the distilling industry. It was noted that the highest wort fermentability usually occurred in malts which were slightly undermodified. In a study on the malt variables which influence the run off from the mash tun, it was found that the relevant variables were fineness of grinding and wort viscosity. Of these, wort viscosity was not found to interfere with run off unless the malt was very undermodified or the mash was very thick. The particle size appears to be the major factor. This is determined by the type of mill used. The use of a Miag Laboratory Disc Mill is recommended for analytical work. Extracts prepared after grinding at setting 2, which ruptures endosperm cells, give an indication of absolute potential extract, whereas extracts determined after milling at setting 7 indicate achievable distilling extract. The difference between the two has been found to correlate well with mash filtration rates.

TABLE 5. Fermentability of unboiled malt worts obtained for malt whisky production (average of 47 analyses). From Dolan (ref 33).

	Mean	SD
Hot Water (Coarse) Extract (% Soluble extract)	74.85	0.51
Fermentability (%)	86.80	0.88
Fermentable extracts (%)	65.82	0.75

Analytical data on malt whisky wort is not readily available. Data on a typical Scotch grain whisky wort was published by Pyke (ref 6). Studies on small scale fermentation carried out in our laboratory have given some information on the levels of different sugars in worts and data on their rate of utilisation has been published by Watson, and Lavery & Berry (refs 37, 38). The sugar composition of first and second extracts and fresh wort are shown in Table 6.

The enzymology of malting and mashing is assumed to be similar to that described for the beer fermentation. However little information is available for whisky production. Preliminary results indicate that α amylase activity persists throughout the mashing process and into the wort (ref 38). However detailed studies on the contribution of different amylolytic enzymes to the breakdown of starch in the malting and mashing stages do not appear to have been published. A list of the enzymes which are involved in the malting and mashing process includes α amylase, β amylase, limit dextrinase, α glucosidases, β fructosidases, phosphorylase, glucosyl transferase, β glucanase and proteinase. One problem which has arisen in the past few years has been the identification of the N nitrosodimethylamine (NDMA) in whisky and other alcoholic beverages produced from malt (ref 39). A detailed study at the FDA Bureau of Foods found that only eleven of forty four whiskies tested were free of nitrosamines (ref 40). Investigations of this problem have indicated that when the kilning of malt has been carried out using natural gas, oxides of nitrogen were sometimes produced in the furnace gases which can interact with amino compounds in the malt to produce carcinogenic nitrosamines, e.g. dimethyl nitrosamine. The level of nitrosamines can be limited by drying at a lower temperature or by the use of fuels which have a high sulphur level. These burn to produce sulphur dioxide which lowers the level of nitrosamines in the malt. Since rootlets of germinating barley have been found to be particularly rich in nitrosamines their efficient separation from the malt also facilitates the reduction in the level of nitrosamines in the malt. Recently specially designed 'No Nox' burners which give a low level of oxides of nitrogen when fitted to kilns which are directly heated with natural gas have been developed (ref 34).

TABLE 6. Composition of first and second extracts and cooled wort from a small scale mashing process. Lavery & Berry (ref 38).

	1st Water	2nd Water	Cooled Wort
Fructose	0.032	0.022	0.029
Glucose	0.509	0.302	0.390
Sucrose	0.250	0.140	0.201
Maltose	4.44	2.51	3.48
Maltotriose	1.39	0.87	1.02
Maltotetraose	0.112	0.130	0.120
Maltopentaose	0.029	0.066	0.016

Peat

One of the unique features of Scotch Malt Whisky is its peaty aroma which is acquired by the use of peat fired kilns for drying the malt. Some 80 aroma constituents have been reported in peated malt, including several phenolic compounds, e.g. phenols, cresols, xylenols, and guaiacol. Although it has not been demonstrated that these phenolic compounds are the main contributors to the characteristic flavour of peated malt whiskies, they have often been used as marker compounds to indicate the degree of absorption of the peat smoke onto the barley (ref 25). The levels of phenolic compounds are frequently measured using the 4 amino phenazone reaction (ref 41); however this is not even a good guide to the phenolic content of the malt since it is not sensitive to several phenols present in the peated malt. Bathgate & Taylor (ref 25) showed that p substituted phenols, e.g. p cresol did not react with 4 amino phenazone. In the assay the phenol is oxidised by alkaline ferricyanide to the corresponding p quinone which then condenses with the four amino phenazone to produce the chromogen measured in the assay. If the phenol is substituted in the p position this oxidation cannot occur so no colour reaction develops. A recent paper has advocated a modified Klebers method which uses p nitroaniline to detect the level of phenols on peated barley (ref 42). Bathgate & Taylor (ref 25) have reported the use of GLC for the estimation of peatiness. The GLC was carried out on a sample which had been prepared by steam distillation of the malt or by ethyl extraction of the distillate. These analyses indicated that peated malt contained phenols and hydrocarbons which were absent from unpeated malt and that the levels of furfurals were also increased. The major phenols present were tentatively identified as guaiacol, phenol, o-cresol, p-cresol, m-cresol, 2,5 xylenol and p ethyl phenol. An ideal procedure for evaluating peatiness would involve the identification of the organoleptic compounds responsible for peatiness and for their direct assay. However until this situation comes about it has been suggested that the term lightly peated be applied to malts containing 1-3ppm phenolic compounds, medium peated for 3-6ppm and heavily peated for malts containing more than 6ppm phenolic compounds.

FERMENTATION

Inoculum

The importance of the choice of yeast strain and the quality of the yeast used for inoculation has been emphasised by Sharp & Watson (ref 43) who have distinguished between primary yeast strains, high yielding fast fermenting strains of yeast which are used by all malt whisky distillers and secondary yeasts which are slower growing strains of brewing yeast and are used by some distillers.

The level of fusel alcohol formed is influenced by the fermentation rate so variations in inoculum size and strain characteristics influenced the organoleptic properties of the spirit. Since the inoculum is not normally produced in sterile conditions on site in the distillery then standards should be established for inoculum purchased from other sources. Those suggested by Sharp & Watson (ref 43) Table 7, indicate the difficulties of studying the microbiology of the malt whisky fermentation. Most malt whisky distillers use the "M" strain of distilling yeast produced by the Distillers Company Limited, Menstrie, Scotland as their main distilling yeast. However, a new strain produced by Maori Products in the U.K. is

TABLE 7. Typical yeast specification from Sharp & Watson (ref 43).

1. The yeast strain must have acceptable fermentation capacity.
2. The yeast must not produce organoleptically unacceptable aromas directly, or indirectly, during fermentation and distillation.
3. Dry matter content should be above 25% (minimum limit 23%).
4. Viability should be above 80% by the lisamine green method (minimum limit 70%) or 75% by the modified methylene blue method (minimum limit 60%). DCL "M" yeast is expected to be over 90%, if not over 95%.
5. Total acidophilic bacteria to be less than 10^6 viable organisms g^{-1} of caked yeast.
6. Total lactic acid group to be less than 5×10^5 viable organisms g^{-1} of caked yeast.
7. Total coliforms to be less than 5×10^5 viable organisms g^{-1} of caked yeast.
8. Total Acetobacter to be less than 5×10^4 viable organisms g^{-1} of caked yeast.
9. Total wild yeast/moulds to be less than 10^4 viable organisms g^{-1} of caked yeast.
10. The yeast must be refrigerated during storage and transportation in the range 1°-5^oC. At no time should the yeast be frozen.
11. The yeast must not contain any enzyme, or fermentable additives.
12. The yeast should be in a stable, easily handled cake form.

being tested by a number of distilleries at the present time (ref 44).

As is the case with many alcoholic beverage fermentations, the whisky fermentation is a very complex process. Because of this there has been a reluctance to study it in detail and to analyse it in a similar manner to other commercial fermentations. The reason for this is not difficult to find. The medium used, wort, is complex in that it contains a variety of sugars and nitrogen sources. It is also difficult to work with in that it contains active amylolytic enzymes and live bacteria which are an inevitable and essential feature of the fermentation. The complexity of the problem may be compounded by the use of two different yeast strains in the inoculum, although this is not practised by all distilleries. The product of the process is also difficult to define. It can be considered under two headings, firstly the major product, ethanol, which must be produced at a high yield and secondly volatile organoleptic compounds which give the unique flavour to the product. Unfortunately for the scientist, but perhaps fortunately for the consumer the precise descriptions of the number and character of compounds contributing to whisky is not known and even if it were, it could be expected to differ for different brands of malt whisky.

However, in the view of the author, these difficulties present another challenge to the essential problem of describing the fermentation and understanding the factors influencing its development. Although, as has been discussed elsewhere, it is accepted that the peating of the malt and the distillation and maturation stages of the process do play an important part in flavour development, it is apparent that the majority of the organoleptic compounds present in whisky are present in the wash at the end of the fermentation and are produced by the yeast during the fermentation.

Industrial Scale Fermentation

An industrial scale fermentation has been described by Watson (ref 37) who emphasised the varied procedures used in the different distilleries and the variation introduced by the use of different wort qualities and the different yeast strains. In the process described, the wort was inoculated immediately the "runnings" from the mash tun had been transferred to the wash back or fermenter. Thus the fermentation was started by inoculating a high gravity wort at a level of 2×10^7 cells ml^{-1}. This was then diluted to the final gravity of 1050-1055 as the second and third washes were run into the fermenters over the next four hours. The fermentation was divided into three phases, an initial active yeast growth phase, a stationary phase and a decline/bacterial phase. The initial growth phase was relatively short and was completed within 12 hours when the yeast concentration had risen to 2×10^8 cells ml^{-1}. During this growth period the temperature of the fermentation rose to 33°C as a result of the heat released by the metabolism of sugars. The fermenters were not fitted at this time with any apparatus for temperature control. The final temperature achieved was influenced by the initial wort temperature which could vary from 20-30°C depending upon the ambient temperature. The sugars were found to be metabolised in the sequence; sucrose, fructose, glucose and maltose during the first 16h. of the fermentation. However, alcohol production continued on to 32h.

In the later stages of the fermentation the viability of the yeast decreased and the level of bacteria in the fermentation rose (ref 45). This rise in the level of bacteria was associated with an increase in lactic acid and a decrease in the pH of the wash. This stage of the fermentation may be the most critical and the most characteristic of the whisky fermentation since the bacteria produce high levels of lactic acid and an acid pH in the wash. The pH of the wash and the level of lactic acid have a marked effect on the nature of the final spirit produced. An increase in the level of ethyl lactate has been reported in fermentations in which there has been a high level of bacterial contamination and it is clear that the more acid the pH of the wash the higher the amount of esters formed during the distillation phase (ref 18). An increased level of furfural has also been reported in distillates from heavily contaminated fermentations by Berry & Ramsay (ref 3, 46). The formation of furfural in the distillation phase from pentoses present in the wort is favoured by the low pH values which occur when high levels of lactic acid are produced by contaminating bacteria.

Laboratory Scale Fermentations

Recently a laboratory scale whisky fermentation has been developed by Ramsay & Berry, and used to study the malt whisky fermentation in greater detail. Results obtained using this technique have indicated that the fermentation can be divided into an initial acceleration phase lasting about 8h, a linear phase lasting a further 20-24h followed by a decline phase (Ref 47). In this it resembled a typical brewery fermentation (ref 48). During the acceleration phase, there was an exponential increase in the number of yeast cells and the fermentation rate. However this phase was soon terminated when the available amino nitrogen in the wort had been assimilated. During the linear phase, ethanol is produced in a linear manner. There was no increase in cell number during this phase however, yeast dry weight did continue to increase. Further investigation indicated that this increase could be attributed in part to the accumulation of glycogen within the yeast (ref 38). Since the period of glycogen accumulation occurs largely when cell division has ceased and the available nitrogen has been utilised, these results would appear to confirm the view that the synthesis of endogenous reserves is stimulated by nitrogen deficiency (Ref 49). The decline in glycogen levels when the wort carbohydrates have been fully utilised indicated that the distilling yeast strains can utilise reserve carbohydrates in anaerobic conditions. In fact the continued production of ethanol after exogenous carbohydrates have been metabolised (ref 47), appears to be at the expense of reserve glycogen. The possibility that certain brewing strains, which may be used as secondary strains in whisky fermentations, cannot metabolise stored glycogen in anaerobic conditions (ref 50) has been considered; however no such strains were identified (ref 51). The utilisation of sugars in the small scale fermentation has been investigated and the sequence of uptake; sucrose, fructose, glucose, maltose and maltotriose (ref 38) is the same as has been reported for the large scale whisky fermentation (refs 6, 37) and for beer fermentations (ref 52).

The small scale fermentation differs from the industrial fermentation in that the level of bacterial contamination remained low, below 100 cells ml^{-1}, throughout the fermentation and the viability of the yeast remained high although a small decrease in yeast viability and in dry weight did occur. This lack of bacterial growth in spite of the use of normal unsterilised, unboiled wort has been attributed to the cleaner procedures and temperature control adopted in the laboratory. Routine fermentations were controlled at 30°C using a temperature controlled water bath. Lower yeast viabilities and higher levels of bacterial contamination were obtained when experiments were carried out in a vessel which had been insulated by surrounding it with polystyrene chips to prevent heat loss. The low bacterial count normally obtained explained the lack of ethyl lactate in the laboratory scale fermentations.

The final yeast cell number has been found to vary directly with the level of α amino nitrogen present in the wort by Berry & Ramsay (Ref 3) (Fig. 1.). This indicated that oxygen or unsaturated fatty acids are not limiting in the whisky fermentation as they can be in the beer fermentation. One important difference in this respect is that inoculum used in whisky fermentations is always grown aerobically. Furthermore, since in malt whisky production adjuncts are not used and all the wort is derived from barley, a high level of unsaturated fatty acids and sterols will be present in the wort (ref 53).

The high rate of biosynthesis during the period of exponential cell replication would be expected to create a high demand for ATP which would utilise the ATP generated during the metabolism of glucose to ethanol by glycolysis. It is not apparent how the cell maintains such a high level of glycolysis and ethanol generation during the linear phase of the fermentation. The major requirement for ATP would be restricted to maintenance requirement (ref 54) and to the use of ATP in glycogen synthesis. The carbohydrate requirement during the maintenance phase has been estimated at 0.1g of glucose h^{-1} g^{-1} dry matter (ref 13).

Although the yield of alcohol obtained in the small scale fermentation was not so high as that obtained in industrial fermentations, the fermentation did proceed until a specific gravity of 1.00 was obtained. Up to 7.4% v/v ethanol was obtained from worts of initial SG = 1.052. The level of ethanol obtained did not appear to be sensitive to variations in α amino nitrogen level between 112 and 298 mg l^{-1}, inoculum level between 0.5 and 0.6 x 10^7 cells ml^{-1} and temperatures between 25° and 30° (ref 47).

<u>Formation of volatile organoleptic compounds</u>

Yeast produces four major groups of the organoleptic compounds which occur in whisky; higher alcohols, fatty acids, esters and carbonyl compounds. These are produced by a variety of metabolic pathways involving different areas of cellular metabolism. A considerable amount of work has been carried out on beer fermentations to investigate the effect of variation of such parameters as wort composition, type and size of inoculum and general fermentation conditions on the level of different organoleptic compounds.

It has been argued that since the process of malt whisky production is tightly controlled by legislation such that no additions can be made to the process other than barley, water and yeast, there is nothing to be gained from studying the fermentation process. This ignores the fact that many parameters can

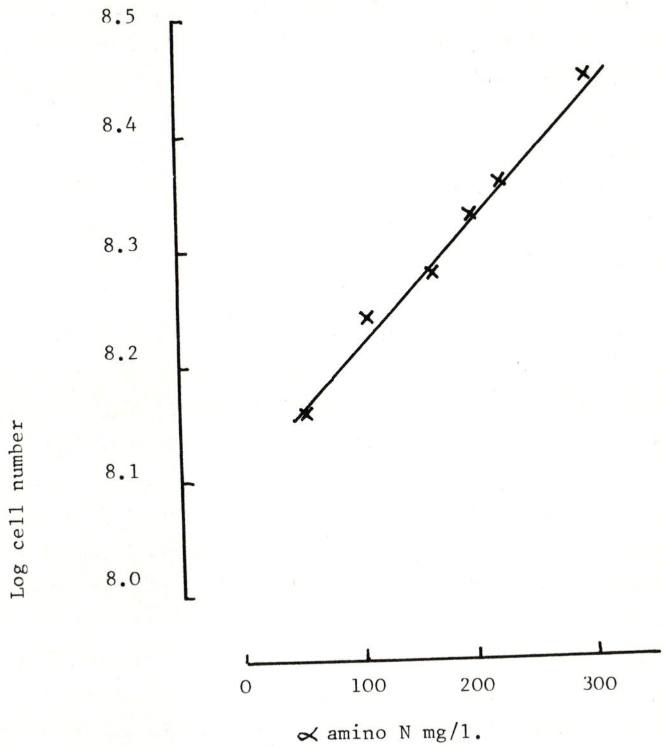

Fig. 1. Relationship between nitrogen content of wort and the final yeast cell number in the fermentation.

be varied, e.g. inoculum size, fermentation temperature and others may vary as a result of changes in plant design or operating procedures, e.g. wort composition, degree of mixing, or level of bacterial contamination (ref 3). The influence of some of these parameters on the formation of different types of volatile compounds of whisky have recently been investigated by Ramsay & Berry (refs 47, 55, 56, 57).

Higher alcohol formation

The biochemical pathways leading to higher alcohol formation are well established and have been reviewed by Webb & Ingraham (ref 58). It is now accepted that the higher alcohols which contribute to whisky flavour can be synthesised either by the catabolism of the corresponding amino acid or from carbohydrates on the biosynthetic pathway leading to that amino acid. The overall mechanism is similar to that involved in the formation of ethanol from pyruvate and the enzyme which is involved in the final step, ADH1, appears to be a non-specific one and is responsible for reducing aldehydes to alcohol during the formation of all the higher alcohols (ref 59).

Using radioactively labelled glucose and amino acids, it has been shown that higher alcohols can be formed from both glucose and amino acids (Ref 60, 61), and that the relative importance of the two routes can vary depending upon the stage of the fermentation and the nutrient status of the medium.

The pattern of higher alcohol formation has been found to follow that of ethanol production closely (Ref 47). An initial period in which amyl alcohol, isobutanol, propanol and 2 phenylethanol increased exponentially was followed by a longer period during which they were produced in a linear manner. There was a close correlation between higher alcohol formation, ethanol formation and attenuation. However after all the wort carbohydrates had been consumed, ethanol and higher alcohols continued to be produced, presumably at the expense of reserve carbohydrates.

Using the small scale fermentation process, the influence of strain on the levels of several higher alcohols has been investigated. Whereas the total level of higher alcohols did not appear to be markedly affected by yeast strain, the relative proportion of each alcohol and in particular the ratio of amyl alcohols to isobutanol were found to be constant for a given strain but differed between strains. In contrast, the use of different grists resulted in changes in the total level of higher alcohols but did not alter the composition of the higher alcohol fractions. Thus the percentage composition of the higher alcohol fraction appears to be a strain characteristic. Several authorities have used the relative abundance of different higher alcohols as indicators of Scotch Whisky, e.g. 2 methyl butanol to 3 methyl butanol or 2 methyl butanol to 2 methyl propanol (Ref 62). This ability to control these ratios by choice of yeast strain could be important. The use of different brewing strains as secondary strains in the inoculation of whisky fermentations could cause variation in the higher alcohol composition of the spirit produced, as they do in beer (ref 63). In view of this, changes in the supply of secondary yeasts should only be made after the effect of the new strain on the quality of the spirit produced has been investigated.

That wort composition is important in controlling the level of higher alcohols is indicated by the high level of variation which occurs when different grists are used. Fuller investigation of the role of wort composition in controlling higher alcohol formation indicated that the highest levels of isobutanol, amyl alcohol and 2 phenylethanol occurred in a wort which had an α amino nitrogen content between 112 and 224mg l^{-1} (ref 55). Increasing or decreasing the nitrogen level beyond this range resulted in a decrease in higher alcohol formation. Since the α amino nitrogen content of grist does not vary very much it is unlikely that the variation observed between grists can be attributed to differences in nitrogen content.

When the wort was enriched for one particular amino acid, the level of the corresponding higher alcohol was stimulated most but higher alcohols corresponding to the other amino acids also showed some stimulation so there was also a general effect on total higher alcohol levels. Addition of threonine however, did not stimulate propanol as anticipated but amyl alcohol and butanol were increased. Since the variation observed between different grists has been shown to be a variation in the total level of higher alcohols rather than in the relative abundance of individual higher alcohols it seems unlikely that it can be attributed to changes in amino acid composition of the wort.

Altering the inoculum size of a fermentation may alter the initial rate of a fermentation but should not alter the amount of growth or the specific growth rate which occurs, since this will be controlled by the availability of the limiting nutrient. Increasing the inoculum size has been found to increase the total levels of higher alcohols, however the composition of the higher alcohol fraction remained unchanged. This is in agreement with the result obtained by Geiger & Piendl, (ref 64) and Van Gheluwe et al., (ref 65) in beer fermentations.

Alterations in the initial pH of the wort had little effect on propanol, isobutanol and 2 phenylethanol, however the level of amyl alcohol increased as the pH was raised from 4.0 to 6.0, resulting in an increase in the overall level of higher alcohols. This effect is again similar to that reported for the beer fermentation (ref 65).

The effect of temperature on the whisky fermentation is of particular interest since the fermentation can vary from ambient temperature, 20-23°C, up to 35°C. Such temperature changes may influence both the rate of formation of higher alcohols and the rate of loss by evaporation. In an attempt to control the loss by evaporation, Ramsay & Berry, (ref 55) carried out control experiments in which evaporation was restricted by the use of a fermentation trap. The results obtained in both open and closed vessels, indicated that the higher alcohol formation increased as the temperature was increased to 30°C, however lower levels were obtained at 35°C. Similar results were obtained in open and closed vessels and no significant change in the compostion of the higher alcohol fraction was observed. The increase in higher alcohol formation up to 30°C has previously been reported in the whisky fermentation by Merritt (ref 66) and in beer (ref 67), however, the differential effect of temperature on individual higher alcohol formation reported by Engan (ref 68) was not confirmed in these experiments.

Studies on agitation were also carried out in open and closed vessels, so that effects of evaporation and the effects of transfer of small quantities of oxygen from air in the headspace of the fermentation into the wash could be distinguished. Stirring was found to stimulate most of the higher alcohols studied, although no affect on propanol was observed. Little difference was observed between open and closed vessels. The effect of agitation resembles the results obtained on beer (ref 69); however the absence of any effect of mild aeration indicates that the whisky fermentation does differ from the beer fermentation in that higher alcohol formation has been reported to be stimulated by mild aeration in beer fermentations (ref 65). This difference may be attributed to the greater demand for oxygen in beer fermentations which use anaerobically grown yeast as inoculum. Any stimulation of growth by raising the fermentation rate could be expected to stimulate higher alcohol formation.

In general, the results obtained in studies on small scale whisky fermentations suggest that factors which increase the rate of the fermentation also increase the level of higher alcohols. This may or may not be accompanied by an increase in growth rate (Ref 55). The increase in higher alcohols at higher inoculum levels is related to increased fermentation rates although the specific growth rate remains constant. The stimulatory effect of increased agitation and elevated temperatures on higher alcohol formation may be attributed to an increase in fermentation rate rather than an increase in yeast growth rate. An increase in fermentation rate and higher alcohol formation caused by the addition of insoluble solids to the wort has been reported by Merritt (ref 9). However, reduced levels of higher alcohols have been reported at elevated wort nitrogen levels (Ref 55) and preliminary results indicate that higher alcohol levels can also be suppressed by the presence of certain long chain fatty acids. These are present in fine solids which occur in the grist and accumulate on the top of the mash bed (Ref 3). It is apparent that, by careful control of process parameters such as temperature, inoculum levels and of wort preparation it is possible to control the level of higher alcohols in a whisky wash.

Fatty acid formation

Volatile fatty acids which are produced during the fermentation and pass over during distillation into the whisky, constitute an important group of flavour compounds in malt whisky and other grain spirits. Acetic acid is quantitatively the most important, comprising between 50-90% of the total acids (ref 70). The most important of the 31 acids reported in whisky include propionic acid (1.5%) isobutyric (4.9%) butyric (1.5%) isovaleric (5.9%) hexanoic (4.2%) Octanoic (26.7) decanoic (31.6%) dodecanoic (16.2%) tetradecanoic (2.2%) and 9 hexadecenoic (2.01%) acids. The values in brackets represent the percentage of the total acids, excluding acetic, in a typical Scotch whisky blend.

Long chain fatty acids are synthesised from acetyl CoA, malonyl CoA and NADPH by the mechanism described by Lynen (ref 71). Medium chain fatty acids appear to arise by leakage from these biosynthetic pathways and can be released from their CoA derivative by hydrolysis (ref 72). Some of the short chain fatty acids e.g.

isobutyric and butyric however, may be derived from the decarboxylation of keto acids produced during amino acid metabolism (ref 73). It has been demonstrated that volatile acids could arise from both carbohydrates and amino acids in a grain whisky fermentation (Ref 60). Acetic acid was mostly derived from carbohydrates but n butyric and isovaleric acids were found to be derived in part from leucine.

It has been reported that medium chain fatty acids are bound to the yeast cell during the fermentation stage and only released during distillation (refs 12, 74). However, recent results obtained by Ramsay & Berry (ref 47) indicate that they are present in the cell free wash and that they are produced throughout the fermentation and not just during the period of yeast growth. In this study, in which the fatty acids were extracted using carbon disulphide rather than concentrated by distillation, octanoic acid (3-5mg l^{-1}) was found to be more abundant than decanoic acid, (2mg l^{-1}) and dodecanoic acid (0.1-0.3mg l^{-1}). These levels are typical of those found in beer (refs 75, 76); however it has been reported that decanoic acid is the most abundant in a variety of whisky brands. This difference may be attributable to the fact that the whiskies are produced from distillates of wash which contained the yeast cells, whereas our extracts were prepared from cell free washes. The relative levels of these fatty acids did not appear to be influenced to any great extent by yeast strain; however it was very sensitive to variations in grist (ref 56). However, the concentration of acids was stable over a wide range of α amino nitrogen levels and only decreased when the α amino nitrogen level was reduced to below 168mg l^{-1}. The level of medium chain fatty acids was relatively unaffected by the addition of amino acids. Unfortunately the level of shorter chain fatty acids, such as isovaleric acid and butyric acid, which might have been affected by the addition of amino acids was not monitored in this study. It is not clear what is responsible for the high level of variation observed in fatty acid composition when different grist preparations were used; however it does not appear to be concerned with nitrogen composition. Other possible causes include variation in the biotin content and the lipid content of wort (refs 77, 78). Increasing the inoculum size was found to cause a marked decrease in the level of all three fatty acids (ref 56). This result is in agreement with those obtained by Krauss & Foch (ref 77) but not with those of Norstedt (ref 76) who reported that inoculum level had little effect on those fatty acids in beer. This effect may be caused by the presence of unsaturated fatty acids in the aerobically grown inoculum, since the presence of unsaturated fatty acids in the medium has been reported to suppress saturated fatty acid formation (refs 78, 79).

When the initial wort pH was varied, it was found to have little effect between pH 4.0 and 5.5 which is the normal pH range for the whisky fermentation; however, acid production could be stimulated by making the pH more alkaline or reduced by reducing the pH to values below pH 4.0.

Fermentation temperature was found to be a critical parameter, since medium chain fatty acids were found to be depressed at higher temperatures, especially when the temperature was raised above 30°C, as it often is in a whisky fermentation (ref 56). This effect of temperature may be the result of changes in membrane permeability, since it has been reported that changes in temperature alter the distribution of fatty acids between the fermentation medium and the yeast cell (ref 80).

The most important factor affecting the level of these fatty acids was found to be agitation in open vessels (Ref 56). Whereas agitation in closed vessels had little effect, agitation in open vessels drastically reduced the level of octanoic, and dodecanoic acids. This change was accompanied by the appearance of 9-decenoic acid, an unsaturated fatty acid which was only present at very low levels in the closed fermentation. Although 9 decenoic acid does not appear to have been reported in commercial whiskies, ethyl decenoic acid has been reported (ref 5). The results have been interpreted in terms of the stirring of the open vessel increasing the rate of diffusion of air into the wort and causing a suppression of fatty acid formation. Aeration has been found to decrease the level of acids in beer (refs 76, 81). The formation of decenoic acid and the suppression of the other acids suggests that it cannot be assumed that the high rate of CO_2 formation is sufficient to prevent small quantities of oxygen entering the headspace of fermentation vessels, through cotton wool plugs in small vessels, or open manholes in larger vessels, and influencing the fermentation. The importance of these results lies not just in the effect of different parameters on the level of fatty acids themselves in the spirit, but on the esters which can be produced from the fatty acids.

Ester Formation

Esters are the most abundant organoleptic compounds in whisky and over a hundred different esters have been identified. Although some of these may be formed during the distillation, the majority of the most abundant esters are produced during the fermentation stage. Although direct ester formation by chemical reaction could occur during the fermentation phase, the rate would be much too low to account for the levels of esters produced, so it seems probable that esters are produced by an enzymatic reaction between acetyl CoA molecules and free alcohols. This mechanism was proposed by Nordstrom (ref 82) and was later confirmed in the case of ethyl acetate by Howard & Anderson (ref 83) who synthesised ethyl acetate using a cell free extract of yeast.

Addition of individual acids and alcohols has shown that there is competition between different alcohols and acids in ester formation (ref 73), so that the most abundant esters are derived from the most abundant acids and alcohols. These are ethanol and acetate in the whisky fermentation. The abundance of the different esters in immature Scotch whisky is shown in Table 1. It is clear that ethyl acetate is the most abundant and the other common esters are either acetate esters of higher alcohols or ethyl esters of medium or long chain fatty acids. Some of the esters present in whisky, especially the ethyl esters of unbranched saturated fatty acids, are not readily excreted into the medium and remain bound to the yeast cell (ref 74). These can be released during the distillation phase of whisky production. When the pattern of formation of esters was investigated in the small scale whisky fermentation, different patterns of formation were observed for acetate esters of higher alcohols and ethyl esters of medium chain fatty acids. The acetate esters followed the same pattern as that of higher alcohol formation, ie synthesis occurred largely in the first 24h of the fermentation and then levelled off, whereas the ethyl esters were typically produced throughout the fermentation as were the corresponding fatty acids (ref 55).

In beer production a correlation has been observed between ester formation and the level of the corresponding higher alcohol. Any change in the process which stimulates the formation of higher alcohols will also result in higher levels of the acetate esters of higher alcohols since acetyl CoA does not appear to be limiting. In these circumstances, the ratio of a given ester to its parent alcohol does not alter; however if the level of acetyl CoA or the rate of ester formation per se is increased, then the ratio of ester to fusel alcohol would increase for all the acetate esters. The reverse is true for the ethyl esters of middle and long chain fatty acids. Throughout most of the fermentation, ethanol is present in excess so the amount of ester produced is dependent upon the availability of the appropriate acyl CoA.

The effect of various parameters on four important acetate esters; ethyl acetate, isobutyl acetate, amyl acetate and 2 phenylethyl acetate was investigated by Ramsay & Berry, (ref 57) in a small scale whisky fermentation. Whereas little evidence of variation between strain was observed, wort composition appeared to be a major factor controlling the level of acetate esters, since the use of different grist samples led to considerable variation in their levels. However, the ratio of esters to alcohols were very similar in both experiments, indicating the effect of different grists was to alter the availability of higher alcohols rather than the ester forming process itself. The level of acetate esters increased with increasing levels of α amino nitrogen up to 150 mg l^{-1} in experiments in which the α amino nitrogen content of wort was varied by either diluting with maltose or adding casein hydrolysate. At higher levels of α amino nitrogen, the level of esters levelled off then decreased. Over a broad range of nitrogen values the ratio of esters to alcohol was constant, indicating that the change could be attributed to changes in the availability of higher alcohols; however at very low levels of nitrogen, below 112 mg l^{-1} and at higher levels, 298 mg l^{-1}, the relative amount of ester produced decreased. This decrease was particularly marked in the case of 2 phenylethylacetate production at high levels of α amino nitrogen. Addition of specific amino acids to the wort stimulated the level of esters of the higher alcohol derived from the added amino acid. The lack of change in the ester to alcohol ratio indicated that this was a direct result of increasing the level of the higher alcohol.

Alteration in the initial pH of the wort had little effect on the level of the acetate esters. Increasing the inoculum level up to 5 x 10^7 cells, approximately twice the normal level, however, increased the level of esters but at higher inoculum levels the esters decreased. This decrease was accompanied by a reduction in the ester to alcohol ratio, indicating that it is caused by a change in the ester forming process rather than the availability of higher alcohols. The optimum temperature for ester formation was found to be 30°C and gentle stirring was found to give a modest stimulation in the level of acetate esters. However, gentle stirring in an open vessel caused a dramatic decrease in the level of acetate esters, which was accompanied by a 50% reduction in the ratio of esters to alcohols (ref 57). It is not known what causes this inhibition of ester formation by oxygen, however it does not appear to be associated with a reduction in higher alcohol formation. Increased biosynthetic activity in the presence of oxygen may reduce the level of acetyl CoA available for ester formation. However, it has also been reported that unsaturated fatty acids repress ester formation in beer fermentation (ref 69, 84). So the repressive effect of trace levels of oxygen may be attributable to the formation of unsaturated fatty acids. High levels of ester produced in high gravity beer fermentations have been controlled by sparging with oxygen (refs 84, 85). In studies on wine and saké (ref 86) and beer (ref 79), unsaturated fatty acids were reported to reduce acetate esters, whereas saturated fatty acids gave a slight stimulation of acetate esters. Preliminary results from our laboratory (Ref 87) suggest that saturated fatty acids may be stimulatory when present at 0.1mM concentration but are inhibitory at higher concentrations.

The effect of fatty acids on ester formation is of considerable importance to the whisky industry since mashing procedures, particularly the degree of raking, can alter the amount of trub, which is rich in fatty acids, entering the fermenter. Factors influencing the formation of the ethyl esters of four medium chain fatty acids; hexanoate, octanoate, decanoate and dodecanoate and three long chain fatty acids; tetradecanoate, hexadecanoate and octadecanoate were investigated by Ramsay & Berry (Ref 57). In this work, one strain was identified which consistently produced low levels of ethyl esters. Since the availability of ethanol was not limiting, this result indicates that the ester producing capacity of strains can vary and be subject to genetic manipulation.

Manipulation of the nitrogen and carbohydrate of the wort had little effect on the formation of these ethyl esters. The results supported the view of Norstedt (ref 76) that ester formation is not influenced by nitrogen level, rather than that of Nordstrom (ref 72) who reported that ethyl octanoate at least was reduced by lowering the nitrogen content of the medium.

The levels of ethyl esters of higher fatty acids decreased with increasing temperature, as did the levels of ethyl hexanoate and octanoate, however the ethyl decanoate level did not appear to be affected by temperature variation. This effect of temperature can probably be attributed to a reduction in the availability of acyl CoA at higher temperatures, since the formation of the corresponding acids was also depressed at higher temperatures (Ref 56). That these results were obtained using distillates prepared from wash samples from which the yeast had not been removed, eliminates the possibility that they can be attributed to changes in cell permeability and excretion of the esters (ref 80). Increasing the inoculum level was also found to depress the levels of ethyl esters of medium and long chain fatty acids. This result is comparable to the effect of the inoculum level on fatty acids. Changes in the initial pH of the wort did not have much effect on the level of ethyl esters, however the highest levels were obtained using a wort pH between 5.0 and 6.0.

The most dramatic effect observed in these experiments occurred when the fermentation was stirred gently in open vessels, that is in most instances a conical flask which was plugged with cotton wool rather than a fermentation trap. Whereas stirring in closed vessels only had a small stimulatory effect on ethyl ester levels, stirring in the open vessels caused a large decrease in their levels. This was accompanied by a dramatic increase in the levels of the ethyl esters of two unsaturated fatty acids, ethyl 9-decenoic acid and ethyl hexadecenoic acid. These results were interpreted as indicating that in the absence of a fermentation trap, high levels of carbon dioxide production do not prevent small quantities of oxygen entering the head space of the fermentation and becoming dissolved in the wort. This resulted in an increase in levels of unsaturated fatty acids and their esters and a depression in the levels of esters of saturated fatty acids. Aeration has been reported to depress ester formation in beer (refs 76, 85, 88).

The addition of unsaturated fatty acids was also reported to depress ester formation in beer fermentation by Äyräpää & Lindstrom (ref 79), however unpublished results obtained in our laboratory suggest that this may not always be so. The effect of adding individual fatty acids was found to be very sensitive to the concentrations added and the results could only be interpreted after consideration of the general inhibitory effect of the fatty acid on the fermentation as a whole. In one experiment, palmitic acid was found to inhibit medium chain ethyl ester formation much more than linoleic acid, which had little effect.

It is appreciated that we do not have sufficient information on the flavour and aroma of Scotch Whisky to be able to interpret the physiological results obtained in terms of the quality of the spirit produced. However, it is considered that sufficient information is available to show that the level of different classes of organoleptic compounds present in the spirit can be influenced by the manipulation of process parameters. It is appreciated that some of the parameters which have been controlled in the experiments discussed in this review, for example α amino nitrogen content of the barley or initial wort pH cannot be manipulated in the production of Scotch Whisky because of legal restrictions. However, other parameters such as temperature, strain, inoculum level, degree of mixing and levels of trub in the wort can be readily controlled. A recent report that different malt whiskies were being produced in the same distillery by Alko in Finland (ref 89) suggests that they may be applying these physiological principles and benefitting commercially from the intense scientific studies which have been carried out by these workers over the past two decades.

Physiological studies of the whisky fermentation are difficult to carry out away from the distillery, because of the need for fresh preparations of wort for each experiment. The procedures described by Ramsay & Berry (ref 47) provide a working system for preparing small quantities of wort in the laboratory, and a background of information on a small scale fermentation which can be used to evaluate new processes and procedures in the whisky industry. Using results obtained in these and other studies it is possible to make certain generalisations concerning the control of the whisky fermentation (Ref 57) (Table 8).

Bacteria in Malt Whisky Production

The traditional techniques of malt whisky production make it inevitable that the fermentation proceeds in non-sterile conditions. The maximum temperatures achieved during wort preparation do not exceed 62.5^{o}C, so bacteria present in the barley are not necessarily killed. The brewing yeast strain, used as the second strain in many distilleries also contains bacteria. Bacterial contamination can also arise through ports in open vessels and as a result of poor plant hygiene. Relatively little work has been carried out on the nature of the bacterial population present in typical and atypical whisky fermentations. A series of strains isolated from whisky fermentations were classified by Bryan Jones (ref 90) as *Lactobacillus fermentum*, *L. brevis*, *L. casei*, *L. delbrueckii*, *L. plantarum* and a variant of *L. collinoides*. However, in view of the problems experienced with the isolation and culture of these organisms (ref 90) and the difficulties experienced in the classification of the genus *Lactobacillus* (ref 91) in general, it would be wrong to place too much emphasis on these names. Since MRS Agar (Oxoid) gave difficulties in maintaining viable cultures, Bryan Jones developed a medium containing malt extract and yeast extract. Dolan & Morgan (ref 92) have reported the use of Wallerstein Laboratories Nutrient Agar containing cycloheximide (ref 93) and have described the development of a selective medium for the enumeration of *Lactobacilli* in the malt whisky fermentation. This involved the addition of cycloheximide (20ppm) and 2 phenyl ethanol (0.3% v/v) to the sucrose

TABLE 8 Summary of Results from 750 ml Fermentations from Ramsay (ref 57)

	Higher alcohols	Medium Chain Fatty Acids	Acetate Esters	Ethyl Esters of Fatty Acids
Yeast Strain	Percentage of each alcohol of the total dependent on yeast strain	Little variation	Variation corresponding to changes in parent alcohols.	As acids
Wort Composition				
(a) Grist variation	Considerable variation	As alcohols	As above	As alcohols
(b) Amino N Concentration	Different alcohols maximum at certain concentration of amino N	Little change except at very low levels of amino N	As above	As acids
(c) Carbohydrate Composition	Little change except for amyl alcohols	Little variation	As acids	As acids
Temperature	Maximum at 25-30°C	Maximum at 20-25°C	As alcohols	As acids
Pitching rate	Concentration increased linearly with log of initial cell number	Reverse of higher alcohols	Little variation	As acids
Initial pH	Little variation	Increased at alkaline pH values	Some variation but little in range 5.0-6.0	As acetate esters
Agitation	Increased on stirring	Small decreases	Small increases	As acids
Aeration	Little effect	Much reduced except for increased 9-decenoic acid	As acids	As acids except for ethyl 9-decenoate

agar medium described by Boatwright & Kirsop (ref 94). A system of identification was devised by Dolan, which reduced the identification time from 6-12 days to 2-4 days. Differentiation between Lactobacilli and Pediococcus was possible using this scheme. Organisms which grow on WLNA plates but not SA_3 were found to be non lactic acid bacteria which were of minimal importance to the distilling industry.

In addition to Lactobacillus, Pediococcus cerevisiae, Streptoccus lactis and Leuconostoc sp. have been identified in whisky fermentations and the presence of a variety of aerobic bacteria which are present in the barley but which disappear early in the fermentation has also been established.

The origin of bacteria in the whisky fermentation has been discussed by Dolan (ref 95). In the order of 10^7 bacteria per gram are present in barley at harvesting; however this value increased to 4×10^9 during barley germination before decreasing to 2×10^8 during kilning (ref 96). The majority of these bacteria were gram negative, and gram positive rods such as Bacillus and Lactobacillus constituted only a small component of the bacterial population. Dolan considers that most of these are killed during the mashing procedure. However, other workers have obtained results which suggest that the malt is one source of bacterial contamination.

Bacterial contamination can also arise from the yeast inoculum. The production of specialised distillery yeast, e.g. M1 strain of Distillers Company (Yeast) Ltd., is a clean rather than sterile process, so the yeast produced may contain some contaminating bacteria. However, the use of secondary brewing strains is more likely to be a greater source of infection. The following values have been proposed as acceptable levels of bacteria in the inoculum:- total bacteria, 1,500, total gram positive bacteria 50, viable lactic acid producing bacteria 10, (per 10^6 cells of the yeast inoculum). Since 1g cake yeast contains in the order of 10^{10} cells, then these values correspond reasonably well with the values quoted by Sharp & Watson (ref 43). (Table 7).

High levels of Lactobacillus may also arise as a result of poor hygiene. The use of stainless steel vessels, and both steam and chemical sterilants to sterilise mash tuns, wort lines and fermenters make it easier to control the level of bacterial growth in the whisky fermentation.

The possible effects of bacteria have been listed by Ramsay (ref 57):-

1. They can use sugars which would otherwise be available to the yeast and so lower the spirit yield.
2. They often grow and excrete by-products, particularly late in the fermentation using substrates which are not assimilable by the yeasts.
3. They can contribute substances which improve flavour.
4. They can produce off-flavours.

5. They can excrete acids or other compounds which affect yeast performance or enzyme activity in the medium.
6. They can provide enzymes which modify the fermentation conditions.

The effect of Lactobacilli on the level of lactic acid in the whisky fermentation was studied by MacKenzie & Kenny, (ref 97) who showed that the suppression of bacterial growth by the addition of penicillin suppressed lactic acid formation. In these experiments, which were carried out in 10 l. fermentations, the main period of lactic acid formation and bacterial growth occurred in the later stages of a 65h fermentation. The release of α amino nitrogen by yeast autolysis after 16h was suggested as being important in stimulating bacterial growth. An increase in the level of Lactobacilli in the later stages of the fermentation has also been reported by Dolan (ref 31) and Berry *et al.*, (ref 45) in commercial fermentations. However, a similar increase was not observed in small scale fermentations which were held at 30°C (refs 47, 57). However, when the fermentation was carried out in a container which was insulated with polystyrene chips in our laboratory, increased bacterial counts were obtained. These were associated with an increase in yeast autolysis. We have interpreted these results in terms of an increase in bacterial growth occurring when the level of α amino nitrogen increases as a result of yeast autolysis. Yeast autolysis occurs in conditions of high temperature in the order of 35-37°C and high ethanol concentrations. However, the observation that the inhibition of bacterial growth by the addition of penicillin also reduced the rate of loss of yeast viability (ref 97). suggests that the increase in bacterial numbers stimulates the rate of yeast death and autolysis.

Different opinions have been expressed on the effect of bacterial contamination on the alcohol yield of whisky fermentations. Harrison & Graham (ref 13) expressed the opinion that 10^8 bacteria per ml would be required to cause a significant reduction in the yield. However, Dolan (ref 31) reported that values as high as 9.6×10^7 Lactobacilli can be found in heavily contaminated fermentations and that such levels can result in a 5% loss of yield. However, precise figures were not presented. Since the metabolism of available carbohydrates is completed long before 30h when most of the ethanol production has occurred, it is possible that a late increase in the number of bacteria occurs at the expense of medium components which are not available to the yeast for ethanol formation (ref 37). The correlation between yeast autolysis, the release of amino nitrogen and the increase in Lactobacilli observed (refs 45, 97) seem to support this view.

Several species of Lactobacilli can form hydroxypropionaldehyde from glycerol. During distillation, this breaks down to form acrolein, a compound which has a pungent, burnt or peppery odour which is unacceptable in whisky. Similarly butyric flavours can be produced by contaminating Clostridia. However, the influence of bacteria on the organoleptic properties of the whisky are not always negative. It has been argued by Berry & Ramsay (ref 3) that the traditional fermentations which continued for three to four days and which used impure inocula in vessels which were not cleaned to present day standards would have had high levels of bacterial contamination and high levels of lactic acid in the wash. The effect of this on the organoleptic properties of the whisky is to increase the level of ethyl lactate and also the level of furfural in the spirit. The ethyl lactate is produced by microbial action in the final stages of the fermentation. However, the furfural is produced in the distillation phase from pentose sugars when the pH of the wash is reduced to pH 4 by the accumulation of acids, normally lactic acid in the wash (Fig. 2).

DISTILLATION

The quality of spirit produced is influenced by the distillation technique in several ways. The pot stills used for the production of malt whisky produce a different product from the continuous Coffey stills used for the production of grain whisky and bourbon. The design and operation of both types of still has been described previously (refs 6, 7, 8), however little work appears to have been carried out on the effect of distillation on the final spirit quality. A small scale study on distillation has suggested that rate of heating, refluxing, atmospheric pressure and cooling rates may all be important in controlling still performance (ref 98).

The distillation of malt whisky is a batch process carried out in two pot stills. Traditionally the pot still is pear or onion shaped and made of copper. It is heated from below, usually with oil by steam heated coils. The use of steam coils is favoured because it minimises pyrolysis of the still contents and reduces the level of furfural in the final spirit. A typical Scotch pot still contains 25-50,000 litres. The distillation is carried out in two stages. The first stage is carried out in the wash still; in this the fermented wort or wash as it is called is boiled over a period of 5-6 hours producing a distillate, the low wines. This represents a three fold concentration of the alcohol to 45° proof. The residue, pot ale has a high BOD and will be discussed in the section on by-products. The second stage is carried out in the spirit still. The distillation is carried out in three phases. The first fraction produced, the foreshots, is rejected and recycled to subsequent distillations; the whisky fraction is taken off at $25\text{-}30^{o}$ over proof. When this fraction is complete, any remaining alcohol in the low wines is distilled over, producing a fraction referred to as feints, which is also recycled to the spirit still.

The composition of the spirit changes as the distillation progresses, so the timing of the cut between foreshots, whisky and feints is critical in determining spirit quality. The ester aroma decreases as the distillation continues, whereas the feints or fusel aroma increases (ref 19). The timing of the cut is determined

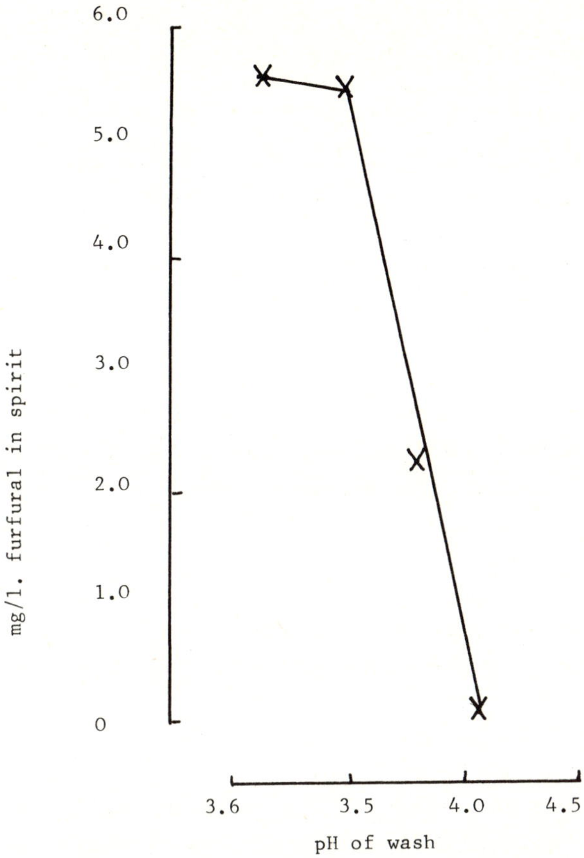

Fig. 2. Effect of wash pH on the formation of furfural

both by changes in hydrometer reading in the spirit chest and also traditionally, by a change in spirit colour which can be recognised by the distiller. Unfortunately, a detailed study of the times at which different organoleptic compounds appear during the distillation of a pot still, does not appear to have been published. However, data on the changes in composition of the vapour at different plates in a Coffey still during whisky production has been presented by Pyke (ref 6).

However, it is not only the timing of the cut which affects spirit quality during distillation. Chemical changes in the spirit also occur. It has been demonstrated that ester formation can occur in the still and that the amount of ester produced increases when the pH of the wash drops below 5.0 (ref 99). The formation of furfural during distillation also increases at lower pH values, so the pH of the wort during the fermentation stage should be carefully controlled (ref 3). Distillation is also important in liberating esters of medium and long chain fatty acids from the yeast. The level of these esters is much lower if the yeast is removed from the wort before distillation (ref 12).

The use of copper stills, although historical, also serves an important function. If whisky wash is distilled in glass apparatus then the distillate has a strong sulphury smell. The absence of this smell in commercial spirit has been attributed by distillers to the action of copper in the still in "fixing" organosulphur compounds. Laboratory studies indicate that the use of a copper still can reduce the level of dimethylsulphide by 70% when compared to glass distillation (ref 18).

MATURATION

Scotch whisky must be warehoused in oak casks for a minimum of three years. In fact, most of the malts used are matured for much longer periods. This period of maturation is not a simple storage period but makes a major contribution to the development of the final flavour. Changes in the concentration of organoleptic compounds in bourbon were investigated over a 12 year maturation period in the laboratory of the Joseph E. Seagram Corporation. During this period, the levels of total solids, colour, total acids, tannins, esters and aldehydes increased (ref 100). In a subsequent series of experiments, it was shown that labelled ethanol was incorporated into acetic acid, acetaldehyde, ethyl acetate and ethyl lignin during a 56 month maturation period. No transfer of label was observed into fusel alcohol or aromatic aldehydes (refs 101, 102). In a study on malt whisky maturation Swan & Burtles (ref 18) reported an increase in isoamylacetate from 7.2ppm to 9ppm in a five year period, an increase of 8 odour units.

Of particular importance to the maturation process is the appearance of three aromatic aldehydes, coniferaldehyde, syringaldehyde and vanillin which are produced during maturation from wood components. The sequence of reactions by which they arise in the spirit has been reported to be:(ref 103)

Wood lignin + ethanol $\rightarrow$ wood lignin + ethanol lignin
Wood lignin + ethanol $\rightarrow$ coniferyl alcohol + sinapic alcohol
Ethanol lignin $\rightarrow$ ethanol + coniferyl alcohol + sinapic alcohol
Sinapic alcohol + $O_2 \rightarrow$ sinapaldehyde
Coniferyl alcohol + $O_2 \rightarrow$ coniferaldehyde
Sinapaldehyde + $O_2 \rightarrow$ syringaldehyde
Coniferaldehyde + $O_2 \rightarrow$ vanillin

Not all materials extracted from the barrel however, have a beneficial effect on the whisky. Oak wood sterols extracted into the whisky have been shown to contain beta sistosterol and beta sistosterol D glucoside. These sterols have been shown to cause floc formation in whisky (ref 104).

Several mechanisms have been proposed as being involved in the maturation process:-

(i) Chemical interactions between ethanol and other distilled components.

(ii) Interactions between ethanol and barrel wood.

(iii) Extraction of wood components from the wood.

(iv) Extraction of components from the barrel which have entered the wood during previous use of the barrel.

An interesting example of the latter mechanism was given by Swan _et al._(ref 5) who noted that although newly distilled grain whisky does not contain alcohols beyond four carbon atoms and esters beyond 3-methyl butyl acetate, matured grain whisky contains many higher boiling components.

It is an interesting point that a 220 l. charred barrel by U.S. law can only be used once for Bourbon production then it has to be sold for use in producing lighter American whisky or Scotch whisky (ref 105). If the wood of the barrel is the source of key flavour compounds, it is not surprising that these will become leached out of the wood as the barrel is used for different distilled beverages. This decrease in the level of materials extracted from the barrel as it is used for successive refills of spirit has been investigated (ref 103). The results showed that after two refills, the level of material extracted had decreased to 20% or less of the value obtained with the original charred barrel.

BY-PRODUCTS

The production of both grain and malt whisky involves the production of large quantities of a wide variety of effluents. The processing of these effluents to produce by-products has led to many ancillary processes which have assumed a major role in the management of whisky production (ref 106).

The effluents are:- draff, the residue from the mashing process; pot ale, the liquor left from the first distillation of malt whisky; and spent wash, the liquor from the first stage of a Coffey still. Carbon dioxide, which is collected from some of the larger distilleries, can also be considered as an effluent which can be converted into a by-product.

Traditionally, the spent grains from the mashing stage were sold in the wet form for cattle and pig feed. The other major effluent, pot ale, was run off into the nearest river or the sea. However, in recent years, environmental pressures have stimulated the development of specific processes to convert these effluents into a marketable form. The first process adopted involved the evaporation of the pot ale to a thick syrup which was then spray dried to produce distillers solubles. However, the most common process at the present time involves the mixing of the evaporated syrup with spent grains. This mixture is then dried out to produce dark grains which can be sold as animal feed (ref 107). Even this does not account for all the effluent; other dilute effluents from spirit stills (spent lees), and from the dark grain plant itself have to be processed by the local sewage treatment plant or by special biotowers (refs 108, 109). The composition and nutritional value of several by-products produced during the whisky process is shown in Table 9.

The recent increase in the price of energy has led to two developments in the field; improvements in the energy efficiency of dark grains plant, and the exploration of new, less energy intensive processes for the conversion of pot ale into valuable by-products (refs 110, 111). One use of pot ale and draff which has been actively considered is their direct use as an animal feed. Their nutritional value for sheep does not appear to be in doubt; however the high copper levels in pot ale, and draff/pot ale mixtures, 17-30mg/kg dry matter, may limit the amount which can be fed to the sheep (ref 112). Alternative uses have involved the application of pot ale to farm land as a fertiliser. The high level of copper in

TABLE 10. Analytical data on pot ale from Old Bushmills Distillery (January, 1977) from Quinn & Marchant (ref 114).

pH	3.20
Carbohydrate	7.0 g/l
Protein	17.5 g/l
Free amino acid	2.8 g/l
Glycerol	4.4 g/l
Titrable acidity (as lactate)	12.5 g/l
BOD	27,500 mg/l
COD	35,500 mg/l
TOC*	21,440 mg/l
Dissolved phosphorus	1310 ppm
Copper	1.9 ppm

*TOC Total organic carbon using Tocsin aqueous carbon analyser.

this context is advantageous, since many Scottish soils are copper deficient.

The use of pot ale and draff as a substrate for anaerobic digestion has also been investigated and has attractions, since the gas produced could contribute to the energy demand of the distillery. The traditional use of distillers solubles was as a substrate for the streptomycin fermentation. Since pot ale is such a rich source of nutrients, the possibility of using it as a substrate for other fermentations has been investigated. Watson (ref 113) reported the use of pot ale as a substrate for growing fresh distillery yeast as the inoculum for the whisky fermentation. The value of this approach was reported to be the high value of the product, aerobically grown distillery yeast, and the use of the product within the company. However this process could only account for a small percentage of the available pot ale.

TABLE 9 Typical analyses of distillers' by products (all values are on an "as produced" basis). From Pass (ref 110)

By-Product	Dry Matter %	Oil %	CP %	DCP %	Fibre %	Lysine %	Av.Lys %	Meth %	M+C %	Threo %	Ca %	P %	DE Mj/kg	ME Mj/kg	MER Mj/kg
Malt distillers dark grains (MDDG)	90.0	7.0	23.0	16.1	10.1	0.77	0.72	0.24	0.47	0.91	0.01	0.80	11.4	8.6	9.7
Malt distillers light grains (MDLG)	90.0	8.0	17.2	12.1	17.3	-	-	-	-	-	0.07	0.42	10.9	7.0	9.6
Malt distillers dried solubles (MDDS)	97.0	0	29.1	23.5	0	1.20	0.96	0.39	0.47	-	3.7	0.90	10.0	8.8	9.0
Malt distillers draff	24.0	2.1	5.0	3.9	4.3	-	-	-	-	-	0.03	0.82	-	-	2.6
Grain distillers dark grains (GDDG)	90.0	10.2	26.3	16.9	6.1	0.73	0.43	0.40	0.63	1.20	0.14	0.75	12.2	9.8	10.3
Grain distillers dried solubles (GDDS)	95.0	8.5	25.6	18.0	4.7	0.85	0.60	0.51	0.58	-	3.58	0.93	13.2	11.1	10.1
Grain distillers maize germ	88.0	18.0	14.5	11.6	7.0	0.78	0.61	0.35	0.92	0.63	0.30	0.50	17.0	15.7	13.2

CP = Crude protein
DGP = Digestible crude protein
Av.Lys = Available Lysine
Meth = Methionine
M+C = Methionine plus cystine
Threo = Threonine
Ca = Calcium
P = Phosphorus
DE = Digestible energy (pigs)
ME = Metabolisable energy (poultry)
MER = Metabolisable energy (ruminants)

Other workers have considered the use of pot ale as a substrate for the production of fungal biomass for use as an animal feed. The most comprehensive piece of work in this area has been carried out using pot ale from Irish malt whiskey distilleries - The Old Bushmills Distillery, Bushmills, County Antrim, Northern Ireland. Analytical data from this pot ale is presented in Table 10. Recent experiments were carried out, using a simple column fermenter and the imperfect fungus Geotrichum. Using batch culture up to 92% reduction in BOD was obtained, leaving a BOD of between 4,500 and 10,000mgl^{-1} which is in the same order as the evaporation condensate effluent (8,500 BOD) produced by dark grain plants. When continuous culture was used, only a 50% reduction in BOD was obtained in a single stage fermentation, however this could be improved to an 87% reduction in a two stage process (ref 114). Growth of Geotrichum in continuous culture was found to be complicated by the formation of two forms of arthrospores; cylindrical and yeast like arthrospores at different dilution rates. The formation of arthrospores was also considered to be stimulated by lactate, which is one of the major carbon sources in pot ale (ref 115). The biomass produced in these fermentations was found to have a true protein content of more than 47% w/w (by Biuret assay) and a crude protein content of more than 53% (N x 6.25). The RNA content of between 5 and 6% is in the range reported for other fungi and the amino acid profile of the protein produced is typical of that produced by other microorganisms in that it contains low levels of sulphur amino acids. However, at 2.0-2.2g/100g protein they fall within the FAO guidelines as do the values for the other essential amino acids (ref 116). However one potential problem of using this material as an animal feed is indicated by the observation of the accumulation of copper in the Geotrichum biomass. Pot ale contains a high level of copper, up to 8.7ppm. These values were obtained when the pH of the pot ale was below pH 3.0. Control of the pH of the final wash to a value greater than pH 4, by reducing the level of bacterial contamination could reduce the level of copper in the pot ale to a value closer to 1ppm. This would facilitate the marketing of products from pot ale and also prolong still life. Much of the copper was found to be present in a complexed form which is associated with the suspended solids of the pot ale. The level could be reduced by a settlement stage. The amount of soluble copper varied between 0.01 and 0.1ppm depending upon the pH of the pot ale. Again the highest level of free copper was obtained at acid pH values (ref 117).

Recent work has concentrated on the use of a mixed population of Geotrichum candidum, G. krusei and Hansenula anomala (ref 118).

Fermentations were carried out in a 100 l. air lift fermenter. This procedure was reported to produce a stable balanced population which gave a similar product and yield to Geotrichum alone. It was also more resistant to bacterial contamination and the reduction in COD obtained, 31.5% in continuous culture and 54.9% in batch culture, was much lower than the values reported for BOD reduction in earlier experiments using Geotrichum alone (ref 114). Again the accumulation of copper in this mixed culture biomass presented problems for utilisation. However, studies on this aspect of the process are continuing (ref 119).

REFERENCES

1. A.H. Rose, in A.H. Rose (Ed.), Economic Microbiology, Vol 1, Alcoholic Beverages, Academic Press, London, 1977., pp 1-14.
2 M.S. Moss and J.R. Hume, The Making of Scotch Whisky, James & James, Edinburgh, 1981.
3 D.R. Berry and C.M. Ramsay, in F.G. Priest and I. Campbell (Eds.), Current Developments in Malting, Brewing and Distilling. Proc. Inst. Brewing (Scottish Section), Aviemore Conference, 1983, pp 45-58.
4 J.H. Khan, J,A.O.A.C., 52 (1969) 1166-78.
5 J.S. Swan, D. Howie, S.M. Burtles, A.A. Williams and J. Lewis in Proc. 2nd Int. Flavour Conf., Athens, July, 1981, Academic Press, pp 201-23.
6 M. Pyke, J. Inst. Brew., 71(1965)209-18.
7 A.C. Simpson, Process Biochem., 3(1968)9-12.
8 T.P. Lyons and A.H. Rose, in A.H. Rose (Ed.), Economic Microbiology, Vol 1, Alcoholic Beverages, Academic Press, London, 1977, pp 635-92.
9 N.R. Merritt, J. Inst. Brew., 73(1967)484-8.
10 K. Jones and R. Wills, J. Inst. Brew., 72(1966)196-201.
11 H. Suomalainen and L. Nykänen, Neeringsmiddelindustrien, 1/2(1970)1-16.
12 H. Suomalainen and L. Nykänen, J. Inst. Brew., 72(1966)469-70.
13 J.S. Harrison and J.C.J. Graham, in A.H. Rose and J.S. Harrison (Eds.), The Yeasts, Vol III, Academic Press, London, 1970, pp 283-348.
14 D.B. Lisle, C.P. Richards and D.F. Wardleworth, J. Inst. Brew., 84(1978)93-6.
15 J. Kaziet and J. Bricourt, Annales de la Nutrition et de l'Alimentation, 32(1978)941-6.
16 M. Lehtonen and H. Suomalainen, Process Biochem., 14(1979)5.
17 R.D. Steinke and M.C. Paulson, J. Agric. Food Chem., 12(1964)381-7.
18 J.S. Swan and S.M. Burtles, Chem. Soc. Revs., 7(1978)201-211.
19 G.W. Shortreed, P. Rickards, J.S. Swan and S.M. Burtles, Brewers Guardian, 108(1979)55.
20 J.R. Piggott, L. Hose and R. Sharpe, Brewing and Distilling International, 10(1980)48-9.
21 A.A. Williams, J. Sci. Food Agric., 26(1975)567-82.

22 M.C. Meilgaard, C.E. Dalgleish and J.F. Clapperton, J. Inst. Brew., 85(1979)38-42.

23 J.R. Piggott and S.P. Jardine, J. Inst. Brew., 85(1979)82-5.

24 J.R. Piggott and P.R. Canaway, in P. Scheir (Ed.), Flavour 81, de Gruyter & Co., Berlin and New York, 1981, pp 33-46.

25 G.N. Bathgate and A.G. Taylor, J. Inst. Brew., 83(1977)163-8.

26 J.S. Swan and D. Howie, In F.G. Priest and I. Campbell (Eds.), Current Developments in Malting, Brewing and Distilling. Proc. Inst. Brewing (Scottish Section), Aviemore Conference, 1983. pp 129-42.

27 P. Salo, L. Nykänen and H. Suomalainen, J. Food, Sci., 37(1972)394-8.

28 G.N. Bathgate, The Brewer, 65(1979)204-7.

29 T.O. Griffin, Process Biochem., 7(1972)17.

30 T.O. Griffin, J. Inst. Brew., 80(1974)236-7.

31 T.C.S. Dolan, J. Inst. Brew., 82(1976)177-181.

32 T.C.S. Dolan, E.T. Dewar and J.D. Gray, J. Inst. Brew., 87(1981)352-355.

33 T.C.S. Dolan, J. Inst. Brew., 89(1983)84-6.

34 D.E. Briggs, J.S. Hough, R. Stevens and T.W. Young, Malting and Brewing Science, Chapman & Hall, London and New York, 1981, p 104.

35 J.R.A. Pollock, Brewing Science, Vol I, Academic Press, London, 1979.

36 A.G. Rutherford, Brewer, 67(1981)82-6.

37 D.C. Watson, in G.G. Stewart and I. Russell (Eds.), Current Developments in Yeast Research, Pergamon Press, 1981, pp 57-62.

38 M. Lavery and D.R. Berry, in F.G. Priest and I. Campbell (Eds.), Current Developments in Malting, Brewing and Distilling, Proc. Inst. Brewing (Scottish Section), Aviemore Conference 1983, pp 293-96.

39 Anon., Science, 25(1979)768-9.

40 D.C. Havery, J.H. Hotchkiss and T. Fazio, J. Food Sci., 46(1981)501-5.

41 C. MacFarlane, J. Inst. Brew., 74(1968)272-5.

42 K. Sugibayashi, H. Sekiya and K. Nishimura, J. Inst. Brew., 89(1983)75-80.

43 R. Sharpe and D.C. Watson, Brewing and Distilling International, 9(1981)50-3.

44 T.C. Seed, Brewing and Distilling International, 13(1983)32-3.

45 D.R. Berry, D.C. Watson and C.M. Ramsay, Brewers Guardian, 108(1979)25-9.

46 C.M. Ramsay and D.R. Berry, Brewing and Distilling International, 1983, In Press.

47 C.M. Ramsay and D.R. Berry, Europ. J. App. Microbiol. & Biotechnol. 1983. Submitted for publication.

48 B.H. Kirsop and M.L. Brown, J. Inst. Brew., 78(1972)51-7.

49 E.A. Dawes and D.W. Ribbons, Bacteriol. Revs., 28(1964)126-49.

50 J.A. Budd, J. Inst. Brew., 80(1974)333-41.

51 M. Lavery, PhD. Thesis, University of Strathclyde, 1983.

52 G.G. Stewart Canad. J. Microbiol. 27(1981)973-90.

53 B.H. Kirsop, J. Inst. Brew., 79(1973)20-5.
54 M.L. Brown and B.H. Kirsop, J. Inst. Brew., 78(1972)39-42.
55 C.M. Ramsay and D.R. Berry, Eur. J. App. Microbiol. and Biotechnol., 1983. Submitted for publication.
56 C.M. Ramsay and D.R. Berry, Eur. J. App. Microbiol. and Biotechnol., 1983, Submitted for publication.
57 C.M. Ramsay, PhD. Thesis, University of Strathclyde, 1982.
58 A.D. Webb and J.L. Ingraham, Adv. in Applied Microbiol. 5(1963)517-53.
59 R. Singh and R.E. Kunkee, Archiv fur Microbiol., 114(1977)255-9.
60 G. Reazin, H. Scales and A. Andreasen, J. Agric. Food Chem., 18(1970)585-9.
61 G. Reazin, H. Scales and A. Andreasen, J. Agric. Food Chem., 21(1973)50-54.
62 P.S. Hall, J. Assoc. Public Analysts, 14(1979)41-5.
63 J.S. Hough and R.S. Stevens, J. Inst. Brew., 67(1961)488-94.
64 E. Gieger and A. Piendl, Ann. Meet. Am. Soc. Brew. Chem., 33(1973)48-54.
65 G. Van Gheluwe, E.C.H. Chen and Z. Valyi, Tech. Quart. Master Brewers Ass. Am., 112(1975)169-75.
66 N.R. Merritt, J. Inst. Brew., 72(1966)374-83.
67 S. Kamiyama and A. Nakagawa, Brewers Digest, 43(1968)60-4.
68 S. Engan, Brygmesteren, 26(1969)23-31.
69 R.G. Anderson and B.H. Kirsop, J. Inst. Brew., 80(1974)48-55.
70 L. Nykänen, E. Puputti and H. Suomalainen, J. Food Sci., 33(1968)88-92.
71 F. Lynen, Biochem J., 162(1967)381-400.
72 K. Nordström, Proc. E.B.C. Congr., Stockholm, 1965, pp 195-208.
73 K. Nordström, J. Inst. Brew., 69(1963)483-95.
74 K. Nordström, J. Inst. Brew., 70(1964)233-42.
75 L.E. Stenroos, K.J. Siebert and M.C. Meilgaard, Am. Soc. Brew. Chem. J., 34(1976)4-13.
76 C. Norstedt, B.O. Bengtsson, I. Bennet, I. Lindström and T. Äyräpää, Proc. E.B.C. Congr., Nice, 1975, pp 581-600.
77 G. Krauss and M. Foch, Proc. Am. Soc. Brew. Chem., 33(1975)37-41.
78 G.T. Taylor, P.A. Thurston and B.H. Kirsop, J. Inst. Brew., 85(1975)219-27.
79 T. Äyräpää and I. Lindström, Proc. E.B.C. Congr., Salzburg, 1973, pp 271-83.
80 H. Suomalainen and M. Lehtonen, J. Inst. Brew., 85(1979)149-56.
81 G. Krauss, M. Foch and H. Holstein, Monats. Brau., 28(1975)229-37.
82 K. Nordström, Svensk. Kemisk. Tidskr., 76(1964)510-43.
83 D. Howard and R.G. Anderson, J. Inst. Brew., 82(1976)70-1.
84 A.K. Palmer and H. Rennie, J. Inst. Brew., 80(1974)447-54.
85 R.G. Anderson, B.H. Kirsop, H. Rennie and R.J.H. Rennie, Proc. E.B.C. Congr., Nice, 1975, pp 243-53.
86 T. Ishikawa and K. Yoshizawa, Agric. Biol. Chem., 43(1979)45-53.
87 C.M. Ramsay and D.R. Berry, unpublished results.

88 D.R. Maule, J. Inst. Brew., 73(1967)351-61.
89 Anon., Kemia-Kemi, 9(1982)126-7.
90 G. Bryan-Jones, in I.G. Carr, C.V. Cutting and G.C. Whiting (Eds.), Proc. Long Ashton Symposium on Lactic Acid Bacteria in Beverages in Food (1975) Academic Press, London pp 165-75.
91 E.A. Barker and F.G. Priest, in F.G. Priest and I. Campbell (Eds.), Current Developments in Malting, Brewing and Distilling, Proc. Inst. Brewing (Scottish Section) Aviemore Conference, 1983, pp 289-91.
92 T.C.S. Dolan and W.J. Morgan, J. Inst. Brew., 84(1978)327-31.
93 D.T. Murphy and L.T. Saletan, Wallerstein Lab. Commun. 29(1966)87-92.
94 J. Boatwright and B.H. Kirsop, J. Inst. Brew., 82(1976)343-6.
95 T.C.S. Dolan, The Brewer, 65(1979)60-4.
96 A. Haikara, V. Makinen and P. Hakulinen, Proc. E.B.C. Congr., Amsterdam, 1977, pp 35-41.
97 K.G. McKenzie and M.C. Kenny, J. Inst. Brew., 71(1965)160-5.
98 S. Forges and J. Flowers, New Scientist 80(1978)942-4.
99 S. Nemoto, Ann. Technol. Agric., 24(1975)397-9.
100 S. Baldwin and A. A. Andreasen, J.A.O.A.C., 57(1973)940-50.
101 G.H. Reazin, S. Baldwin, H.S. Scales, H.W. Washington and A. A. Andreasen, J.A.O.A.C., 59(1976)770-6.
102 G.H. Reazin, Brewers Guardian, 108(1979)31-4.
103 G.H. Reazin in F.G. Priest and I. Campbell (Eds.), Current Developments in Malting, Brewing and Distilling, Proc. Inst. Brewing (Scottish Section) Aviemore Conference, 1983, pp 205-29.
104 K.J. Byrne, G.H. Reazin and A. A. Andreasen, J.A.O.A.C., 64(1981)181-5.
105 Anon., Brewing Distilling International, 12(1982)22-3.
106 G.D. Wilkin, Process. Biochem., (1977)22.
107 A.C. Simpson in A.H. Rose (Ed.), Economic Microbiology Vol. 1, Alcoholic Beverages, Academic Press, London, 1977, pp 537-94.
108 D. McDonald, Brewing Distilling International, 7(1977)28-9.
109 D. Brown, R. McKay and W. Weir, Prog. Water Technol., 8(1976)291-300.
110 R.T. Pass, Brewers Guardian, Sept (1981) 17-21.
111 D. Ashboren, Brewing, Distilling International, 13(1983)36-9.
112 E.W. Wainman and P.J.S. Dewey, Anim. Prod., 34(1982)325-8.
113 D.C. Watson, unpublished results presented at University of Strathclyde Distilling Symposium, Dunblane, Scotland, 1981.
114 J.P. Quinn and R. Marchant, Water Res., 14(1980)545-51.
115 J.P. Quinn and R. Marchant, Europ. J. Applied Microbiol. and Biotechnol. 6(1979)251-61.
116 J.P. Quinn and R. Marchant, Europ. J. Applied Microbiol. and Biotechnol., 6(1979)263-70.

117 J.P. Quinn, T.W. Barker and R. Marchant, J. Inst. Brew., 86(1980)274-6.
118 T.W. Barker, J.P. Quinn and R. Marchant, Europ. J. Appl. Microbiol. and Biotechnol., 14(1982)247-53.
119 J.P. Quinn, T.W. Barker and R. Marchant, J. Appl. Bacteriol., 51(1981)149-57.

MICROORGANISMS AND THEIR ENZYMES IN THE MATURATION OF CHEESES

B.A. LAW
National Institute for Research in Dairying
Shinfield
Reading, U.K.

CONTENTS

I. INTRODUCTION

Cheese maturation can be regarded as the slow, controlled decomposition of the fats, proteins and carbohydrates of bland lactic curds, mediated by both the starter bacteria which the manufacturer has added deliberately, and the secondary flora whose composition varies widely between different cheese types. When it is considered how few species of starter bacteria are used in the basic curd-forming stage of cheese manufacture (see reference 2) it seems remarkable that as many as 400 different varieties are claimed to result from ripening this starting material. However, close examination of the properties of these cheeses shows that many differ only in size, shape, packaging and name, the latter often coming from small regions or towns in their country of origin (1). The list of cheese types is less formidable and more logical if they are grouped according to their moisture content (governed by manufacturing conditions) and secondary flora (Table 1 and ref. 2). High moisture cheeses (50-80%) are termed 'soft'; some are consumed fresh (e.g. Cottage cheese, Mozzarella) while others develop a surface flora of moulds and yeasts (e.g. Brie, Camembert). Semi-hard and semi-soft cheeses have moistures in the range 45-50% and include such varieties as mild-flavoured Caerphilly and Edam, with relatively simple lactic floras, contrasting with

TABLE 1

Major cheese categories, their starter compositions and secondary microfloras

Cheese category	Example varieties	Moisture content (%)	Starter composition	Starter function	Secondary flora
Unripened, soft	Cottage	not >80	*Streptococcus diacetylactis* *Leuconostoc* spp.	Acid and diacetyl production	none
	Mozzarella	>50	*Strep. thermophilus* *Lactobacillus bulgaricus*	Acid production	none
Ripened, soft	Camembert Brie	48) 55)	*Strep. cremoris* *Strep. lactis*	Acid production	*Penicillium caseicolum*, yeasts
Semi-soft	Caerphilly	45	*Strep. diacetylactis* *Leuconostoc* spp. *Strep. cremoris* *Strep. lactis*	Acid and diacetyl production	Lactobacilli
	Limburg	45	*Strep. lactis* *Strep. cremoris*	Acid production	Yeasts, *Brevibacterium linens*
Semi-hard	Gouda	40	*Strep. cremoris* *Strep. lactis* *Strep. diacetylactis* *Leuconostoc*	Acid and carbon dioxide production	Propionibacteria[a]
Blue-vein	Roquefort Gorgonzola Stilton	40-45	*Strep. lactis* *Strep. diacetylactis* *Strep. cremoris* *Leuconostoc* spp.	Acid, carbon dioxide production	*Penicillium roqueforti*, yeasts, micrococci
Hard	Cheddar	<40	*Strep. cremoris* *Strep. lactis* *Strep. diacetylactis*[b] *Leuconostoc* spp.	Acid production	Lactobacilli, pediococci

Emmental	38	*Strep. thermophilus* *Lactobacillus helveticus* *L. lactis* *L. bulgaricus* *Propionibacterium shermanii*	Acid, carbon dioxide and propionic acid production	*Propionibacterium shermanii*, group D streptococci
Gruyere	38-40	*Strep. thermophilus* *Lactobacillus helveticus* *L. lactis* *L. bulgaricus* *Propionibacterium shermanii*[c]	Acid, carbon dioxide and propionic acid production	*Propionibacterium shermanii*, group D streptococci plus yeasts and coryneforms including *B. linens*

[a] Sometimes present as adventitious bacteria but not vital to typical cheese characteristics (Ref. 3)

[b] Not always included

[c] Introduced with the starter but have no lactic acid-forming function. They grow as a secondary flora

Limburger and Munster with complex surface floras and very strong flavours. Blue vein cheeses are in the semi-hard category, being distinguished by the growth of *Penicillium roqueforti* within the cheese matrix during maturation. These are not chance contaminants, but are specially-selected pure strains, added as spores to the cheese milk. The hard cheeses are made by processes which include a heating ('scalding') stage applied to the curds. This may be up to 40°C for cheeses made with mesophilic starters (*Str.cremoris* and *lactis*) or 50°C for those made with thermophilic cultures (*Str.thermophilus* and lactobacilli; see Table 1). Secondary floras range from non-starter lactic acid bacteria (lactobacilli and pediococci) to propionibacteria, surface yeasts and brevibacteria. Although the mechanisms of ripening are not precisely known for all cheese varieties, it is generally accepted that the range of cheese flavours reflects the diversity both of physico-chemical conditions in cheese (due to differences in salting methods, moisture levels, ripening temperatures and cheese pH) and the biochemical and enzymic properties of the secondary microfloras. The role of the starter bacteria can be seen as one of providing a highly acid environment which suppresses the growth of pathogens and spoilage organisms, while permitting the slow growth of desirable floras. This also ensures that most of the enzymes in cheese operate at sub-optimal pH so that flavour develops gradually and the cheese remains palatable for a reasonable time.

A fundamental understanding of the microbiological and biochemical changes which occur during cheese maturation is essential to the cheese industry if it is to maintain traditional cheese characteristics and standards in the face of new manufacturing technologies, whose primary aim is to reduce costs by increasing unit output and turnover. This chapter attempts to explain in detail the microbiology and enzymology of cheese ripening in terms of general principles and using selected examples of varieties from the broad categories listed in Table 1. For a detailed catalogue of the manufacturing methods and equipment used in the production of the worldwide range of cheeses the reader is referred to the works of Davis (4) and Kosikowski (5).

II. MICROBIAL RENNETS

The proteolytic cleavage of the Phe_{105} - Met_{106} bond of κ-casein, which releases a charged macropeptide, destabilizing the micellar casein suspension in milk, is the basis of curd formation during cheesemaking (6). Traditionally, this process is catalysed by calf rennet (chymosin) but shortages of this enzyme have stimulated the search for substitutes, many of which are of microbial origin. This subject has been so extensively reviewed (7-12) that a detailed account of the origins, properties and utilisation of microbial rennets would be superfluous in this chapter. However, an outline of the action of these proteinases is included for the sake of completeness.

TABLE 2

Ratios of milk-coagulating activities to proteolytic activities for chymosin, bovine pepsin and two microbial rennets

Enzyme	Coagulation:Proteinase[a]
Chymosin	1.40
Bovine pepsin	0.04
Mucor rennet	0.52
Endothia rennet	0.15

[a]Data adapted from reference 7

The main difference between microbial rennets and chymosin is in the extent to which they catalyse general proteolysis in addition to the specific and desired breakdown of κ-casein. Because chymosin is very specific, it shows a very high ratio of milk-coagulating activity to proteolytic activity (Table 2). Bovine pepsin, a digestive proteinase, is less specific and shows a much lower ratio. In general, bacterial milk-clotting enzymes (e.g. from *Bacillus subtilis, B.polymyxa* or *B.mesentericus*) are too proteolytic and cause bitterness and body defects in hard and semi-hard cheese. However, the fungal rennets, although showing lower clotting:proteolytic ratios than chymosin, have been used successfully for the manufacture of these types of cheese. The enzyme from *Endothia parasitica* is too proteolytic (Table 2) for ripened cheese and shows relatively broad bond specificity. However, it can be used if curd scalding temperatures are normally greater than 50°C, so that the enzyme is inactivated. It may therefore find application in Emmental and Gruyère manufacture. Rennets from both *Mucor pusillus* var. *Lindt* and *Mucor miehei* are available commercially and have relatively high clotting:proteolytic ratios (Table 2). They are typical acid aspartyl proteinases (like chymosin) and show strong bond specificity for aromatic and hydrophobic side chains. Preparations based on *Mucor miehei* proteinases are generally preferred because their Ca^{2+} requirements are similar to that of chymosin; increased Ca^{2+} concentrations are required for good clotting activity with *Mucor pusillus* rennet. The *Mucor* rennets have the disadvantage of being relatively heat stable so that they persist in cheese whey and preclude the use of this by-product in other foods unless it is subjected to suitable pasteurisation (7). However, a new generation of modified heat labile *Mucor* coagulants have been developed and several recent publications have reported successful cheesemaking with them (13, 14).

III. THE STARTER BACTERIA

A. Fermentation Stage

The lactic fermentation of starters has effects on cheese maturation in addition to the more obvious influence on curd syneresis and bacteriological stabilization. For example, their growth and metabolism confers a relatively low redox potential on the cheese (15, 16). This is now thought to be important for the preservation of volatile sulphur compounds such as hydrogen sulphide and methanethiol in their flavourful, reduced state during maturation (16, 17). Starter metabolism also generates specific compounds (e.g. acetic acid, acetaldehyde, ethanol, diacetyl) which may subsequently contribute synergically to the overall flavour profile of mature cheese. These compounds are products of carbohydrate metabolism, formed by alternative pathways of pyruvate utilization (see reference 2). Such pathways have been demonstrated in *Str.diacetylactis* when it ferments citrate (18) and in *Str.lactis* and *cremoris* when they come under the influence of various growth limitations (19, 20). Induction of NADH oxidase in some group N streptococci also allows alternative pathways of pyruvate metabolism to be expressed (2, 21, 22). Their importance was demonstrated by Czulak (23) who showed that the pyruvate dehydrogenase of group N streptococci was inhibited in cheese made from milk containing high concentrations of polyunsaturated fat. The cheeses developed rather bland flavours attributable to low concentrations of diacetyl and acetaldehyde. Manning and Robinson (24) had earlier suggested that diacetyl formed part of the spectrum of compounds responsible for Cheddar aroma.

Str.thermophilus and the homofermentative lactobacilli, comprising the thermophilic starters for high-cook cheese, have not been studied in the same detail as have the mesophilic streptococci. However, the evidence presently available indicates that they do not contribute significantly to the flavour volatiles of cheese. *Lactobacillus bulgaricus* for example is capable of phosphoroclastic conversions to CO_2, acetate and formate but does so only under alkaline conditions (25). Its normal metabolism is homolactic, with lactic acid as the sole end product. The same appears true of *Str.thermophilus;* it does not metabolise citrate nor produce pyruvate dehydrogenase (26). However, this does not mean that the thermophilic starters are not important in cheese maturation; their rate of growth and lactose utilization is critical in the production of good quality Emmental and Gruyere cheese (Section VI, E).

B. Maturation

1. Proteolysis. This process is essential for the conversion of the springy curd to typically inelastic textures of mature cheeses. Body/texture changes are chiefly attributable to α_{s1}-casein breakdown because this protein forms strong interactions with other caseins to form a structural network, which is weakened as the α_{s1}-casein bonds are broken (27). General casein breakdown can also be linked with increased brittleness in maturing hard cheese because cleaved peptide bonds expose new ionic groups which bind water and render the cheese proteins less soluble. The later stages of proteolysis produce small peptides and amino acids which are involved in the development of cheese flavours, the buffering of cheese and the nutrition and growth of secondary floras. The contribution to cheese proteolysis by enzymes of starter bacteria varies in significance, depending on

the extent to which these secondary floras produce proteinases and peptidases during cheese ripening. Microbial proteolysis is also superimposed on the action of chymosin, the coagulant used in the cheese vat to destabilize casein micelles and form the gel from which the cheese curds are made. In the short term, chymosin is very bond specific (Section II), but during the following weeks and months of maturation this enzyme contributes to gross proteolysis in cheese. It is thought to produce a large proportion of the larger peptides to a lower limit of 1400 molecular weight (28-31). Early rennet proteolysis is typified by the hydrolysis of the Phe_{23} - Phe_{24} bond (32, 33) or the Phe_{24} - Val_{25} bond (34) of α_{s1}-casein. Beta-casein degradation occurs only slowly in cheese and its products appear later in cheese maturation; the most sensitive bonds are Ala_{189} - Phe_{190} and Leu_{191} - Tyr_{192}. Endopeptidases from the starter bacteria also contribute to gross casein hydrolysis (i.e. hydrolysis leading to increased pH 4.6 - soluble N). For example an intracellular endopeptidase from *Str.diacetylactis* rapidly hydrolyses α_{s1}-casein (but not whey proteins) and appears to be specific for peptide bonds involving the α-amino group of hydrophobic residues (e.g. X-Leu or X-Phe; (35)). Zevaco and Desmazeaud (36) later demonstrated that this proteinase also hydrolysed Pro_{186} - Ile_{187} and Ala_{189} - Phe_{190} in β-casein although activity was low. On the other hand the enzyme efficiently degraded peptides derived from β-casein by chymosin action, by attacking the Lys_{176} - Ala_{177}, Lys_{119} - Val_{120} and Pro_{206} - Ile_{207} bonds. These authors concluded that the starter proteinases functioned in cheese chiefly by degrading those peptides released from casein by chymosin. This is consistent with reports by several research groups that cheeses made with chymosin alone contain pH 4.6-soluble N but very little peptide and amino acid N, whereas cheese made with starter and chymosin contain relatively large amounts of free amino N. Several independent studies (29, 30, 37) have established that in both Gouda and Cheddar cheeses, the proteinases of starter bacteria can slowly degrade whole casein to low mol. wt. (<1400) peptides and amino acids. In normal cheese, chymosin-mediated gross proteolysis is so rapid (especially of α_{s1}-casein) that it is doubtful whether this activity is significant. The most important role of starter proteinases and peptidases appears to be the degradation of large rennet-derived peptides to small peptides and amino acids. Recent evidence from studies with *prt*$^-$ variants suggests that the cell-bound extracellular proteinases of starters are significant in cheese proteolysis (38) but a significant proportion of the proteinase activity, and most of the peptidase activity, is intracellular and only released into the cheese matrix during the early ripening stage when the cells lyse (39).

These intracellular peptidases have been widely studied and described in numerous papers. For example, Sorhaug & Solberg (40) used starch gel zymograms to demonstrate the presence of dipeptidases in *Str. lactis* capable of hydrolysing a wide range of substrates containing alanine, leucine, phenylalanine, tryptophan, histidine and glycine. Similar techniques with PAGE revealed the presence of further amino- and dipeptidases in *Str. lactis, diacetylactis* and *cremoris* active against peptides and peptide derivatives containing arginine, methionine, tyrosine and proline (41, 42). Yamauchi *et al.,* (43) reported the isolation of an aminopeptidase from *Str. cremoris* whose hydrolytic activity was greatest on peptides released by chymosin from α_{s1}-casein. Some of these peptidases have been purified and characterized (42, 44, 45). Collectively, the peptidases of Group N streptococci are probably capable of completely hydrolysing casein to free amino acids after initial proteolysis by chymosin and starter proteinases (46). Despite their predominantly neutral or alkaline pH optima, these enzymes retain sufficient activity in cheese to function during maturation and can be recovered in active form from extracts of ageing cheese (47).

The contribution of thermophilic starter bacteria to cheese proteolysis has not been studied to the same extent as that of mesophilic starters, but they are known to produce a range of proteinases and peptidases. Lactobacilli tend to be more proteolytic than *Str. thermophilus*, though the activity of both groups is rather weak (48). Richter & Gill (49) suggested that there may be a multiple proteolytic system in *Lb. bulgaricus* on the basis of its complex pH and temperature profile. Little is known about proteinase specificity; the surface bound proteinase is most active against β-casein (50, 51). Proteinases and peptidases are produced by *Lb. casei* and the specificities of the peptidases are known (52-54). They hydrolyse a variety of di- and tripeptides containing lysine, leucine, tryptophan, phenylalanine, histidine, methionine and proline. Both the carboxypeptidase and endopeptidase have very restricted specificities for synthetic substrates and it is impossible at present to assess their contribution to cheese proteolysis. The endopeptidase has subsequently been identified as a ribosomal arylpeptidyl amidase (55) and such enzymes have now been detected in many other species of *Lactobacillus* (56).

2. Lipolysis. The release of flavourful free fatty acids from milk triglycerides in maturing cheese has an obvious function in the ripening of Italian hard cheeses and blue-vein cheese where these compounds are an easily-identifiable part of the typical flavour profile. However, these types contain exogenous lipases and lipolytic moulds, respectively, to ensure high levels of lipolysis and the contribution of starter bacteria is extremely limited since these organisms only hydrolyse milk fat very slowly (57). Cheese varieties with more subtle, less easily defined flavour profiles probably owe part of their characteristics to low concentrations of volatile free fatty acids (5-10μg g^{-1} cheese), released by the enzymes of starter and non-starter lactic acid bacteria. Stadhouders & Veringer (58) showed that mesophilic starters hydrolysed mono- and diglycerides, but their activity on triglycerides was very weak. This explained discrepancies

in previous results (59, 37) indicating that particular starters could cause different degrees of lipolysis in different cheeses. Presumably the extent of partial triglyceride breakdown by milk lipase (lipoprotein lipase) and heat-resistant psychrotroph lipase governs the amount of substrate available to the starter lipase. Nakae and Elliot (60) showed that volatile fatty acids are also produced by *Str.diacetylactis* from amino acids *via* oxidative deamination, but it is not certain whether such pathways operate in cheese since Dulley and Grieve (61) demonstrated that milk fat was necessary for production of volatile fatty acids other than acetate in Cheddar cheese.

Thermophilic starters are sufficiently lipolytic to contribute to fatty-acid formation in cheese. Paulsen *et al.*, (62) showed that, although volatile fatty acids in Swiss cheese were produced chiefly by carbohydrate metabolism of the lactic and propionic-acid bacteria, the same organisms also produced longer chain fatty acids during ripening by breaking down cheese fat.

IV. THE NON-STARTER FLORA

The lactic starter cultures have metabolic and enzymic activities which can combine to give fresh curds a basic "cheesy" taste as it matures into cheese. The subsequent growth and metabolism of secondary, non-starter floras is now considered as the means by which different varieties acquire their distinctive characteristics.

A. Soft cheeses

These high-moisture cheeses range from unripened, mild-flavoured varieties such as Cottage or Mozzarella, to surface-ripened cheeses with very distinctive aromas and flavours. Most ripened soft cheeses are made with a surface flora of the white mould, *Penicillium camemberti* (otherwise known as *caseicolum* or *candidum*) though Romadour, made in Austria, has a surface smear chiefly composed of *B.linens*, giving it a characteristic odour made up of fatty acids, ammonia, methanethiol and other unidentified aroma compounds. However, the mould-ripened soft cheeses typified by Camembert and Brie are quantitatively the most important. Temperature and humidity control are vital to the correct development of a balanced surface flora of these cheeses (1). In addition to the *P.camemberti* spores which are sprayed onto the cheese surface, a flora of yeast usually develops. Species of *Geotrichium, Saccharomyces, Kluyveromyces, Torulopsis* and *Candida* predominate (63, 64) but although their presence in high numbers (up to 10^9 cfu g^{-1}) and their proteolytic and lipolytic activities suggest that they may contribute to flavour, specific metabolic pathways leading to flavour compounds have not yet been identified. The yeasts are salt and acid tolerant and ferment lactic acid, so their function may be to raise surface pH before the germination and growth of *Penicillium* spores. However, this is not a universally-held view and some workers consider that the colonization of the cheese surface and development of typical aroma is possible without the intervention of the yeasts (65).

Penicillium camemberti is both lipolytic and proteolytic; although the volatile fatty acids released by the lipases from milk fat contribute to the aroma profile, it is the action of the proteinases and peptidases in texture formation and amino acid liberation which have received more research attention (31, 66-69). Two extracellular proteinases (acid aspartyl and neutral metalloproteinase) have been purified and characterized; the peptidases include two distinct carboxypeptidases and an aminopeptidase. The *P.camemberti* proteinases cleave bonds in α_{s1}- and β-casein, producing large peptides. Even though few points in the polypeptide chains of these proteins are cleaved (4 in α_{s1}-casein and 3 in β-casein) the effect on cheese texture is profound because the network of associating casein molecules is broken down very effectively (27). Trieu-Cuot & Gripon (70) have proposed a sequence for proteolysis in camembert cheese which involves chymosin, plasmin (native milk proteinase) and the two *P.camemberti* proteinases. Early α_{s1}-casein proteolysis is thought to be chymosin-mediated but the similar actions of the *Penicillium* acid enzyme and chymosin prevent an assessment of their relative importance beyond the 7th day of maturation, when mould growth and enzyme secretion become significant near the cheese surface. Because chymosin-mediated hydrolysis of β-casein is slow in cheese, its degradation by the mould enzymes is more easily observed from the 7th day of maturation. Lenoir & Auberger (71) suggested that the acid proteinase activity was highest in young cheese from 6 to 16 days after manufacture, but it disappeared after 22 days. The neutral proteinase, on the other hand, became active from the 13th day and persisted at a high level for at least one month. However, from their detailed observations of casein-derived peptides, Trieu-Cuot & Gripon (70) have come to the opposite conclusion, suggesting that the aspartyl acid proteinase is more important than the neutral metalloproteinase for most of the maturation period. Since they have directly observed the appearance of specific products of both *Penicillium* proteinases, Trieu-Cuot & Gripon's evidence is the more convincing at present. Plasmin also hydrolyses β-casein but its action is not seen until 21-35 days, after the pH of the cheese has risen to nearer the neutral/alkaline optimum of this enzyme.

The relationship between the proteinases and the peptidases in the sequential breakdown of caseins to amino acids in the cheese is not clear because relatively little is known about the specificities of the latter group of enzymes. Both extra- and intracellular enzymes are produced; presumably the intracellular peptidases only become significant in older cheeses in which the *Penicillium* mycelia become moribund and leak their enzymes into the cheese matrix. In any case the pH optima of these peptidases are relatively high (72) so that they could only be effective in the later stages of maturation. Despite the lack of agreement on the role of yeasts in the microflora development and ripening of

Camembert cheese Schmidt (73) has shown that many isolates produce a range of intracellular proteolytic enzymes capable of liberating small peptides and amino acids from casein. They may therefore contribute to the final stages of protein breakdown together with the *P.camemberti* peptidases.

Amino acid catabolism by the surface flora yields a variety of flavour compounds and precursors, thought to be important to the development of the subtle, distinct flavour and aroma of Camembert cheese. For example, ammonia contributes to the aroma profile and results from amino acid deamination by the contaminating yeast microflora, particularly species of *Geotrichium* (74). Cheeses which also develop a growth of *Brevibacterium linens* are likely to take on a stronger ammoniacal aroma since this organism deaminates the amino groups of most amino acids including the side chain groups (74, 75). General flavour characteristics common to many surface ripened varieties of cheese have been attributed to 3-methyl-1-butanol, phenylethanol and phenol; though direct evidence is lacking for mechanisms operating in cheese, it is known that many microorganisms, including *B.linens* can convert leucine, phenylalanine and tyrosine respectively into these three compounds. Volatile sulphur compounds make up another important group in soft cheese flavour and aroma. *P.camemberti* itself was reported to produce H_2S, dimethylsulphide and methanethiol from methionine (76) by a combination of oxidative deamination and demethiolation and *B.linens* also has this capability (77). Methanethiol has no proven direct role in Camembert aroma but it may be involved in non-enzymic reactions such as that with formaldehyde to yield bis (methylthio) methane, thought to be important to the aroma profile (78, 79). *Brevibacterium linens* also appears to interact with micrococci to produce a variety of thioesters, but these are thought to be of greater significance in surface smear cheese (79).

Although much emphasis has been placed on proteolysis in the maturation of mould-ripened soft cheeses, lipolysis is also important. *Penicillium camemberti* produces extracellular lipases and oxidises the liberated fatty acids to methyl ketones (Section IVB for pathways). Although these compounds are a more obvious part of the flavour profile of blue-vein cheese, their presence is also desirable for the correct flavour balance in Camembert cheese (80). Oxidative activity among strains of *P.camemberti* is very variable (81) and the main products are 2-nonanone and 2-undecanone. The unsaturated alcohol, oct-1-en-3-ol attenuates the harsh character of the ketone flavours with a softer, mushroom-like note but the origin of this compound is unknown (82).

B. Blue-vein cheeses

Varieties within this category are usually classified as semi-soft or semi-hard but are described separately because their flavour is generated as a result of mould ripening; Danish Blue, Roquefort, Gorgonzola and Stilton are typical examples. The milk or curds of these varieties are inoculated with spores of blue-green *Penicillium* moulds (*P.roqueforti*). When the freshly-made cheeses are 'spiked' to admit air, the spores germinate and the resulting mould growth spreads throughout the inside of the cheese, aided by an open texture conferred by the inclusion of gas-forming leuconostocs in the starter culture (83). Mould growth is partly controlled by the NaCl concentration in the cheese (84). Salt is normally applied dry to the outer surface from which it diffuses to the cheese interior. Godinho & Fox (85, 86) showed that *P.roqueforti* grew most quickly in the middle depth region, rather than the rind or centre regions because high and low salt concentrations were inhibitory, whereas 1-3% (w/w) was stimulatory. Mesophilic starters are used to form the curds so that the general considerations of proteolysis and lipolysis have already been outlined. The spores of *P.roqueforti* germinate early in maturation and mycelia are visible after 8-10 days. Maximum growth is reached by one to three months. The mycelia and spores are strongly lipolytic and proteolytic and their action in the cheese is superimposed on that of the starters. The extracellular acid and neutral proteinases of *P.roqueforti* have been shown to act synergically with the starter proteinases and peptidases (87, 88). Although the relative quantitative contributions of the two mould enzymes in cheese are not known, both are actively secreted by the *P.roqueforti* mycelium (89). These proteinases liberate peptides with molecular weights >1000 from which the starters produce small peptides and amino acids. *Penicillium roqueforti* also contributes to peptide breakdown by producing intra- and extracellular amino- and carboxypeptidases (90). Proteolysis is extensive in mature blue-vein cheese and its rate is inversely proportional to NaCl concentration (91). As much as 10% of the cheese N can be present as free amino acids and this high level is thought to act as a buffer, keeping the cheese pH around 6.5, thus aiding other enzyme activities (84).

Blue-vein cheeses normally have a surface slime or crust containing moulds, yeasts and micrococci. Devoyod and Bret (92) showed that these three groups became dominant, at the expense of coliforms and staphylococci, as the cheese was acidified by the lactic flora early in maturation. Changes occur in the composition of the yeast flora during ripening in response to surface salting (93). *Candida lipolytica* is a common species which is thought to contribute directly to flavour development by its lipolytic activity (84). Salt is also assumed to be the source of micrococci in the surface slime and, although their role in ripening is not known, they tend to be more lipolytic than starters and may contribute through such action (94). The microbiology of surface slime is dealt with in more detail in Section IV, C.

The overriding flavour notes in blue-vein cheese are those of fatty acids and methyl ketones, both classes of compounds being produced by *P.roqueforti*. Two lipases are secreted by the mould, but only the acid enzyme is significant in cheese ripening (95). As was the case with proteolysis, high NaCl concentrations inhibit lipolysis (96).

The rate of release of fatty acids by the lipases of *Penicillium* sp. governs the rate of ketone formation (84). Methyl ketones are formed from fatty acids, by partial oxidation *via* the fatty acid β-oxidation pathway (Fig. 1), which occurs in both spores and mycelium. The β-oxo-acyl-CoA, formed by enzymic dehydrogenation of fatty-acyl β-hydroxyacyl-CoA derivatives is deacylated by thiohydrolase to form free β-keto acid. The decarboxylase which then catalyses formation of corresponding methyl ketones is most active in mycelia and β-oxolauric acid (yielding 2-undecanone) is its preferred substrate, but the preponderance of heptan-2-one in cheese is due to the preference of the preceding thiohydrolase for β-oxo-octanyl-CoA.

MILK TRIGLYCERIDES

↓ Lipases

FATTY ACIDS (C_n)

↓ β-oxidation

OXO-ACYL-CoA

↓ thiohydrolase

β-KETO ACID

↓ decarboxylase

METHYL KETONE (C_{n-1})

Fig. 1. Metabolic pathway for formation of methyl ketones by mycelium and spores of *Penicillium roqueforti*; *n* has a value of 6, 8, 10 or 12.

Flavour-simulation studies suggest that δ-lactones contribute to cheese flavour (97), and Jolly and Kosikowski (98) increased the quality of blue-cheese flavour by increasing the concentrations of δ-tetradecalactone and δ-dodecalactone. Two mechanisms for δ-lactone formation are possible. Traces of δ-hydroxy acid may be present in milk glycerides, and they can be released by lipases during cheese ripening. The free acids may spontaneously undergo ring closure to form lactones, or they may be converted enzymically (99). Alternatively, esterified δ-keto acids in milk glycerides may be released by lipases, reduced to hydroxy acids and then converted into lactones (Fig. 2). Boldingh and Taylor (99) cited examples of yeasts and moulds which could produce lactones by this pathway.

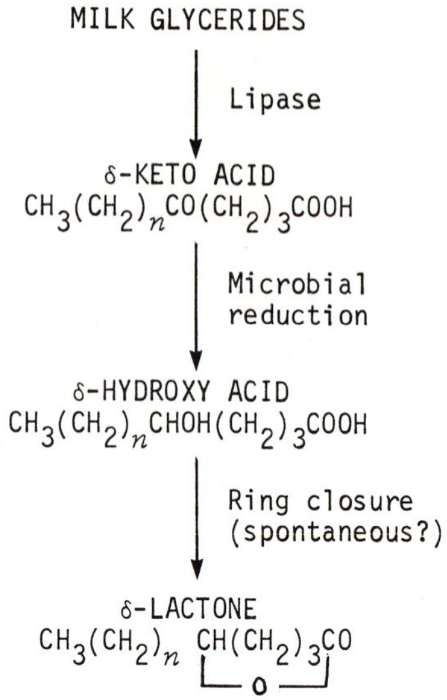

Fig. 2. Metabolic pathway for formation of δ-lactones in cheese; *n* has a value of 4 or 6.

C. Semi-soft cheeses

These varieties are usually made with mesophilic starters at low inoculation rates (<1%, v/v) and full acidity develops during both the vat stage and the light pressing stage. Typical examples include Limburg, Munster and Tallagio; for details of differences between these varieties the reader is referred to reference 1.

These cheeses are initially dry-salted on their surface, then after 1-2 days, wiped with brine-soaked cloths. This second salting has the effect of transferring a heterogeneous yeast and bacterial microflora to the cheese surface. The yeasts are typically salt-tolerant, lactate-utilizing strains of *Geotrichium candidum* whose growth and metabolism raises the surface pH to >6.5 and produces low mol.wt. growth factors for subsequent growth of *B.linens* and micrococci. On Tellagio cheese both micrococci and coagulase-negative staphylococci form the major pigmented flora (100) but *B.linens* is important on Limburg (101, 102). Suitable conditions of air flow, humidity and temperature for the proper development of surface smear floras have been reviewed (103, 104). The yeasts and bacteria are proteolytic and catabolise amino acids to ammonia and reduced sulphur compounds, giving the cheese surface its typically putrid aroma.

D. Semi-hard cheeses

Varieties in this group range from English territorial cheeses such as Lancashire and Leicester to the well-known and widely-made "Dutch" cheeses, Edam and Gouda. There have been no recent definitive studies of the microbiology of the English territorials and the reader is referred to the works of Davis (4) and Chapman & Sharpe (1) for general descriptions of their manufacture and ripening.

Edam and Gouda are brine-salted cheeses whose curds are washed to remove some of the acidity produced during manufacture. Thus the early part of the ripening process involves rapid growth of starter streptococci with the production of typical fermentation volatiles and lactic acid. A secondary non-starter lactic flora of lactobacilli may subsequently develop but this is not known to be essential for the production of typical flavour in Edam. Propionibacteria may develop in Gouda cheese to produce small eyes. The chief compositional difference between Edam and Gouda lies in the higher fat and lower moisture content of the latter. Although the flavour compounds in these varieties are not well defined, it is generally agreed that limited proteolysis and lipolysis are responsible for producing their relatively mild organoleptic characteristics.

E. Hard cheeses

1. Cheddar. The secondary flora of Cheddar cheese comprises organisms which gain access accidentally. This adventitious flora includes thermoduric bacteria carried over from the cheese milk (most Cheddar is made from milk pasteurized at approx. $71^{o}C$ for 15-17 sec) and contaminants of the cheese vat from the creamery environment. A wide variety of micro-organisms have been isolated from Cheddar cheese but great care is required if false results are to be avoided; surface contaminants must be excluded from the cheese samples by aseptic techniques (105). Lactobacilli and pediococci represent the major non-starter bacteria in Cheddar; they may be present in fresh curds in numbers as low as 100 g^{-1} but are capable of multiplying during the first weeks of maturation to reach as high as 100,000,000 g^{-1} (Fig. 3). Since the starter streptococci die out during the early ripening period the non-starter lactic acid bacteria (NSLAB) are the dominant flora in the cheese for the greater part of its storage life. Other types of bacteria may be present as stable or declining populations; typical genera include group D streptococci (enterococci), micrococci, staphylococci, Gram-negative rods, coryneforms, aerobic and anaerobic sporeformers, but the combined effects of acid and salt prevent them from multiplying significantly in normal cheese (Section V,C).

The relative importance of starter and non-starter bacteria in the development of typical flavour in Cheddar cheese has been the subject of considerable past speculation and although it has been established that the cheese ripens in a satisfactory way when the secondary flora is excluded experimentally (59, 106) the growth of NSLAB can influence the rate and quality of flavour formation. In New Zealand Cheddar the NSLAB are regarded as potentially detrimental to the production of clean-flavoured cheese and their growth requires control to avoid fermented flavour defects (Section V,C and VII,A). In an attempt to define more

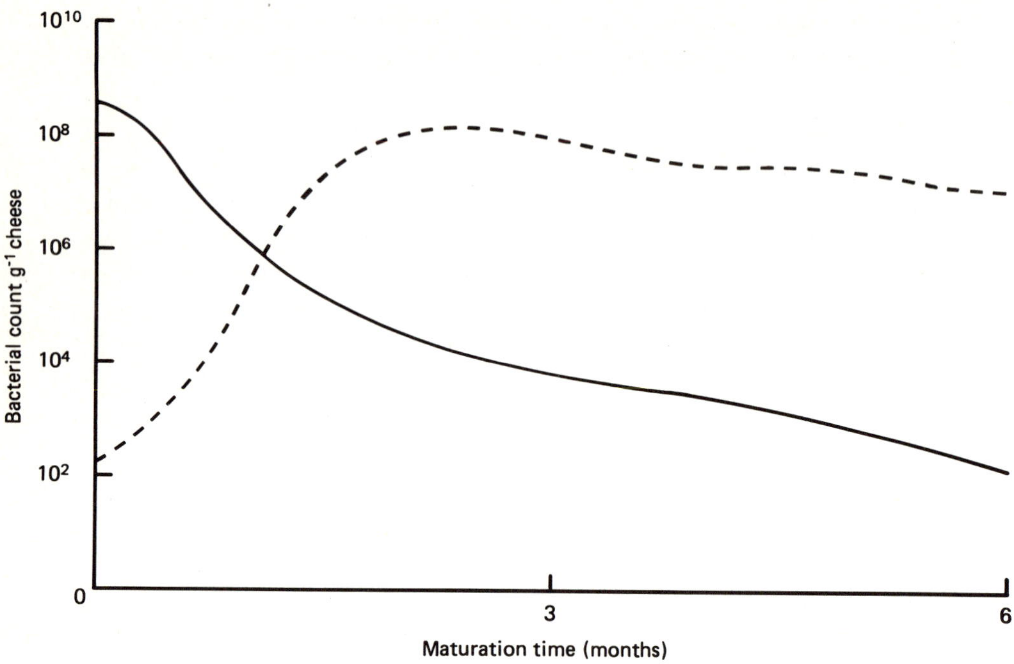

Fig. 3. Changes in the lactic flora of ripening Cheddar cheese: , starter streptococci; -------- non-starter lactic acid bacteria (lactobacilli or pediococci).

closely the role of secondary floras, Reiter *et al.* (59) isolated groups of adventitious bacteria from commercial creameries (reference floras) and added them back to experimental cheeses under bacteriologically controlled conditions. They reproduced the flavours of the original commercial cheeses from which the reference floras were derived and intense flavour developed more rapidly than it did in "starter-only" cheeses. However, this flavour intensification was accompanied by the development of flavour defects (yeasty, fruity and sour) thought to have been due to heterofermentative lactobacilli (Section V,C). Subsequent attempts to define the most important components of the reference floras have failed (106) even though the inclusion of different reference-flora groups affected formation of several potential flavour compounds in different ways; concentrations of volatile fatty acids, higher fatty acids and hydrogen sulphide were higher in reference-flora cheeses than in starter-only cheese. The principal source of volatile fatty acids were NSLAB (acetate) and the lactobacilli and Gram-negative rods (butyrate). Umemoto & Sato (107) also observed that lactobacilli from Cheddar cheese were lipolytic and that they reach sufficiently high numbers in maturing cheese to influence free fatty acid concentrations. However, concentrations of fatty acids in cheese do not appear to be critical in relation to typical flavour. Indeed, Lawrence (108) showed that 23 out of 41 good quality New Zealand Cheddar cheeses did not contain any detectable free fatty acids higher than C_4 (butyrate). Manning and Price (107) reported that volatile fatty acids did not influence Cheddar aroma at low pH values, but they did not rule out a role in the general background taste of cheese. Higher free fatty acids do not themselves appear to be directly beneficial to Cheddar flavour; indeed, at the high concentrations produced by heat-resistant psychrotroph lipases, they can be detected organoleptically as rancid, soapy flavour defects (110). Also, high concentrations of butyrate and hexanoate, combined with high concentrations of ethanol from some starters, lead to fruity flavours often associated with lipolytic rancidity.

Levels of hydrogen sulphide in reference-flora cheeses are highest in the presence of lactobacilli, some strains being able to desulphurylate cysteine (111). Hydrogen sulphide is a desirable component of Cheddar aroma (109) and it may be involved in the generation of methanethiol but, unlike methanethiol, its concentration in cheese does not appear to be critical. In a later study, Law and Sharpe (112) showed that only Gram-negative rods and coryneform bacteria produced methanethiol from methionine, but the enzymes involved in the reaction were too unstable, and unsuited to conditions in cheese, to influence thiol concentrations during ripening. These data support the conclusion that this important compound is produced by non-enzymic reactions and they strengthen the hypothesis (113) that typical flavour develops in Cheddar cheese independently of its viable microflora.

2. Emmental and Gruyere cheese. In contrast to Cheddar cheese, the growth and metabolism of the starter and non-starter flora during maturation is a vital part of the ripening process (114). The secondary flora consists mainly of a growing population of *Propionibacterium shermanii* which is added by the cheesemaker with the starter to the cheese milk. However, its growth and metabolism have to be considered together with that of the starter flora because they are interdependent. The *Str.thermophilus*/*Lb.helveticus* starter grows and produces lactic acid both in the cheese vat and during pressing, but the most important stage is recognized as the 24h draining and pressing time during which the cheese cools from 50°C to 20°C. Accolas *et al.* (115) and Moquot (116) have described the effects of the temperature gradient in cooling Emmental cheese on the acid production, first by *Str.thermophilus*, then by *Lb.helveticus* such that after 24h all the lactose is fermented to lactic acid, more being produced near the cool periphery than in the warmer centre of the cheese. This lactate gradient, together with the salt gradient which results from brining the young cheese, ensures that the propionibacteria grow best near the centre since they are inhibited by high salt and high lactate concentrations. The rate of growth and CO_2 production of this secondary flora is very

critical because it governs the size and distribution of holes ("eyes") in the cheese and hence its market value. Emmental should have larger holes than Gruyere, an objective reached by using higher ripening temperatures and lower salting rates for the former variety (116).

Fermentation of lactate and residual sugars by propionic acid bacteria is thus a vital stage in Swiss-cheese ripening, and follows initial lactic fermentation by the starters. Hettinga and Reinbold (117) have reviewed the extensive literature on the propionate fermentation, which involves a complex double cycle *via* pyruvate, catalytic amounts of methylmalonyl-CoA, and an apparently unique biotin-dependent transcarboxylation reaction (Fig. 4).

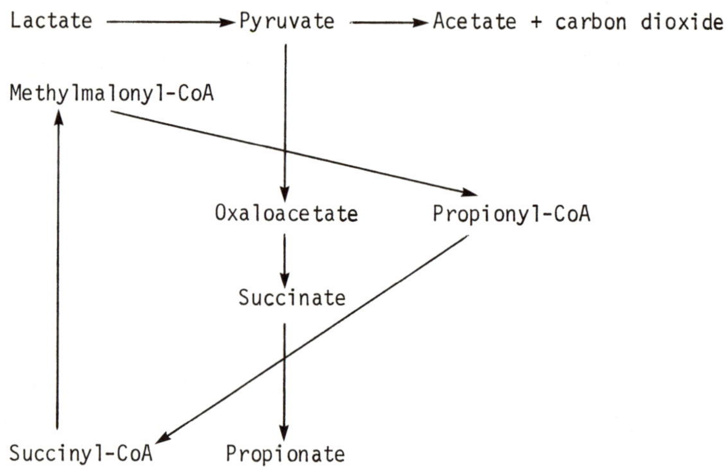

Fig. 4. Pathways leading to propionate, acetate and carbon dioxide production in *Propionibacterium shermanii*

The products are carbon dioxide (which forms the 'eyes'), and propionic and acetic acids which, with lactic acid, dominate the flavour of Emmental cheese in particular. Eye formation begins after the propionibacteria reach maximum population (from 25-40 days and onwards) while the texture of the cheese is sufficiently plastic to deform, rather than crack. So called "secondary fermentation" causes excessive CO_2 formation in older cheeses, and leads to such texture defects. It appears to be caused by late stimulation of propionibacteria by amino acids released by the proteolytic action of certain starter cultures or by undesirable contaminants (116, 118).

Production of volatile fatty acids by lipolysis is not significant in the best flavoured cheeses but higher fatty acids, produced by the starter and propionic-acid bacteria, may contribute to typical flavour. Lipolytic rancidity develops if *Clostridium tyrobutyricum* (derived from feed silage) is present in the cheese (118).

Protein breakdown is important in Swiss cheese not only for texture development but also for the production of small peptides and amino acids which apparently contribute 'nutty', 'brothy' and sweet tastes (120). Chymosin contributes to gross proteolysis as it does in other cheese varieties. Peptides and amino acids are formed through the further action of peptidases released from cells of the starter streptococci and lactobacilli. El Soda *et al.* (53) separated and identified an endopeptidase, dipeptidase, aminopeptidase and carboxypeptidase from *L.casei* and Desmazeaud and Juge (52) detected dipeptidase, aminopeptidase and proteinase activities in *Str.thermophilus*. Propionibacteria also produce intracellular peptidases which are released by autolysis into the cheese (121). Extracts of *Propionibacterium shermanii* contained 12 proteinase and 7 peptidase components, separable by electrophoresis, and earlier observations by Langsrud *et al.* (122) had suggested that proline-releasing peptidases predominate in this organism. Proline is reported to have a sweet taste and its presence in particularly high concentrations in Emmental cheese (123) led Langsrud and Reinbold (124) to suggest that it was the basis of this aspect of Swiss-cheese flavour. *Propionibacterium shermanii* may also contribute to sweet flavour by producing dimethyl sulphide (125, 126).

Although the characteristic sweet, nutty flavour of Emmental cheese is regarded as manifesting a complete, balanced state of maturation, mountain varieties of Gruyere cheese develop additional flavour due to growth of surface flora. This consists initially of lactate-utilizing yeasts whose growth raises the surface pH value towards 6 so that *Brevibacterium linens* can also grow as a secondary flora (127). These orange-pigmented coryneform bacteria are probably stimulated by vitamins and amino acids produced by the yeasts (cf. Limburg cheese flora; 101, 102). The role of the surface microflora in flavour development is not clearly understood but both yeasts and *B.linens* are proteolytic and the latter produces a range of peptidases (128) of which the aminopeptidase is the most thoroughly studied (129). The aroma and taste near the cheese surface is mildly putrid, partly due to formation of methanethiol by *B.linens* (77, 78, 130). Esterification by the concerted action of *B.linens* and micrococci (79) of methanethiol and acetic or propionic acids produces thio-esters which have a cheesy aroma. However, the same authors found many more volatile compounds in Gruyere cheese (esters, alcohols, carbonyls and hydrocarbons) whose origins and contributions to flavour are impossible to assess at present. The surface flora probably adds to the antimicrobial action of the lactic flora (e.g. *Str.thermophilus*; 131) by producing fatty acids which are inhibitory to *Clostridium botulinum* (132). This is a common surface contaminant and would normally be able to germinate and grow because the surface pH value is raised by yeast growth.

V. MICROBIOLOGICAL DEFECTS

The starter lactic acid bacteria suppress the growth of pathogens and most spoilage organisms in cheese. They do this chiefly by their acidifying effect with lactic acid, though other factors may also be involved; production of nisin and related antibiotics, hydrogen peroxide, low redox potential (133, 134). However, defects which either reduce the value of the cheese or render it inedible can arise from the use of poor quality milk, starter failure due to bacteriophage or antibiotics (previous chapter), poor choice of starters, the development of atypical secondary floras, or surface contamination with moulds.

A. Effect of raw milk flora

1. Thermoduric bacteria. Most cheese manufacturers heat treat their milk using high temperature short time pasteurization (HTST; approx. 70°C for 15-17 sec) which destroys most of the bacteria present in the raw milk. Some thermoduric lactobacilli and streptococci may remain but they are not generally regarded as major sources of defects in cheese made under good operating conditions. However, a recent study of cheese milk pasteurizers in the Netherlands (135) showed that plant which is not cleaned regularly may accumulate thermoduric milk streptococci (resembling *Str. thermophilus*) which are transferred downstream to the cheese vat. These organisms sometimes multiplied to $>10^{7}$ cfu ml^{-1} in Gouda cheeses and caused "unclean" and "yeasty" off-flavours as well as open texture defects.

Thermoduric catalase-positive cocci (micrococci and staphylococci) and microbacteria are common Cheddar cheese contaminants but have little influence on its flavour; they normally represent only a small proportion of its total bacterial population (10^{3} - 10^{4} cfu g^{-1}). Only when they are added experimentally in very high numbers (10^{8} g^{-1}) can they be associated with changes in cheese quality, usually negatively (136). Thermoduric sporeformers, other than the clostridia (mentioned in connection with Emmental cheese) do not appear to have any significant effect on cheese quality (Section V,C).

2. Psychrotrophic bacteria. This group is not directly significant in the spoilage of ripened cheeses but large populations of Gram-negative psychrotrophs can build up in stored, refrigerated bulk raw milk and cause defects by producing heat-resistant enzymes which carry over into the cheese after the bacteria themselves have been destroyed (137, 138). The main genera involved are *Pseudomonas, Aeromonas, Flavobacterium* and *Acinetobacter;* although their optimum temperatures for growth may be between 20 and 30°C, they can multiply significantly at 4°C and above. These bacteria contaminate the milk during milking from a variety of sources such as bedding, soiled teat surfaces, wash water (139). Most workers agree that they must be present in numbers $>5 \times 10^{6}$ ml^{-1} milk before manufacturing problems arise (137, 138). Proteinases and lipases are economically the most

significant groups of enzymes which may influence the yield (as weight of cheese solids recovered relative to milk solids) and grade of cheese. There are many reports of lipolytic flavour defects in Cheddar, Emmental and soft cheeses made with heat-treated milks which have supported the multiplication of lipase-producing psychrotrophs before pasteurization (110, 140, 141). Lipases of psychrotrophic bacilli and pseudomonads have also been implicated in contributing to yield reductions by hydrolysing milk fat and causing poor fat retention in curds (142). However, yield losses are usually associated with increased loss of non-protein N into whey, related to the action of proteinases on casein during raw milk storage. Most reports of significant yield reductions concern soft cheese manufacture where curds and whey are separated over a protracted period at temperatures conducive to enzymic action (142-145). Evidence for measurable effects on hard cheese yields is less convincing because conditions of study have been atypical (146, 147) and Law *et al.* (148) showed that even when psychrotroph populations in raw milk were high enough to cause casein breakdown which was detectable by gel electrophoresis, Cheddar cheese yield reduction could not be demonstrated.

Although there is some evidence for adverse effects on soft cheese yields by psychrotrophs, their proteolytic action appears to have a stimulatory effect on the rate of lactic acid production by starter cultures. The streptococci and lactobacilli used in starters are nutritionally fastidious and require some preformed amino acids for growth, although they can utilize small peptides as amino acid sources (149). Cousin & Marth (150) considered that the marked stimulation of acid production in *Streptococcus cremoris* and *Str. lactis*, in response to the addition of cultures or culture filtrates of *Pseudomonas* spp., was due to proteolytic activity on casein, providing the streptococci with higher concentrations of peptides and amino acids than would normally be nutritionally available in milk. Their results confirm earlier reports of starter culture stimulation in high-count milks and in milks containing culture filtrates of *Ps. fluorescens* (151-153). Collectively these observations may account for the shorter manufacturing times for Cheddar cheese made from stored market milks which had contained up to 10^6 cfu/ml of caseinolytic psychrotrophs (144, 154). The specific consequences of such effects for the quality of cheese are difficult to assess but in general terms any factor which unpredictably changes the rate of growth and acid production of a "familiar" starter is likely to lead to a lack of uniformity in curd texture and moisture content.

Quality control for cheese milk relies on total bacterial counts, which are assumed to represent the psychrotrophic counts in refrigerated milk that has been stored before use. Many new rapid methods for the enumeration of bacteria in raw milk are in use or under development (155), but there are at present no proven rapid methods available for the direct determination of potential spoilage enzymes in milk for cheesemaking. The difficulties inherent in devising such assays have been discussed at length by Law (137), though techniques for measuring proteinases and lipases are under development (156-158). Of these, the Hide Powder Azure (HPA) assay (156) is discussed here since its practical potential has been investigated and it could usefully be incorporated in future surveys of product quality in relation to milk quality. This method detects the presence of psychrotrophic proteinases in raw milk by the enzymic release of a soluble blue dye from an insoluble protein substrate (HPA). Relevant criteria of sensitivity for such a method are difficult to establish when little is known about the proteinases it must detect, but the assay will show the presence of proteolytic activity in milk before that activity is sufficient to measurably degrade the casein (159). Earlier work with stored milks has shown that only those in which casein has undergone degradation are likely to have high enough enzyme loads to cause manufacturing problems (148, 160). By inference therefore, the use of the HPA assay provides a safe basis for the acceptance or rejection of milks intended for processing to any stored product, dependent simply on a pass (no blue colour in the milk) or a failure (blue colour visible in the milk) after incubation with HPA under standardized conditions. Low numbers of enzyme-producing bacteria can only be ensured at source by observing high standards of milking hygiene (139), but some new methods designed to prevent bacterial growth in contaminated milk have recently been discussed (155).

B. Defects produced by starter cultures

Defects in ripened cheeses which are directly attributable to starter cultures only occur in hard and semi-hard varieties.

1. Bitterness. Bitter taste is caused by an accumulation of peptides with molecular weights ranging from around 1,000 to 12,000 which characteristically contain a high proportion of hydrophobic amino-acid residues (e.g. leucyl, prolyl, phenylalanyl; 161, 162). Richardson and Creamer (162) showed that bitter peptide in New Zealand Cheddar cheese originated from near the chymosin-sensitive bond of α_{s1}-casein, supporting the idea that peptides released by rennet were the substrate for starter endopeptidases. These hydrophobic bitter peptides tend to accumulate in cheese, probably because they are degraded only slowly by the peptidases of starter bacteria. Their rates of degradation are likely to be influenced by the peptidase activity in the starter cells, which is strain dependent (163) and by the rate of release of peptidases from lysing cells.

Opinions differ as to the importance of proteolysis by mesophilic starters in the production of bitter defects in cheese. Early hypotheses (e.g. reference 164) suggested that bitter peptides were produced by chymosin and that the so-called 'bitter' starters were those which had insufficient peptidase activity to break down the bitter peptides to non-bitter peptides and amino acids. However, the situation is more complex than this. While it is true that chymosin produces bitter peptides from casein, the starter proteinases can also do this and, indeed, can produce small bitter peptides from non-bitter, casein-derived peptides (Fig. 5). Lowrie and his co-workers (165, 166) suggested that this latter process is the single most important determinant in bitterness development and that starters which multiplied at relatively high cooking temperatures during Cheddar manufacture (the 'fast' starters) were the most likely to give bitter cheese, simply because the resultant high cell numbers contributed large quantities of bitter peptide-producing proteinases. This hypothesis was supported with experimental evidence showing that 'bitter' starters could be made to produce non-bitter cheese if their number in curds was restricted by controlled bacteriophage infections or by higher cooking temperatures. Conversely, the slow 'non-bitter' starters made bitter cheese if they were allowed to multiply to high cell numbers by altering the manufacturing process (166). Direct evidence for the involvement of starter cell-wall proteinases in the development of bitterness was provided recently by the observation that proteinase-deficient variants of 'fast' starters produce less bitterness in cheese than their parent strains, even when total starter cell populations are high (38).

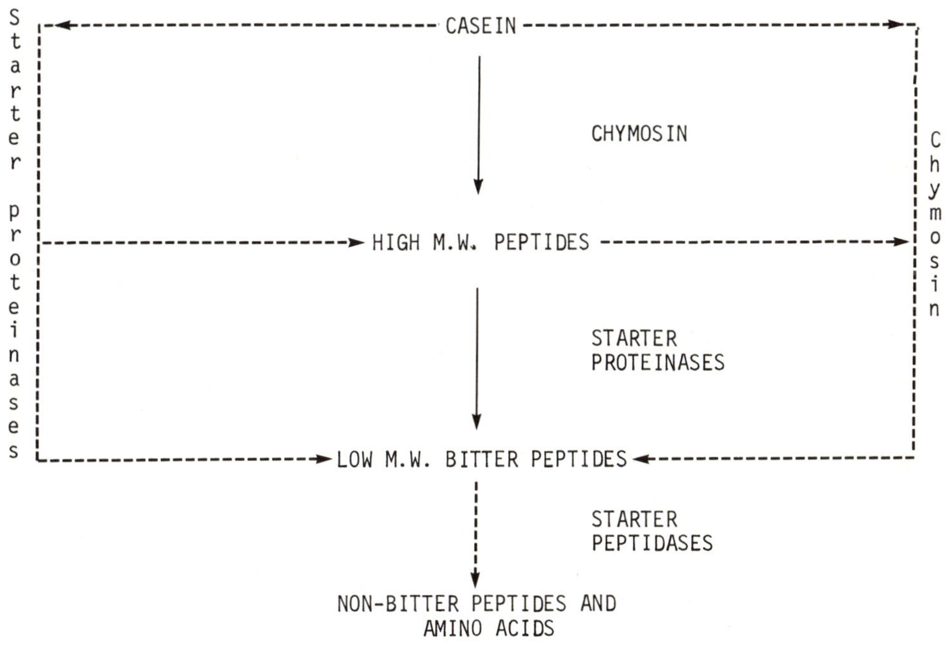

Fig. 5. Lowrie and Lawrence (165) model for bitterness development in Cheddar cheese. Bold arrows indicate the important stages (1 and 2). Broken lines indicate stages of lesser importance.

The factors controlling bitter defects in Gouda cheese appear to be more complex since the starters generally reach high populations in curd at the relatively low cooking temperatures used for this variety. Stadhouders and Hup (167) showed that factors influencing the retention of chymosin in Gouda curd (e.g. cooking temperature, initial milk pH) also influence the tendency of the cheese to become bitter. They emphasised that some starter strains produce more bitter peptide-degrading peptidases than others. It is not known whether these are specific peptidases confined to non-bitter strains or general peptidases present at different levels. Chiba and Sato (168) identified both di-peptidase and amino-peptidase activity in fractions of cell-free extracts from starter streptococci capable of reducing bitterness but individual enzymes were not isolated. It appears, then, that proteolysis by mesophilic starters is important in producing the bitter defect in cheese but its contribution depends on the cheese variety in question.

2. "Fermented" flavours. In addition to bitterness, other off-flavours are associated with the metabolism or enzymes of starters. For example, starters that utilize citrate or have induced NADH oxidase, and *Leuconostoc cremoris* strains with high alcohol dehydrogenase activity, are more likely to produce ethanol (among other alternatives to lactate) which, over long storage periods in cheese, reacts with butyric or hexanoic acids (lipolysis products) to form fruity-flavoured esters (169). Other starter strains (*Str.lactis* var *maltigenes*) can transaminate and decarboxylate leucine to 3-methylbutanal which is the basis of an unpleasant malt-like flavour defect in cheese (170). So-called "fermented flavour" is variously described as buttery, green or acetic acid-like and is caused by *Str.diacetylactis* strains which produce excessive quantities of carbonyls from lactose (171).

C. Defects attributable to secondary floras

The secondary bacterial flora of most hard and semi-hard cheeses is dominated by non-starter lactic acid bacteria (NSLAB; pediococci, lactobacilli) which usually grow in maturing cheese without causing obvious defects. However, if the flora becomes predominantly heterofermentative the residual lactose will be fermented by pathways which generate CO_2 (hexose monophosphate pathway). If large numbers are present (10^7 - 10^9 g^{-1} cheese), sufficient CO_2 can be formed to cause cracked, open-textured cheese and "blowing" of film-wrapped blocks. Heterofermentative lactobacilli belong to the species *brevis* and may be derived from poor quality raw milk as heat-resistant strains or from an insufficiently hygienic factory environment. *Lb.casei* can also cause CO_2 blowing in cheese by fermenting citrate, but if it is known to be a common component of the secondary flora, the cheese-maker can use citrate-fermenting starters so that the organism has little residual substrate to metabolise in the maturing cheese.

The role of pediococci in producing cheese defects is poorly documented but Fryer (172) suggested that cheese in which their growth is restricted has improved quality and lower incidence of fermented off-flavours.

Defects caused by other bacteria are rare in hard cheese, though coliforms can produce gassing and "faecal" flavours if they are allowed to contaminate the cheese vat or the curds. Late blowing by *Cl. tyrobutyricum* in Emmental has already been mentioned and it should be noted that a similar defect can occur in the washed-curd cheeses such as Edam and Gouda. The growth of this organism can be prevented by adding nitrate (0.02% w/v) to the cheese milk, but there is increasing speculation concerning the role of this compound and nitrite in the formation of carcinogenic nitrosamines in protein-containing foods under some manufacturing conditions, or after consumption (173-175). Devoyod (176) considered that Edam and Gouda cheese made with nitrate-treated milk was not a hazard but that the nitrate content of the whey should be controlled because of its own toxicity to infants at high concentrations and also because it may be involved in nitrosamine formation during whey processing in the presence of other food ingredients. For these reasons, alternatives to nitrate are under investigation, of which lysozyme has received the most attention. Both pure egg white lysozyme and fresh egg white itself are very effective inhibitors of late blowing bacteria in cheese (177) but dried egg white is ineffective (178). At present, therefore, there is no cheap, convenient method for using lysozyme (treatment with the pure enzyme would cost approximately £250 $tonne^{-1}$ cheese) but circumstances may arise in which its use may be favoured instead of nitrate.

Although the low moisture, high acid cheeses are particularly stable to internal spoilage, they are susceptible to undesirable surface mould and yeast growth which is traditionally controlled with potassium sorbate. More recently, the mould inhibitor natamycin (pimaricin) has been gaining acceptance as an alternative. It is a colourless, odourless complex tetraene, produced by *Streptomyces natalensis* and selective against yeasts and moulds. Depending on the method of application, pimaricin can be used at concentrations between 80% and 99% lower than sorbate to achieve similar mould control (179). Minimum inhibitory concentrations for pimaricin and sorbate are compared in Table 3. It is most commonly applied to hard cheese as an emulsion incorporated into the plastic coat or wrapping, though it can be included in brine vats or painted onto the cheese surface. In countries which do not yet permit the use of pimaricin, mould growth can be minimized by vacuum packaging, gas flushing and tight wrapping techniques (180).

Secondary flora defects in mould-ripened cheeses usually relate to poor manufacturing techniques or incorrect ripening conditions, leading in turn to poor growth of desirable moulds. This allows the colonization of the cheese by wild moulds, yeasts and possibly bacteria. Such defects, their causes and remedies are already well documented (1, 4, 5).

Table 3.
Activities of Pimaricin and Sorbate against food spoilage moulds and yeasts (surface grown)

	M.I.C. ($\mu g\ ml^{-1}$)	
	Pimaricin	Sorbate
Aspergillus niger	5	1000
Fusarium sp.	10	500
Mucor mucedo	5	500
Oidium lactis	10	500
Penicillium digitatum	5	500
Rhizopus sp.	5	500
Saccharomyces cerevisiae	1-5	500
Torulospora rosei	10	500
Zygosaccharomyces barkeri	1	500

Data from Klis (179)

VI. MYCOTOXINS

The moulds which grow in and on cheese, whether after deliberate addition or as a result of contamination, can produce a series of toxic substances collectively called mycotoxins. For example, *P.roqueforti* strains used in blue cheese manufacture can synthesize penicillic acid, roquefortine, isofumigaclavines A and B, PR toxin, mycophenolic acid and siderophores; *P.camemberti* (Camembert cheese) produces cyclopiazoic acid (181). Penicillic acid is a toxic antibiotic with possible, though not confirmed, carcinogenic properties in rodents (182) but is not thought to represent a human health hazard since it is unstable in cheese and cheese is not a good substrate for its production (183, 184). The roquefortines and isofumigaclavines A and B are toxic alkaloids which are commonly isolated from blue cheese (185) but their toxicity is low and on present evidence their production by *P.roqueforti* should not cause undue concern (186). However, further toxicological studies would clearly be desirable. Although many blue cheese strains of *P.roqueforti* produce PR toxin, it has never been detected in cheese itself, probably because it is unstable (187) and the conditions in cheese do not favour its production (188). Mycophenolic acid is a mildly toxic antibiotic with some history of causing chromosome damage in mammalian cells (189, 190). However, its presence in blue cheese can be avoided because it is relatively simple to screen for producer strains and avoid them (181). The siderophores are natural sequestering agents for iron and are found in blue cheese (191) but their health significance has not been assessed. They may interfere with iron absorption in the gut or be converted to carcinogenic alkyl hydroxylamine, but there is no experimental evidence of such phenomena at present. *P.camemberti* produces cyclopiazonic acid which is acutely toxic to rats (181) and has been isolated from the crust of Camembert cheese (192, 193). However, concentrations in cheese are such that consumers are unlikely to ingest enough to induce toxicity symptoms in the course of eating normal portions of the product (192).

Spoilage moulds on hard and semi-hard cheeses are chiefly species of *Penicillium* and some are toxigenic. In a recent review, Bullerman (194) cites several surveys of mouldy cheese from which were isolated *P.cyclopium, P.viridicatum, P.brevi-compactum, A.flavus, A.repens, A.versicola, Cladosporium* spp. and *Alternaria* spp. Between 1.8 and 12% isolates produced mycotoxins, identified as penicillic acid, patulin, ochratoxin A, citrinen and aflotoxins. However, the extent to which these contaminants represent real health hazards to consumers is not known; the potential hazard is probably small because low carbohydrate foods, especially cheese, are poor substrates for toxin production, even though the moulds may grow prolifically. Also, toxin production is usually low or non-existent below $10^{\circ}C$ so that refrigeration of cheese, especially after it has been cut, is an important preventive measure against toxigenesis.

Aflotoxins constitute a group of toxic and carcinogenic secondary metabolites produced by some strains of *Aspergillus flavus* and *A.parasiticus*. The major form is hydroxylated aflotoxin B_1, designated aflotoxin M_1 (AFM_1). It has been isolated from many different cheeses, though concentrations tend to be lower in soft cheese than in other types (195). Although contaminating *Aspergillus* moulds cannot be ruled out as a source of AFM_1, these organisms do not produce aflotoxin optimally under the conditions prevailing in cheese (196). Contaminated feed in which aspergilli have grown represents a more likely route of access, *via* the digestive tract of dairy cattle to their milk. AFM_1 is concentrated into curds during cheesemaking because it associates with casein and only a proportion of the toxin is destroyed by heat treatment (197). The chief concern over aflotoxins lies in their long-term carcinogenicity rather than their toxicity, but there are insufficient definitive data available to fully assess their public health significance. Based on experimental trial data, the U.S.A. Food and Drugs Administration regulations permit a maximum level of 0.5 ng ml^{-1} AFM_1 in milk. By comparison, Stolloff (198) found a variety of cheeses to contain the toxin at levels ranging from 0.02 - 1.3 ng g^{-1} during a year (1977) of high milk contamination in the U.S.A. He later concluded that in a normal year AFM_1 levels in dairy products should not exceed the equivalent of 0.1 ng ml^{-1} milk (199).

Methods for determining AFM_1 in milk are well documented, reflecting the concern for long-term surveys of its importance. The most common methods use thin-layer or high performance liquid chromatography (196).

VII OCCURRENCE AND BEHAVIOUR OF PATHOGENS

Raw milk may contain a variety of human pathogens ranging from *Mycobacterium tuberculosis* and *Brucella abortus* (where they have not been eradicated from cattle) to *Staph.aureus, E.coli, Streptococcus* spp. *Salmonella* spp. and *Campylobacter jejuni,* the second group being associated with bovine mastitis. Many of these are heat labile and efficient pasteurization kills them before milk is used for cheesemaking. However, varying degrees of post-pasteurization contamination of milk and curds are possible, mainly with staphylococci, streptococci, and coliforms; their possible presence emphasizes the need for active starter cultures with their excellent antibacterial properties which have already been discussed (Section V). The relative microbiological safety of cheese is reflected in the low incidence of disease associated with its consumption. A recent worldwide survey (200) showed that only 22 outbreaks connected with cow's milk cheese have been reported between 1951 and 1965 and only one between 1965 and 1972. The majority of cases involved *Staph.aureus* in Cheddar cheese, though Rumanian, Stilton, canned American and Cheshire cheeses have also caused disease *via* this organism. Other organisms isolated after cheese-induced illness include *Salmonella typhimurium, Strep.bovis - morbificans,* enterococci (group D) and Proteus. Salmonellae and *E.coli* usually decline in numbers over the first months of ripening in hard cheese, though they can persist in soft cheese (201, 202).

There is no universally-accepted microbiological standard for cheese, different countries or companies laying down their own requirements as they see fit. Specifications used by large retailers tend to be very stringent, allowing <100 yeasts and moulds g^{-1}, no coagulase-positive staphylococci in 10g and no coliforms in 0.1g. Such requirements are unrealistic, especially for soft cheese, and farmhouse cheese made with raw milk. Concern over enterotoxin-producing staphylococci, for example, appears to be exaggerated in the light of evidence that in normal hard cheese at least 2.8×10^7 organisms g^{-1} must be reached before enterotoxin production can be detected (203, 204). Even in cheese subjected to bacteriophage-induced starter failure, no toxin was detected with staphylococcal populations below 4×10^6 g^{-1}. Recent advances in the use of relatively simple staphylococcal thermonuclease assays to estimate growth and toxin levels of these organisms should improve screening of potentially harmful cheese (205).

Group D streptococci are potentially dangerous in cheese because they generate amines by decarboxylating free amino acids. Amines are significant in two separate ways; they can combine with nitrogen oxides (especially in nitrate-treated foods) to form carcinogenic nitrosamines (206, 207) or they can act directly by producing symptoms of hypo- and hypertension (208). Tyramine (from tyrosine), histamine (from histidine) and tryptamine (from tryptophan) are the food-related biogenic amines. The amounts of amines which can safely be ingested before the onset of toxic symptoms varies from person to person depending on their efficiency of detoxification. Patients using anti-depressant drugs based on mono-amine oxidase inhibitors are particularly at risk by eating cheese. Cases of apparent amine poisoning have usually involved cheeses containing in the order of 100mg for every 100g cheese (208)- normal cheeses contain between 1 and 30 mg per 100g. It should be noted that other amine-producing bacteria have been isolated from cheese, including *Propionibacterium* spp., streptococci of the *viridans* group, *Lb.bulgaricus* and *Lb.plantarium*. Cheese starter streptococci are not active amine producers. Cheese appears to be a good substrate for amine production but the low incidence of cheese-related poisoning suggests that good cheesemaking practice minimises contamination by producer organisms.

For a detailed analysis of the occurrence and behaviour of pathogens in cheese the reader is referred to Report Document 122 of the International Dairy Federation (209).

VIII ACCELERATED RIPENING

Hard and semi-hard cheese varieties represent a means by which the excellent nutritional properties of milk can be preserved in a safe, palatable form. While cheesemaking remained a domestic process or a farmhouse industry, the rate at which cheese acquired its typical flavour was relatively unimportant, compared with its shelf life as a food. Today, though, cheesemaking is a capital-intensive process which benefits from high turnover rates, and considerable sums of money are tied up in cheese in the cheese store. Also, in the UK it has become less easy to sell bland young cheese as consumers develop more sophisticated tastes resulting from their introduction to a wider range of cheese varieties from overseas. The need for accelerated ripening to produce mature cheeses in a short time arises from these considerations. As an example, the aim for Cheddar cheese would be a reduction from 6-9 months storage to 2-3 months.

A wide variety of methods have been proposed to achieve reduction in ripening times and these have been reviewed and assessed in a Bulletin commissioned by the IDF (210). They include temperature control, use of exogenous enzymes (chiefly proteinases and lipases), supplementation with attenuated or genetically modified starter cultures, alterations to cheesemaking technology, and very rapid fermentation of semi-liquid cheese slurries. The present discussion will be confined to those techniques which show the most immediate prospect of practical application and which have received the greatest attention from experimenters.

A. Temperature control

The technical simplicity and absence of legal barriers associated with this method recommend it to many manufacturers. It has long been known that cheese matures more quickly at higher temperatures but it has not been possible to consistently maintain flavour quality against off-flavours under these circumstances.

A general guide to the flavour enhancement attainable by high-temperature ripening of Cheddar cheese under very hygienic experimental conditions is shown by the data in Table 4. These time savings can be brought into perspective by noting that with interest rates based on an MLR of 12%, cheese costs about £25 tonne^{-1} month^{-1} to store.

TABLE 4

Accelerated Cheddar cheese ripening by elevated storage temperature

Normal temperature (^{o}C)	Experimental temperature (^{o}C)	Ripening period (weeks)	*Advancement (weeks)
6	12	8	8
6	13	24	24
6	18	8	12

* Based on extrapolation of data for low-temperature cheese to estimate time taken to reach a similar flavour intensity score as high-temperature cheese. Data source; Law *et al.* (230), Law & Wigmore (223).

In a recent study of New Zealand Cheddar cheese, Fryer (172) suggested that rapid ripening by temperature control could be achieved in commercial creameries, provided that standards of production hygiene were high (see ref. 211 for guidelines in this area). After extensive factory-scale trials a temperature regime has been evolved with an accompanying set of bacteriological criteria. The method depends on cheeses being pressed quickly and transferred to blast-cooled stores so that the cheese temperature falls rapidly to <10°C. If the cheeses are then held for 14 days, there is continuous slow growth of non-starter lactic acid bacteria (NSLAB) in such a way that they ferment residual lactose only to lactic acid and not to any of the alternative products which could produce off-flavours. The conditions should be such that the count of NSLAB is $<10^{3}$ g^{-1} at hooping and remains $<10^{6}$ g^{-1} after 14 days. Fryer (172) claims that after this initial period the balanced flavour of the cheese can be encouraged to develop more rapidly at a relatively high storage temperature, without fear of secondary bacterial metabolism causing flavour defects. The method has the advantage of simplicity, provided the manufacturer has the necessary equipment for rapid pressing, and should be applicable to any hard or semi-hard cheese whose flavour development depends on enzymic processes, as distinct from the metabolism of a viable secondary flora. However, the assumption is made that only NSLAB can grow and influence the quality of the cheese; this requires verification at elevated temperatures and it would be prudent to determine whether normally dormant microorganisms begin to multiply in warmer cheeses.

B. Enzyme addition

There have been numerous reports of accelerated cheese ripening involving enzyme addition to milk or curds (review; 212). In general, lipase addition is useful in strong-flavoured cheeses such as Italian hard cheese (213), Ras cheese (214), Domiati cheese (215) and blue-vein cheese (216). Blue cheese benefits from *Aspergillus* lipase directly and indirectly; the enzyme increases the liberation of fatty acids to enhance one flavour note, and concomitantly releases methyl ketone and δ-lactone precursors in the form of β-keto acids and δ-hydroxy acid respectively. There is no general agreement on the efficacy of microbial lipases in accelerating the formation of typical Cheddar cheese flavour; Sood and Kosikowski (217) claimed favourable results with low-level additions to American cheese but this has not been confirmed by other workers. Indeed, it is the author's experience that exogenous lipases either have no effect at all, or induce rancid off-flavours, depending on how much enzyme is used (unpublished data). However, proper data comparisons have been hampered by the lack of simple reliable methods for the determination of individual free fatty acids in maturing cheese.

Proteolysis is considered by most investigators to be so important to cheese maturation that much of the effort towards achieving accelerated cheese ripening has centred around methods to increase the rate of this process. Notable examples are the studies of Nakanishi & Itoh (218, 219), Kosikowski & Iwasaki (220) and Malkii *et al.* (221). The more recent results of Sood & Kosikowski (217) indicate that suitably-screened food grade proteinases can halve normal American Cheddar ripening times without off-flavour formation. Data from trials with English Cheddar cheese, in which food grade proteinases were added to curds are summarised in Table 5. Enzymes were added as powders using salts as a diluent; the proteinases must not be added to milk as they will produce soft curds, lead to large reduction in yields, and contaminate the whey (222, 223). The aspartyl acid proteinase of *Aspergillus oryzeae* always produced bitter cheese without significant enhancement of typical flavour, even when addition levels were varied over a range of 2 orders of magnitude. Bacterial neutral proteinase (*B.subtilis*) also produced bitter cheese at high levels of addition but an optimum amount has been determined (2×10^{-5} Anson units g^{-1} cheese) equivalent to 0.00125% (w/w) of Novo Neutrase (1.5S) or Sturge neutral proteinase, which significantly enhances the intensity of flavour in Cheddar cheese without producing flavour defects. There seems no reason why this type of enzyme should not be effective in any hard or semi-hard cheese whose maturation is partially proteolytic in nature. The *B.licheniformis* alkaline proteinase (subtilisin) produced very bitter cheese at 2×10^{-5} units g^{-1} and was not investigated further because of excessive cost (£15 $tonne^{-1}$ cheese); its commercial equivalent is Alcalase (Novo) but this is not at present available as a food grade powder. Pronase is a broad specificity proteinase from *Streptomyces griseus*. It produced strong flavours in cheese at

TABLE 5

Effect of commercial proteinases on flavour development in Cheddar cheeses after two months' maturation

Proteinase	Concentration (* units g^{-1})	Cheddar intensity (0-8 scale)	Bitter intensity (0-4 scale)	Other off-flavours (0-4 scale)
Acid	50	1.8ϯ	3.3	1.1
(A.oryzae)	10	2.6	2.8	0.7
	20	2.7	2.6	0.4
	0.4	2.1	2.2	0.3
	0.08	2.2	0.7	0.2
Neutral	50	3.1ϯ	1.6	0.9
(B.subtilis)	10	3.4ϯ	0.7	0.4
	2	3.2ϯ	0.1	0.2
	0.4	2.7	0.1	0.2
Alkaline *(B.licheniformis)*	2	2.9	3.5	0.0
Pronase *(Str.griseus)*	2	4.1ϯ	0.5	0.6
Untreated	-	2.4	0.1	0.2

* 1 unit = amount of enzyme required to bring about an increase in A_{595} of 0.5 in 15 min using the Hide Powder Azure (HPA) assay of Cliffe & Law (156) 1 HPA unit = 10^{-5} Anson units.

ϯ significantly different from untreated cheese at $P < 0.05$

(Data from ref. 223)

the 2 x 10^{-5} units g^{-1} but the cheese was bitter. It may have potential if used at lower levels of addition since this would bring the cost of application closer to that of Neutrase (approx. £0.12 $tonne^{-1}$ cheese).

All proteinase treatments resulted in soft-bodied, crumbly cheese, though this was least noticeable in Neutrase cheeses containing only 2 x 10^{-5} units g^{-1}. For example, a subjective panel assessment (based on the procedure of Green *et al.*, ref. 224) suggested that only one of 3 replicates treated with Neutrase was significantly more crumbly than untreated control cheese (222). The textural differences appeared to be due to excessive breakdown of β-casein by the exogenous proteinases.

Despite this progress in assessing and using enzymes many problems remain to be overcome before they are applied commercially. For example, the weak body of proteinase-treated cheese may be unacceptable to the retailer because it could lead to excessive losses during cutting and packaging. Enzyme preparation mixtures, with high ratios of exo- to endopeptidase activity may overcome this problem but none are available commercially at present.

The technical difficulties associated with the addition of enzyme preparations to cheese safely and efficiently require fundamental work for their solution; enzymes added to cheese milk are largely lost in the whey. Proteinases in particular cause weakening of the coagulum or curd, losses of soluble N in whey, and the whey itself is rendered unsuitable for many food applications unless the contaminating enzymes can be easily inactivated by heating. On the other hand, powdered enzymes added to cheese curds are difficult to distribute evenly and may present a health hazard to creamery personnel. Different cheese varieties may require different methods of addition matched to their unique manufacturing process. Research is needed on the suitability of various encapsulation methods (fats, proteins, phospholipids) using criteria such as degree of capture and retention, enzyme stability in capsules and mechanisms of release within treated cheese. Ultimately, the goal is to provide "packaged" enzyme systems for complete control of cheese ripening, even to the extent of changing flavour characteristics and decreasing the levels of flavour in older cheese, as the market demands. Reactions which are not normally possible in cheese may be achieved in encapsulated enzyme/substrate systems operating under optimal conditions. Magee & Olson (225, 226) described the encapsulation of carbohydrate-metabolising enzymes and their substrates in milk fat. Their system has application for low-cook soft cheese since the enzymes are released when the fat melts (>33°C) but most hard and semi-hard cheeses require a higher manufacturing temperature. Hardened fat may give the opposite problem that the enzymes would be difficult to release into the cheese, a necessary step in "delivering" enzymes in such a way that they can act on their substrates. Phospholipid liposomes or geletin capsules may be more appropriate since these compounds can be degraded in cheese to allow enzyme release.

C. Attenuated starters

Although evidence for the direct involvement of starter bacteria in the formation of flavour compounds in any cheese type is incomplete, it remains the widely-held view that in most hard and semi-hard varieties they contribute to flavour development at least by limited lipolysis and by their ability to degrade large chymosin-derived peptides to small peptides and amino acids as flavour precursors. It follows therefore that higher numbers of starter cells in cheese may accelerate flavour development and several reports support this conclusion (227-229). The basic requirement is for a starter culture whose acid-producing activity has been impaired sufficiently to prevent it interfering with normal cheesemaking (e.g. lac^{-} mutants, heat shock, lysozyme, solvent treatment) but whose cheese-ripening enzymes remain active, or are enhanced. Three important advantages can be gained from using modified starters. First, they contain a natural balance of ripening enzymes which is less likely to produce flavour defects in comparison with exogenous enzyme preparations. Second, where legal barriers to enzyme addition in traditional varieties exist, they may be overcome by adding extra enzymes in the form of starter cells. Third, the cells, containing their enzymes, are entrapped intimately within the curd matrix in the same way as normal starter cells. Their major potential disadvantage is considered to be the high cost of producing and treating sufficient starter culture to increase the total numbers of starter cells significantly above those normally present in cheese curd; for example the cells from a 200 litre constant-pH culture would be needed to double the starter population of 1 tonne of cheese. Perhaps the application of modern genetic techniques will provide ways of enhancing specific desirable enzyme activities in starter cultures; strains which "over-produce" peptides for example, or lipolytic starters for Italian hard cheese.

REFERENCES

1. H.R. Chapman and M.E. Sharpe, in R.K. Robinson (Ed.), Dairy Microbiology, Vol. 2, The Microbiology of Dairy Products, Applied Science, London, 1981, Ch. 5, p.158.
2. B.A. Law, in A.H. Rose (Ed.), Economic Microbiology, Vol. 7, Fermented Foods, Academic Press, London, 1982, Ch. 5, p.148.
3. G. Kleter, Neth. Milk Dairy J., 30 (1976) 254-270.
4. J.G. Davis, Cheese, Vol. 3, Churchill Livingstone, London, 1976, pp.487-920.
5. F.V. Kosikowski, Cheese and Fermented Milk Foods, 3rd ed., F.V. Kosikowski and Associates, N.Y., 1982.
6. D.G. Dalgleish, in P.F. Fox and J.J. Condon (Eds.), Food Proteins, Applied Science, London, 1982, Ch. 8, pp.155-178.
7. K. Aunstrup, in A.H. Rose (Ed.), Economic Microbiology, Vol. 5, Microbial Enzymes and Bioconversions, Academic Press, London, 1980, Ch. 2, p.50.
8. M.L. Green, J. Dairy Res., 44 (1977) 159-188.
9. P.J. de Konings, Int. Dairy Fed. Bull., 126 (1980) 11-23.
10. R. Martens and M. Naudts, Int. Dairy Fed. Bull., 91 (1976) 1-9.
11. R. Martens and M. Naudts, Int. Dairy Fed. Bull., 108 (1978) 51-56.
12. J. Phelan, Dairy Inds. Int., 42 (1977) 50-54.
13. J.P. Ramet and F. Weber, Le Lait, 61 (1981) 381-392.
14. J.P. Ramet and F. Weber, Le Lait, 61 (1981) 458-464.
15. Th. E. Galesloot, Neth. Milk Dairy J., 14 (1950) 176-214.
16. B.A. Law, M.J. Castanon and M.E. Sharpe, J. Dairy Res. 43 (1976) 301-311.
17. D.J. Manning and J.C. Price, in Brief Communications, XXI Int. Dairy Congr., Moscow, Vol. 1, Book 1, 1982, pp.507-508.
18. T.M. Cogan, Dairy Inds. Int. 41 (1976) 12-16.
19. T.D. Thomas, Appl. Env. Microbiol. 32 (1976) 474-478.
20. T.D. Thomas and K.W. Turner, Appl. Env. Microbiol. 41 (1981) 1289-1294.
21. J.C. Bruhn and E.B. Collins, J. Dairy Sci. 53 (1970) 857-860.
22. R.F. Anders, D.M. Hogg and G.R. Jago, Appl. Microbiol. 19 (1970) 608-612.
23. J. Czulak, L.A. Hammond and J.F. Horwood, Austr. J. Dairy Technol. 29 (1974) 124-128.
24. D.J. Manning and H.M. Robinson, J. Dairy Res. 40 (1973) 63-75.
25. S.K. Rhee and M.Y. Pack, J. Bacteriol., 144 (1980) 217-221.
26. W. Tinson, A.J. Hillier and G.R. Jago, Austr. J. Dairy Technol. 37 (1982) 22-25.
27. L.K. Creamer and N.F. Olson, J. Food Sci. 47 (1982) 631-646.
28. M.J. Desmazeaud and J.-C. Gripon, Milchwissenschaft. 32 (1977) 731-737.
29. F.M.W. Visser, Neth. Milk Dairy J., 31 (1977) 210-239.
30. R.B. O'Keefe, P.F. Fox and C. Daly, J. Dairy Res., 43 (1976) 97-107.
31. F.M.W. Visser and A.E.A.de Groot-Mostert, Neth. Milk Dairy J., 31 (1977) 247-264.
32. R.D. Hill, E. Lehar and D. Givol, J. Dairy Res., 41 (1974) 147-153.
33. J.-C. Gripon, M.J. Desmazeaud, D. Le Bars and J.-L. Bergère, Le Lait, 55 (1975) 502-516.
34. L.K. Creamer and B.C. Richardson, New Zealand J. Dairy Sci. Technol. 9 (1974) 9-13.

35. M.J. Desmazeaud and C. Zevaco, Ann. Biol. anim. Bioch. Biophys., 16 (1976) 851-863.
36. C. Zevaco and M.J. Desmazeaud, J. Dairy Sci., 63 (1980) 15-24.
37. B. Reiter, Y. Sorokin, A. Pickering and A.J. Hall, J. Dairy Res., 36 (1969) 65-76.
38. O.E. Mills and T.D. Thomas, New Zealand J. Dairy Sci. Technol. 15 (1980) 131-141.
39. B.A. Law, M.E. Sharpe and B. Reiter, J. Dairy Res., 41 (1974) 137-146.
40. T. Sørhaug and P. Solberg, Appl. Microbiol. 25 (1973) 388-395.
41. M.J. Desmazeaud and L. Vassal, Le Lait, 59 (1979) 327-344.
42. T. Sørhaug and J. Kolstad, Neth. Milk Dairy J., 35 (1981) 338-343.
43. K. Yamauchi, S. Kaminogawa and I.K. Hwang, Vth Int. Congr. Food Sci. Technol. Abstr. (1978), p.173.
44. I.K. Hwang, S. Kaminogawa and K. Yamauchi, Agric. Biol. Chem., 45 (1981) 159-165.
45. M.J. Desmazeaud and C. Zevaco, Ann. Biol. anim. Bioch. Biophys., 17 (1977) 723-736.
46. L. Mou, J.J. Sullivan and G.R. Jago, J. Dairy Res., 42 (1975) 147-155.
47. A.J. Cliffe and B.A. Law, J. Appl. Bacteriol. 47 (1979) 65-73.
48. C. Tourneur, Le Lait, 52 (1972) 149-174.
49. T.C. Gehrig, R.L. Richter and C.W. Dill, Texas Agric. Exp. Stu. Prog. Rep., PR3733 (1980) 57-63.
50. P.J. Argyle, G.E. Mathison, and R.C. Chandan, J. Appl. Bacteriol. 41 (1976) 175-184.
51. R.C. Chandan, P.J. Argyle and G.E. Mathison, J. Dairy Sci. 65 (1982) 1408-1413.
52. M.J. Desmazeaud and M. Juge, Le Lait, 56 (1976) 241-260.
53. M. El-Soda, M.J. Desmazeaud and J.-L. Bergère, J. Dairy Res., 45 (1978) 445-455.
54. B. Eggimann and B. Bachmann, Appl. Env. Microbiol., 40 (1980) 876-882.
55. M. El-Soda and M.J. Desmazeaud, Agric. Biol. Chem., 45 (1981) 1693-1700.
56. M. El-Soda, M.J. Desmazeaud and N. Zeyada, Neth. Milk Dairy J., 35 (1981) 349-352.
57. R.C. Lawrence, T.F. Fryer and B. Reiter, Nature, 213 (1967) 1264.
58. J. Stadhouders and H.A. Veringa, Neth. Milk Dairy J., 27 (1973) 77-91.
59. B. Reiter, T.F. Fryer, A. Pickering, H.R. Chapman, R.C. Lawrence and M.E. Sharpe, J. Dairy Res., 34 (1967) 257-272.
60. T. Nakae and y.A. Elliot, J. Dairy Sci., 48 (1965) 287-292.
61. J.R. Dulley and P.A. Grieve, Austr. J. Dairy Technol., 29 (1974) 120-123.
62. P.V. Paulsen, J. Kowalewska, E.G. Hammond and B.A. Glatz, J. Dairy Sci., 63 (1980) 912-918.
63. J.L. Schmidt and J. Lenoir, Le Lait, 58 (1980) 272-282.
64. J.L. Schmidt and J. Lenoir, Le Lait, 58 (1980) 343-350.
65. J. Auclair, J.-P. Accolas, L. Vassal and G. Moquot, Int. Dairy Fed. Bull., 136 (1981) 20-26.
66. J.-C. Gripon and B. Debest, Le Lait, 56 (1976) 423-438.
67. J. Lenoir, B. Auberger and J.-C. Gripon, Le Lait, 59 (1979) 244-268.
68. P. Trieu-Cuot, M.-J. Archieri-Haze and J.-C. Gripon, J. Dairy Res., 49 (1982) 487-500.
69. P. Trieu-Cuot, M.-J. Archieri-Haze and J.-C. Gripon, Le Lait, 62 (1982) 234-249.
70. P. Trieu-Cuot and J.-C. Gripon, J. Dairy Res., 49 (1982) 501-510.
71. J. Lenoir and B. Auberger, in Brief Communications, XXI Int. Dairy Congr., Moscow, Vol. 1 Book 1, 1982, p.336.
72. B. Auberger, M. Montals and J. Lenoir, in Brief Communications, XXI Int. Dairy Congr., Moscow, Vol. 1, Book 1, 1982, p.276.
73. J.L. Schmidt, in Brief Communications, XXI Int. Dairy Congr., Moscow, Vol. 1, Book 1, 1982, p.365.
74. D. Hemme, C. Bouillanne, F. Métro and M.J. Desmazeaud, Sci. Aliment., 2 (1982) 113-123.
75. C. Bouillanne, F. Métro, D. Hemme and M.J. Desmazeaud, in Brief Communications, XXI Int. Dairy Congr., Moscow, Vol. 1, Book 1, 1982, p.286.

76. T. Tsugo and H. Matsuoko, in XVI Int. Dairy Congr., Copenhagen, B, 1962, pp.385-394.
77. M.E. Sharpe, B.A. Law, B.A. Phillips and D.G. Pitcher, J. Gen. Microbiol., 101 (1977) 345-349.
78. J.-P. Dumont, G. Pradel, S. Roger and J. Adda, Le Lait, 56 (1976) 18-38.
79. A. Cuer, G. Dauphin, A. Kergomard, J.-P. Dumont and J. Adda, Agric. Biol. Chem., 43 (1979) 1783-1784.
80. J.-P. Dumont and J. Adda, in D.G. Land and H.E. Nursten (Eds.), Progress in Flavour Research, Applied Science, London, 1979, Ch. 21, p.255.
81. G. Lamberet, B. Auberger, C. Canteri and J. Lenoir, Rev. Lait Francaise, 406 (1982) 13-19.
82. J.-P. Dumont, S. Roger, C.Cerf and J. Adda, Le Lait, 54 (1974) 501-516.
83. C. Moreau, Le Lait, 60 (1980) 254-271.
84. J.E. Kinsella and D. Hwang, Biotechnol. Bioeng., 18 (1976) 927-938.
85. M. Godinho and P.F. Fox, Milchwissenschaft, 36 (1981) 205-208.
86. M. Godinho and P.F. Fox, Milchwissenschaft, 36 (1981) 329-333.
87. M.J. Desmazeaud, J.-C. Gripon, D. Le Bars and J.-L. Bergère, Le Lait, 56 (1976) 379-396.
88. J.-C. Gripon, M.J. Desmazeaud, D. Le Bars and J.-L. Bergère, J. Dairy Sci., 60 (1977) 1532.
89. D. Le Bars and J.-C. Gripon, J. Dairy Res., 48 (1981) 479-487.
90. J. Paquet and J.-C. Gripon, Milchwissenschaft, 35 (1980) 72-74.
91. M. Godinho and P.F. Fox, Milchwissenschaft, 37 (1982) 72-75.
92. J.J. Devoyod and G. Bret, in XXVII Int. Dairy Congr., Munich, D, 1966, p.585.
93. M. Galzin, P. Galzy and G. Bret, Le Lait, 50 (1970) 1-37.
94. J.J. Devoyod, Le Lait, 49 (1969) 20-39.
95. A. Monassa and G. Lamberet, in Brief Communications, XXI Int. Dairy Congr., Moscow, Vol. 1, Book 1, 1982, p.509.
96. M. Godinho and P.F. Fox, Milchwissenschaft, 36 (1981) 457-520.
97. N.P. Wong, R. Ellis, D.E. La Croix and J.A. Alford, J. Dairy Sci., 56 (1973) 636.
98. R.C. Jolly and F.V. Kosikowski, J. Agric. Food Chem., 23 (1975) 1175-1176.
99. J. Boldingh and R.J. Taylor, Nature, 194 (1962) 909-913.
100. R. Todesco, R. Lodi, G. Mucchetti and S. Carini, Latte, 6 (1981) 741-751.
101. M. Purko, W.O. Nelson and W.A. Wood, J. Dairy Sci., 34 (1951) 699-705.
102. M. Purko, W.O. Nelson and W.A. Wood, J. Dairy Sci., 34 (1951) 874-878.
103. S. Bouman, Zuivelzicht, 69 (1977) 1130.
104. F. Emch, P. Munch and U. Glunk, Schweiz. Milchwirt. Gorschung, 7 (1978) 55.
105. B.A. Law, M.E. Sharper, L.A. Mabbitt and C.B. Cole, in R.G. Board and D.W. Lovelock (Eds.), Scoiety for Applied Bacteriology Technical Series, Vol. 7, Sampling - Microbiological Monitoring of Environments, Academic Press, London, 1973, Ch. 1, pp.1-9.
106. B.A. Law, M.J. Castanon and M.E. Sharpe, J. Dairy Res., 43 (1976) 117-125.
107. Y. Umemoto and Y. Sato, Agric. Biol. Chem., 39 (1975) 2115-2122.
108. R.C. Lawrence, New Zealand J. Dairy Sci. Technol., 2 (1967) 55-57.
109. D.J. Manning and J.C. Price, J. Dairy Res., 44 (1977) 357-361.
110. B.A. Law, M.E. Sharper and H.R. Chapman, J. Dairy Res., 43 (1976) 459-468.
111. M.E. Sharpe and J.G. Franklin, in XVI Int. Dairy Congr., Copenhagen, B, 1962, p.46.
112. B.A. Law and M.E. Sharpe, J. Dairy Res., 45 (1978) 267-275.
113. B.A. Law and M.E. Sharpe, Dairy Inds. Int., 45 (1980) 5, 15, 17, 19-20, 22.
114. J.P. Accolas, D. Hemme, M.J. Desmazeaud, L. Vassal, C. Buillanne and M. Veaux, Le Lait, 60 (1980) 487-524.
115. J.P. Accolas, M. Veaux, L. Vassal and G. Moquot, Le Lait, 58 (1978) 118-132.
116. G. Moquot, J. Dairy Res., 46 (1979) 113-160.
117. D.H. Hettinga and G.W. Reinbold, J. Milk Food Technol. 35 (1972) 358-372.
118. C. Steffen, Milk Ind., 81 (1979) 24-27.
119. T. Langsrud and G.W. Reinbold, J. Milk Food Technol., 37 (1974) 26-41.
120. S.L. Biede and E.G. Hammond, J. Dairy Sci., 62 (1979) 227-237.
121. T. Langsrud, G.W. Reinbold and E.G. Hammond, J. Dairy Sci., 61 (1978) 303-308.

122. T. Langsrud, G.W. Reinbold and E.G. Hammond, J. Dairy Sci., 60 (1977) 16-23.
123. A.I. Virtanen and M. Kreula, Meijeri. Aikakaus., 10 (1948) 13-23.
124. T. Langrud and G.W. Reinbold, J. Milk Food Technol., 36 (1973) 593-609.
125. T.W. Keenan and D.D. Bills, J. Dairy Sci., 51 (1968) 797-799.
126. G.J. Dykstra, D.L. Drerup, A.L. Branen and T.W. Keenan, J. Dairy Sci., 54 (1971) 168-172.
127. J.P. Accolas, D. Melcion and L. Vassal, in Brief Communications, XX Int. Dairy Congr., Paris, E, 1968, p.762.
128. H. Torgersen and T. Sørhaug, in Brief Communications, XX Int. Dairy Congr., Paris, E, 1978, pp.478-479.
129. H. Foissy, Milchwissenschaft, -3 (1978) 221-223.
130. J.-P. Dumont and J. Adda, J. Agric. Food Chem., 26 (1978) 364-367.
131. S.R. Pulsani, D.R. Rao and G.R. Sunki, F. Food Sci., 44 (1979) 575-578.
132. N. Grecz, R.O. Wagenaar and G.M. Dack, Appl. Microbiol., 7 (1979) 228.
133. F.J. Babel, J. Dairy Sci., 60 (1977) 815-821.
134. G.F. Ibrahim, A.K. Baldock, D.R. Radford and L.B. Ireland, J. Food Protect. 44 (1981) 263-267.
135. F.M. Driessen and S. Bouman, Nord. Mejerit., 47 (1981) 39-44.
136. A.J. Gillies, Austr. J. Dairy Technol., 26 (1971) 145-149.
137. B.A. Law, J. Dairy Res., 46 (1979) 573-588.
138. M.A. Cousin, J. Food Protect., 45 (1982) 172-207.
139. C.M. Cousins and A.J. Bramley, in R.K. Robinson (Ed.) Dairy Microbiology Vol. 1, The Microbiology of Milk, Applied Science London, 1981, Ch. 4, pp.119-163.
140. J.P. Dumont, G. Delespaul, G. Mignot and J. Adda, Le Lait, 57 (1977) 619-630.
141. R. Mourques, J.P. Accolas and J. Auclair, Revue Lait. Fr., 243 (1967) 289-293.
142. C.L. Hicks, M. Allauddin, B.E. Langlois and J. O'Leary, J. Food Protect., 45 (1982) 331-334.
143. M. Feuillat, S. Le Guennec and A. Olsson, Le Lait, 56 (1976) 521-536.
144. M.A. Cousin and E.H. Marth, Cultured Dairy Products J., 12 (1977) 15-18, 30.
145. F.O. Mohamed and R. Bassetti, J. Dairy Sci., 62 (1979) 222-226.
146. T.A. McCaskey and F.J. Babel, J. Dairy Sci., 49 (1966) 697.
147. A.R. Yates and J.A. Elliot, Can. Inst. Food Sci. Technol. J., 10 (1977) 269-271.
148. B.A. Law, A.T. Andrews, A.J. Cliffe, M.E. Sharpe and H.R. Chapman, J. Dairy Res., 46 (1979) 497-509.
149. B.A. Law, J. Gen. Microbiol., 105 (1978) 113-118.
150. M.A. Cousin and E.H. Marth, J. Food Protect., 40 (1977) 406-410.
151. T.J. Claydon and H.C. Fryer, Appl. Microbiol., 8 (1960) 278-281.
152. T.J. Claydon and J.A. Koburger, Appl. Microbiol., 9 (1961) 117-121.
153. J.A. Koburger and T.J. Claydon, J. Dairy Sci., 44 (1961) 1811-1817.
154. M.A. Cousin and E.H. Marth, J. Dairy Sci., 60 (1977) 1048-1056.
155. B.A. Law and L.A. Mabbitt, in T.A. Roberts and F.A. Skinner (Eds.), S.A.B. Symposium Series, Vol. 11, Food Microbiology: Advances and Prospects, pp.131-150.
156. A.J. Cliffe and B.A. Law, J. Dairy Res., 49 (1982) 209-219.
157. A.T. Andrews and M.D. Taylor, FEBS Letters, 141 (1982) 207-209.
158. D. Stead. J. Dairy Res., 50 (1983) in press
159. D. Kalogridou-Vassiliadou and B.A. Law, J. Dairy Res., 49 (1982) 511-514.
160. B.A. Law, A.T. Andrews and M.E. Sharpe, J. Dairy Res., 44 (1977) 145-148.
161. V.R. Harwalkar, J. Dairy Sci., 55 (1972) 735-741.
162. B.C. Richardson and L.K. Creamer, New Zealand J. Dairy Sci. Tehcnol. 8 (1973) 46-51.
163. J.J. Sullivan, L. Mou, J.I. Rood and G.R. Jago, Austr. J. Dairy Technol. 28 (1973) 20-26.
164. J. Czulak, Austr. J. Dairy Technol. 14 (1959) 177-179.
165. R.J. Lowrie and R.C. Lawrence, New Zealand J. Dairy Sci. Technol. 7 (1972) 51-53

166. R.J. Lowrie, R.C. Lawrence and M.F. Peberdy, New Zealand J. Dairy Sci. Technol., 9 (1974) 116-121.
167. J. Stadhouders and G. Hup, Neth. Milk Dairy J., 29 (1975) 335-353.
168. Y. Shiba and Y. Sato, Jap. J. Dairy Fd. Sci., 29 (1980) 161-167.
169. D.D. Bills, M.E. Morgan, L.M. Libbey and E.A. Day, J. Dairy Sci., 48 (1965) 1168-1173.
170. R.M. Sheldon, R.C. Lindsay, L.M. Libbey and M.E. Morgan, Appl. Microbiol., 22 (1971) 263-266.
171. W.E. Sandine, C. Daly, P.R. Elliker and E.R. Vedamuthu, J. Dairy Sci., 55 (1972) 1031-1039.
172. T.F. Fryer, in Brief Communications, XXI Int. Dairy Congr., Moscow, Vol. 1, Book 1, 1031-1039.
173. A. Oterholm, Meieriteknikk, 6 (1980) 33-39.
174. M. Terral, Ann. Nutr. Aliment., 34 (1980) 865-874.
175. A. de Vuyst, W. Vervack and M. Foulou, Le Lait, 56 (1976) 414-422.
176. J.J. Devoyod, Annal. Nutr. Aliment., 30 (1976) 789-792.
177. S. Carini and R. Lodi, Ind. Latte, 18 (1982) 35-48.
178. F. Wasserfall, E. Voss and D. Prokopek, Keiler Milchwirt. Forschung. 28 (1976) 3-16.
179. W.S. Klis, Food Technol. 13 (1960) 124-132.
180. R.C. Chandan, Dairy Rec. 83 (1982) 141-144.
181. P.M. Scott, J. Food Protect. 44 (1981) 702-710.
182. P.M. Scott, in T.D. Wyllie and L.G. Morehouse (Eds.), Mycotoxic fungi, mycotoxins, mycotoxicosis., Marcel Dekker, New York, 1977 pp.283-356.
183. F.Y. Lien and L.B. Bullerman, J. Food Sci., 42 (1977) 1222-1224, 1228.
184. W.T. Stott and L.B. Bullerman, J. Food Sci., 41 (1976) 201-203.
185. P.M. Scott and B.P.C. Kennedy, J. Agric. Food Chem., 24 (1976) 856-868.
186. C. Moreau, Le Lait, 60 (1980) 254-271.
187. P.M. Scott and S.R. Kanhere, J. Assoc. Off. Anal. Chem., 62 (1979) 141-147.
188. J. Guirard, M.-T. Piva, J. Crouzet and P. Galzy, Ann. Nutr. Aliment., 31 (1977) 551-555.
189. R. Marinari, R. Fleischmajer, A.H. Schragger and A.L. Rosenthal, Arch. Dermatol., 113 (1977) 930-932.
190. M.T. Umeda, T. Tsutsui and M. Saito, Gann, 68 (1977) 619-625.
191. S.A. Ong and J.B. Neilands, J. Agric. Food Chem., 27 (1979) 990-995.
192. J. Le Bars, Appl. Environ. Microbiol., 38 (1979) 1052-1055.
193. P. Still, C. Eckardt and L. Leinster, Fleischwirtschaft, 58 (1978) 876-877.
194. L.B. Bullerman, J. Dairy Sci., 64 (1981) 2439-2452.
195. K. Polzhofer, Z. Lebensm, Unters. Forsch., 163 (1977) 175-179.
196. R.S. Applebaum, R.E. Brackett, D.W. Wiseman and E.H. Marth, J. Food Protect., 45 (1982) 752-777.
197. F. Kiermeier and M. Buchner, Z. Lebensm, Unters. Forsch., 164 (1977) 87-91.
198. L. Stoloff, J. Food Protect., 43 (1980) 226-230.
199. L. Stoloff, G. Wood and L. Carter, Jr., J. Dairy Sci., 64 (1981) 2426-2430.
200. N.S. Galbraith, P. Forbes and C. Clifford, Brit. Med. J., 284 (1982) 1761-1765.
201. T.W. Dommett, Austr. J. Dairy Technol., 29 (1974) 198-203.
202. R. Mourques, L. Vassal, J. Auclair, G. Moquot and J. Vandeweghe, Le Lait, 57 (1977) 131-149.
203. S.R. Tatini, J.J. Jezeski, J.C. Olsson and E.P. Casman, J. Dairy Sci., 54 (1971) 312-320.
204. S.R. Tatini, W.D. Wesala, J.J. Jezeski and H.A. Morris, J. Dairy Sci., 56 (1973) 429-435
205. G.I. Ibrahim and A.K. Baldock, J. Food Protect., 44 (1981) 655-660.
206. M. Castegnaro and E.A. Walker, in E.A. Walker (Ed.), N-Nitroso-Compounds: Analysis, Formation and Occurrence, IARC Scientific Publications, Lyon, 1980, p.445.
207. E. Pedersen, in E.A. Walker (Ed.), N-Nitroso-Compounds: Analysis, Formation and Occurrence, IARC Scientific Publications, Lyon, 1980, p.493.
208. S.T.Edwards and W.E. Sandine, J. Dairy Sci., 64 (1981) 2431-2438.
209. Int. Dairy Fed. Bull., 122 (1980) 4-23.
210. B.A. Law, Int. Dairy Fed. Bull., 108 (1978) 40-48.

211. H. Luck, in R.K. Robinson (Ed.), Dairy Microbiology, Vol. 2, The Microbiology of Milk Products, Applied Science, London, 1981, Ch. 7, pp.279-324.
212. B.A. Law, Dairy Inds. Int., 45 (1980) 15, 17, 19, 20, 22, 48.
213. R.V. Chaudhari and G.H. Richardson, J. Dairy Sci., 54 (1971) 467-471.
214. M.H.A. El-Salam, S. El-Shibiny, E. El-Bagoury, E. Ayad and N. Fahmy, J. Dairy Res., 45 (1978) 491-495.
215. A.A. Neshawy, A.A. Abdel-Baky and S.M. Faharat, Dairy Inds. Int., 47 (1982) 29-31.
216. R.C. Jolly and F.V. Kosikowski, J. Agric. Food Chem., 23 (1975) 1175-1176.
217. V.K. Sood and F.V. Kosikowski, J. Dairy Sci., 62 (1979) 1865-1872.
218. T. Nakanishi and M. Itoh, Jap. J. Dairy Sco., 22 (1973) A110.
219. T. Nakanishi and M. Itoh, Jap. J. Dairy Sci., 23 (1974) A121.
220. F.V. Kosikowski and T. Iwakaki, J. Dairy Sci., 58 (1975) 963-970.
221. Y. Malkki, L. Rouhianeu, R. Mattson and P. Markkonen, United States Pat., 1977, 4,062,730.
222. B.A. Law and A.S. Wigmore, J. Dairy Res., 49 (1982) 137-146.
223. B.A. Law and A.S. Wigmore, J. Soc. Dairy Technol., 35 (1982) 75-76.
224. M.L. Green, A. Turvey and D.G. Hobbs, J. Dairy Res., 48 (1981) 343-355.
225. E.L. Magee and N.F. Olson, J. Dairy Sci., 64 (1981) 600-615.
226. E.L. Magee and N.F. Olson, J. Dairy Sci., 64 (1981) 616-621.
227. H.E. Petersson and G. Sjostrom, J. Dairy Res., 42, 313-326,
228. M.P. Thompson, G.A. Somkuti, J.F. Flanagan, M. Bencivengo, D.P. Brower and D.H. Steinberg, J. Dairy Sci., 62, Suppl. 1 (1979) 68.
229. J.R. Dulley, D.E.J. Brooks and P. Grieve, in Brief Communications, XX Int. Dairy Congr., Paris, E, 1978, p.468.
230. B.A. Law, Z.D. Hosking and H.R. Chapman, J. Soc. Dairy Technol., 32 (1979) 87-90.

APPLICATION OF BIOTECHNOLOGY TO THE ANIMAL HEALTH INDUSTRY

William J. Marsheck, Jr.
Diamond Shamrock Corporation
T. R. Evans Research Center
Painesville, Ohio 44077

1. INTRODUCTION

The objective of food animal health overall is to provide methods to reduce economic losses due to disease and improve animal production efficiencies. This will ensure supplies of animal protein at reasonable prices to consumers and ensure profit to producers and processors. Although a shift may occur in the type of meat protein desired, man's desire to obtain protein and energy from animal sources will not change and will grow with the population.

The contributions of biotechnology to the food animal health industry have been very significant. Antibiotics produced by fermentation comprise most of the products. Antibiotic technology applied to food animal health was first used as a spin-off from human health application of antibiotics, but increasingly restrictive government regulations and special requirements for specific properties have forced the industry to look for new compounds not used in human medicine. Recombinant DNA technology promises to provide new peptide materials which will increase the importance of microbial products in the growth and well-being of food animals (1).

1.1 Background and Issues

The application of biotechnology to the animal health industry goes beyond disease prevention, reaching back to the production of agricultural feed. About 50% of the U.S. produced grain crop is used for food animal production. This provides 25% of the total livestock feed, with the remainder coming from roughage and agro-byproducts. Biotechnology needs to overcome diminishing gains in agricultural productivity via improved nitrogen fixation and photosynthesis efficiency, growth regulation, and resistance to environmental stress factors. Since agriculture and ultimately food animal production is dependent on petroleum and petrochemicals, biotechnology's role will be to provide the means to produce more food with less energy.

The use of drugs for various animal health applications increased significantly in the early 1970's and will continue as a major force in food animal production. With new applications of biotechnology employing recombinant DNA techniques, new and improved vaccines will be generated. The sale and use of animal vaccines which have shown little growth in the last few years should increase with efficacious new products. Antimicrobials, antiparasiticides, growth promoters, and vaccines are used in cattle, both beef and dairy, poultry, swine, and sheep production worldwide to address a variety of animal propagation problems (2). In the U.S. and Europe, modern practices of animal husbandry necessitate that large numbers of animals be brought together in close quarters, causing stress on the animal enhancing the disease risk. Foreign diseases such as African Swine Fever in pigs, Foot and Mouth Disease in hoofed animals, Rinderpest in cattle, and Newcastle Disease in chickens, to name a few, pose a constant threat to U.S. and European animal herds from areas where the causative agents are less rigorously controlled. In other areas of the world where herds are allowed to pasture, these and additional disease problems may arise. The underutilization of the pasture lands in Central Southern Africa and South America for domestic cattle production is attributed to tick and tsetse fly borne diseases (3). The eradication of trypanosomiasis alone could increase, by 120 million head, the number of cattle raised in this area.

Beef and dairy cattle are the major component of the agricultural economics in many societies. Transportation of cattle in the U.S. over long distances and confinement in large feedlots lead to respiratory tract and enteric diseases. Although there is a trend toward increased consumption of meat with lower fat content, i.e. poultry, beef still remains a popular source of protein. In areas where land has little agricultural value, the potential is usually present for some cattle production. Dairy cattle provide the most efficient source of protein. In increasing productivity by breeding programs, the stress of higher milk production in lactating cows has lead to diseases of the reproductive system contributing to infertility and fetal mortality.

In swine, the trend is also toward the concentration of animal populations for more effective utilization of feed and land. This leads to animal stress and disease susceptibility. Swine producers are major users of antibiotics and nutritionals in feed.

For poultry, the per capita consumption of chicken is increasing as people become more conscious about the fat content in their diet. Vaccines and feed nutritionals are important considerations for poultry production. (Table 1)

TABLE 1

Some key conditions for food producing animals

CATTLE

- Brucellosis (*Brucella abortus*)
- Colibacillosis
- Salmonellosis
- Shipping Fever Complex (*Pasteurella* and Infectious Bovine Rhinotracheitus)
- Feedlot Bloat
- Rabies
- Black Leg (*Clostridium chauvoci*)
- Anaplasmosis
- Leptospirosis
- Helminths
- Efficient Use of Feed and Growth

SWINE

- Atrophic Rhinitis (*Bordetella bronchiseptica*)
- Brucellosis (*Brucella suis*)
- Colibacillosis
- Transmissible Gastroenteritis
- Hemorrhagic Dysentery
- Leptospirosis
- Helminths
- Nutritional

POULTRY

- Chronic Respiratory Disease (*Mycoplasma gallisepticum*)
- Coccidiosis
- Salmonellosis
- Newcastle Disease
- *Pasteurella multocida*
- Nutritional

1.2 Regulation and Drug Characteristics

The approval process for a new animal health drug is long and costly. New drugs are evaluated for toxicity, drug residue, efficacy, and environment impact. The introduction in the U.S. of new animal drug products has decreased over the last ten years. In the 1967-71 period, an average of 4.2 new products per year were approved by the U.S. Food and Drug Administration, while in the 1972-76 period, this figure slipped to 2.4. Finally, in the 1977-81 span, less than one new product was approved per year resulting in what some have termed a "drug lag" (4) in the U.S. In an effort to speed up the approval process, FDA's Bureau of Veterinary Medicine has been reorganized (5) to consist of:

- Division of Biometrics and Production Drugs
- Division of Drug Manufacturing and Controls

- Division of Therapeutic Drugs for Food Animals
- Division of Therapeutic Drugs for Non-Food Animals

This reorganization abolishes animal species divisions, replacing these with separate divisions for food animal and non-food animal drugs, manufacturing and quality control, and production drugs, including mathematical and computer support.

In addition to the reorganization, the FDA has added a "fast track" status for innovative new products which have new and outstanding features offering significant health benefits. Diamond Shamrock's tiamulin for Swine Dysentery and Merck's ivermectin, an antiparasiticide, have been accepted by the FDA on fast track status. Both are produced, at least in part, by fermentation. At present, the U.S. Department of Agriculture has been responsible for approval of biologicals including bacterins and vaccines.

Any drug intended for use in food animals should exhibit low plasma levels and short biological half-life in the animal. Rapid clearing of the compound from the body after drug withdrawal is essential in permitting undelayed animal slaughter or product generation. An antibiotic used as a feed additive for growth promotion should not be absorbed from the gastrointestinal tract.

2. ANTIBIOTICS

Antibiotics, specifically antibacterials, are used extensively in the livestock and poultry industry, not only for prevention and treatment of disease, but also at subtherapeutic levels as feed additives for growth promotion and increased feed efficiencies (weight gain per weight feed consumed). In recent years, new antibiotics developed exclusively for animal applications have appeared in an attempt to create products with specialized activity and prevent overlap with compounds used for human medicine.

2.1 Antibacterials

Antibiotics used as feed additives are generally known as growth permittants. Their mode of action in promoting increased weight gain with less feed consumption is believed to be via interaction with the bacteria in the gut of the animal. The following have been proposed as possible modes of action for antibiotics increasing animal growth:

- Decrease in number of intestinal bacteria
- Influence metabolism of these bacteria

- Nutrient sparing effect (amino acids, vitamins, etc.)
- Influence thickness of intestinal wall and increase absorption
- Prevention of disease which stresses animal via toxin production, etc.

The amount of antibiotic employed in animal feeds is appreciably below the therapeutic level. Sublethal amounts of antimicrobials are known to influence the ability of certain bacteria such as Streptococci and E. coli to adhere to epithelial cells which contributes to colonization and resultant pathogenicity by the microorganism. The attachment via pili or fibrillae and subsequent colonization of bacteria in the gastrointestinal tract could influence nutrient absorption by the animal. Vaisanen et al. (6) found at antimicrobial concentrations 12.5-50 percent of the minimal inhibitory concentration, cells of E. coli were unable to agglutinate human erythrocytes and the ability to adhere to human buccal cells was reduced. In a separate study, Vosback (7) using E. coli strains demonstrated reduced adhesion to intestine tissue culture cells in the presence of antibacterials at 25% of the MIC concentration. Both reports cited the reduction in the bacterial protein structures which participate in the E. coli adhesion process. Studying the transit time of nutrients through the epithelium from small intestines of germ-free and re-infected animals, Savage et al. (8) found in mice and rats that the enzymes, alkaline phosphatase and disaccharidase, in germ-free animals were two to three times higher than in animals associated with indigenous microorganisms.

The total world animal health market for antibacterials (including the synthetic sulfas and nitrofuran compounds) is approximately $2 billion. Table 2 indicates some of these antibiotics, their application, and manufacturers. Chlortetracycline (Fig. 1) was the first antibiotic to be used for growth promotion in food animals (9). The compound produced by Streptomyces aureofaciens exhibits broad spectrum antibacterial activity as well as having activity against some large viruses and rickettsia. Included at low levels in feed, chlortetracycline reduces the incidence of liver abscesses in feedlot cattle and has been used extensively in swine production. Chlortetracycline, tetracycline, and oxytetracycline comprise the largest selling antibacterial antibiotic class for animal health. Sales of the tetracycline-based antibiotics are lead by Pfizer

TABLE 2

Some antibacterial antibiotics for animal health applications

COMPOUND	MANUFACTURER	ACTIVITY	STRUCTURAL TYPE
Avoparcin	American Cyanamid	gram-positive	glycopeptide
Bacitracin	A. L. Laboratories International Minerals and Chemicals	gram-positive	alkaline polypeptide
Bambermycin	Hoechst A.G.	gram-positive	glycolipid
Destomycin	Meiji Seika	broad-spectrum	glycoside similar to Hygromycin B
Erythromycin	Abbott	gram-positive	macrolide
Hygromycin B	Eli Lilly	broad-spectrum	glycoside
Lincomycin	Upjohn	gram-positive	amino acid and sugar
Macarbomycin	Meiji Seika	gram-positive	glycolipid
Neomycin	Upjohn	broad-spectrum	glycoside
Spectinomycin	Abbott	broad-spectrum	actinamine-deoxyketose
Spiramycin	Rhone-Poulenc	gram-positive	macrolide
Tetracyclines - chlortetracycline - oxytetracycline	American Cyanamid Diamond Shamrock Rachelle	broad-spectrum	most widely used antimicrobials in U.S.
Thiopeptin	Fujisawa	gram-positive	sulfur-containing peptide
Tylosin	Eli Lilly	gram-positive, some gram-negative	macrolide
Virginiamycin	Smith Kline Beckman	gram-positive	macrocyclic lactone peptolide

(oxytetracycline) and American Cyanamid, Rachelle, and Diamond Shamrock for chlortetracycline.

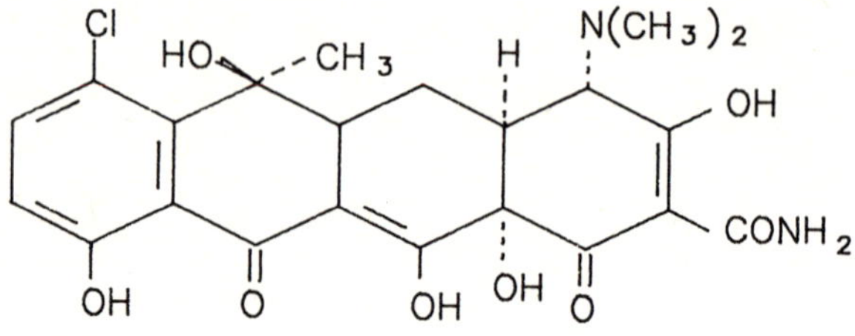

Fig. 1. Chlortetracycline

Tylosin (Fig. 2), sold by Eli Lilly, is the second largest selling antibacterial antibiotic for animal health applications. The compound is sold as a feed additive or a therapeutic injectable against gram-positive bacteria and some gram-negative bacteria, Spirochetes, some Vibrios, and Mycoplasma. It is the only antibacterial approved by the FDA for use in combination with the rumen metabolic growth promotant, Rumensin®, as feed additives for feedlot beef cattle. In addition, tylosin has been used for Swine Dysentery in piglets, growth promotion in pigs, and mycoplasma infections in a variety of animals. Tylosin, a 16-member macrolide antibiotic containing three sugar moieties, was found to be produced by Streptomyces fradiae NRRL 2702 (ATCC 19609) (10) in a mixture of components including lactenocin, desmycosin, macrocin, and relomycin (11). The latter compound, relomycin, has reduced antibiotic activity. In studies on the biosynthesis and control of tylosin formation, investigators have found that the two enzymes involved in the condensation of two acetates, five propionates, and one butyrate to form the lactone ring, propronyl CoA carboxylase and methylmalonyl CoA carboxytransferase (12), are both depressed by glucose and inorganic phosphate (13).

In an elegant study on the analysis of pathway mutants and use of cofermentation and bioconversion of intermediates, Baltz and Seno (14) at Eli Lilly showed that the tylosin sugars, mycarose, mycaminose, and mycinose are derived from a common intermediate. The O-methylation of the 3"-position of macrocin appears to be a key rate limiting step (15) in that a number of mutants selected for increased production of tylosin also had increased levels of macrocin O-methyl transferase activity (16).

Modifications to the tylosin structure, possibly in part encouraged by the expiration of the tylosin fermentation patent in April, 1982, have resulted in a number of interesting derivatives. By hydrolyzing tylosin under mild acidic conditions, 5-0-mycaminosyl tylonide was produced (17). The compound is said to have high antibacterial activity against Pasteurella multocida

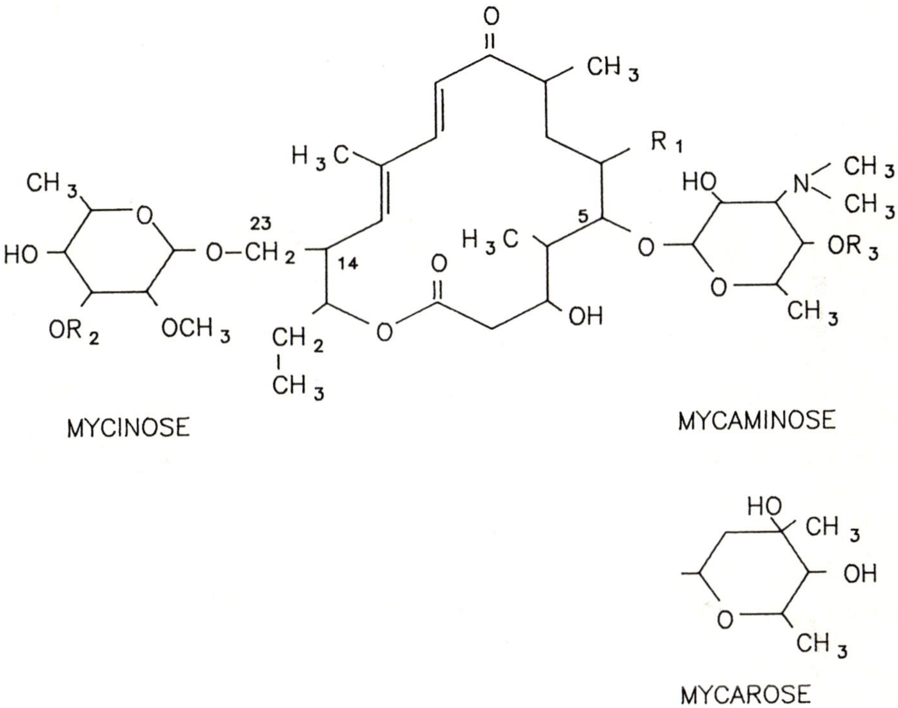

	C–23	R1	R1	R3
TYLOSIN	MYCINOSE	CHO	CH_3	MYCAROSE
DESMYCOSIN	MYCINOSE	CHO	CH_3	H
LACTENOCIN	3–DEMETHYLMYCINOSE	CHO	H	H
MACROCIN	3–DEMETHYLMYCINOSE	CHO	H	MYCAROSE
RELOMYCIN	MYCINOSE	CH_2OH	CH_3	MYCAROSE

Fig. 2. Tylosin and derivatives produced during the fermentation.

and P. haemolytica, the causative agents of respiratory diseases in sheep, cattle, swine, and fowl. Further chemical modification by Sakakibara and coworkers (18) generated 19-deformyltylosin derivatives with superior antibacterial activity against macrolide-resistant bacteria. The addition of isovaleryl moiety to the 4"-hydroxy of mycarose increased the effectiveness of the derivative to marcolide-resistant strains of Staphlococcus aureus (19).

The first antibiotic developed exclusively for animal health applications is the Hoechst A.G. compound, bambermycin (20). Representative of a group of complex phosphorus-containing glycolipid antibiotics, the antibiotic is formed by a group of gray-green streptomyces including Streptomyces bambergiensis ATCC 13879, S. ghanaensis ATCC 14762, S. geysiriensis ATCC 15503, and S. ederensis ATCC 15304. The compound was originally described as moenomycin in 1965 for animal health applications and ten years later approved by the FDA for use as a dried mycelial feed supplement to improve feed efficiency. Indications of biological activity have been cited in poultry, cattle, and swine (21,22). Bambermycin is primarily effective against gram-positive bacteria with minimal inhibitory concentrations against Streptococci and Staphylococci in the 0.001-0.005 mg/ml range. As with other members of the class of antibiotics, bambermycin inhibits the growth rate of some gram-negative (E. coli) bacteria possessing certain plasmids, including resistance transfer factor. Macarbomycin, produced by Streptomyces phaeochromogenes (23), is another member of this interesting group of phosphorus-containing glycolipid antibiotics. Studies with macarbomycin showed direct inhibition of peptidoglycan synthesis in the cell wall. The data suggests direct binding of the antibiotic to the synthetase enzyme. Investigators at Sankyo (Japan) have described the newest member of this group, pholipomycin (24). Produced by Streptomyces lividoclavatus, the antibiotic is isolated by methanol extraction of the mycelial cake. As with bambermycin, growth promoting activity is claimed in broiler chickens and swine, with increased feed efficiencies. Interestingly, pholipomycin exhibits three to nine times higher activity against E. coli than other members of this family of antibiotics.

Another new animal health antibiotic, tiamulin (Fig. 3), is being developed in the U.S. and Canada by Diamond Shamrock and in Japan by Sankyo on licenses from Sandoz. Tiamulin is a semi-synthetic antibiotic generated by a chemical modification of pleuromutilin. First isolated in 1951 by Kavanagh et al. (25), the tricyclic diterpene pleuromutilin was rediscovered as a weak gram-positive antibacterial by investigators at Sandoz Research Institute, Vienna, in fermentation broths from the basidiomycete Clitopilus passeckerianus NRRL 3100 (26). Pleuromutilin, 14-acetylmutilin and mutilin (14-hydroxyl form) are all produced during the fermentation with the composition of products dependent upon fermentation medium and conditions. The synthesis of derivatives differing in the acyl portions attached to the 14-hydroxyl of mutilin led to tiamulin. Tiamulin has excellent activity against Spirochetes and Mycoplasma and also good activity against some other gram-positive bacteria including Staphlococci, and Streptococci (27). Moderate to no gram-negative activity is seen with the compound. Tiamulin can be used in drinking water or as a premix feed additive against Treponema hyodysenteriae, the causative agent of dysentery, and Mycoplasma (pneumonia) in swine. In comparisons with tylosin against field strains of mycoplasma, tiamulin was more effective (28). Unfortunately, its application is limited for mycoplasmal infections in some animals due to problems with incompatibility with monensin.

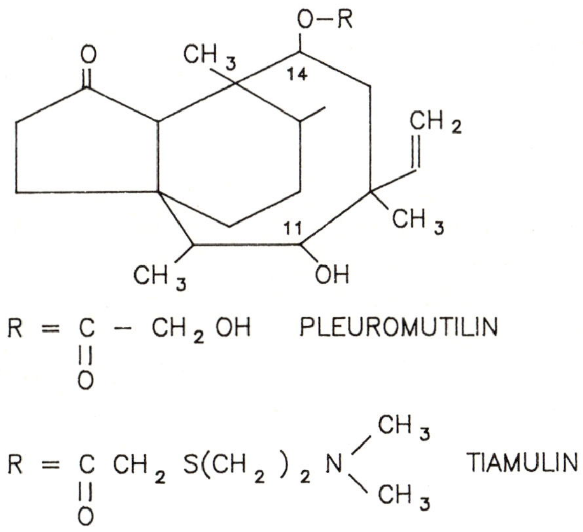

Fig. 3. Pleuromutilin and semi-synthetic derivative tiamulin.

2.2 Coccidiostats

Coccidia are protozoan parasites belonging to the genera _Eimeria_ and _Isospora_ and infect vertebrates as well as invertebrates. The _Eimeria_ are economically most important due to their infection of poultry, cattle, and sheep. Coccidiosis of chicken is widespread and any one of nine species of _Eimeria_ may infect. This disease is estimated to cost the U.S. poultry industry almost $300 million per year (29). Until the late 1960's synthetic agents were used as drug therapy. For the most part, rapid resistance arose to each compound. In the early 1970's, with the introduction of the polyether ionophore antibiotic monensin, by Eli Lilly, the economic plight caused by coccidiosis in the broiler industry was reversed and a dramatic increase in production realized.

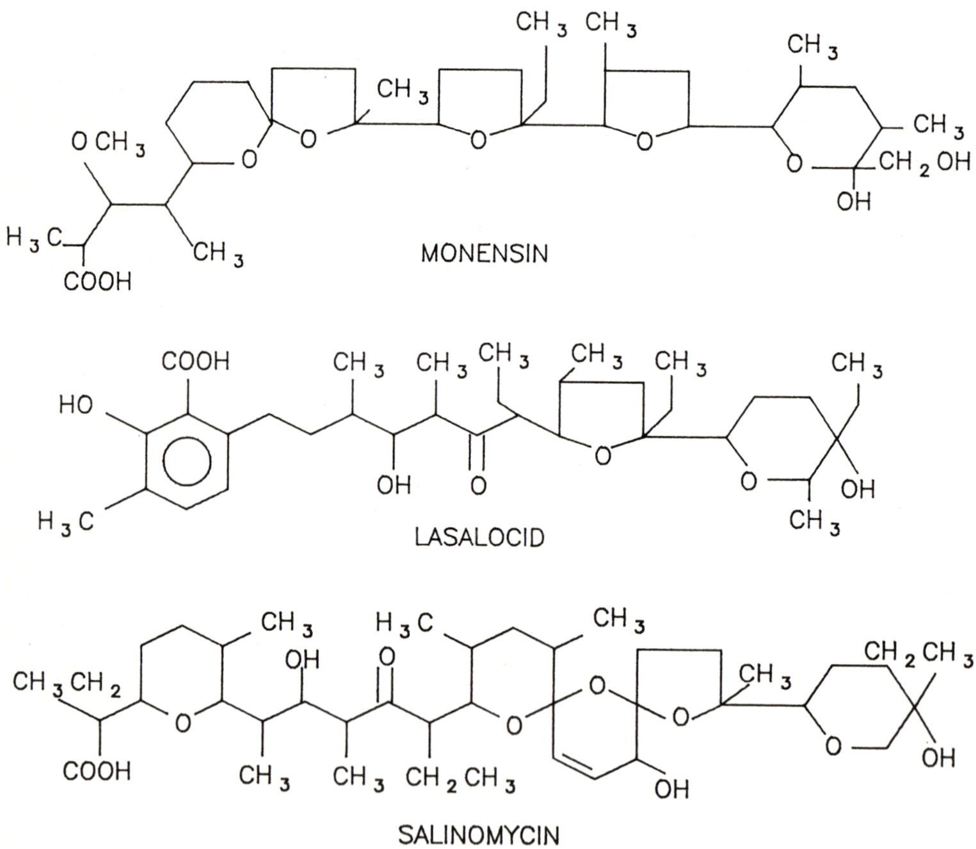

Fig. 4. Polyether ionophore antibiotics.

Monensin (Fig. 4) was isolated from Streptomyces cinnamonensis ATCC 15413 about 25 years ago (30). Isolated from the culture filtrate detected by its gram-positive antibacterial activity, monensin was identified among the complex of four closely related compounds. As with all other polyether ionophores, toxicity was and continues to be a problem. The LD_{50} (oral) in mice is 125 mg/kg and in chickens 200 mg/kg (31). In spite of this problem, monensin is highly effective against strains of Eimeria and dominates the world-wide marketplace.

Over 20 distinct polyether antibiotics have been isolated from various streptomycetes. Compounds of this class cause cations to become lipid-soluble, enabling them to pass through membranes and have become a valuable tool in the study of membrane transport and secretion of inorganic cations. Recently other polyether ionophores are being introduced in an attempt to challenge monensin's market share. Kaken (Japan) has licensed salinomycin to Pfizer, Hoechst, and A. H. Robin for world-wide marketing. Lasalocid, produced by Streptomyces lasaliensis, is marketed by Hoffman-LaRoche. All of the above polyether antibiotics may act as growth promotants by controlling subclinical levels of coccidiosis.

In an effort to have a back-up product for monensin, Eli Lilly has begun development of narasin. Isolated initially from Streptomyces lydicus NRRL 12034, narasin differs from salinomycin only by substitution of a proton by a methyl group on the first ring. Utilizing strain Streptomyces aureofaciens NRRL 5758, Boeck et al. (32) reported on a strain and media improvement for the fermentation from 4.3 to 1690 µg/ml in shake flask and 2430 µg/ml in stirred fermentor.

2.3 Anthelmintics

The infection of sheep, cattle, and swine by parasitic worms can cause considerable economic loss in meat production. Short of the death of animals in serious infections, subclinical infections can adversely affect weight gain and production efficiency. Helminths of major impact are the gastrointestinal round worms and tape worms, lung worms, and liver flukes. In cattle alone, an economic loss of $100 million per year to internal parasitic worms has been estimated (33). At present, synthetic anthelmintic compounds consisting of benzimidazoles, levamisole (imidazothiazole), and organophosphates have been used. Thiabendazole (Merck) was the first broad spectrum anthelmintic introduced almost twenty years ago. In recent years, increased resistance to some synthetic anthelmintics has created a need for new classes of compounds with anthelmintic activity.

The first fermentation produced anthelmintic, hygromycin B (34), (Eli Lilly) has had limited applications in the treatment of nematode infections in swine and poultry. With the discovery of avermections (35) (Fig. 5), these new structures produced by fermentation have the potential to capture a major portion and probably increase the current worldwide animal anthelmintic market of $455 million. During the isolation of new soil microorganisms at the Kitasato Institute, Tokyo, Japan (36), a streptomycete, later designated Streptomyce avermetilis NRRL 8165, was isolated and later found by investigators at Merck to produce a complex of closely related compounds possessing broad spectrum nematicidal activity. This new family of 16-membered macrocyclic lactone derivatives is isolated from the mycelium of the microorganism and lacks detectable antimicrobial activity. One of the semi-synthetic derivatives, 22,23-dihydroavermectin B_1, has been given the generic name, ivermectin.

The avermectins have been found to be effective against most mature and larval forms of gastrointestinal and systemic nematodes

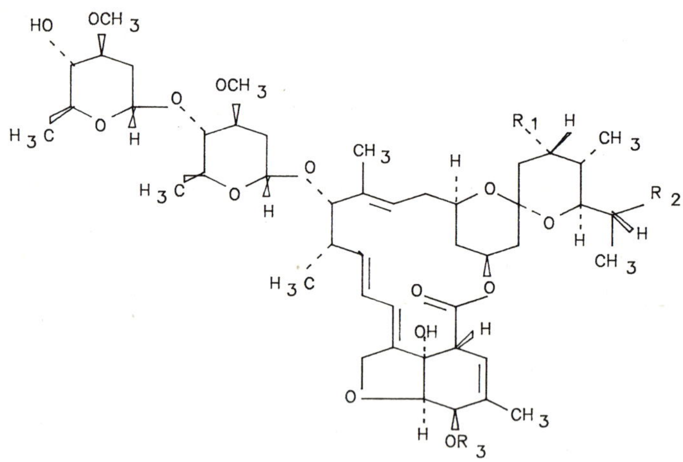

Fig. 5. Structural type of the avermectins.

in sheep and cattle, including the benzimidazole-resistant strains (37,38,39). Their mode of action is believed to be via blockage of neurotransmission, resulting in rapid and irreversible nematode paralysis. In addition to the anthelmintic activity, the avermectins are active against ectoparasitic arthropod pests, a major cause of economic loss in livestock product. Applied topically or systemically, ivermectin is effective against mites, ticks, and larvae of biting flies.

An additional complex of thirteen compounds termed the milbemycins (40), produced by *Streptomyces hygroscopicus* subsp. *aureolacrim*, has been described by investigators at Sankyo. Very structurally similar to the avermectins, the milbemycins lack the sugar moieties present in the avermectins. Like the avermectins, the milbemycins have ascaricidal and insecticidal activity.

Additional fermentation produced compounds possessing anthelmintic activity will most likely be identified as fermentation cultures are examined for non-antimicrobial activity. Additional macrolide compounds and possibly other structural types will be discovered using rapid screening tests. One such test uses the free-living nematode, *Caenorhabditis elegans* (41). The assay is sensitive to benzimidazole anthelmintics, avermectin, and levamisole and provides a convenient *in vitro* test system avoiding the use of animals and large amounts of compound.

3. GROWTH PROMOTION

3.1 Rumen Metabolic Regulators

Ruminant animals, cattle and sheep, are capable of converting solar energy fixed in plants not available to man for food, to a form which is nutritionally available to man. Cellulose makes up approximately 50 percent of the carbon in plant vegetation. This and other ingested plant polymers (hemicelluloses, pectins, etc.) are fermented by the microbial ecosystem of the rumen stomach to short-chain volatile fatty acids, acetate, propionate, and butyrate, and to the gasses, methane and carbon dioxide (42). The fermentation produces excess hydrogen which is removed by conversion with CO_2 to methane and is released by the animal. The volatile fatty acids are absorbed from the rumen stomach into the blood stream to be converted by the body to glucose and fat. Only propionate is gluconeogenic, being converted to glucose in the liver. Since more energy is recoverable from propionate, a shift of the fermentation to increased propionate at the expense of acetate and butyrate would result in increased feed efficiency.

Production of methane and its removal also represents an energy loss and a reasonable reduction in its production would also increase feed utilization efficiency.

Monensin, a fermentation polyether antibiotic produced by Streptomyces cinnamonensis and used as a coccidiostat, is also used extensively as a feed additive for cattle (43). Administered at a rate of 30 g/ton/feed, monensin improves feed efficiency by shifting the volatile fatty acid synthesis to propionate production. The production of methane is also decreased (44). The animals eat less feed, but gain the same weight. The rate of weight increase is not affected.

Other ionophore polyether antibiotics, including salinomycin, lasalocid, and narasin, are also effective as rumen metabolic regulators. The mode of action of these compounds is believed to be via specific inhibition of certain gram-positive bacteria and Ruminococci of the ruminant flora. Lasalocid has recently been introduced in the U.S. by Roche for growth promotion and improved feed efficiency in feedlot cattle.

3.2 Hormonal

In contrast to antibiotics which act as growth permittants and rumen metabolic regulators which affect the volatile fatty acid composition resulting in nutritional effects, hormonal growth promoters affect cellular mechanisms in the animal. This cellular control may regulate the rate of synthesis and amount of protein in the muscle fibers. Considerations of the role of hormonal growth promoters in enhancing feed efficiency and promoting more rapid growth include the interaction of muscle protein and amino acid metabolism with organs such as the pituitary gland, liver, etc., and the important role of protein breakdown which may be independent of protein synthesis. The age and sex of the animal will greatly influence the effect of a particular growth promoter (45).

Steroids have been employed as anabolic agents in beef production. Synovex S® (estradiol and progesterone) and Synovex H® (estradiol and testosterone) marketed by Syntex, and trenbolone acetate not approved in the U.S. but used in Europe, are representative of these compounds. The most widely used anabolic agent, the non-steroidal estrogenic compound diethylstilbesterol (DES) was added as a feed additive or as an implant in beef cattle and sheep. The compound is currently banned in the U.S. and other parts of the world as potentially carcinogenic.

Filling the void left by the removal of DES from the market, International Minerals and Chemicals has seen its fermentation-produced zeranol (Fig. 6) rapidly grow in sales during the past six years (46). Investigators at Purdue University and Commercial Solvents Corporation in the late 1950's determined that resorcyclic acid lactones were the causative agents in moldy corn producing estrogenic toxicity in swine (47). Substances isolated by ethanol extraction from the growth medium of cultures identified as *Gibberella zeae* NRRL 2830 (perfect stage of *Fusarium roseum*) were estrogenic. The major component, zeralenone (2,4-dihydroxy-6(10-hydroxy-6-oxo-trans-1-undecenyl)-benzoic acid lactone, when reduced to zeranol in which the double bond and the 6-ketone in the lactone ring are reduced is capable of growth promotion in cattle. The initial production of zearalenone took place on a surface grain fungal growth medium from which zearalenone was extracted. The later use of vermiculite and a defined glucose medium yielded 15-17 g/liter after a 5-week incubation period (48). A mutation and selection procedure using *G. zeae* ATCC 20028 produced a major breakthrough when yields on surface medium with *G. zeae* ATCC 20273 were increased to 30 g/L.

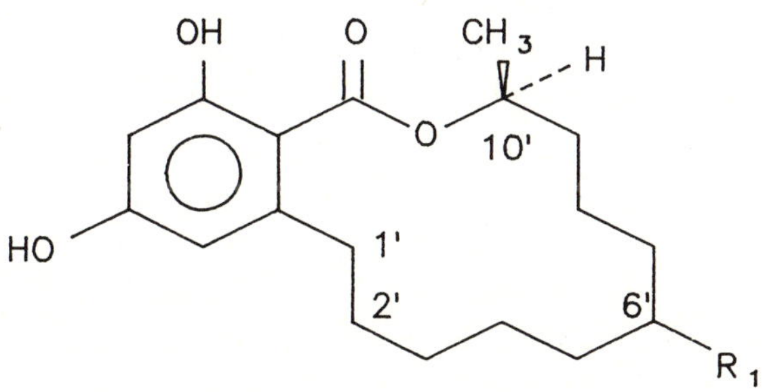

		R_1	C 1' – 2'
I	ZEARALENONE	=O	CH=CH
II	ZEARALANOL, ZERANOL	OH	CH_2-CH_2

Fig. 6. Resorcyclic acid lactones.

More importantly, the new strain was found to produce zeralenone in submerged fermentation. The mutant isolated in 1971 allows submerged production in excess of 30 g/L after a period of approximately 10-14 days. The use of zeranol was approved by the FDA in 1969 as an implant in the ear for cattle (3-12 mg/animal) and sheep (1-12 mg/animal). Claims have been made of 10-15 percent weight increase with a 10-12 percent feed efficiency improvement.

Recombinant DNA technology is being applied to the area of growth promotion by the cloning, expression, and production of bovine growth hormone. Growth hormone, produced by the anterior lobe of the pituitary, is a protein having a molecular weight of 22,000 and containing 191 amino acids (49). The amino acid sequence varies with the growth hormone source. The hormone affects protein anabolism and metabolism of lipids, carbohydrates, and minerals. This, in turn, has its consequences in milk production, wool growth, animal growth rate, and fat deposition. Several joint ventures combining new biotechnology with the commercial expertise of large chemical companies have been announced to clone, express, and eventually market growth hormone. Molecular Genetics, Inc., with American Cyanamid, Genentech with Monsanto, and Biogen with International Minerals and Chemicals are expected to commercialize bovine growth hormone. Using recombinant DNA based bovine growth hormone developed by Genentech, researchers at Cornell University reported a 12% increase in milk production when daily injections were given to dairy cattle (50). Another report has cited a net increase of 18% and a reduction in feed per kg of milk of 29% (51).

Keshet et al. (52) at the Weizmann Institute of Science have published their procedure for cloning of the bovine growth hormone gene:

1) Total RNA extraction from anterior lobes of bovine pituitary glands followed by isolation of messenger RNA (poly A tailed) using oligo (dT)-cellulose chromatography.
2) Synthesize double-stranded cDNA (copy DNA) using reverse transcriptase enzyme.
3) Linearize plasmid pBR322 with restriction enzyme PstI (ampicillin gene site) and add approximately 20 guanosine base residues to 3′ end using terminal deoxynucleotidyl transferase enzyme.
4) Add cytosine residue to cDNA.
5) Double stranded cDNA fractionated by electrophoresis to

isolate species of about 500 base-pairs to match protein product size.

6) cDNA mixed with equimolar amounts of dG-tailed linearized plasmid vector and allowed to anneal.

7) Plasmid transformed into Escherichia coli and transformants screened for tetracycline resistance and ampicillin sensitivity (inactivational insertion of cloned fragment at PstI site).

Because the bovine growth hormone was inserted into ampicillin gene, a fused protein product of bovine growth hormone with β-lactamase was expected. A radioimmunoassay procedure was used to analyze the clones for presence of BGH protein. Mini-cell preparations were prepared by culturing in the presence of ^{35}S-methionine and incubating extracts with ^{125}I-labelled BGH and rabbit antiserum raised against BGH. One clone D-4 having 820 base pairs was determined to have the sequence corresponding to the known amino acid sequence of BGH. The clone contained information at the 5′ end for a 23 amino acid signal peptide. Use of the cloned fragment as a probe for other BGH genes resulted in the conclusion that only one bovine growth hormone gene exists.

The availability of growth hormone from biotechnology for increased milk and beef production raises problems of delivery of the protein hormone. Slow release technology for these types of compounds needs to be developed as daily injection is an impractical route of administration. Controlled-release technology developed for steroids and other anabolic agents may not be applicable to the larger protein molecules (53).

3.3 Nutrition

To obtain the best quality product at the highest feed efficiency, the animal production industry has employed a variety of nutritional supplements in animal feed. Many of these factors are produced by fermentation and with the application of recombinant DNA methodology for strain development and improvement, additional nutritional factors and supplements will be generated by microbial processes. Essential amino acid supplementation to animal diets permits greater weight gains and higher feed efficiency. Although ruminants, cattle and sheep, obtain their amino acid requirements via their normal diet and the synthesis of amino acids by the rumen microbial flora, monogastric animals such as swine and poultry require ten essential amino acids.

The addition of lysine increases the quality of cereal protein, allowing for more of the amino acid content to be utilized in the synthesis of animal protein. The most utilized amino acid is methionine which has generally been supplied as a dl-synthetically prepared mixture. L-lysine produced by bacterial fermentation began with efforts at Pfizer in the mid-1950's. Since that time, a number of patents have been issued in the United States (Table 3). The most recent patent, 4,346,170 (54), covers the use of recombinant DNA technology to produce strains of E. coli carrying plasmids containing genes of the lysine biosynthetic pathway. Having "decontrolled" the gene expression to feedback regulation by isolation of mutants resistant to L-lysine analogs, the genes are cloned and expressed (55). Although L-lysine yield is low compared to those cited for nutritional and analog resistant mutants of Corynebacterium and Brevibacterium (56,57), the trend is set to utilize gene amplification and expression in amino acid producers. Tryptophan is another amino acid potentially produced by fermentation which has been shown to be important as a feed supplement. Several recent U.S. patents indicate high interest in the production of this amino acid using indole and serine or pyruvic acid as precursors in an enzymatic process (58,59). A recent study in swine indicates that the inclusion of tryptophan to the corn-soybean meal diets will allow the reduction of soybean meal by approximately 4 percent (60).

In one of the boldest ventures in biotechnology, Imperial Chemicals Industry (U.K.) started production in 1979 of their single-cell protein fermentation process (61). Designed as an animal feed substitute for soybean meal and fish meal, the product, Pruteen is the culmination of development covering ten years. In the fermentation process, Methylophilus methylotrophus is grown on methanol, salts, and ammonia. The bacterium, consisting of 70 percent protein, is an ideal supplement to the diets of swine, poultry, and calves. Current economics due to methanol costs limit further expansion and utilization of the material. Current fermentation production at ICI capacity is rated at 50,000 metric tons per year.

Bacterial preparations are manufactured and sold for feed supplementation and improvement of feed utilization. Dried preparations of Bacillus subtilis are sold for use as swine and calf milk replacers and feeds. Cultures of Lactobacillus acidolophius supplement normal microbial populations in the gastrointestinal

TABLE 3

Significant U. S. patents for L-lysine production by fermentation

U.S. PATENT NUMBER	DATE	COMPANY	ORGANISM	APPROACH
2,979,439	1961	Kyowa Hakko Kogyo	*Micrococcus glutamicus* ATCC 13286, 13287	nutritional requirements for homoserine, threonine and methionine threonine and cystathione threonine and homocysteine
3,527,672	1970	Ajinomoto	*Brevibacterium lactofermentum* ATCC 21986	nutritional requirements for threonine, isoleucine, and valine
3,616,218	1971	Ajinomoto	*Brevibacterium flavum* ATCC 21117-19	nutritional requirements for threonine
3,687,810	1972	Kyowa Hakko Kogyo	*Corynebacterium flavum* ATCC 21513-16 *Brevibacterium flavum* ATCC 21128	resistance to cell-wall antibiotics, i.e., penicillin, cycloserine, etc.
3,707,441	1972	Ajinomoto	*Brevibacterium flavum*, *Corynebacterium acetoglutamicum*	resistance to S-(2-amino-ethyl)-L-cysterine and nutritional requirements
3,708,395	1973	Kyowa Hakko	*Brevibacterium flavum* *Corynebacterium glutamicium*	resistance to variety of amino acid analogues and nutritional requirements
4,346,170	1982	Ajinomoto	*Escherichia coli*	cloning into *E. coli* genes from strains resistant to L-lysine analogues

tract. Advertisements from Prairie States Enterprises, St. Joseph, Missouri, U.S., claim that the bacterial preparation "lowers intestinal pH to remove conditions that promote rapid proliferation of pathogenic bacteria" (62). The Microbial Genetics Division, Pioneer Hi-Breed International, Inc., markets Probios® brand of microbial cultures which are claimed to establish and maintain balanced microflora in the animal gut. Related to the use of microbial culture supplements is the use of animal feed enzymes. Investigated in the early 1960's, bacterial and fungal amylases and proteases were studied to improve feed efficiencies. Interest in this approach has waned and there is currently little activity in this area. A combination of microbial culture addition to feed and the concept of increased enzyme activity to improve feed efficiency might be merged via the use of recombinant DNA technology and the cloning into these cultures useful enzyme activity.

4. BIOLOGICALS (VACCINES AND BACTERINS)

The world-wide sales of animal health biologicals including those for companion animals is approximately $500 million. The U.S. market for these products is about $150 million with about 60 percent devoted to vaccines and bacterins for food-producing animals. The term biologicals includes preparations of viruses and bacteria used to elicit an immune response when administered to an animal to protect against disease caused by future exposure to the infectious agent. Viral vaccines are directed against a variety of virus diseases in cattle, swine, and poultry. Viral vaccines may be live attenuated strains weakened by serial passage in tissue culture or a non-host organism. Vaccines are also manufactured by utilizing chemicals such as formalin or β-propiolactone to produce killed virus preparations. Bacterins are formulations of specific bacterial cells, killed by chemical or physical agents. Attenuated or avirulent bacterial strains or even spores of an avirulent strain may also be employed in generating humoral immunity against the disease.

Virus produced diseases (Table 4) cause major economic losses in the production of animal protein. Vaccines directed against these diseases provide, in many cases, an efficacious method for control of these infections, while in others there are no vaccines or those available are not effective. Some vaccines are used with great care because they, in fact, may cause the infection they

TABLE 4

Important viral diseases of economic animals

RNA VIRUSES	
Myxoviruses	Medium-sized, enveloped viruses with fringe of projections on surface; generally cause respiratory diseases via affinity for mucins in respiratory tract.
examples:	Swine Influenza, Avian Influenza, Newcastle Disease, Parainfluenza associated with shipping fever in cattle. Pseudomyxovirus - Rinderpest "Cattle Plague".
Rhabdoviruses	Includes rabies virus--highly fatal in all warm-blooded animals, and vesicular stomatitis virus-relatively benign vesicular infection.
Coronaviruses	Similar to Myxoviruses except surface projections are petal or club shaped and more widely spaced.
examples:	Transmissible Gastroenteritis of swine (TGE)-highly fatal in piglets. Avian Infectious Bronchitis--also highly fatal disease in young chickens.
Reoviruses	Double-stranded RNA virus.
examples:	Blue Tongue, Neonatal Calf Diarrhea (Rotaviruses).
Togaviruses	Includes arthropod borne viruses found in warm parts of world.
examples:	Nairobi Sheep Disease, Bovine Virus Diarrhea, and Swine Fever (hog cholera).
DNA VIRUSES	
Herpesviruses	Major disease-causing virus; protein capsid usually covered by outer envelope.
examples:	Infectious Bovine Rhinotracheitis, Pseudorabies in swine, Infectious Laryngotracheitis in chickens, Marek's Disease in chickens, Malignant Catarrhal Fever, Cytomegaloviruses.
Indoviruses	African Swine Fever - mortality rate approximately 100% in endemic regions of Africa.

were designed to prevent. For some viruses, only live vaccines are sufficiently immunogenic, however, a weakened animal or reversion of the attenuated strain may lead to disease. In addition, the growth of the virus from a live vaccine may produce an immunosuppressive effect. Killed virus vaccines can cause accentuation of a subclinical disease. Another problem with classical vaccines is the need for refrigeration and a limited shelf-life of the product.

Transmissible Gastroenteritis (TGE) of swine is a serious problem in piglets leading to a high death frequency. Current vaccines consist of live attenuated tissue culture generated vaccines given to pregnant sows. The piglets then receive antibody via the colostrum. One of the major problems with the vaccine is the potential of introducing TGE virus on farms with no prior history of the disease and the danger of spread of the infection to adjacent farms. In the viral diseases of African Swine Fever, Bovine Leukemia, and Malignant Catarrhal Fever, there are no vaccines currently available. Of the three, African Swine Fever, caused by a large icosahedral DNA virus, possesses the most serious threat world-wide (63). The disease produces an often fatal disease in swine and all infected or exposed herds are slaughtered. Cases of the disease are beginning to appear in chronic or subclinical form leading to lowering of weight gains. Infection by ASFV causes immunosuppression in the animal which can be correlated with the virulence of the strain. It has been estimated that movement of the disease into U.S. swine herds could result in economic losses as high as $300 million.

In poultry, Marek's Disease causes lymphoid tumors in young chicks. Attenuated virus grown in chicken kidney cells is used as a vaccine. The commercial vaccine is not always effective, causes local toxicity, and is relatively expensive. In the United Kingdom a policy of slaughter rather than vaccination is carried out in the control of Newcastle's Disease in poultry. Again, toxicity caused by the vaccine is economically intolerable.

The use of recombinant DNA technology holds great promise to provide greatly improved vaccine replacements for those currently available and to generate vaccines against diseases where there currently is no prophylactic treatment. Application of this technology to viral vaccine research and development provides the means, for the first time, of locating in the viral genome the coding sequence responsible for the protective antigenic determinant. A genetic analysis of genes responsible for encoding the viral capsid proteins is technically possible, and amplification of the antigens to determine which one produces neutralizing antibody can take place. Cloning and expression of the gene in a suitable microbial system leads to the production of large amounts of peptide antigen for use as the vaccine against the virus.

There are two major problems in the generation of a recombinant DNA based vaccine. First, the gene coding for the protective antigenic determinant must be identified. If little is known about the capsid peptide structure of the virus, investigations must begin at this level. The virus may be grown in tissue culture and viral protein isolated. Following partial enzymatic digestion, the protein is fractionated and the fractions tested for the strongest immunoprecipitation response against antiserum raised against the virus. The amino acid sequence of the peptide(s) giving the strongest immunological response is determined and synthetic gene(s) synthesized. Alternatively, virus generated messenger RNA (m-RNA) can be isolated from tissue culture, double-stranded copy DNA (ds cDNA) generated and clones prepared. If the size of the peptide is known, m-RNA species are fractionated according to size before cDNA is made. In the case of some positive-strand RNA viruses, the viral genome can be used directly for reverse transcription to cDNA because of the presence of a polyadenylated (poly A tail) portion at the 3′-end of the genome. Expression of the cloned fragments in *E. coli* mini-cells or a *in vitro* translation system is carried out to generate enough peptide to determine immunogenicity with antiserum against the virus. Following detection of the proper clone, the cDNA clones could be sequenced, edited for optimum antigenicity, and expressed.

Optimal expression of a cloned gene is critical for the successful production of any rDNA based product including vaccines. Beyond the initial problems of attaining the proper reading frame of the inserted DNA fragment which is ultimately tested by the antigenicity of the peptide, are the concerns around producing high yields of the cloned product. Placing the reading of the cloned sequence in the correct orientation under the control of a highly efficient gene promoter to insure efficient transcription, preventing the action of host proteases to degrade the peptide after it is biosynthesized, and transporting the peptide to the periplasmic space in *E. coli*, or even outside the cell, are all factors in the high expression of an r-DNA vaccine. Beyond the manufacture of the antigenic peptide, there are further questions around delivery of the vaccine. Classical adjuvants, such as aluminum hydroxide or mineral and vegetable oil emulsions, used in the past to heighten the antigenicity of attenuated or kill viruses may not be applicable for r-DNA subunit vaccines. New carriers and delivery systems will need to be devised to protect and produce efficient presentation of the antigen to the immune system.

A recombinant DNA based vaccine against the dreaded Foot-and-Mouth Disease promises to be the first major contribution of the technology to animal vaccines. The disease caused by a single-stranded RNA picornavirus affects approximately 30 species of cloven foot animals and is endemic in every continent except North America and Australia. Seven serotypes of the virus are known and the differences are such that recovery from infection from one does not produce protection to the others. Protection is provided only for reinfection with the same serotype (64). Subserotypes exist showing poor cross reactivity within each group. An inactivated vaccine exists for the disease but improper killing has caused outbreaks of the disease. Other problems with the current vaccine approach include low titer of the antigen, instability, and need for refrigeration. Genentech Inc. has announced (65) the successful isolation and cloning of the cDNA sequence responsible for the immunizing antigen of the viral structural capsid. Using the tryptophan promoter-operator system in E. coli, approximately 17 percent of the total cellular protein was found to be the VP3 capsid protein which elicited protective antibody against Foot-and-Mouth Virus challenge. Cloning and subsequent sequencing of an cDNA insert has permitted the identification of a central 20 amino acid portion of the 213 amino acid VP1 fragment which is responsible for animmunoglobulin response (66,67). Synthesis of this peptide has produced the first synthetic peptide vaccine possibility. This capsid protein and the smaller peptide portion is from only one of the seven serotype viral strains and will not confer protection to the other six strains.

Additional clonings of cDNA inserts encoding viral capsid proteins have been reported. Rose and Gallione (68) at the Salk Institute generated and cloned cDNA responsible for the G protein of Vesicular Stomatitis Virus. The strategy used by these investigators to obtain full length cDNA clones involved careful optimization of the reverse transcription of isolated m-RNA to generate complete copies, and chromatographic purification of the first cDNA strands before complementary strand production was attempted. After ds cDNA was produced, the cDNA was purified before cloning. Cloning and sequencing of the haemagglutin gene code for human Influenza Virus (69) and Hepatitis B (70) also signal the potential of the technology in developing subunit and synthetic vaccines. Transgene, a French genetic engineering company, using a viral gene for rabies surface antigen supplied by the Wistar Institute, has cloned and expressed the antigen (71). Numerous other new genetic engineering companies have cited animal vaccines as initial targets due to the need for new approaches in this area, the potential of the technology, and the relative ease (compared to human health) for government approval of animal biologicals.

Bacterial vaccines or bacterins have played an important role in the control of bacterial diseases in economic animals. Effective bacterins have been employed against potential infections caused by Brucella, Clostridia, Bacillus anthracis, and to a lesser degree, Pasteurella and Haemophilus. New technology needs to be applied for the development of protective preparations against salmonellosis, colibacillosis, Swine Dysentary (Treponema hyodysenteriae) and Moraxella. Due to their increased cellular complexity when compared to viruses, a variety of antigenic sources are available from the bacterial cell. Somatic and capsular antigens of protein and mucoprotein composition may be used. Flagellar or piliate antigens resulting from proteinaceous surface projecting structures in Salmonella, Escherichia, and Proteus may be highly antigenic. Other possibilities for eliciting humoral immunity come from the use of spore antigens from Clostridia or Bacillus, and soluble antigens such as inactivated toxins (toxoids).

Bacteria which produce piliated protein surface antigens are being studied in an attempt to generate vaccines by applying new non-destructive isolation procedures or recombinant DNA technology for production of the protein. Infections of Escherichia coli (colibacillosis) in calves and piglets cause cholera-like diarrhea in these animals. Neonatal enteric colibacillosis in piglets causes severe economic losses in swine production. The disease causes death or runting of 10 percent of newborn calves and piglets in the U.S. Current vaccines are based upon inactivated enterotoxin or killed bacterial cells. The *E. coli* bacterium produces piliated surface projections on the cell surface which enables the colonization of the mucosal surface of the small intestine by means of specific adhesion to epithelial cells. The K-88 antigen from *E. coli*, which is swine-specific, and the K-99, which is found in organisms from calves, is encoded by genes on conjugative plasmids (72,73). In addition to the episome-carried surface antigens, enterotoxins are also plasmid-borne. The gene responsible for the K-88 antigen has been cloned and mapped (74). A number of laboratories, including Cetus, Philom Bios, and Intervet, have announced their intention of producing a vaccine against coliform enteritis based on cloning and expression of the K-antigens. In one approach, the plasmid containing K-99 surface antigen was transferred from an enteropathogenic *E. coli* strain into *E. coli* minicells (ATCC 31563) to produce an anucleated live *E. coli* vaccine (75). These organisms, unable to propagate, are able to produce antigen which is capable of generating antisera reactive against enteropathogenic strains of *E. coli*. The vaccine is administered to the cow or sow before parturition so that antibodies are passed to the young before birth. Utilizing a monoclonal antibody approach, Molecular Genetics, Inc. (76) has successfully produced antibody against K-99 antigen for use in passive immunization in calves. In field trials, the monoclonal was administered orally to newborn calves and provided protection against the life-threatening *E. coli* infection.

Surface antigens, including pilus-associated antigens, are being investigated for other vaccine approaches. Antigens from *Pasteurella multocida* have been isolated utilizing potassium thiocyanate, which disrupts hydrogen bonding near the surface membrane, and isoelectric focusing electrophoresis in an attempt to purify antigens to identify immunologically-protective antigens (77,78). The protective antigen from *Bacteroides nodosus*, the causative agent for foot rot in sheep, is also associated with surface pili.

Efficacious and safe vaccines against a variety of infectious organisms, including the *Mycoplasma*, *Anaplasma*, *Lieshmania*, *Schistosoma*, and *Trypanosoma*, have eluded development. The application of recombinant DNA and monoclonal antibody technology will lead to a better understanding of the surface antigens of these organisms and the genetic determinants which control their synthesis.

African trypanosomes, a major constraint to livestock production in over 10 million square kilometers of semi-humid tropics of Africa (79), are able to evade the host's humoral immune system by varying the surface coat antigen (80). This coat glycoprotein, denoted VSG, can exist in over 100 variations caused by alterations in the sequential expression of genes specifying the surface antigens. The complete nucleotide sequence including restriction map of cDNA coding for one of these variant glycoproteins has been determined after cloning in E. coli plasmid pBR322. This surface antigen variation coupled with a better understanding of the differentiation of the organism, as in the case of T. brucei which requires non-dividing stumpy forms from slender, rapidly-dividing forms for the onset of host humoral immunity (81), will hopefully result in prophylactic treatment procedures.

Hybridoma techniques are being used to help determine the key protective antigenic determinant from coccidia Eimeria tenella (82). Identification of the antigen would form the basis for a vaccine for a disease which is controllable now only by chemotherapy.

5. IMMUNOSTIMULATORS

Vaccines cause the production of specific immunoglobulins by B-cells which, in the presence of the infectious agent presenting the surface antigen, bind to the bacterial, viral, or protozoal organism allowing for the destruction of the foreign organism by components of the complement system and macrophages. Beyond the humoral system the cell-mediated response, including macrophages, cytotoxic T-cells, and natural killer cell activity, plays an equally important role in immune defense mechanisms.

The cellular immune response, as well as the humoral response in cattle, are believed to work together in cattle against Salmonella infections. Robertsson et al. (83) showed a delayed-type hypersensitivity skin reaction in Salmonella infected cattle following intracutaneous injection of crude extracts of S. typhimurium and S. dublin. Working with protozoan infections in chickens caused by Eimeria maxima (coccidia), investigators concluded that both antibody and cell-mediated responses were necessary for inhibition of development of the parasite (84). A deficiency of immune cells in the gut allows for localized infection and ultimate circulation of the organism. Cellular mediated immunity is also operative in amoebic infections in providing host defense against establishment of amoebic liver abscesses (85).

Viral infections can repress the cellular immune response and seriously jeopardize the host's defense. Infectious Bovine Rhinotracheitis Virus, one of the infectious agents isolated from cattle suffering from respiratory shipping fever complex, has been shown to reduce alveolar macrophage function so that bacteria including Pasteurella haemolytica could not be cleared from the lungs (86). In addition, the antibody-dependent T-cell cytotoxicity response was also reduced. Depression of cell-mediated immunity has also been shown as a serious side effect of vaccines and other types of stress on the animal caused by physical, chemical, and nutritional factors.

Recently, interest has increased in the search for compounds which stimulate a depressed immune system. Although the major emphasis is on human applications, it is expected that products for animal applications will be developed. Genentech, Inc. has announced the production of bovine interferon alpha-1 using r-DNA technology for preventative and therapeutic treatment of viral related diseases such as shipping fever complex (87). Resulting from stress caused by shipment to feedlots, shipping fever causes a $500 million loss in the U.S. from sickness and death in 12 million cattle (88). Operating in part by stimulating natural killer cell activity (89), bovine interferon may be effective against a variety of virus infections. Being more effective in the earlier stages of virus infection, interferon would be best used prophylactically.

The anthelmintic compound, levamisole, has proven effective in trials for prophylactic treatment of shipping fever in cattle. The activity of the compound was noticed by its ability to augment the response of cattle to Brucellosis vaccine. This imidazothiazole compound represents a first try at utilization of immunopotentiators in stress situations. The compound stimulates cellular immunity by stimulating T-lymphocytes, increasing phagocytosis, and restoring defective leucocyte migration inhibition in response to antigen stimulation (90). A number of recently discovered compounds from microbial fermentation studies have been shown to produce immunostimulating effects. Streptoverticillium sp. ATCC 31499 produces a pentapeptide which possesses thymic hormone-like activity and induces release of interferon-type substances in mice spleen (91).

Screening from fermentation broths for inhibitors of enzymes found on the cellular surface of lymphocytes and macrophages, workers at the Institute of Microbial Chemistry, Tokyo discovered several immuno-enhancing agents including bestatin (92). Bestatin, [2S,3R-3-amino-2-hydroxy-4-phenylbutanoyl]L-leucine, is an inhibitor of aminopeptidase B and leucine aminopeptidase and was initially detected by its ability to inhibit the hydrolysis of arginine β-naphthylamide by rat macrophages, lymphocytes, and mouse spleen cells. The compound, at low concentrations in vitro, has mitogenic effects by increasing ^{3}H-thymidine incorporation into lymphocytes (93). It has been shown to enhance the recognition by the immune system to tumor cells, and seems to generally stimulate components of both the cellular and humoral immune system possibly via increased macrophage activation. Cyclic dipeptides (diketopiperazines) have been claimed capable of immunostimulative properties in certain systems. Cyclo (L-valine-L-proline) and cyclo (L-tyrosine-L-proline) have been isolated from culture broth of a Streptomyces (94). Several synthetic diketopiperazines cyclo glutamine-lysine, glutamine-ornithine, asparagine-lysine, asparagine-ornithine have been claimed in a recent U.S. patent as immunoregulatory (95). A sesquiterpene, diketocoriolin B produced by a strain of Coriolus consors, is able to stimulate mouse antibody formation to sheep red blood cells (96). An anthracycline, aclacinomycin A, was found to augment both humoral and cell-mediated response possible through inhibition of T-suppressor cells (97).

6. SUMMARY

This overview of biotechnology as applied to problems and opportunities in the production of economic animals has considered products from classical fermentation processes as well as the potential of recombinant DNA technology. Secondary metabolites produced by fungi and streptomyces will continue to be new sources of products as antimicrobials or parasiticides. The time seems right for biologicals, specifically vaccines, generated by recombinant DNA technology to increase in importance and expand market sales for these products. Other peptide-based products, i.e., growth hormones, immunostimulators, etc., should create new market potentials. Adding this new technology to the more traditional fermentation technology, the animal health industry will continue to play a critical role in providing products for the efficient production of animal protein.

7. REFERENCES

1 Business Week, June 21, 1982, 130.
2 N.B. King, Am. J. Vet. Res., 42 (1981) 1092-1108.
3 D.M. Jones and G.V.R. Born, Parasitol., 81 (1980) 383-387.
4 Animal Nut. Health, July/August, (1982) 17.
5 Biotech. News, 2 (1982) 4.
6 V. Vaisanen, K. Lounertmma, and T.K. Korhonen, Antimicrob. Agents Chemother., 22 (1982) 120-127.
7 K. Vosbeck, H. Mett, U. Huber, J. Bohn, and M. Petignet, Antimicrob. Agents Chemother., 21 (1982) 864-869.
8 D.C. Savage, J.E. Siegel, J.E. Snellen, and D.D. Whitt, App. Environ. Microbiol., 42 (1981) 996-1001.
9 J.H. Jukes, Antibiotics in Nutrition, Medical Encyclopedia, Inc., New York, 1955, p. 18.
10 J.M. McGuire, W.S. Boniece, C.E. Higgins, M.M. Hoehn, W.M. Stark, J. Westhead, and R.N. Wolfe, Antibiot. Chemother., 2 (1961) 320-327.
11 R.L. Hamill and W.M. Stark, J. Antibiot., 17 (1964) 133-139.
12 S. Omura, A. Nakagawa, H. Takeshima, J. Miyazawa, and C. Kitao, Tetrahedron Letters, 50 (1975) 4503-4506.
13 K. Vu-Trong, S. Bhuwapathanapun, and P.P. Gray, Antimicrob. Agents Chemother., 17 (1980) 519-525.
14 R.H. Baltz and E.T. Seno, Antimicrob. Agents Chemother., 20 (1981) 214-225.
15 E.T. Seno and H. Baltz, Antimicrob. Agents Chemother., 20 (1981) 370-377.
16 E.T. Seno and R.H. Baltz, Antimicrob. Agents Chemother., 21 (1982) 758-763.
17 H.A. Kirst, U.S. Patent 4,341,771 (1982).
18 H. Sakakibara, T. Fujiwara, O. Okegawa, E. Honda, S. Watanabe, and T. Matsuda, U.S. Patent 4,345,069 (1982).
19 M. Tsuchiya, M. Hamada, T. Takeuchi, H. Umezawa, K. Yamamoto, H. Tanaka, K. Kiyoshima, S. Mori, and R. Okamoto, J. Antibiot., 35 (1982) 661-672.
20 F. Bauer and G. Dost, Antimicrob. Agents Chemother., 5 (1966) 749-752.
21 H.J. Lantzsch and K.H. Menke, Arch. Gefluegelkd., 38 (1974) 82-85.
22 G.V. Davis, D.W. Arnett, and A.B. Erhart, Cattle Feeder's Day, Sum. Feed Trials, 1969-1972, June 8, 1973, 3-6.
23 J. Suzuki, M. Hori, T. Saeki, and H. Umezawa, J. Antibiot., 25 (1972) 94-104.
24 A. Torikata, H. Yoshikawa, and T. Katayama, J. Antibiot., 30 (1977) 1060-1063.
25 F. Kavanagh, A. Hervey, and W.J. Robbins, Proc. Natl. Acad. Sci., 37 (1951) 570-574.
26 F. Knauseder and E. Brandl, J. Antibiot., 24 (1976) 125-130.
27 H. Egger, G. Laber, and E. Schutze, Osterreichische Apotheker-Zertung, 35 (1981) 934-937.
28 E.M. Allan and H.M. Pirie, Res. Vet. Sci., 31 (1981) 174-176.
29 Industrial Chem. News, 3 (1982) 11.
30 M.E. Haney and M.M. Hoehn, Antimicrob. Agents Chemother., 7 (1967) 349-352.
31 Monensin, Biologics Technical Information, Calbiochem-Behring Corp., La Jolla, CA.
32 L.D. Boeck, M.M. Hoehn, R.E. Kastner, R.W. Wetzel, N.E. Davis, and J.E. Westhead, in L. A. Underkofler (Ed.), Develop. Indust. Microbiol. Vol. 18, Society Industrial Microbiology, Washington, 1977, pp. 471-485.
33 Animal Nut. Health, Jan/Feb., (1982) 10.

34 R.L. Mann and W.W. Bromer, J. Am. Chem. Soc., 80 (1958) 2714.
35 R.W. Bing, B.M. Miller, E.E. Baker, J. Birnbaum, S.A. Currie, R. Hartman, Y. Kong, R.L. Monaghan, G. Olson, I. Putter, J.B. Tunac, H. Wallick, E.O. Stapley, R. Oiwa, and S. Omura, Antimicrob. Agents Chemother., 18 (1979) 361-367.
36 H.B. Woodruff, ASM News, 48 (1982) 520-524.
37 C.C. Wang, Ann. Reports Med. Chem., 16 (1981) 269-279.
38 J.R. Egerton and D.A. Ostlind, Antimicrob. Agents Chemother., 18 (1979) 372-278.
39 T.A. Yazwinski, M. Williams, T. Greenway, and W. Tilley, Am. J. Vet. Res., 42 (1981) 481-482.
40 Y. Takiguchi, H. Mishima, M. Okuda, M. Teras, A. Aoki, and R. Fukuda, J. Antibiot., 33 (1980) 1120-1127.
41 K.G. Simpkin and G.C. Coles, J. Chem. Techn. Biotech., 31 (1981) 66-69.
42 M.J. Wolin, Science, 213 (1981) 1453-1468.
43 T.W. Perry, W.M. Beeson, and M.T. Mohler, J. Anim. Sci., 42 (1976) 761-765.
44 S. Nissen and A. Trenkle, Iowa State Univ. Vet., 1 (1976) 10-13.
45 V.R. Young, F.C. Lu, and J. Rendel (Eds.), Anab. Agents in Animal Prod., Georg Thieme Publishers, Stuttgart, 1976, pp. 20-42.
46 Chemical Week, March 17, 1982, 44-48.
47 F.N. Andrews and M. Stob, U.S. Patent 3,196,019 (1965).
48 P.H. Hidy, R.S. Baldwin, R.L. Greashman, C.L. Keith, and J.R. McMullen in D. Perlman (Ed.), Adv. Appl. Microbiol., Academic Press, New York, 1977, pp. 59-82.
49 L.J. Machlin in F. C. Lu and J. Rendel (Eds.), Anab. Agents in Animal Prod., Georg Thieme Publishers, Stuttgart, 1976, pp. 43-53.
50 Biotech. News, 2 (1982) 2.
51 L.J. Machlin, J. Dairy Sci., 56 (1972) 575-580.
52 E. Keshet, A. Rosner, Y. Bernstein, M. Gorecki, and H. Aviv, Nucleic Acid Res., 9 (1981) 19-30.
53 R. Langer, Chem. Eng. Comm., 6 (1980) 1-48.
54 K. Sano and T. Tsuchida, U.S. Patent 4,346,170 (1982).
55 R.W. Old and S.B. Primose, Studies in Microbiology, Vol. 2, Univ. California Press, Berkeley, 1980, pp. 73-88.
56 K. Nakayama and K. Araki, U.S. Patent 3,708,395 (1973).
57 A.K. Misra, J. Dasgupta, and V.C. Vora, J. Chem. Tech. Biotechnol., 30 (1980) 453-457.
58 A. Mimura, Y. Takahaski, K. Yuasa, and M. Shibukawa, U.S. Patent 4,349,627 (1982).
59 Y. Asai, M. Shimada, and K. Soda, U.S. Patent 4,335,209 (1982).
60 A. Bertelsen, Feedstuffs, May 4, 1982, p. 14.
61 E.R. Howells, Chem. Ind., 15 (1982) 508-512.
62 Feedstuffs, 54 (1982) 50.
63 A.L. Lunder, Animal Nut. Health, March, (1982) 6-8.
64 A. Buxton and R. Fraser, Animal Microbiol., Vol. 2, Blackwell Scientific Publications, Oxford, 1977, p. 611.
65 J.L. Bittle, R.H. Houghten, H. Alexander, T.M. Shinnick, J.G. Sutcliffe, R.A. Lerner, D.J. Rowlands, and F. Brown, Nature, 298 (1982) 30-33.
66 D.G. Kleid, D. Yansura, B. Small, D. Dowbenko, D.M. Moore, M.J. Moore, M.J. Grubman, P.D. McKercher, D.O. Morgan, B.H. Robertson, and H.L. Bachrach, Science, 214 (1981) 1125-1128.
67 F. Brown, Trends Biochem. Res., December, 1981, 325-330.
68 J.K. Rose and C.J. Gallione, J. Virol., 39 (1981) 519-528.
69 I. Heiland and M.J. Gething, Nature, 291 (1981) 851-852.

70 G.R. Dressman, Y. Sanchez, I. Ionesco-Matin, J.T. Sparrow, H.R. Six, D.L. Peterson, F.B. Hollinger, and J.L. Melnick, Nature, 295 (1982) 158-160.
71 Chemical Week, August 18, 1982, 23.
72 I. Orskov and F. Orskov, J. Bacteriol., 91 (1966) 69-75.
73 H.W. Smith and M.A. Linggood, J. Med. Microbiol., 5 (1972) 243-250.
74 F.R. Mooi, F.K. de Gaaf, and J.D.A. van Embden, Nucleic Acids Res., 6 (1979) 849-865.
75 G.G. Khachatourians, U.S. Patent 4,311,797 (1982).
76 SIM News, 32 (1982) 9.
77 K.L. McKinney and P.A. Rebes, Can. J. Microbiol., 28 (1982) 511-521.
78 D.J. Stewart, B.L. Clark, and J.E. Peterson, Res. Vet. Science, 32 (1982) 140-147.
79 A.C. Allison and E.M. Enger in J. E. Butler (Eds.), Adv. Exp. Med. Biol., Vol. 137, Plenum Press, New York, 1981, pp. 225-240.
80 G. Allen, L.P. Gurnett, G.A.M. Cross, J. Mol. Biol., 157 (1982) 527-546.
81 C.N. Sendashonga and S.J. Black, Parasite Immunol., 4 (1982) 245-257.
82 Ind. Chem. News, 3 (1982) 11.
83 J.A. Robertsson, S.B. Svenson, and L.H.M. Renstrom, Res. Vet. Science, 32 (1982) 225-230.
84 M.E. Rose and P. Hesketh, Parasite Immunol., 4 (1982) 171-185.
85 E. Ghadirian and E. Meerovitch, Parasite Immunol., 4 (1982) 219-225.
86 A.J. Forman and L.A. Babiuk, Infect. Immunity, 35 (1982) 1041-1047.
87 Biotech. News, 2 (1982) 1.
88 Animal Nut. Health, Jan/Feb (1979) 19-20.
89 R.M. Krause, ASM News, 45 (1979) 358-366.
90 P.N. Plowman, Trends Pharm. Science, 2 (1981) 6-7.
91 SCRIP, 567 (1981) 15.
92 H. Umezawa, M. Ishizuka, T. Aoyagi, and T. Takeuchi, J. Antibiot., 29 (1977) 857-859.
93 M. Ishizuka, J. Sato, Y. Sugiyama, T. Takeuchi, and H. Umezawa, J. Antibiot., 32 (1980) 653-662.
94 M. Munekata and G. Tamura, Agric. Biol. Chem., 45 (1981) 2613-2618.
95 G. Heavner, F. Tjoeng, and G. Goldstein, U.S. Patent 4,331,595 (1982).
96 M. Ishizuka, T. Takeuchi, and H. Umezawa, J. Antibiot., 34 (1981) 95-102.
97 M. Ishizuka, T. Takeuchi, T. Masuda, S. Fukawawa, and H. Umezawa, J. Antibot., 34 (1981) 331-340.

PRODUCTION, PROPERTIES AND APPLICATIONS OF XANTHAN

JOHN F. KENNEDY and IAN J. BRADSHAW
Research Laboratory for the Chemistry of Bioactive Carbohydrates and Proteins, Department of Chemistry, University of Birmingham, Birmingham B15 2TT, U.K.

1. INTRODUCTION

Xanthan (or xanthan gum) is the common name for the microbial exopolysaccharide produced by fermentation with the bacterium Xanthomonas campestris. Since substantial commercial production of xanthan commenced in 1964, this biopolymer has become established as a thickening, suspending, stabilising and gelling agent of major importance in both food and industrial applications. Surprisingly, it is only more recently that the molecular basis for the remarkable and sometimes unique physical properties of xanthan have been explained [1].

Naturally, the success of xanthan has induced widespread screening programs in recent years for microorganisms which can produce biopolymers of equal or even greater commercial interest. However, it is apparent from various reports [2-4] that a polysaccharide will only display solution properties of commercial interest if it can adopt secondary and tertiary structures in an aqueous environment. This can only take place if the nature of the chemical (or primary) structure of the polymer does not impose restrictions on the formation of such ordered structures. Although relatively few of existing monosaccharides are commonly found in the composition of microbial polysaccharides, the various permutations available for linkage between the monomer units via glycosidic linkages is large. When the further variations of sugar ring conformation and glycosidic linkage configuration are considered, it is clear that the variety of possible chemical structures for biopolymers produced by microorganisms is enormous. Not surprisingly, few of microbial polysaccharides investigated actually possess useful solution properties. Indeed, it is possible that the chemical structure of the polymer may have to be slightly unusual in some respect. In the case of xanthan, many of the unique solution properties stem from the cellulose-like backbone [5,6] of the molecule.

A major obstacle to the introduction of potentially useful biopolymers onto the commercial market is the development work to scale-up from the production of a polymer by bacteria on solid media in the laboratory to that of an economic large-scale industrial fermentation. For instance, the microorganism must not only be capable of synthesising the biopolymer in liquid media, but also be capable of converting the carbon source (e.g. D-glucose) to the polymer with reasonable efficiency. However, there are various microbial polysaccharides available with potential for commercialisation such as scleroglucan [7], pullanan [8], microbial sodium alginate [9], Zanflo [10] and PS-60 polysaccharide [11].

It would seem likely that an established product such as xanthan will continue to lead the commercial application field of biopolymers for some years to come. It is anticipated that the most important and financially lucrative future use for xanthan will be as a mobility control agent [12-14] in Enhanced

Oil Recovery (EOR). However, the utilization of xanthan in EOR, in addition to the expected increase in other commercial areas, would require a significant increase in the present levels of production. Thus, it is pertinent at this point in time to review the current methods of xanthan production, as well as the solution properties and commercial applications of this unique gum.

2. ORIGIN OF XANTHAN

The interest in the polysaccharide xanthan was instigated by the Northern Regional Research Laboratories (N.R.R.L.) of the U.S. Department of Agriculture who conducted an extensive screening program of their wide collection of cultures during the 1950's for microorganisms which could biosynthesise water-soluble gum products of likely commercial importance. This particular study was carried out for two main reasons. Firstly, it was considered that there was a market in the United States for new water-soluble gums to supplement or replace the existing ones derived from plant sources such as seaweed (alginates), trees (e.g. gum arabic), seeds (e.g. locust bean gum) and cereals (starch). Since the supply and quality of the majority of plant derived gums is subject to change due to seasonal variations (alginates), adverse climatic conditions (e.g. gum arabic, alginates and starches) and skilled labour dependency (e.g. gum arabic, gum karaya and gum tragacanth); it was recognised that the production of gums by fermentation might be more reliable. Secondly, due to the overproduction of corn starch the U.S. Department of Agriculture had a plentiful supply of a carbon source which could be readily converted to D-glucose for use in the fermentation of microbial gums.

Of the various microorganisms which were tested, the bacterium *Xanthomonas campestris* NRRL B-1459 was found to produce the polymer with the most interesting solution properties [15]. In particular, solutions of polysaccharide B-1459 (later named as xanthan gum) were found to exhibit (a) high viscosity at low concentration (b) high pseudoplasticity levels (i.e. reversible decrease in viscosity with increase of shear rate) (c) atypical insensitivity to a wide range of salt concentrations, pH and temperature. (This is a simplified version of the unique solution properties of xanthan, but further details will be discussed later). Equally importantly, it was found that the biopolymer could be produced in good yield from D-glucose by fermentation with *X. campestris* [16]. This work was responsible for the subsequent venture into the commercial production of xanthan by Kelco Co. Inc. in America. A further significant event in the development of xanthan was the acceptance of the polymer as a food additive by the USA Federal Food and Drug Administration in 1969. To meet the growing demand for xanthan, the polymer is now produced by other companies such as Rhone-Poulenc (France) and Pfizer Inc. (USA). In addition, it is expected that Kelco Co. will start commercial production of xanthan in the U.K. in the near future.

X. campestris, the bacterium responsible for the biosynthesis of xanthan, is a naturally occurring microorganism, originally isolated from cabbage plant where it causes black rot disease [17]. Although the exact origin of the X. campestris strain NRRL B-1459 is unknown, it was subsequently deposited in the American Type Culture Collection as ATCC 13951. Production of xanthan with X. campestris in the laboratory was first reported by Lilly et al [18] in 1958, who used only the natural carbohydrates, proteins and trace minerals of the cabbage plant as the source of fermentation substrate. Evidence by Sutton and Williams [19] has shown that the chemical composition of xanthan produced by commercial fermentation is similar to the polymer biosynthesised on living cabbage tissues under natural conditions. In addition, both the polysaccharide fractions from culture filtrates and that from infected cabbage leaves reacted in immunodiffusion tests to give similar precipitin bands with an antiserum to the polysaccharide from culture filtrates. Thus, it seems that xanthan produced by fermentation is identical to the polymer produced on cabbage leaves.

A number of other Xanthomonas spp. and related strains of X. campestris are known to exude a gum during fermentation [20,21]. Originally, it was considered that there was a large number of distinct Xanthomonas species. However, only four species of Xanthomonas are now recognized (viz. X. campestris, X. albilineans, X. axonopodis and X. fragaria) [22]. A common feature of Xanthomonas species and related strains is that they pathogenize specific plant hosts. Some examples of which are cabbage (X. campestris), sugar cane (X. albilineans and X. vasculorum), strawberry (X. fragaria) and walnut (X. juglandis).

3. STRUCTURE OF XANTHAN

3.1 Basic Repeating Unit

Xanthan was initially shown by Sloneker and Jeanes [23] to be composed of D-glucose, D-mannose, D-glucuronic acid, acetic acid and pyruvic acid. Based on the Smith degradation analysis and methylation analysis of Sloneker et al [24] and Siddiqui [25] respectively, the primary structure of xanthan was originally proposed to consist of a 14 or 16 sugar unit repeat sequence. These early studies failed probably through the lack of suitable techniques to give accurate representation of the polymer molecule. Due mainly to the biosynthetic complexity which would be required, the tentative structures [24,25] proposed for xanthan were regarded with scepticism. Several sets of workers [5,6,26] were, thereby, inspired to reinvestigate the structure of the polysaccharide.

The accepted structure of xanthan (Fig. 1) was first proposed by Jansson et al [5]. Elucidation of the structure was obtained by using more sophisticated techniques of methylation analysis [27] and a specific degradation scheme for polysaccharides containing uronic acid residues [28]. The structure consists of a (1 → 4) β-D-glucopyranose or cellulose-like backbone to which trisaccharide side-chains are attached to alternate sugar residues on the main chain at the C-3 position. The side-chain consists of two D-mannose residues and a D-glucuronic acid residue. The terminal β-D-mannopyranose residue is (1 → 4)-linked to the β-D-glucuronic acid residue, which in turn is (1 → 2)-linked to the non-terminal α-D-mannopyranose residue. In addition, the methylation evidence of Jansson et al [5] indicated that the acetic acid was O-6 linked to the non-terminal D-mannopyranose residue and that the pyruvic acid was acetal linked through O-4 and O-6 to the terminal D-mannopyranose residue. However, both these constituents occur in non-stoichiometric amounts. The 'S' configuration of the pyruvic acid acetal had been established earlier by Gorin et al [29], when it was still considered [30] that this constituent was attached to a D-glucopyranose residue. Support for the above proposed repeating unit of xanthan came from Melton et al [6] and Lawson and Symes [26] who used partial acid hydrolysis and acetolysis respectively to determine the structure.

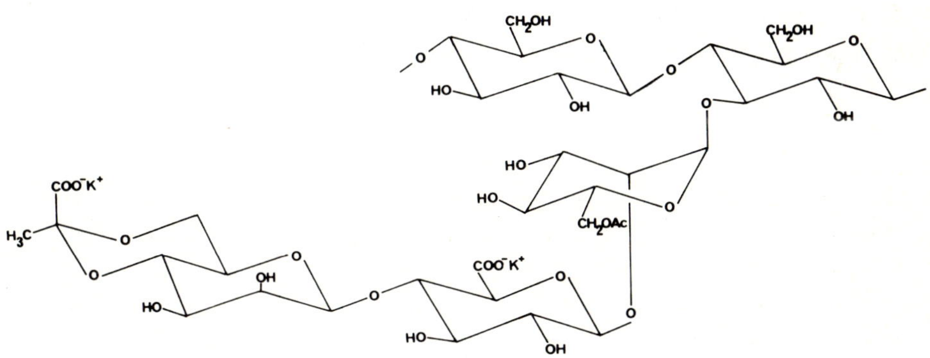

Fig. 1. Structure of Xanthan Repeating Unit.

A range of values have been reported for the molecular weight of the xanthan polymer. Using classical light-scattering techniques, Dintzis et al [31] obtained a molecular weight $\bar{M}_w = 2 \times 10^6$ for xanthan in 4 M aqueous urea, which had been heated to 90°C prior to collection of data. In contrast, urea solutions of two xanthan fermentation broths which had undergone no thermal treatment gave molecular weights of $\bar{M}_w = 13 \times 10^6$ and 50×10^6. Using similar techniques, Rinaudo and Milas [32] obtained a value of $\bar{M}_w = 2 \times 10^6$ for xanthan in aqueous NaCl solutions. Both of the above studies disagree with the work of Holzwarth [33], who performed band-sedimentation analysis of xanthan solutions (in 0.75 M NaCl) in which the macromolecules had been tagged with a fluorescent group. The fluorescent derivatives of xanthan were prepared by isocyanide coupling of fluoresceinamine to the carboxyl groups of the polymer. This study produced values of $\bar{M}_w = 15 \times 10^6$ and 62×10^6 for commercially purified xanthan and untreated fermentation broth respectively. Recent observations of a time-dependent aggregation of xanthan in aqueous solution by Southwick et al [34], using quasi-elastic light scattering, may have provided an explanation for the discrepancies in the reported values for the molecular weight of the polymer. Since no associations were found with xanthan in 4 M urea solutions, this implies that hydrogen bonding is important in stabilizing the aggregates in water. Variation of the fermentation conditions used in the production of xanthan is a further factor which could influence the molecular weight of the polysaccharide.

3.2 Effect of Variation of *Xanthomonas* Strain and Fermentation Conditions on Xanthan Repeating Unit

Although the main structural features of xanthan have now been established, numerous studies have shown that variation in *Xanthomonas* strain and culture conditions gives rise to some variation in the composition of the polymer molecule. In this respect xanthan is somewhat unusual, since a large number of other microbial exopolysaccharides show uniformity in their chemical structure [35]. While the variation in xanthan structure tends to be mainly due to change in pyruvate and acetate levels, variation in carbohydrate composition can also occur. The variation in degrees of acetylation and pyruvylation in several commercial samples of xanthan are illustrated in Table 1 [36]. Variations of this nature must be expected for several reasons. Firstly, different commercial companies producing xanthan may use slightly differing strains of *X. campestris*. Further, different conditions for maintaining stock cultures of the microorganism may be employed. Lastly, it is probable that other fermentation parameters such as medium composition and culture time are different. As shown (Table 1), variations of 0.32 - 0.44 and 0.34 - 0.90 have been observed for degrees of pyruvylation and acetylation respectively in various commercial xanthan samples. Interestingly, xanthan products from the same company can even show differences in their chemical composition. This emphasises the difficulty in maintaining consistency during the microbiological production of xanthan.

TABLE 1

Degrees of acetylation and pyruvylation for several commercial xanthan samples. Data of Smith *et al* [36]*

Xanthan sample	Degree of pyruvylation	Degree of acetylation
Keltrol[a]	0.39	0.90
Keltrol[a]	0.32	0.35
Kelzan[a]	0.40	0.34
Rhodopol[b]	0.44	0.67
Rhodigel[b]	0.39	0.53
Viskotan[c]	0.34	0.79

Sources of samples were [a]Kelco Co. Inc. (San Diego, USA), [b]Rhone-Poulenc Industries (Paris, France), [c]Lohmann Fermentations, GmbH, (Cuxhaven, W. Germany.) The latter company no longer produces xanthan.

As mentioned above, proper maintenance of *X. campestris* stock cultures is important for consistency in the production of xanthan, as variation is a recognised characteristic in *Xanthomonas* species [37]. Indeed, recommended procedures [38] have been reported which allow the physiological state and synthetic activity of the *X. campestris* NRRL B-1459 strain to be maintained constant and invariable. However, when colonial variation does occur with *X. campestris*, changes in the level of pyruvate substitution have been observed [39-41]. In this case the variants were substrains found in certain sub-cultures of *X. campestris*. Despite the loss in pyruvate, the chemical composition of the xanthan produced by normal and variant strains of *X. campestris* was almost identical (Table 2).

TABLE 2

Comparison of the composition of xanthan obtained from normal and sub-strains of X. campestris. Data of Sandford et al [41] reproduced with permission*

Type of X. campestris strain	Composition (g/100g)				
	D-glucose	D-mannose	D-glucuronic acid	Pyruvate	Acetate
normal	37.0	43.4	19.5	4.4	4.5
sub-strain	37.7	42.9	19.3	2.5	3.7

A wide variation in degrees of pyruvylation and acetylation of xanthan samples was also observed [42] when numerous strains of Xanthomonas were used to produce the polymer (Table 3). Fermentation conditions for the production of the xanthan samples were kept constant, except for a few strains of Xanthomonas which required addition of amino acid supplement to the medium for satisfactory growth. Nevertheless, the carbohydrate composition of the polymer samples showed little variation with D-glucose, D-mannose and D-glucuronic acid occurring in the approximate molar rations 2:2:1; consistent with the accepted structure of xanthan. However a range of pyruvate and acetate contents was noticeable in the different preparations. In particular, two strains (556 and 1128) of X. phaseoli produced polymers with very low acetate and pyruvate contents respectively. Variation of acetate and pyruvate levels was also found [42] in xanthan samples when a laboratory strain of X. campestris was grown on alternative carbon sources (alanine, glutamate, glycerol and pyruvate) to D-glucose. Again, the carbohydrate composition of the polymer was consistent with the structure of xanthan.

The above studies were all performed using batch fermentation techniques. However, the variation in composition of xanthan produced by X. campestris [43] and X. juglandis [44] using continuous culture with various growth-limiting nutrient conditions has also been studied. Davidson [43] found that the carbohydrate composition of xanthan was more or less identical when X. campestris was grown under various nutrient limitations in a chemostat culture (Table 4). Although levels of acetate showed only a small variation (3.1 - 5.2%), pyruvate content showed considerable differences (0.9 - 8.5%). Particularly low levels (0.9 and 1.1%) of pyruvate in xanthan were observed when X. campestris was produced under phosphorus and magnesium limitations respectively, suggesting that these two elements have a direct effect on the phosphoenolpyruvate and the pyruvylation mechanism of the biosynthesis.

TABLE 3

Chemical composition of various *Xanthomonas* polysaccharides.

Data of Sutherland [42] reproduced with permission

Strain	D-Glucose	D-Mannose	D-Glucuronic acid	Pyruvate	Acetate
X. campestris	30.1	27.3	14.9	7.1	6.5
X. albilineans 1738[a]	30.2	27.5	13.9	5.1	5.8
X. albilineans 2257[a]	25.5	26.1	14.2	5.8	5.1
X. axonopodis 457	30.6	28.6	13.7	7.2	5.5
X. fragaria 1469[a]	32.5	28.5	15.1	7.0	3.9
X. fragaria 1822	24.6	26.1	14.0	4.9	5.5
X. gummisudans 2182	34.8	30.7	16.5	4.7	10.0
X. juglandis 411	33.2	30.2	16.8	6.9	6.4
X. manihotis 1159	33.3	28.7	16.7	7.6	6.8
X. phaseoli 556	29.1	34.2	17.8	7.7	1.8
X. phaseoli 1128	30.9	28.6	15.3	1.8	6.4
X. vasculorum 702	34.9	30.2	17.9	6.6	6.3
X. vasculorum 796	33.2	26.2	15.3	5.4	8.8

[a] These preparations also contained D-galactose

TABLE 4

Production and composition of xanthan produced by *X. campestris* when grown in continuous culture under various growth limitations

Data of Davidson [43] reproduced with permission.

	Limiting nutrient					
	Glucose	Nitrogen	Sulphur	Magnesium	Potassium	Phosphorus
Cell yield ($g\ l^{-1}$)	1.1	1.6	1.8	1.5	1.7	2.9
Polymer yield ($g\ l^{-1}$)	2.7	7.0	6.4	5.7	5.3	5.0
Ratio of mannose: glucose	1.10	1.08	1.09	1.20	1.13	1.18
Uronic acid (%)	21	21	18	16	23	17
Acetate (%)	4.9	3.1	-	5.2	4.5	-
Pyruvate (%)	5.8	6.6	8.5	1.1	5.5	0.9

When X. juglandis was grown under various growth limitation conditions in a chemostat culture two polysaccharides were produced [44]. While one polymer was similar in composition to xanthan, the other one was composed of glucose and rhamnose. The composition of crude polysaccharide isolated by acetone precipitation of the fermentation medium is illustrated in Table 5. Material produced from sulphur-limited cultures of X. juglandis was very similar in composition to xanthan. In contrast, the use of carbon-limited cultures gave rise to "xanthan" material heavily contaminated with the other polymer. Interestingly, the composition of the material produced under potassium and magnesium limitation showed a large increase in glucose content and corresponding decrease in mannose and glucuronic acid, compared to the other nutrient limitation conditions employed. Since this material was only poorly soluble, it seems possible that it was a polymer with a cellulose backbone containing a much reduced side-chain substitution compared to the normal xanthan molecule.

The production of xanthan material which has reduced side-chain substitution has also been observed [45] during batch fermentation of certain mutant strains of a wild-type X. campestris (T646). Trace amounts of the microbial exopolysaccharide which was produced by the non-mucoid mutant (T646NM2) gave a similar analysis (Table 6) to the wild-type polysaccharide except for a much reduced pyruvate content. (The acetate content was very low for both the wild-type and non-mucoid mutant derived polysaccharide). However, the analysis of the polymer material produced by a 'crenated' mutant of X. campestris revealed that the ratio of glucose to mannose was not 1:1, as usually occurs in xanthan. Although these polymer preparations were contaminated with a lipopolysaccharide (composed of galactose and rhamnose), the reduced amounts of mannose and glucuronic acid suggests that some regions of the "xanthan" molecular backbone lacked sidechains or parts thereof.

A further point of interest is in the nature of distribution of the pyruvate groups in the xanthan molecule. Typical pyruvate contents indicate that 1 in 2-3 terminal D-mannopyranose residues are substituted with this constituent. However, the question arises whether or not some strands of xanthan molecules contain greater substitution than others or alternatively that the pyruvate is more randomly distributed between molecular chains. Two lines of evidence suggest that the former is the more likely, since xanthan can be separated into fractions of various pyruvate contents. Sandford et al [46] achieved this separation by fractional precipitation of a xanthan sample with ethanol. Alternatively, Sutherland [47] observed that an affinity column consisting of Rhizobium strain TAI antiserum coupled to Sepharose gel was able to separate xanthan into pyruvylated and non-pyruvylated molecular strands. An earlier

TABLE 5

Composition of xanthan produced by *X. juglandis* XJ 107 in continuous culture when grown under various growth limitations

Data of Evans *et al.* [44] reproduced with permission.

Limiting nutrients	Dilution rate (h^{-1})	Percentage composition of crude polysaccharide					
		Glucose	Mannose	Glucuronic acid	Rhamnose	Pyruvate	Acetate
Nitrogen[a]	0.03	34.3	23.2	18.0	5.0	1.6	3.0
Nitrogen	0.03	34.0	26.3	18.0	5.7	1.7	3.0
Carbon	0.045	38.0	24.1	15.2	15.0	1.8	0.5
Potassium	0.03	75.0	5.6	2.0	7.5	-	-
Magnesium	0.03	63.3	10.7	7.1	7.5	-	-
Phosphorus	0.03	29.9	29.0	17.0	2.0	3.7	3.0
Sulpher	0.035	34.3	30.8	18.1	1.4	1.5	1.6-3.2

[a] Complex medium used. In all other cases a chemically defined medium was employed.

TABLE 6

Comparison of the composition of exopolysaccharides produced by wild-type, "non-mucoid" and "crenated" mutant strains of *X. campestris*.

Date of Whitfield *et al* [45] reproduced with permission.

Strain	Glucose (%)	Mannose (%)	Uronic acid (%)	Galactose (%)	Rhamnose (%)	Pyruvate [mmol. g polymer^{-1}]	Acetate [mmol. g polymer^{-1}]
T646	42	38	20	0	0	24	0.7
T646NM2[a]	40	38	22	0	0	1.9	0.8
T646D[b]	22	19	13	24	22	<0.1	0.1
T646KR[b]	29	21	11	27	12	<0.1	0
T646E[b]	37	12	9	28	14	<0.1	0.1

T646 was obtained as *X. campestris* ATCC 13951 from the American Type Culture Collection

[a]Non-mucoid mutant which formed smaller colonies than the wild-type bacteria and lacked the characteristic glossy mucoid appearance.

[b]Class of mutants of unusual colonial morphology referred to as crenated mutants.

observation [48] that the antibodies against the polysaccharide from Rhizobium strain TAI were directed against pyruvate acetal groups formed the basis for this separation technique.

It would appear that under standardised fermentation conditions, numerous Xanthomonas species or related strains will produce a polymer of similar carbohydrate composition. However, one strain (S19) of Xanthomonas is known to produce a polymer which contains galactose [49]. It has been observed that the solution properties of X. campestris and X. phaesoli exopolysaccharides are similar [50, 51]. Thus, it seems likely that the majority of Xanthomonas polysaccharides will resemble "normal" xanthan in not only carbohydrate composition, but also physical properties in solution. Certainly, one difference between "xanthans" is the varying degrees of pyruvylation and acetylation present in the molecule. Since these two parameters also vary in commercial samples of xanthan, it is of interest to know what influence these groups have on the solution properties of the polymer. Conflicting views exist [41, 52] on this point and will be discussed in more detail later. Even the same strain of Xanthomonas can produce variation in acetate and pyruvate content, when grown under different fermentation conditions. Moreover, under certain growth conditions it appears that the xanthan which is produced has side-chain residues, partly or wholly, missing.

4. CONFORMATIONAL ASPECTS OF XANTHAN IN SOLUTION

Until recently, structural knowledge of polysaccharides had been confined solely to their chemical (or primary) sequence. However, the increasing diverse commercial use of polysaccharides, particularly xanthan, has created the need for an explanation of the solution properties of these gums at a molecular level. As mentioned earlier, polysaccharides which display interesting solution properties tend to form ordered (secondary, tertiary or quaternary) structures in an aqueous environment. Conveniently, various physical methods such as nmr, circular dichroism and optical rotation are capable of detecting order-disorder transitions of polysaccharides in solution. These techniques are useful, therefore, in not only correlating the solution properties of polysaccharides such as xanthan with their molecular structure, but also in the screening of new polymers for potentially useful properties.

Evidence for the existence of an ordered form of xanthan in solution came initially from Rees [53] who observed a sigmoidal change in the optical rotation of the polymer with variation of temperature (see Fig. 2). This was interpreted as an indication that a conformational change (order-disorder transition) was taking place in solution. Observation of this conformational transition has since been confirmed by further optical rotation work [50,54,55], viscosity measurements [50, 54], nmr relaxation data [50] and circular dichroism [50]. It has been found [54, 55] that an increase in ionic strength causes an increase in the transition midpoint temperature (Tm) of the conformational change which occurs with xanthan in solution. Indeed, at the salt levels encountered in most commercial applications the ordered form of xanthan is stable to above 100°C. The transition mid-point temperature is also sensitive to variation in xanthan composition, chemical removal of pyruvate and acetate results in increases and decreases respectively in Tm [56]. Further, reduction in the molecular weight of the polymer by partial cellulase hydrolysis was observed [57] to cause a decrease in the melting temperature (Tm) of the polymer.

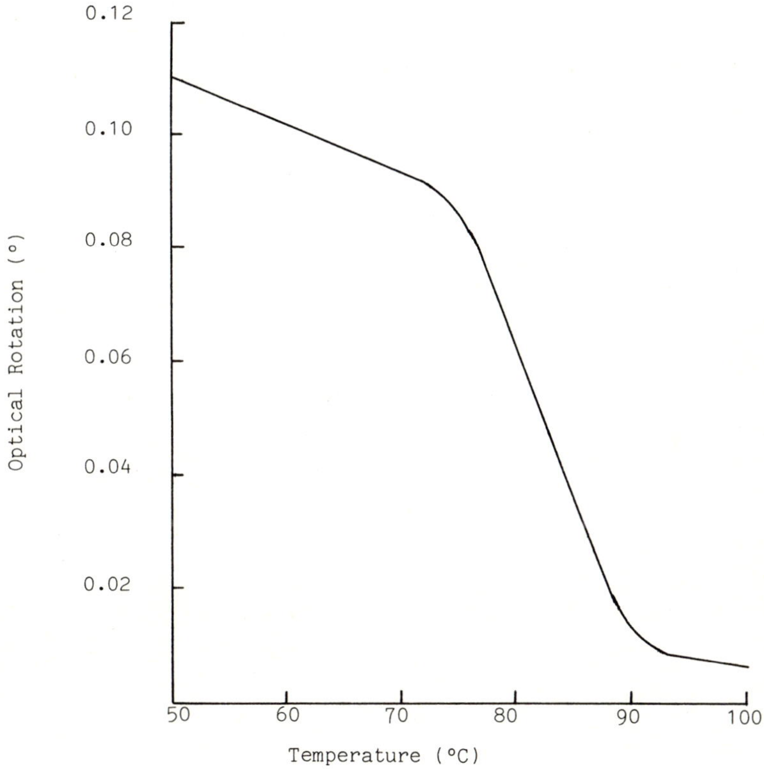

Fig. 2 Effect of temperature on optical rotation of xanthan solutions (1%). Data of Rees and Welsh [4] reproduced with permission

X-ray diffraction studies by Moorhouse et al [58] on xanthan fibres showed that in the solid state and probably solution, the polymer adopted a right-hand helical conformation having five-fold symmetry and a pitch of 4.7nm. The X-ray pattern was interpreted in terms of a single-stranded structure, but a double-stranded helix was not ruled out. Subsequently, from various lines of evidence two conflicting models, single and double-stranded structures, have been proposed for the conformation of xanthan in solution.

Based on nmr relaxation, optical rotation and circular dichroism studies, Morris [1] proposed that xanthan existed as a single-stranded ordered structure in solution in which the charged trisaccharide side-chain folds back along the cellulose backbone, to give a rigid rod-like molecule. Support for this model has come from a variety of physical techniques. Firstly, the thermally induced transition of xanthan observed by optical rotation was found to be independent of polymer concentration [50, 55]. Thus, suggesting that the order → disorder transition of xanthan is either unimolecular or extremely cooperative. While the latter alternative does occur with DNA, it is unlikely for xanthan since the transition is rather broad. Secondly, polarimetric stop-flow studies showed that the salt-induced disorder → order transition of xanthan possessed first order kinetics [59]. Lastly, low angle light scattering studies revealed that no increase in the molecular weight of the polysaccharide took place until well below the transition mid-point [60]. Again, suggesting that an intra- and not inter-molecular conformational transition was occurring. The observed increase in the molecular weight of xanthan at low temperature was attributed to an entropically favoured aggregation of the ordered chains [60]. Clearly, the existence of xanthan in solution as an ordered single-stranded helix would require stabilisation by intra-molecular bonding. While no behaviour of this type has been observed for a polysaccharide, ordered structures of this nature occur with the protein α-helix.

In contrast to the single-stranded model for xanthan, it has also been proposed that the polymer exists as a double-stranded ordered structure in solution. Initial evidence from optical rotation studies by Holzwarth [54] on the thermally induced transition associated with xanthan, suggested that multi-stranded breakdown might be involved with the order → disorder conformational change. Following electron micrograph studies on very thin films of xanthan, Holzwarth and Prestridge [61] proposed that the native structure of the polymer was multi-stranded with two (or perhaps three) strands arranged in a right-handed helix. Support for the double-stranded model has come from hydrodynamic [33] and low angle light scattering [62] evidence. These studies gave rise to values of ~ 1900 and ~2000 daltons/nm respectively for the linear mass density of xanthan, which are approximately twice the value for the single-stranded structure proposed from X-ray work.

Thus, the exact nature of the secondary structure of xanthan in solution has not been clarified and further work will be required to establish which of the two models (single- or double-stranded) is the more valid. Another aspect of the conformation of xanthan in solution which remains uncertain is the rigidity of the molecule. Morris [1] originally proposed that xanthan was a rigid rodlike molecule in solution. Rinaudo and Milas [32] have also adopted this model to fit their intrinsic viscosity and sedimentation data. Holzwarth [63] interpreted intrinsic viscosity evidence to suggest that xanthan was composed of stiff, but worm-like chains in solution. An alternative to these views was put forward by Whitcomb and Macosko [64], who deduced from their intrinsic viscosity data that xanthan existed in solution as rodlike molecules, which possessed some degree of flexibility. The hydrodynamic length of the molecule was estimated to be ~ 1.5 μm.

Regardless of the nature of the conformation adopted by xanthan, it seems likely that the polymer molecules show some form of association in solution. This has been exemplified by observations of the unusual behaviour of xanthan solutions to increasing ionic strength. Normally, polyelectrolytes such as xanthan experience a drop in viscosity on addition of inorganic salt, as the highly expanded conformation which is present at low ionic strength adopts a more compact random coil due to charge screening [65]. However, the solution viscosity of xanthan is not only stable to salt addition, but can actually show an increase with polymer concentrations above ~ 0.2-0.5% [15]. Recent data [66] on the flow of xanthan solutions through Millipore filters, and quasi-elastic light scattering measurements of hydrodynamic radii suggest that an increase in the hydrodynamic size of the polymer takes place with added electrolyte above a critical value (~ 0.02M NaCl). Southwick et al [66] have proposed that this increase in hydrodynamic dimensions at >0.02 M NaCl is due to screening of the intermolecular, electrostatic repulsions by the excess of salt ions leading to a self-association of xanthan chains.

It appears that the pyruvate groups of xanthan play some role in this macromolecular association. Smith et al [36] observed that the fractional change in viscosity on addition of KCl to salt-free (1%) solutions of purified xanthan was positive only when the degree of pyruvate substitution exceeded ~ 0.31. They suggested that the macromolecular association of xanthan solutions was promoted by the intermolecular apolar interactions of pyruvate methyl groups, which are suitably located on the periphery of the helical polymer structure. Further evidence that pyruvate was involved in the macromolecular association was obtained through Millipore filtration experiments on xanthan solutions [61]. Native xanthan solutions showed an increasing resistance to flow through a 3 μm filter with increasing concentration of NaCl (Fig. 3). In comparison, pyruvate-free xanthan prepared by very mild TFA hydrolysis [52] was able to flow through the Millipore filter with comparative ease over the full range of NaCl concentrations studied. Pyruvate-free xanthan prepared from a mutant strain of X. campestris has also been reported to show improved filtration properties [68]. No evidence was found by Nisbet et al [67] that deacetylation of the native xanthan enhanced passage through Millipore filters and thus acetate is unlikely to be involved in the macromolecular association process. This is in agreement with the view of Smith et al [36] who concluded that the acetate groups were located too close to the centre of the xanthan helix to be involved in the intermolecular interaction.

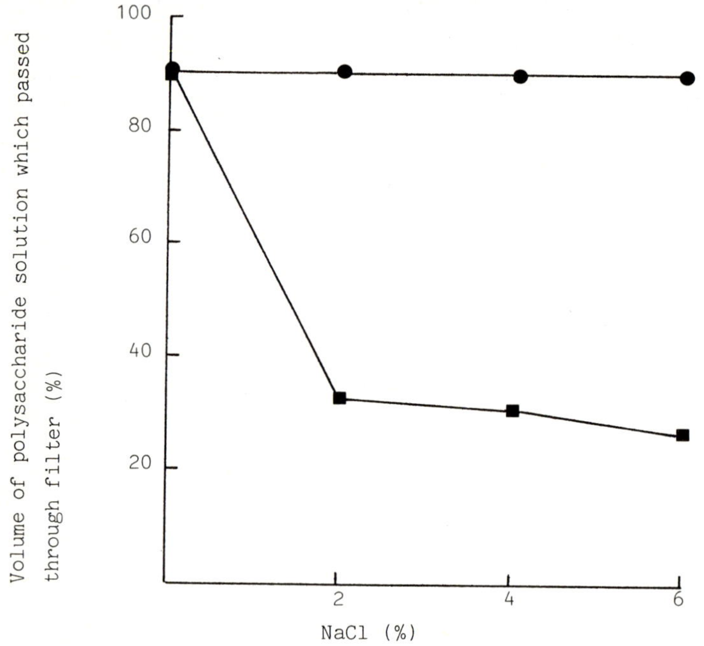

Fig. 3 Comparison of flow of solutions (0.3%) through a 3 μm filter for xanthan (●) and pyruvate-free xanthan (■). Data of Nisbet et al (67).

The loss of entropy which occurs with the aggregation of molecules in solution should be less for rigid rods than for random coils. It follows that the salt-induced macromolecular association of xanthan chains should be energetically favourable with respect to entropy losses. Thus, shielding of the interchain, electrostatic repulsions by salt ions will allow the favourable enthalpic interactions (such as hydrophobic forces and hydrogen bonding) to take place, resulting in the xanthan chain associations. Evidence that the xanthan can form parallel, side-by-side arrays has been provided by Rinaudo et al [69] who observed formation of cholesteric liquid crystals in polymer solutions using birefringence, optical rotation and laser light diffraction measurements.

5. MOLECULAR BASIS FOR SOLUTION PROPERTIES OF XANTHAN

Although the precise nature of the xanthan conformation has not been resolved, the inter- and intramolecular interactions between the polymer molecules, arising from the secondary and tertiary structures, can satisfactorily account for the unusual and sometimes unique physical properties of this polysaccharide in solution. The most basic property of xanthan is the ability to form solutions of very high viscosity at-rest or low-shear rates even with low polymer concentrations (Fig. 4). This arises from the rigidity of the ordered structure, intermolecular interactions and high molecular weight of the polymer.

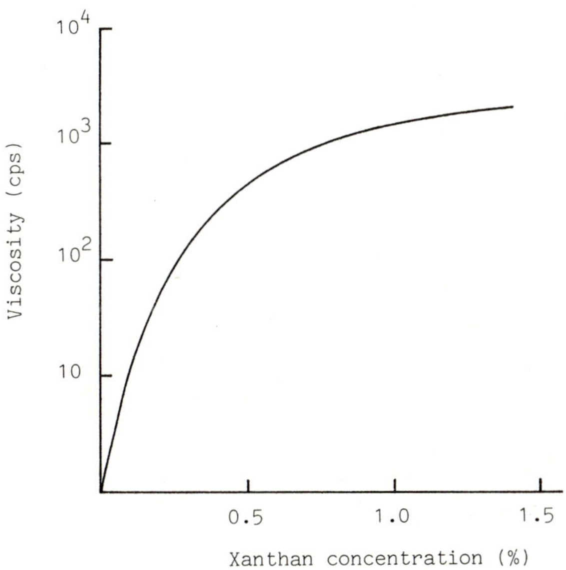

Fig. 4 Effect of concentration on viscosity of xanthan. Measurements are shown for a shear rate of 10 s^{-1}

Since the extent of pyruvate and acetate substitution in xanthan samples can vary considerably (Section 3.2), it is of interest to know what effect these groups will have on viscosity and other solution properties of the polymer. This is particularly important if consistency is to be maintained in xanthan solutions used in commercial applications. Sandford et al [41] suggested that the solution viscosity of xanthan was greater for polymer material with high pyruvate content (4.0 - 4.8%) than that with low pyruvate (2.5 - 3.0%). They concluded that the high and low pyruvate xanthan samples interacted differently in solution, thus yielding the various viscosities when tested at low shear rate ($\sim 4\ s^{-1}$). However, the polymer samples of low pyruvate content were derived from a variant strain of X. campestris and, thus, the differences observed could be due to changes in other characteristics such as molecular weight as well as acetal groups. In contrast, Bradshaw et al [52] suggest that the pyruvate content of xanthan does not directly affect the solution viscosity of the polymer. Native xanthan and pyruvate-free xanthan were found to show similar viscosity at 10 s^{-1} over a range of concentrations (0.2 - 1.5%). The pyruvate-free xanthan was prepared from the native sample using very mild TFA hydrolysis and differed only from the original material in respect of acetal group content. This infers that comparative studies of biopolymers should probably be made where possible, when variation in the extent of acetal substitution has been achieved by gentle chemical procedures in preference to material from different fermentation conditions as greater heterogeneity of polymer molecules may then result. In support of this, chemically prepared pyruvate-free xanthan showed significantly higher solution viscosity than pyruvate-free material obtained from a mutant strain of X. campestris [70]. Thus, indicating that some change in xanthan characteristic (but not pyruvate content) has occurred which influences solution viscosity.

Under the influence of small applied forces (very low shear rates) solutions of xanthan may display a yield point, thus showing total resistance to flow. The yield point can be considered as a measure of the force required to dissociate the intermolecular association between xanthan molecules. Once the yield point has been passed, xanthan solutions are extremely pseudoplastic i.e. the measured viscosity decreases rapidly as the shear rate is increased (Fig 5). This is an instantaneous process and the solutions should return to their original viscosity when the shearing force is removed and the molecular associations of xanthan molecules can then reform. However, Chen and Sheppard [71] observed that xanthan solutions (0.1%) present in the disordered form (60°C, distilled water) were susceptible to some degree of degradation (loss of viscosity) when sheared at 5000 s^{-1} for 30 min. In contrast, xanthan solutions (0.1%) in the ordered form (60°C, 10% NaCl) showed negligible degradation when treated to the same shear stress conditions.

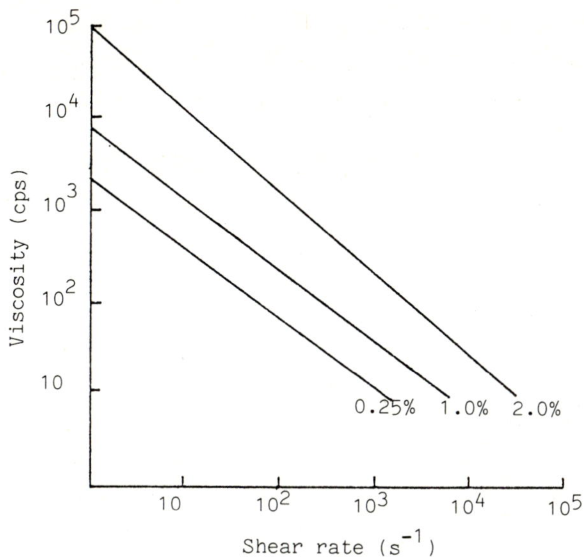

Fig. 5 Effect of shear rate on viscosity of xanthan solutions.

As mentioned earlier xanthan solutions display good compatibility with salts and actually show an increase in viscosity with salt addition (0.1% NaCl) to polymer concentrations >0.2 - 0.5%. This is as a result of the stabilisation of the xanthan structure in the ordered form giving rise to intermolecular association. Below concentrations of ~ 0.2%, addition of salt causes a slight depression of xanthan solution viscosity. Many polysaccharides change markedly in solution viscosity as the pH is altered. However, in the presence of moderate amounts of salt (0.1% NaCl) the solution viscosity of xanthan is essentially independent between the pH range 1-11. Xanthan is gradually deacetylated above pH >9, but loss of acetate should have no effect on solution viscosity [52].

While most polysaccharide solutions show a sharp decrease in viscosity with increase of temperature, xanthan with added salt can maintain its ordered structure and thus viscosity up to 100°C. Furthermore, salt-free xanthan solutions show an anomalous increase in solution viscosity on heating, before the expected decrease [15]. This behaviour is indicative of the unwinding of an ordered conformation such as a helix into a random coil with subsequent increase in effective hydrodynamic volume and, therefore, viscosity. Solutions of xanthan are also resistant to degradation by heat; they can be held at elevated temperatures for long periods of time without any major change in viscosity [72].

6. INTERACTION OF XANTHAN WITH CERTAIN PLANT POLYSACCHARIDES

Although solutions of xanthan alone will not form gels, they do display the relatively unusual property of a synergistic interaction with certain substituted (1 → 4)β -D-glycans. The best known examples [73-75] are the interactions shown with plant galactomannans such as locust bean gum (from *Ceratonia siliqua*), tara gum (from *Caesalpinia spinosu*) and guar gum (from *Cyanopsis tetragonolobus*). These plant gums consist of a (1 → 4)β -D-mannan backbone substituted to varying extents with (1 → 6)α -D-galactose residues. The addition of any of these plant galactomannans to a solution of xanthan at room temperature causes a synergistic viscosity increase to take place. Further, if combinations of xanthan and locust bean gum (or tara gum) are heated to 60-65°C and then cooled, cohesive gels are formed. These are true gels, which do not flow or recover from mechanical damage. The gels are also thermally reversible and show sharp melting and setting behaviour over a narrow range of temperature [51].

Although considerable commercial use has been made of xanthan-galactomannan mixed gels, the molecular origin of this interaction remained obscure until relatively recently. In general, the strength of interaction between xanthan and galactomannans can be correlated with the degree of substitution of the mannan chain. Thus, locust bean gum which has a D-mannose:D-galactose ratio of ∼4:1 gives far stronger and more rigid gels than tara gum [∼3:1 (Man:Gal)]. While guar gum with a ratio of ∼2:1 displays no gelling interaction. Based on the structural evidence available [76-79], which indicated that the D-galactose substituents in locust bean gum occur in blocks of the linear mannan chain, it was proposed [50,51,75] that gel formation was due to cooperative association between xanthan and the unsubstituted ("smooth") regions of the galactomannan to form a 3-dimensional network in solution. Further, it was suggested that these "junction zones" were linked by the substituted "hairy" regions of the mannan chain to provide the network with its overall solubility. More specifically, the non-associated regions would remain extensively hydrated compared to the associated regions which resemble the crystalline or solid state. A similar approach was used by Dea *et al* [80] to explain the gelling interactions of galactomannans with agar and carrageenan in solution. Interestingly, both agar and carrageenan possess double helical structures [81-83].

Minor revision of the proposed model for the association between xanthan and galactomannans may be necessary to explain the interaction of the microbial polymer with plant polysaccharides of similar D-galactose content [84]. In particular, guar gum and the polysaccharide from *Leucaena leucocephala* both have Man : Gal ratios of ~ 2:1, but differ in the degree of interaction shown with xanthan. The polysaccharide from Leucaena leucocephala shows a significant gelling interaction with xanthan, unlike guar gum. Nevertheless, guar gum does display a viscosity enhancement effect with the microbial polymer, which results in a doubling of solution viscosity relative to that of either polymer alone, at an optimum guar gum: xanthan ratio of ~ 10:1 [85]. Further, the presence of guar gum in xanthan solutions stabilises the ordered conformation of the polymer to temperatures above those it will persist in isolation [51]. Clearly, guar gum is capable of showing some form of interaction with xanthan. However, the addition of guar gum to a mixed gel of xanthan: locust bean gum weakens considerably the gel structure [85]. This is indicative of guar gum blocking potential sites in xanthan which could otherwise associate with locust bean gum.

It appears that this difference in the interaction with xanthan between galactomannans of similar D-galactose content is due to the distribution of the sidechain residues. This has arisen from examination of the fine structure of galactomannans using α-D-galactosidases and β-D-mannanases [84]. In contrast to previous evidence [76] which suggested that guar gum was composed mainly of the repeating unit $[(\mathrm{Gal} \rightarrow) \mathrm{Man} \rightarrow \mathrm{Man}]_n$, a more random distribution of D-galactose residues was observed. However, the polysaccharide from Leucaena leucocephala was found to possess significant proportions of the above repeating unit. Since the preferred conformation of the mannan chain in aqueous solution is ribbonlike [86-87], the D-galactose residues located at the C-6 position on adjacent D-mannose units would lie on opposite sides of the main chain. It follows that a galactomannan polysaccharide with significant regions of the [(Gal→) Man→Man] repeat unit will have regions of D-galactosyl groups located on one side of the mannan backbone. Thus, McCleary [84] has suggested that the gelling interaction between xanthan and galactomannans does not absolutely require long "blocks" of unsubstituted residues, but rather sections where all the D-galactosyl residues are located on one side of the main chain and therefore can participate in "junction zones".

Gelling interaction with galactomannans is a common feature of the various "xanthans" excreted by the Xanthomonas species [88]. Pyruvate-free xanthan has also been found to form gels with galactomannans [52]. Thus, it is unlikely that the pyruvate groups play a significant role in the gelling process. However, it is necessary for xanthan to exist in the ordered form for gelation to take place. Interestingly, xanthan shows even stronger interaction with plant polysaccharides such as Konjac mannan, which incorporate a part cellulose backbone [(1→4)-β-D-glucan] into their structure, compared to galactomannans [51]. Further, other microbial polysaccharides such as Arthrobacter polymers show no interaction whatsoever with plant polysaccharides, despite possessing an ordered structure [51]. Dea et al [51] have suggested that the strength and specificity of the interaction shown by xanthan with plant polysaccharides incorporating a cellulose backbone in their structures, may be related to the parasitic action of Xanthomonas bacteria. Since this type of backbone is characteristic of plant cell walls, it is considered that this interaction, which is an example of the formation of quaternary structure by polysaccharides, may be involved in the recognition of appropriate sites on the host plant to colonise [51].

7. PRODUCTION OF XANTHAN

7.1 Natural Function and Biosynthesis

Extracellular polysaccharides may be classified into two types depending upon their arrangement in relation to the microbial cell (i) polymers which are physically attached to the microbial cell in the form of capsules (ii) polymers which are excreted by the microbial cell as loose slime into the environment. Xanthan is of the latter type of extracellular polysaccharide.

The natural function of xanthan and other microbial polysaccharides is not fully recognised, but various suggestions have been put forward to explain their existence. One view is that the polysaccharide layer surrounding the microbial cell protects the microorganism from various environmental factors [89]. In particular, the water-retaining capacity of polysaccharides provides the microorganism with a protective layer under dry environmental conditions, thus providing some degree of desiccation resistance. Also, the polysaccharide layer may act as a barrier against attack from bacteriophage. As mentioned earlier (Section 6), a more specific function has been suggested for xanthan viz. recognition of appropriate sites on the host plant for colonisation of the bacteria to take place [51].

Most of the reports on the synthesis of xanthan by X. campestris have been concerned with optimisation of conditions for the industrial production of the polymer. Until recently only limited information on the biosynthesis of xanthan and other commercially important microbial polysaccharides was available. However, it is certain that the biosynthesis of microbial heteropolysaccharides such as xanthan is a complicated process, involving a multienzyme system. The majority of studies on the biosynthesis of microbial polysaccharides have been carried out with microorganisms of immunological and clinical interest. Thus, most of the present ideas [90,91] on the biosynthesis of xanthan have been derived from related studies. Since an increased knowledge of the control mechanisms acting on polysaccharide intermediates and the specific enzymes involved in the synthesis may lead to improved polymer production, the biosynthesis of xanthan is likely to receive greater interest in the future. Indeed, some recent reports [92-94] have provided useful information on lipid-linked intermediates involved in the biosynthesis and of the mode of pyruvylation and acetylation of xanthan.

The first step [90] in the biosynthesis of a microbial polysaccharide is uptake of the substrate (usually carbohydrate), which may occur by active transport, group translocation or facilitated diffusion [95]. A substrate that has entered the cell is phosphorylated either during the active transport process or via a hexokinase enzyme which utilises adenosine 5'-triphosphate (ATP). The presence of hexokinase activity in X. campestris has been demonstrated [96]. After phosphorylation, the substrate is subjected to either catabolic processes for cell energy or anabolic processes leading to the formation of polymers (exopolysaccharides, lipopolysaccharides, cell-wall polysaccharides).

The next step in the biosynthesis involves conversion of the phosphorylated substrate to the various sugar nucleotides (usually nucleotide monosaccharides) required for assembly of the polysaccharide repeating unit via enzymes such as UDP-Glc pyrophosphorylase [97]. These activated monosaccharides are the immediate precursors of the repeating unit of the polymer and their presence will be essential for the overall synthesis. Indeed, "in vitro" studies of the biosynthesis of xanthan by X. campestris have shown that UDP-glucose, GDP-mannose and UDP-glucuronic acid are necessary for the synthesis of polymer with the correct repeating unit [92]. Only small amounts of polymer were produced when any of these nucleotide monosaccharides were missing. This work also showed that UDP-glucose, GDP-mannose and UDP-glucuronic acid sequentially donate their sugar moieties to a lipid acceptor (isoprenoid alcohol phosphate) to form a pentasaccharide-P-P-lipid. Subsequent polymerisation occurs by transfer of the pentasaccharide unit from the lipid carrier to the growing polysaccharide chain.

Further work has shown that the pyruvic acid acetal residues in xanthan are formed by transfer from phosphoenolpyruvate (PEP) to the pentasaccharide-P-P-lipid [93]. Although it had been suggested previously [90] that PEP was the likely donor, this evidence is the first confirmation of this occurrence in the biosynthesis of xanthan. Since xanthan with very little or no pyruvate can be obtained, this step is clearly not critical to the overall biosynthetic process. Recently, it has been reported that the addition of acetyl residues also occurs at the lipid intermediate stage and that the donor is Acetyl-CoA [94]. Again, the preparation of xanthan lacking O-acetyl residues is indication that the acetylation step is not essential to the overall biosynthesis.

The regulatory mechanisms controlling the release of the polysaccharide from the cell, and also the size of the polymer chain are largely unknown.

Presumably, some specific enzyme is present which cleaves the terminal monosaccharide residue from the isoprenoid alcohol phosphate, to allow return of the carrier lipid for further polymer synthesis. This is another important step as it is considered that the availability of lipid isoprenoid alcohol phosphate is a critical factor in the biosynthesis of such exopolysaccharides as xanthan [98,99]. It has been suggested [90] that the chain length (molecular weight) of microbial polysaccharides may be dependent upon the growth rate of the microorganism. In particular, a higher growth rate may lead to a more rapid turnover of the carrier lipid and therefore earlier release of the polymer with a shorter chain length (or molecular weight).

There are still some uncertainties surrounding the biosynthesis of xanthan and other microbial polysaccharides. However, it is likely that our knowledge in this field will be increased in the future, since this could lead to improvement in the commercial production of these biopolymers.

7.2 *Xanthomonas* species

As mentioned earlier (Section 2), various *Xanthomonas* species are known to produce a polysaccharide during fermentation. However *X. campestris* is the species which has received the maximum attention, since it was found to be the most efficient producer of xanthan. It is well known that the characteristics of microorganisms are inherently unstable [100-102]. Spontaneous mutation can take place resulting in changes in the genetic information stored in the cell. Such changes can affect the regulatory mechanisms and enzyme levels involved in any cell process including polysaccharide biosynthesis. One example of undesired mutation is the degeneration of a high-yielding strain to a less efficient producer of biopolymer.

It follows that it is important to preserve strains in a stable form. Since the number of spontaneous mutants is a function of the number of replications, this parameter should be kept to a minimum by using appropriate methods of preservation which maintain the microorganism viable, but biologically inactive. For long-term preservation of microbial cells lyophilization, freezing and "L-dried" (dried from the liquid state) are the main methods employed [103], with the former technique being the most favoured. Short-term storage methods require the microbial cells in an accessible form, from which a high recovery of viable cells can be obtained which grow with a minimum lag-phase. Storage on agar slants is the most popular short-term procedure, but it has been suggested [102] that other non-propagative methods are superior.

It is important that the conditions of such short term storage favour the desired characteristics of the microorganism, so that unwanted mutants are not selected. In particular, problems with the development of variant strains during short term storage on agar slants have been observed with X. campestris [39]. This resulted in changes in the yields of xanthan produced and the viscosity of the gum. However, it was suggested that the rate of appearance of the variant forms was diminished when the D-glucose content of the stock medium was sufficient to avoid depletion and when transfer to fresh medium was reduced to 14 days. It was also suggested that a fresh lyophil culture should be used to initiate a new series of agar slant cultures every 3 months. As an alternative to storage on agar slants, drying of microbial cells on paper strips has been proposed as a simple, convenient technique for short-term non-propagative maintenance of X. campestris cultures [104].

Although degeneration of a microorganism to an altered form is usually undesirable, induced mutagenesis by chemical means or uv or γ irradiation is a useful method for strain development. In particular, induced mutation with suitable selection procedures may lead to a microorganism with improved characteristics or a biopolymer with improved solution properties. For instance, it may be possible to improve the yield of polymer produced or increase the viscosity of the product. In the case of X. campestris, however, it may be difficult to further improve on such parameters, since a good deal of development work has been performed on this species. Nevertheless, various altered strains of X. campestris have been reported which possess a particular desired characteristic. This includes a strain which is capable of converting NO_3^- to NH_4^+ and NH_3, thereby avoiding the need for pH control by base addition during fermentation [105]. A strain which produces colourless xanthan, instead of the usual yellow pigmented polymer has also been reported [106]. As mentioned earlier (Section 4), it is possible to obtain pyruvate-free xanthan from a mutant strain of X. campestris [68]. Degeneration of X. campestris has been observed during continuous culture studies (see Section 7.5), besides during storage on agar slants [39]. Thus, a degenerative resistant strain of X. campestris has been developed for long term continuous fermentation to produce xanthan [107]. One of the disadvantages of induced mutation is that unwanted mutations are also produced. Thus, difficulty can often occur in the isolation of mutants altered only in the desired parameters. Future methods for improving microorganisms for biopolymer production will perhaps involve genetic transfer procedures, which will alleviate this problem.

7.3 Media Composition

In the biosynthesis of xanthan on the cabbage plant by X. campestris the cabbage provides the carbohydrate substrates, proteins, minerals and buffering capacity necessary for cell growth and polymer production. In laboratory scale or commercial fermentation, carbon sources, nitrogen sources, trace minerals and pH buffering capacity are supplied in a way that simulates the natural conditions. Since xanthan is presently produced commercially by a batch fermentation, most studies on the production of this polymer relate to this type of procedure.

In batch fermentation, a set amount of a pure culture of X. campestris is inoculated into a medium containing all of the necessary ingredients. Preparation of the inoculum involves growth of the microorganism in several stages and identification tests before introduction into the final fermentation medium. The relative size of the inoculum can affect the production of xanthan. In one study the volume of inoculum added was 7.5% (v/v) of the fermentation medium [40]. This was necessary to achieve optimum yields of xanthan. In a commercial process, the scale-up time required to produce sufficient inoculum for the fermentation may be several days. Thus, one method has been reported which requires only an inoculum volume of 0.01 - 0.5% of the total fermentation medium volume [108]. This was achieved by controlling the levels of Fe, Mn and Zn in the inoculum culture. It appears that the composition of the inoculum medium can also affect the physical properties of the final polymer product, since it has been reported that concentrations of sulphate greater than 0.2% reduced the viscosity of xanthan [109]. After the inoculum stage, the fermentation is continued (48 - 96h) under aseptic conditions until one of the substrates (usually carbohydrate) is exhausted.

For optimal production of xanthan, it is necessary to establish the optimal medium composition and fermentation parameters. The carbohydrate concentration affects the efficiency of carbon source conversion into biopolymer. D-glucose concentrations in the range 1-5% were found to produce the best yields of xanthan [18]. The conversion efficiency of D-glucose to polymer by X. campestris decreases when higher D-glucose concentrations are used [16]. However, this problem can be overcome by gradual addition of the carbohydrate substrate to the fermentation broth [110]. In this way, 7% (w/v) D-glucose was converted to a final xanthan concentration of 5% (w/v), representing a yield of 70%. One disadvantage of this procedure is the time (136 h) required for the fermentation.

Various carbohydrate sources have been used as an alternative to D-glucose for the production of xanthan, including sucrose [111], hydrolysed cereal grains [112, 113], molasses [114] and hydrolysed acid whey [116]. Using these carbon sources, 40-85% yields of xanthan have been reported, variations being due to the type and quantity of carbohydrate substrate used. One study using chemically defined media found that 4% D-glucose or sucrose were the optimum initial carbohydrate levels to employ [111]. While X. campestris can produce xanthan from enzymically hydrolysed lactose (acid whey) [115], polymer formation does not result when the microorganism is grown on whey itself [116]. Interestingly, the D-glucose and D-galactose from the hydrolysed lactose were used at approximately the same rates by X. campestris and no diauxic effect was observed [115]. Further, it was suggested that both sugars had been used for polymer production. The fermentation of hydrolysed acid whey by X. campestris for the production of xanthan has the advantage that utilization of a waste product from cottage cheese manufacture is achieved, which would otherwise require disposal by some costly procedure. Acid whey also contains adequate nitrogen and other growth factors required for the fermentation. Similarly, glucose syrups prepared from cereal grains have been reported to contain some nitrogenous component, which partly fulfilled the nitrogen requirement for xanthan production [113].

A nitrogen source in the fermentation medium is required for protein synthesis including that of the enzymes involved in polymer formation. However, in general, efficient conversion of carbon source to polymers such as xanthan requires a high carbon:nitrogen ratio; an excess of nitrogen often [111] causes inhibition of polymer formation. Simple inorganic materials such as ammonium or nitrate salts may be used as the nitrogen source. Sodium nitrate and ammonium nitrate have been reported to be superior to ammonium sulphate [111] for cell growth and polymer production. Certain amino acids such as L-glutamic acid can also be used effectively by X. campestris as nitrogen sources. A wide variety of more complex nitrogen sources such as dried distillers solubles [117], yeast autolyzate [118], peptone [119], meat peptone [120], soy peptone [112], soybean whey [121] have also been used in xanthan production. Thus, the bacterium X. campestris shows versatility in respect of carbon and nitrogen sources which can be utilised.

Addition of certain organic acids to the fermentation medium has been shown to have a stimulatory effect on the production of xanthan [111, 122-124]. Citrate at an optimum concentration of 4.7 - 9.4 mM was found to stimulate xanthan formation in pH uncontrolled fermentations [122,123]. However, pH control of the media eliminated this beneficial effect. The authors thus suggested that the metabolism of citrate by X. campestris provides a favourable pH pattern for xanthan production [122,123]. Other organic acids such as pyruvate, succinate and α-ketogluturate were also found to stimulate xanthan production [123]. Concentrations of these organic acids greater than the optimum inhibited polymer formation, but not growth of X. campestris. Addition of deoxycholate or cholate to the fermentation medium also results in increases in the production of xanthan [125].

The nutritional requirements of X. campestris also necessitates the presence of phosphorous and this is usually added in the form of a phosphate buffer. Magnesium is usually added to the fermentation medium as the sulphate salt. Various trace minerals are also required in the fermentation medium, some of which may be present as part of the added organic components.

7.4 Fermentation Conditions

Various fermentation parameters can affect the production of xanthan, one of which is the pH of the fermentation medium. Without control, the pH decreases during the fermentation due to the formation of organic acids and acidic functions contained on xanthan. If the pH falls below 5, the formation of xanthan drastically decreases. Thus, it is necessary to control the fermentation medium at the optimum pH of 7 [117], by the use of a buffer or addition of base during the process. Fermentation temperature is another critical parameter and optimum xanthan production is obtained at 28°C [126]. Further increase in temperature reduces the viscosity of the xanthan product [127]. The intrinsic viscosity of xanthan produced at 35°C was 11.5 dL/g compared to a value of 25 dL/g when the polymer was obtained at the optimum fermentation temperature. This variation may be due to a difference in the molecular weight of xanthan.

Since production of microbial polymers such as xanthan is always performed under aerobic conditions, it is necessary to control the aeration and agitation of the fermentation medium. As the fermentation proceeds the viscosity of the broth will increase, causing problems with oxygen limitation which will in turn affect the quantity and perhaps quality of the product. In small scale shake-flask cultures oxygen limitation is almost certain to occur. With larger scale fermentations, some form of forced aeration using sterile air is likely to be employed. Further, adequately designed agitation is necessary to disperse the introduced air evenly throughout the medium. Agitation of the medium is also useful for increasing the rate of transport of nutrients across the cell membrane, which in turn enhances the growth rate of the microorganism.

7.5 Continuous Culture

In continuous fermentations addition of fresh medium is made continually to the culture, with simultaneous harvesting of the desired product. The fermentation is thus performed under steady state conditions and growth is controlled through one limiting nutrient which is present in a sufficiently low concentration such that other nutrients are present in excess. Thus, greater control of the overall process is obtained than in batch culture and a better understanding of the microbial physiology of the fermentation is obtained. It is also possible to utilise the nutrients in a more efficient manner for production.

Initial work on the continuous culture of X. campestris to produce xanthan showed that polysaccharide production was a function of the dilution rate [128,129]. While xanthan could be produced at 0.84 $gh^{-1}kg^{-1}$ with a dilution rate (D) of 0.15 h^{-1}, at the highest dilution rate studied (0.196 h^{-1}) the rate of polymer production fell to 0.69 $gh^{-1}kg^{-1}$. This work, which was performed using complex fermentation media, indicated that yields of xanthan in the range of 81-89% (theoretical) of D-glucose utilised were obtained. However, problems with strain stability were obtained and after 6.5 - 8.7 culture residence times, variant (small colony type) forms of X. campestris appeared and a drop in polymer production resulted [129]. This is related to the observation of degeneration of X. campestris during storage [39] and is a reflection of the reversion of some microorganisms from a high-yielding to a low-yielding strain under certain conditions.

When xanthan was produced using X. juglandis in continuous culture in a simple salts medium, the nutrient limitation conditions had some effect on strain stability [44]. Culture variation was evident under carbon, magnesium and potassium limitation, while xanthan could be produced at a consistent level in yields of 60% under sulphur limitation for 12 weeks. Considerably higher viscosities were achieved under sulphur or nitrogen limitation compared to carbon, magnesium and potassium limitation. As discussed earlier (Section 3.2) variation in nutrient limitation conditions was found to affect the chemical composition of the polymer [44]. Using sulphur limited conditions, 0.05 h^{-1} was found to be the optimum dilution rate for xanthan production. This is equivalent to a mean residence time of 20 h and thus compares favourably with batch culture time. Slight differences in physical properties of the polymer were observed with samples produced at different dilution rates.

A method to control the viscosity of the xanthan produced in continuous culture has been reported [130]. The "viscostat" consisted of a single stage 8L fermentor fitted with a Bendix Ultraviscoson to sense the broth viscosity which was linked to the control which actuated the feed medium pump. In this way addition of fresh medium (0.25% D-glucose; complex salts) was made to the fermentor to maintain the viscosity at a constant level. Although the authors reported problems with the accuracy of the viscosity sensor, it was suggested that this method provided a means of detecting bacterial contamination in continuous culture systems. The number of recent reports in the patent literature concerned with continuous fermentation using X. campestris to produce xanthan suggests that industry is considering the use of this process [131-135]. This has included the description of a method to produce xanthan with a low pyruvate content (0.1 - 1.0%) [134]. Also, a method to increase the productivity of the fermentation and enhance the rheological properties of xanthan by control of certain trace elements (Fe, Mn, Zn, Cu, Co and B) in the culture medium has been reported [135].

A semi-continuous method for production of xanthan with X. campestris has been described [136]. The fermentation in a simple salts medium along with D-glucose (2.2%) was initiated in a batch-type process. Once the fermentation was complete (24h), 10% of the original broth was used to inoculate fresh medium. In this way fresh medium was inoculated every day for 14 days with 10% broth from the previous fermentation. No problems with culture deterioration were observed due to the use of a degenerative resistant strain of X. campestris [107] in these experiments. Relatively low yields were obtained, but it was suggested that the procedure was more efficient than a normal batch type process. In particular, the preparation time required for the culture to inoculate each medium in the batch process could be avoided.

Two-stage [137] and multi-stage [138] continuous culture processes have also been developed for the production of xanthan. In the former process, the first stage is performed in a low dextrose concentration (0.8 - 1.0%) and is designed to allow growth of X. campestris cells with minimal production of polymer. The bacteria are contained in an inner membranous chamber which allows passage of nutrients but not intact cells. Flow of the enzymes necessary for polymer synthesis from the inner to the outer chamber also takes place. The second stage of the process is conducted in the absence of any intact cells. This is accomplished by allowing the exhausted liquor, containing the polymer synthesising enzymes to flow continuously into the second fermentor which contains sufficiently high concentrations of dextrose (2-5%) to enable the formation of xanthan in relatively high yields (65-70%) to take place. Residence times of approximately 6h and 20h respectively were required for the first and second stages of the fermentation process. A clear advantage of such methods is that problems with cell removal from viscous broths can be avoided.

7.6 Industrial Scale Production

Since the industrial processes used by individual companies are proprietary, little information is available on this aspect of xanthan production. But it is likely that different fermentation conditions, strains of X. campestris and fermentor design will be employed. Further, the process will be carried out using the cheapest substrates and nutrients available. However, cheaper substrate materials such as hydrolysed cereal grain often contain components which are beneficial to production. Although xanthan is probably still produced in batch-scale operations, a switch to continuous fermentation may occur in the future. Indeed, if the applications for xanthan continue to grow and the anticipated large-scale use of the polymer in Enhanced Oil Recovery occurs, a considerable increase in production of xanthan will have to take place. The use of continuous fermentation would help, since the time of incubation is reduced in this process compared to batch-type operations. Further, more efficient use of the substrates is theoretically possible and this could help in reducing the costs of xanthan production. However, the problem of avoiding contamination in continuous fermentation is one of the main reasons why industry has not adopted this method as yet.

Yield and quality of product obtained in a laboratory scale fermentation are not necessarily reproduced when the conditions are scaled-up to an industrial process. Thus, pilot plant production is a necessary intermediate step between the laboratory study and the industrial process. However, with a few exceptions [16,40,139,140], little information has been reported on semi-pilot or pilot plant production of xanthan. One study [139] suggests that larger savings on production costs of xanthan can be made by optimization of operational variables such as mixing, aeration and fermentation time compared to economics over the choice of substrate. Careful consideration needs to be given to mixing and power requirements, since the solution properties of the fermentation medium change from Newtonian to pseudoplastic behaviour as the xanthan is formed [141]. Oxygen mass transfer in the culture medium also requires consideration and is affected by the ratio of impeller size to fermentor size [141]. Power-law models to determine the effect of shear viscosity of xanthan solutions on the various physical parameters of the fermentations have been developed recently [142,143]. One method to reduce the energy input and increase the yield of xanthan obtained involves emulsion fermentation [144]. This batch process involved addition of a water-immiscible non-polar oil during the fermentation to reduce the viscosity of the medium.

8. RECOVERY METHODS USED IN THE PRODUCTION OF XANTHAN

8.1 General

Following the fermentation with *X. campestris* to produce xanthan, the polymer product may be isolated from the broth by a variety of methods. The aim of the recovery process is to obtain the xanthan in a suitable form (either as solid material or in a broth solution) with minimum degradation and free from other fermentation products and fermentation medium constituents. The final fermentation broth contains, besides xanthan, *X. campestris* cells, insoluble material, unused carbohydrate substrate and various inorganic salts. To comply with toxicity regulations, it is essential to obtain a product free from *X. campestris* cells and any reagents employed in the recovery process, when food-grade xanthan is required. Generally, the intended application of the xanthan product defines the grade of purity required which in turn governs the choice of recovery method which is used.

On a laboratory scale, the isolation of xanthan from X. campestris fermentation broth can be achieved relatively easy. Microbial cells can be removed from the broth solution by ultracentrifugation. Further, separation of xanthan from low molecular weight carbohydrates and inorganic salts can be obtained by dialysis. The polymer can be further purified and isolated in the solid form using various precipitation procedures (see later). If very high degrees of purity are desired, some form of ion-exchange or affinity chromatography could be used to free the xanthan from traces of other biopolymers. However, on a commercial scale, most of these steps (except precipitation) are not feasible.

The commercial isolation of xanthan is generally a difficult and expensive process, which is reflected in the high cost of the polymer. In particular, the high viscosity of X. campestris fermentation broths makes the recovery difficult from several aspects. Microbial cell removal by mechanical means such as centrifugation or filtration is not efficient. Also, high power requirements are necessary to obtain satisfactory mixing of the broth solutions with any recovery reagents used. Even, general handling of the viscous solutions is difficult and, therefore, dilution is usually required at some stage of the process.

The principal steps in most commercial recovery processes for xanthan involves (i) deactivation and removal (or degradation) of X. campestris cells (ii) precipitation of the polymer to obtain xanthan in the solid form. (iii) dewatering, drying and milling. Various treatments can be used to obtain further purification e.g. deactivation of contaminating enzymes and also to enhance the solubility of the xanthan when rehydration is required. It follows the greater the degree of purity required, the higher the cost of the isolated xanthan product. The use of clarified fermentation broths may have increasing importance in future applications of xanthan. This would avoid the cost of precipitation reagents and any problems associated with rehydration of the xanthan.

8.2 Prerecovery methods

Some form of prerecovery treatment is usually necessary to optimise the performance of xanthan in specific applications. In particular, the use of xanthan in tertiary oil recovery will require polymer solutions free from particulate matter such as cells and microgels, which could otherwise cause plugging of the rock pores [12,145]. Many of the reported methods for clarification of xanthan solutions are equally applicable to fermentation broths or redissolved solutions of the polymer. However, the first step in the isolation process is usually pasteurization of the fermentation broth to kill the

X. campestris cells. Various conditions [146,147] have been reported for this step e.g. 98.9°C for 3 mins [147]. It has even been suggested [147] that these conditions enhance the viscosity of xanthan solutions.

Clarification of X. campestris fermentation broth may be performed chemically using alkali [146,148] or hypochlorite [149,150]. One disadvantage is that deacetylation and degradation of xanthan can occur if the polymer is allowed to stand in alkaline conditions for prolonged periods. Hypochlorite also causes bleaching of the yellow pigmented fermentation broth and may even be used for deactivation [151] of cellulases present in the fermentation broth. Alternatively, such enzymes may be deactivated by treating wet xanthan, obtained by precipitation from the fermentation broth, with various chemicals including propylene oxide [152] at 40-70°C for 1-5 h.

As mentioned above, efficient clarification of viscous X. campestris fermentation broths by centrifugation or filtration is difficult. However, it has been reported that increasing the temperature of the fermentation broth to 124°C enhances the filtration considerably [153]. Presumably, xanthan would be in the disordered form at this temperature and thus show a lowered viscosity which allowed more efficient filtration to take place. Alternatively, coagulation with salts may be used to aid filtration of X. campestris fermentation broths [154]. After the fermentation, the broth is diluted with an aqueous solution of an alkali metal salt such as NaCl and agitated at 95°C for 2 h. A coagulant (potassium aluminium sulphate, 2-250 ppm) is then added to the fermentation broth and further agitation carried out at 95°C for 2 h. The resulting clarified polysaccharide solution, obtained by passage through a pre-coated filter at 75°C, was claimed to be suitable for direct use in enhanced oil recovery applications. Adsorption of the particles in the fermentation broth to added clay particles [155], followed by coagulation, has also been used in the clarification of xanthan.

A variety of methods [156-161] have been reported for degradation of X. campestris cells with enzymes to obtain clarified fermentation broths or redissolved solutions of the polymer. Acid, neutral and alkaline proteases have all been used to this effect and sources of the enzymes have included Bacillus subtilis [156,157], Aspergillus niger [157,159] and Pellicularia filamentosa (sasaki) [161]. A disadvantage of alkaline proteases is that deacetylation or degradation of the polymer may occur; one particular method using such a type of enzyme required a pH of 12.5-13.0 for optimal activity. A wide variety of enzyme concentrations, temperatures and incubation periods are encountered in these protease clarification procedures. One method reported by Colegrove [156] using an alkaline protease from Bacillus subtilis required enzyme treatment (25-2,000 ppm) for at least six hours at temperatures of up to 60°C. Alternatively, Radar and Wong [161] used an enzyme mixture (25 ppm) from Pellicularia filamentosa and Pellicularia sasaki for 15 min at 24°C. In addition to protease activity, the enzyme mixture contained cell lytic and β -(1 $\longrightarrow$ 3) glucanase activity and was claimed to improve the filterability, flow rate and viscosity of xanthan solutions.

An alternative approach to clarification of xanthan broths, involving solubilisation of the X. campestris cells, was adopted by Schroek [162]. After the initial process, a second fermentation was performed under acidic conditions with a fungus such as Trichoderma viride which is capable of consuming the X. campestris cells. Since fungal cells of Trichoderma viride are larger and therefore easier to filter than the microbial cells of X. campestris, more efficient clarification of the fermentation broth could be achieved.

Prerecovery procedures may also be used to improve the reconstitution of the final xanthan product. Treatment of xanthan with glyoxal [163] prior to precipitation from the fermentation broth was claimed to improve dispersion and also the viscosity of xanthan. Chemical modifications of xanthan such as mild alkaline [15] and acid hydrolysis [52] to remove acetate and pyruvate respectively could also be carried out at this stage of the process.

8.3 Precipitation procedures

Following prerecovery treatment, xanthan can be precipitated from the clarified fermentation broth by a variety of procedures. The process favoured commercially is addition of a water miscible organic solvent such as a lower alcohol in sufficient concentration to effect precipitation of xanthan from solution. Isopropanol tends to be the solvent which is used most frequently for this process. In an early process reported by O'Connell [146], the fermentation broth was pasteurized and diluted with 3-4 vol. of water. Following bleaching of the broth with hypochlorite, isopropanol (55% w/v) was added to effect precipitation of the polymer. Various attempts [164-167] were subsequently made to improve the xanthan recovery process using isopropanol. Reduction in the amount of isopropanol required can be achieved by addition of KCl to the fermentation broth prior to the precipitation step [164]. While another procedure reported [165] an improvement in the filtration of the fermentation broth by addition of quantities of isopropanol insufficient to precipitate xanthan. Following removal of solids at high temperatures, further isopropanol was added to precipitate the xanthan. It has also been suggested [166] that the residue from the alcohol distillation, following precipitation of xanthan, can be reused in the production of the polymer due to the presence of large amounts of phosphate salts and smaller amounts of carbohydrates and nitrogenous material.

The main disadvantage of the isopropanol precipitation process for recovery of xanthan is the cost of the operation, which forms a considerable part of the total production expenditure. Large quantities of the expensive solvent are required for this process. Also, distillation costs for recovery of the isopropanol for reuse are high. Further, losses of solvent during the overall process are to be expected. Another factor is the expensive fire safety precautions which are required for the production plant, due to the highly inflammable nature of isopropanol.

Advantage has been taken of the acidic groups in xanthan in several recovery processes which used precipitation with multivalent metal ions [168-170]. Precipitation of xanthan from the fermentation broth with calcium ions in alkaline pH formed the basis for two earlier methods of this nature [168,169]. The pH is critical for this reaction, since the divalent ion can only precipitate xanthan when the acidic groups are fully ionised. Although the exact mechanism of the reaction is unknown, some form of cross-linking is likely between polymer molecules. In a later process, Towle [170] describes the use of certain aluminium salts (chlorides, nitrates and acetates) to achieve recovery of xanthan. Initially, the fermentation broth was diluted with H_2O (2 vol.) and the aluminium salt added to reduce the solution to pH 3. After centrifugation of the broth, precipitation of xanthan was accomplished by increasing the pH into the range 4-4.5. The pH appears to be critical in this procedure, since outside of this range the material was found to be marginally soluble or else gelling occurred. One problem with metal ion precipitation methods is that the xanthan is recovered in an insoluble form and difficulty can occur when cation-exchange is attempted to obtain a soluble polymer product. In the case of the latter method, the xanthan precipitate (Al^{3+} form) was repeatedly slurried in a mixture of isopropanol:HCl (20:1) to convert the polymer to the H^+ form. A process involving addition of both calcium ions and isopropanol to the fermentation broth to precipitate xanthan has recently been reported [171].

The reaction of organic cations, particularly quaternary ammonium salts [172-177], with xanthan has been used in several processes for recovery of the polymer. Addition of quaternary ammonium salt to X. campestris fermentation broth results in precipitation of the xanthan due to the hydrophobic nature of the organic cation. The mechanistic aspects of this precipitation reaction has recently been studied in detail by Kennedy and Bradshaw [178,179]. In the original process [172-174], a commercial mixture of cetyltrimethylammonium chloride was used for precipitation of xanthan. The formed xanthan-quaternary ammonium salt complex was recovered by filtration and washed in several changes of methanol containing 0.05-0.20% KCl. After recovery of the xanthan precipitate (K^+ form), the supernatant was distilled to achieve recovery of the solvent and the quaternary ammonium salt residue for reuse. The obvious disadvantage of this method is the necessity to achieve high recovery of solvent and quaternary ammonium salt. Polyethoxylated quaternary ammonium compounds [176] have also been used for isolation of xanthan and in a recent study, cetyltrimethylammonium bromide [177] was found to be more efficient than the commercial mixture, Cetavlon, for the recovery of the polymer. Quaternary ammonium salts may have some use for isolation of industrial grade xanthan material, but essentially quantitative removal of the precipitant from the polymer would have to be obtained before food use could be considered.

Other precipitants which have been reported for the isolation of xanthan include polyamine compounds [180] and poly(hexamethylenebiguanide hydrochloride) [181]. The latter reagent possesses the ability to precipitate xanthan, yet leaves the majority of X. campestris cells in the supernatant. Again, quantitative recovery of expensive reagent is an important factor if these methods were to be used. As an alternative to precipitation processes for recovery of xanthan, direct conversion may be made from the fermentation broth to the solid product by evaporation. On a laboratory scale, freeze-drying is a convenient method of achieving this aim. However, on an industrial scale, spray or drum drying would be the analogous technique employed. Although two methods [182,183] for a drum-drying process of xanthan were reported a few years ago, it does not appear that this procedure is favoured commerically.

8.4 Post Precipitation Processing

After precipitation the xanthan will contain large amounts of liquid and thus centrifugation or pressing may be used to remove the excess liquid. Final drying of the polysaccharide precipitates can be obtained using batch or continuous driers [164]. This may be performed under vacuum or with forced air or inert gas if the polymer contains isopropanol. Care has to be taken in the drying process to avoid impairing the quality or rehydration properties of the final product. Most commercial xanthans have a final moisture content of about 10%. After drying, the xanthan can be milled to the required mesh size, which is sometimes the only difference between some commercial grades of the polysaccharide produced by one company [184].

8.5 Dissolution of Xanthan

Most commercial samples of xanthan are supplied to the user in powdered form. Problems can often occur in the uniform dispersion and subsequent dissolution of polysaccharide powders in aqueous solution. In particular, clumping of individual particles of xanthan can occur in solution which can effect the rehydration time of the product. Even the mesh size of the polymer can affect its rehydration properties [185]. To overcome rehydration problems, the manufacturers of xanthan recommend several methods to achieve uniform dispersion and thus rapid preparation of polymer solutions. These include the use of high-shear mixing, educator mixing and "wetting" the product with miscible solvents such as lower alcohols. As mentioned earlier, complexation of xanthan with glyoxal improves dispersion of the polymer [163]. Treatment of xanthan with borate can also produce similar results [186].

Following preparation, solutions of xanthan may be further processed to optimise the performance for a particular application. One example is the treatment of xanthan solutions with ultrasonic waves to improve injectability for Enhanced Oil Recovery use without impairing viscosity [187]. Treatment of xanthan solutions with enzymes possessing β-D-(1$\rightarrow$3) glucanase activity can also improve the flow properties of the polymer through porous material [188,189]. It was suggested that the enzyme cleaved some of the glycosidic linkages in xanthan, while leaving the viscosity of the solution intact. Alteration of the molecular weight of xanthan using cellulase should also be possible, if required for a specific application [57]. Under the appropriate conditions (salt-free), this enzyme can be used for the controlled degradation of the main polymer chain.

8.6 Use of Clarified Xanthan Broths

Some of the problems associated with the recovery process could be avoided if clarified fermentation broths were used in applications instead of solid xanthan materials. In particular, the cost of precipitation reagents and their subsequent recycling could be avoided. Also, the possibility of denaturing of xanthan by the commercial processing could be reduced. Due to the complex nature of the higher levels of structure adopted by xanthan in solution, it would seem likely that commercial processing could affect solution properties. Indeed, Duda *et al* [190] suggested that a commercial xanthan in powder form (Kelzan) was slightly denatured compared to the broth form (Flocon 1035) and thus existed in the disordered form when dissolved in H_2O (salt-free). The drying process was probably responsible for this alteration of the powder form of xanthan. The implications of these findings was that the broth form of xanthan showed better injectivity performance for Enhanced Oil Recovery applications. The use of broths also avoids problems with rehydration of the product.

One disadvantage in the use of clarified xanthan broths is the transportation costs which would arise, if the solution was not intended for applications close to the site of production. In this case, some form of concentration of the broth solutions might be required. One report has suggested [191] that xanthan solutions of up to 15% can be obtained using ultrafiltration. Precautions to avoid bacterial growth in the xanthan broth solutions by addition of such additives as formaldehyde or sodium azide would also have to be considered.

9 COMMERCIAL APPLICATIONS OF XANTHAN

9.1 General

The commercial applications of polysaccharides such as xanthan are normally sub-divided into food uses and industrial (or non-food) uses. Due to the potential of xanthan for Enhanced Oil Recovery, a separate section on this particular application has been included in this review (Section 9.5). The commercial success of xanthan results from its valuable and sometimes unique solution properties. In summary, they are the ability of xanthan to form solutions of high viscosity at-rest or low shear forces, which are also highly pseudoplastic and display a yield value. Further, the viscosity of xanthan solutions is stable over a wide range of salt concentration, temperature and pH (stability to the variation of temperature and pH does require the presence of a small amount of salt). These solution characteristics of xanthan give rise to the functional properties such as thickening and stabilising ability which are used in a diverse range of applications. In addition, xanthan shows a synergistic interaction with various galactomannans which results in enhanced solution viscosity or formation of gels, depending upon the concentration and type of plant polysaccharide used.

Thickening ability is related to the viscosity of a gum solution, which can be defined as the "resistance to flow" of a liquid. Different gums and different concentrations of a particular gum will display a characteristic viscosity behaviour i.e. show variation in "resistance to flow" with variation of applied shear force. Since a particular application will require a set viscosity at a specified shear rate, careful choice over both gum and concentration has to be made. Due to its high solution viscosity at low concentration and shear thinning behaviour, xanthan is a particularly useful thickener.

Xanthan is also used for the stabilisation of suspensions, emulsions and foams which are solids, liquids or gases respectively dispersed in an aqueous medium. This functional property is enhanced by a high solution viscosity and existence of a solution yield value. Thus, if the suspended particles exert a force less than the yield value, they will be dispersed effectively. Again, the shear thinning behaviour of xanthan is important, since some stabilised products require a low viscosity during their application.

All water-soluble polysaccharides possess some degree of thickening ability, but comparatively few have gelling properties. The characteristics of existing gelling systems tend to be unique and, thus, the interchange of gums in specific applications is impossible. Although xanthan by itself does not possess the functional property of gelation, it will form thermoreversible gels with certain galactomannans. The phenomenon of gelation arises from inter-molecular associations ("junction zones") to form a three-dimensional continuous network which traps and immobilizes the water within it to form a rigid structure, which shows resistance to flow under pressure. The gelling and melting of xanthan-galactomannan gels involves the formation and destruction of the "junction zones" which non-covalently link the polymer chains. Essentially, gels possess properties of both liquids and solids.

Various derivatives of xanthan have been prepared in attempts to modify the solution properties of the polymer for particular applications. This has included the carboxyalkyl ether [192], hydroxyalkyl ether [193], dialkylaminoalkoxy ether [194] and alkylene glycol ester [195] derivates of xanthan. Methods [196,197] for the preparation of the sulphate ester derivative of xanthan have also been reported. Also, various graft copolymers of xanthan have been prepared using acrylamide or acrylamide derivatives [198, 199]. However, it does not appear that any of the above derivatives of xanthan are produced commercially.

In addition to possessing useful solution properties, it is also necessary for the polymer to be compatible with a wide range of materials encountered in commercial applications. As already discussed, xanthan shows good compatibility with high concentrations of a range of salts. Xanthan is compatible with many acid solutions such as 5% sulphuric acid, 5% acetic acid and 25% phosphoric acid. The viscosity of such solutions is reasonably stable provided storage is at ambient temperatures. Xanthan is also compatible with many basic solutions, but alkaline degradation is likely at pH > 10, even at ambient temperature. Oxidising or free radical reagents are also likely to cause degradation of xanthan. Further information on the compatibility of xanthan with various chemicals can be found elsewhere [184]. Xanthan is also compatible with a wide range of other carbohydrate polymers such as cellulosic materials, alginates, carrageenans and starch. For food use, it is sometimes necessary for xanthan to be compatible with a complex mixture such as proteins, fats, low molecular weight materials (including carbohydrates), besides carbohydrate polymers.

It is possible that xanthan could come into contact with enzymes in some applications. Although xanthan is not susceptible to degradation by many common enzyme preparations such as amylases, pectinases and proteases [184], recent reports suggest that the polymer is not as stable to enzymic attack as had been considered. In particular, xanthan can be degraded by enzyme preparations from *Bacillus* spp. [200-202] and a *Corynebacterium* strain (20/122) [203].

Further, a cellulase preparation has been observed to degrade xanthan when the polymer is in the disordered form under salt-free conditions [204].

Figures for the commercial consumption of xanthan are not readily available and estimates have to be regarded with caution. But one estimate suggests that 10 and 12 million pounds (total) of xanthan were used in the USA and Worldwide respectively in 1979 [205]. Further, 30% of the total USA consumption was used for food applications.

9.2 Safety Aspects and Regulatory Status of Xanthan

The safety properties of xanthan for food and pharmaceutical applications have been investigated extensively. Short-term feeding studies (12 weeks) of xanthan to dogs indicated that no acute toxicity or growth-inhibiting activity took place [206]. Long-term toxicological evaluation using two year feeding studies to rats and dogs showed that xanthan had no effect on growth or survival rate nor was any incidence of tumour detected [207]. Also, xanthan had no effect on the three-generation reproduction of rats [207]. Tests have also shown that xanthan is a non-sensitizer and that it does not cause skin or eye irritation [184].

To gain acceptance as a food additive a polysaccharide has to satisfy the stringent government regulations of individual countries or groups of countries (such as the E.E.C.). However, the successful toxicological testing of xanthan has resulted in a widespread acceptance of its suitability for inclusion in foods. Clearance for the general food use of xanthan where Standards of Identity regulations do not preclude such use was cleared by the United States Food and Drug Administration (FDA) in 1969 [208]. Approval for use of xanthan in specified foodstuffs in this country was later obtained [209-213]. Xanthan is also on the FDA list of GRAS (generally regarded as safe) substances [214]. The polymer has also been cleared for general food use in Canada [215] and is on Annex II of the European Economic Community list of emulsifiers, stabilisers etc. [216].

The major advantage of xanthan as a food gum additive is that it does have such widespread acceptance. Since the regulatory restrictions for food use and the cost of complying with them has been growing over the last few years, it will be difficult to introduce a new microbial gum onto the approved food additive list, even if it has superior rheological properties to xanthan [217]. One estimate suggests that the testing required by the USA FDA to gain food clearance approval for a new additive, would cost about $2,000,000 and take up to 7½ years to achieve [217].

9.3 Food Uses of Xanthan

Polysaccharides such as xanthan are incorporated into foods to control the rheology of the final product. In performing this functional task, the polymer has a major effect on various sensory properties such as texture, flavour-release and appearance, which contribute to the acceptability of the product for consumer use. Due to its solution properties, xanthan can be used to provide a desired functionality at a lesser concentration than most gums and therefore may be cheaper to use on a cost effective basis. Further, polysaccharides such as xanthan which display pseudoplastic properties in solution have less "gummy" mouthfeel than gums with more Newtonian solution behaviour.

Some examples [218-232] of the application of xanthan for the food industry are as follows. One application which utilises several of xanthans' properties is French dressing compositions [218]. The high at-rest viscosity and yield value of xanthan results in stability of the oil-in-water emulsion present in the product. Further, the high pseudoplasticity of xanthan solutions allows easy pouring of the product, yet provide good cling to salads. The stability of xanthan at low pH and high viscosity at low concentrations are particularly relevant for this application. Similar to other hydrocolloids, xanthan imparts good freeze-thaw stability to products thus rendering the polymer suitable for frozen foods [219]. Inclusion of low concentrations of xanthan in juice drinks stabilises the fruit pulp to provide a uniform product with acceptable mouthfeel [220]. Inclusion of xanthan in bread has also been suggested [221-223]. In particular, the polymer has been cited as a replacement for gluten in bread [221, 222]. Xanthan is also used in the preparation of canned foods where it provides viscosity control during processing and improves pumping and filling operations due to the pseudoplastic behaviour of the polymer [184]. A relatively unexplored potential for xanthan is in the preparation of fabricated foods. But one report has indicated the use of xanthan as a binder for protein fibres in meat analogues [224].

Numerous food applications of xanthan make use of the synergistic viscosity and gelling effects which take place with galactomannans. In such applications, xanthan is often used in blends with guar gum and/or locust bean gum. Inclusion of xanthan and locust bean gum in cottage cheese creaming emulsions provides viscosity stability to the product [225]. This prevents drainage from the product and allows a more uniform consistency to be maintained. Gelling of instant desserts [226] and meat products [227] can be accomplished using the xanthan/locust bean gum system. A particular advantage of this system is the low concentrations of the two polymers which can be used to provide effective gels. A combination of xanthan/locust bean gum can be used as a partial replacement for starch to provide products with better texture and less pasty mouth-feel.

The xanthan/locust bean gum gelling system has been reported to improve production of starch jelly candies by decreasing the time for the product to set [228]. A mixture of these two gums can also be used to stabilise margarines [229]. Combinations of xanthan/guar gum and xanthan/guar gum/carrageenan/ locust bean gum are used to stabilise ice-cream [230] and milk-shakes [231] respectively. Xanthan has also been used in combination with carboxymethylcellulose as a food stabilising agent [232].

The application of xanthan is not limited to foods for human consumption. It has been used in pet foods as a thickening agent and binder [233]. Xanthan has also been included in liquid cattle feed supplements [234] and calf milk replacers [234], where it prevents separation of active ingredients from the suspension.

9.4 Non-Food Uses of Xanthan

Xanthan is used in a diverse range of industrial (or non-food) applications, particularly as a suspending or thickening agent. The polymer is used in the suspension of agricultural chemicals where it controls drift during spraying and provides good cling to crops [235]. Inclusion of xanthan in paints results in effective dispersion of the pigments and provides desirable application of the product on surfaces [236]. Xanthan can be used as an additive in metal pickling baths due to its stability in strong acid and alkaline solutions [237]. Xanthan is often used to suspend heavy particles in products such as ceramic glazes [238]. The ability of xanthan to impart the desired dispersion stability at rest, and low viscosity under application is used in such applications as toothpaste [239].

The ability of xanthan to control the viscosity of solutions is used in a variety of applications. This includes the petroleum industry where it is used in oil well drilling fluids [240], well completion and fracturing fluids [241] and as a mobility control agent in Enhanced Oil Recovery (see next section). In such applications the ability of xanthan to thicken dense salt solutions is of particular importance.

Various gelling systems of xanthan are used for industrial applications. A mixture of xanthan/locust bean gum can be used in deoderant gels [184]. Under certain conditions of concentration and pH, xanthan will form gels with di- and trivalent metal ions. An example of such a gelling system is xanthan/ iron which is used to reduce the level of glue in particle board manufacture [242]. Xanthan will also gel with borate and this combination has been used for preparation of explosives [243].

9.5 Use of Xanthan in Enhanced Oil Recovery (EOR)

After completion of the conventional methods (primary and secondary production) of oil recovery, more than 50% of the oil initially present in the reservoir rocks remains unrecoverable. For economic and energy conservation reasons, there is widespread interest in the development of tertiary (or enhanced) production procedures to increase the efficiency of oil recovery. A wide variety of methods are available for this specific purpose [244]. Water-flooding of the oil rock reservoirs is used in secondary production methods, but due to the relative mobilities of the injected fluid and the more viscous oil, by-passing of the hydrocarbon occurs. However, the addition of polymers with thickening ability to the injected fluid, reduces the mobility of water and improves the sweep efficiency of oil displacement. A wide variety of polymers have been used as mobility control agents for Enhanced Oil Recovery. Xanthan is among those polymers which is considered to have the greatest potential for this application [12-14]. The use of xanthan may also be considered for micellar-polymer flooding of oil reservoirs. The inclusion of the surfactant would reduce the immiscibility of oil and water and thus further increase the sweep efficiency of hydrocarbon displacement.

Xanthan possesses many of the basic properties required for EOR use. In particular, the ability of xanthan to form solutions of high viscosity at low concentrations is necessary for the physical requirements of the application and also for cost effectiveness. Further, injection of an EOR polymer into the rocks requires pumping and thus the pseudoplastic behaviour of xanthan solutions is essential for this operation. It is also necessary for the polysaccharide to flow through the relatively small rock pores without causing blockage or problems with injectability. Xanthan shows molecular associations in salt solution and simulated tests have shown that this could lead to decreased passage through small pores under highly saline conditions [67]. Pore blockage of xanthan in tests has also been attributed to poorly dissolved polymer aggregates (microgels) [245,246] and residual bacterial cell debris [247,248]. The use of clarified fermentation broths rather than xanthan powders may help to eliminate microgels [249], while enzyme treatments can be used to alleviate problems with bacterial cell debris (see Section 8.2).

It is also important for EOR applications that the polymer can withstand the conditions of the oil reservoir for the period ($\sim$two years) which it will be underground. In some cases these conditions will be extreme, for instance, in North Sea oilfields temperatures of 80-102°C, salinities up to 10% and pressures of 200-400 bar can be expected. Temperature stability is a particularly critical requirement and several tests have been performed on xanthan [250,251,252]. When suitable precautions are taken xanthan shows reasonable stability at high temperatures. In one study the viscosity of xanthan solutions was preserved at 97°C for one year using sodium sulphite, thiourea and isopropanol as oxygen scavenger, radical transfer agent and readily oxidizable agent respectively [251]. The xanthan solutions were deoxygenated prior to the addition of the antioxidant additives. One problem with the use of such additives is that they may become separated from the polymer by hydrodynamic chromatography effects in the rock pores.

Due to the wide variation of physical conditions which are found in oil reservoirs, it is possible that no mobility control polymer agent will be found which can be applied worldwide in EOR. However, at present, xanthan still appears to be the leading contender for this application. The increase in efficiency of oil recovery employing polymer flooding of reservoirs may only be 5%. Thus, under present economic conditions, this tertiary oil recovery process may only be marginally profitable when applied under true conditions. However, it is considered that the process would be more efficient, and more profitable, if initiated earlier in the life of a reservoir when oil saturations are higher [253].

9.6 Analysis of Xanthan

Different grades of xanthan are supplied by the commercial producers of the polymer for the various areas of application such as food, industrial and Enhanced Oil Recovery. It follows that different levels of purity can be expected between the different xanthan products. Clearly, in the interests of both the producer and consumer there is a need to monitor the quality of the xanthan. Thus, a variety of analytical tests such as viscosity, carbohydrate content, ash content and powder colour are available [254] to control the quality of xanthan produced. For some application areas this information may not be sufficient and other characteristics such as filtration behaviour might be required.

Also, information on the content of non-carbohydrate constituents (acetate and pyruvate) in xanthan may be required. Colorimetric procedures are available for the assay of acetate [255] and pyruvate [256] content in xanthan. The latter constituent may also be determined by an enzymic method involving lactate dehydrogenase [257]. However, these procedures are not entirely satisfactory and, in particular, pyruvate determination requires an initial acid hydrolysis which may destroy material. Alternatively, acetate and pyruvate contents may be performed on intact xanthan samples using NMR [36,258]. One method involves recording the ^{1}H NMR spectra of xanthan on a 60 MHz instrument using Fourier Transformation at 90°C [258]. At this temperature the proton signal from the methyl groups in acetate and pyruvate can be detected and quantified by addition of an internal standard to the xanthan solution.

For the comparison of xanthans from different fermentation conditions or *Xanthomonas* strains, more detailed analysis such as sugar composition or sequence might be useful. Sophisticated techniques such as glc/mass spectrometry can be used for this purpose, but they tend to be time-consuming. Alternatively, it has been suggested that certain enzyme preparations may be used for a more convenient comparison of different xanthans [42].

10 CONCLUSIONS

This review has attempted to cover the more important aspects of xanthan. Further information on xanthan can be found in a wide selection of other reviews devoted entirely or partly to the biopolymer [254,259-262]. More detailed information on aspects such as the production [164,263,264], recovery [265] and applications [261,262,266,267] of xanthan is also available.

The past few years have seen some valuable contributions to our knowledge of xanthan, notably the correlation of structure with solution properties. However, there still exists some areas where our knowledge of xanthan is not complete. This includes some aspects of the biosynthesis of xanthan and the exact conformation of the biopolymer in solution. There is also plenty of scope for new work on the production of xanthan. In particular, the successful scale-up of a continuous fermentation process for production of xanthan on a commercial basis would be a significant breakthrough.

The areas for the commercial application of xanthan should continue to expand and demand for the biopolymer in existing applications is also likely to increase. A particularly interesting area is the use of xanthan in EOR and the next few years will be crucial in determining if the biopolymer can fulfil its promise in this application. The use of xanthan in fabricated foods is another area which will be watched closely in the future. Xanthan is likely to be the leading biopolymer for the forseeable future, but it may act as a precedent for the introduction of new ones which are discovered to have interesting solution properties.

11 REFERENCES

1 E.R. Morris, in P.A. Sandford and A. Laskin (Eds.), A.C.S. Symp. Ser., Vol. 45 (Extracellular Microbial Polysaccharides), American Chemical Society, Washington D.C., 1977, pp. 83-89.
2 E.R. Morris, in J.M.V. Blanshard and J.R. Mitchell (Eds.), Polysaccharides in Food, Butterworths, London, 1979, pp. 15-31.
3 D.A. Powell, in R.C.W. Berkeley, G.W. Gooday and D.C. Ellwood (Eds.), Microbial Polysaccharides and Polysaccharases, Academic Press, London, New York and San Francisco, 1979, pp. 118-160.
4 D.A. Rees and E.J. Welsh, Angew. Chem. Int. Ed. Engl., 16 (1977) 214-224.

5 P.E. Jansson, L. Kenne and B. Lindberg, Carbohydr. Res., 45 (1975) 275-282.
6 L.D. Melton, L. Mindt, D.A. Rees and G.R. Sanderson, Carbohydr. Res., 46 (1976) 245-257.
7 F.E. Halleck, U.S. Pat. 3,301,848 (1967).
8 S. Yuen, Process Biochem., 9 (1974) 7-9,22.
9 L. Deavin, T.R. Jarman, C.J. Lawson, R.C. Righelato and S. Slocombe, Ref 1, pp. 14-28.
10 K.S. Kang, G.T. Veeder and D.D. Richey, Ref 1, pp. 211-219.
11 R. Moorhouse, G.T. Colegrove, P.A. Sandford, J.K. Baird and K.S. Kang, in D.A. Brant (Ed.), A.C.S. Symp. Ser., Vol. 150 (Solution Properties of Polysaccharides), American Chemical Society, Washington D.C., 1981, pp. 111-124.
12 E.I. Sandvik and J.M. Maerker, Ref. 1, pp. 242-264.
13 E. Unsal, J.L. Duda and E. Klaus, In R.T. Johansen and R.L. Berg (Eds.), A.C.S. Symp. Ser., Vol. 91 (Chemistry of Oil Recovery), American Chemical Society, Washington D.C., 1979, 141-170.
14 A. Gabriel, Ref. 3, pp. 191-204.
15 A. Jeanes, J.E. Pittsley and F.R. Senti, J. Appl. Polymer Sci., 5 (1961) 519-526.
16 S.P. Rogovin, R.F. Anderson and M.C. Cadmus, J. Biochem. Microbiol. Technol. Eng., 3 (1961) 51-63.
17 J.C. Walker and M.E. Gallegy, Amer. J. Bot., 38 (1951) 663-665.
18 V.G. Lilly, H.A. Wilson and J.G. Leach, Appl. Microbiol., 6 (1958) 105-108.
19 J.C. Sutton and P.H. Williams, Can. J. Bot., 48 (1970) 645-651.
20 H. Stolp and M.P. Starr, Phytopathol. Z., 51 (1964) 442-478.
21 H. Stolp, M.P. Starr and N.C. Baigent, Ann. Rev. Phytopathology, 3 (1965) 231-264.
22 R.E. Buchanan and N.E. Gibbons (Eds.), Determinative Bacteriology, 8th Edition, Williams and Wilkins, Baltimore, 1974.
23 J.H. Sloneker and A. Jeanes, Can. J. Chem., 40 (1962) 2066-2071.
24 J.H. Sloneker, D.G. Orentas and A. Jeanes, Can. J. Chem., 42 (1964) 1261-1269.
25 I.R. Siddiqui, Carbohydr. Res., 4 (1967) 284-291.
26 C.J. Lawson and K.C. Symes, Ref. 1, pp. 183-191.
27 H. Bjorndal, C.G. Hellerqvist, B. Lindberg and S. Svensson, Angew. Chem. Int. Ed. Engl., 9 (1970) 610-619.
28 B. Lindberg, J. Lonngren and J.L. Thomson, Carbohydr. Res., 28 (1973) 351-357.
29 P.A.J. Gorin, T. Ishikawa, J.F.T. Spencer and J.H. Sloneker, Can. J. Chem., 45 (1967) 2005-2008.
30 J.H. Sloneker and D.G. Orentas, Can. J. Chem., 40 (1962) 2188-2189.
31 F.R. Dintzis, G.E. Babcock and R. Tobin, Carbohydr. Res., 13 (1970) 257-267
32 M. Rinuado and M. Milas, Biopolymers, 17 (1978) 2663-2678.
33 G. Holzwarth, Carbohydr. Res., 66 (1978) 173-186.
34 J.G. Southwick, H. Lee, A.M. Jamieson and J. Blackwell, Carbohydr. Res., 84 (1980) 287-295.
35 I.W. Sutherland, Adv. Microb. Physiol., 8 (1972) 143-208.
36 I.H. Smith, K.C. Symes, C.J. Lawson and E.R. Morris, Int. J. Biol. Macromol., 3 (1981) 129-134.
37 M.P. Starr, Ann. Rev. Microbiol., 13 (1959) 211-238.
38 A. Jeanes, P. Rogovin, M.C. Cadmus, R.W. Silman and C.A. Knutson, Polysaccharide (Xanthan) of Xanthomonas campestris NRRL B-1459: Procedures for Culture Maintenance and Polysaccharide Production, Purification and Analysis, U.S. Agric. Res. Serv. North Cent Reg, ARS-NC-51, 1976.
39 M.C. Cadmus, S.P. Rogovin, K.A. Burton, J.E. Pittsley, C.A. Knutson and A. Jeanes, Can. J. Microbiol., 22 (1976) 942-948.
40 M.C. Cadmus, C.A. Knutson, A.A. Lagoda, J.E. Pittsley and K.A. Burton, Biotech. Bioeng., 20 (1978) 1003-1014.

41 P.A. Sandford, J.E. Pittsley, C.A. Knutson, P.R. Watson, M.C. Cadmus and A. Jeanes, Ref. 1, pp. 192-210.
42 I.W. Sutherland, Carbohydr. Polymers, 1 (1981) 107-115.
43 I.W. Davidson, FEMS Microbiol. Lett., 3 (1978) 347-349.
44 C.G.T. Evans, R.G. Yeo and D.C. Ellwood, Ref. 3, pp. 51-68.
45 C. Whitfield, I.W. Sutherland and R.E. Cripps, J. Gen. Microbiol., 124 (1981) 385-392.
46 P.A. Sandford, P.R. Watson and C.A. Knutson, Carbohydr. Res., 63 (1978) 253-256.
47 I.W. Sutherland, J. Chromatogr., 213 (1981) 301-306.
48 W.F. Dudman and N. Heidelberger, Science, 164 (1969) 954-955.
49 V.S. Fareed and E. Percival, Carbohydr. Res., 49 (1976) 427-438.
50 E.R. Morris, D.A. Rees, G. Young, M.D. Walkinshaw and A. Darke, J. Mol. Biol., 110 (1977) 1-16.
51 I.C.M. Dea, E.R. Morris, D.A. Rees, E.J. Welsh, H.A. Barnes and J. Price, Carbohydr. Res., 57 (1977) 249-272.
52 I.J. Bradshaw, B.A. Nisbet, M.H. Kerr and I.W. Sutherland, Carbohydr. Polymers, 3 (1983) 23-38.
53 D.A. Rees, Biochem. J., 126 (1972) 257-273.
54 G. Holzwarth, Biochemistry, 15 (1976) 4333-4339.
55 M. Milas and M. Rinaudo, Carbohydr. Res., 76 (1979) 189-196.
56 G. Holzwarth and J. Ogletree, Carbohydr. Res., 76 (1979) 277-280.
57 M. Milas and M. Rinaudo, Ref. 11, pp. 25-30.
58 R. Moorhouse, M.D. Walkinshaw and S. Arnott, Ref. 1, pp. 90-102.
59 I.T. Norton, D.M. Goodall, E.R. Morris and D.A. Rees, J. Chem. Soc. Chem. Commun., (1980) 545-547.
60 D.A. Rees, Pure Appl. Chem., 53 (1981) 1-14.
61 G. Holzwarth and E.B. Prestridge, Science, 197 (1977) 757-759.
62 G. Paradossi and D.A. Brant, Macromolecules, 15 (1982) 874-879.
63 G. Holzwarth, Ref. 11, pp. 15-23.
64 P.J. Whitcomb and C.W. Macosko, J. Rheol., 22 (1978) 493-505.
65 O. Smidsrod and A. Haug, Biopolymers, 10 (1971) 1213-1227.
66 J.G. Southwick, A.M. Jamieson and J. Blackwell, Carbohydr. Res., 99 (1982) 117-127.
67 B.A. Nisbet, I.J. Bradshaw, M. Kerr and I.W. Sutherland, in 2nd European Symposium Enhanced Oil Recovery, Technip., Paris, 1982, pp. 71-78.
68 W.C. Wernau, G.B. Pat, 2,008,600A (1979).
69 G. Maret, M. Milas and M. Rinaudo, Polymer Bulletin, 4 (1981), 291-297.
70 I.J. Bradshaw, B.A. Nisbet, M.H. Kerr and I.W. Sutherland, Unpublished Results.
71 C.S.H. Chen and E.W. Sheppard, Polym. Eng. Science, 20 (1980) 512-516.
72 P. Kovacs, Food Trade Review, 43 (1973) 17-22.
73 J.K. Rocks, Food Technol., 25 (1971) 22-31.
74 P. Kovacs, Food Technol., 27 (1973) 26-30.
75 I.C.M. Dea and E.R. Morris, Ref. 1, pp. 174-182.
76 C.W. Baker and R.L. Whistler, Carbohydr. Res., 45 (1975) 237-243.
77 T.J. Painter, J.J. Gonzalez and P.C. Hemmer, Carbohydr. Res., 69 (1979) 217-226.
78 J.E. Courtois and P. Le Dizet, Bull. Soc. Chim. Biol., 50 (1968) 1695-1709.
79 J.E. Courtois and P. Le Dizet, Bull. Soc. Chim. Biol., 52 (1970) 15-22.
80 I.C.M. Dea, A.A. McKinnon and D.A. Rees, J. Mol. Biol., 68 (1972) 153-172.
81 N.S. Anderson, J.W. Cambell, M.M. Harding, D.A. Rees and J.W.B. Samuel, J. Mol. Biol., 45 (1969) 85-99.
82 S. Arnott, W.E. Scott, D.A. Rees and C.G.A. McNab, J. Mol. Biol., 90 (1974) 253-267.
83 S. Arnott, A. Fulmer, W.E. Scott, I.C.M. Dea, R. Moorhouse and D.A. Rees, J. Mol. Biol., 90 (1974) 269-284.
84 B.V. McCleary, Carbohydr. Res., 71 (1979) 205-230.
85 E.R. Morris, D.A. Rees, G. Robinson and G.A. Young, J. Mol. Biol., 138 (1980) 363-374.

86 P.R. Sundarajan and V.S.R. Rao, Biopolymers, 9 (1970) 1239-1247.
87 E.D.T. Atkins, E.D.A. Hopper and D.H. Issac, Carbohydr. Res., 27 (1973) 29-37.
88 H.R. Schuppner, Aust.Pat.401, 434 (1966).
89 J.F. Wilkinson, Bacteriol. Rev., 22 (1958) 46-73.
90 I.W. Sutherland, Ref 1, pp. 40-57.
91 I.W. Sutherland, Adv. Microbiol. Physiol., 23 (1982) 79-150.
92 L. Ielpi, R.O. Couso and M.A. Dankert, FEBS Letters, 130 (1981) 253-256.
93 L. Ielpi, R.O. Couso and M.A. Dankert, Biochem. Biophys. Res. Commun., 102 (1981) 1400-1408.
94 L. Ielpi, R.O. Couso and M.A. Dankert, Biochem. Int., 6 (1983) 323-333.
95 S. Roseman, in L.E. Hopkin (Ed.), Metabolic Pathways, Academic Press, London and New York, 1972, pp. 41-89.
96 C. Whitfield, I.W. Sutherland and R.E. Cripps, J. Gen. Microbiol., 128 (1982) 981-985.
97 R.L. Bernstein and P.W. Robbins, J. Biol. Chem., 240 (1965) 391-397.
98 I.W. Sutherland (Ed.), Surface Carbohydrates of the Prokaryotic Cell, Academic Press, London, New York and San Francisco, 1977.
99 I.W. Sutherland, Ref 3, pp. 1-34.
100 C.H. Clarke and A.W.B. Johnson, Mutat. Res., 36 (1976) 147-164.
101 C.H. Clarke and D.M. Shankel, Bacteriol. Rev., 39 (1975) 33-53.
102 J.W. Drake, The Molecular Basis of Mutation, Holden-Day, San Francisco, 1970.
103 S.P. Lapage, J.E.Shelton, T.G. Mitchell and A.R. MacKenzie, in J.R. Norris and D.W. Ribbons (Eds), Methods in Microbiology, Vol. 3A, Academic Press, New York, 1970, pp. 135-228.
104 D.K. Kidby, P.A. Sandford, M.C. Cadmus and A. Herman, Appl. Environ. Microbiol., 33 (1977) 840-845.
105 W. Bechstedt, U. Behrens, E. Stolle and M. Wagner, Ger. (East) Pat. 121, 954 (1975).
106 M. Wagner and E. Stolle, Ger. (East) Pat. 129, 802 (1978).
107 W.P. Weisrock and E.F. McCarthy, Eur. Pat. 46,007A (1982).
108 T.R. Jarman, Eur. Pat. 66, 957A (1982).
109 W.P. Weisrock, U.S. Pat. 4,377,637 (1983).
110 W.C. Wernau, U.S. Pat. 4,282,321 (1981).
111 P. Souw and A.L. Demain, Appl. Environ. Microbiol., 37 (1979) 1186-1192.
112 W.H. McNeely, U.S. Pat. 3,433,708 (1969).
113 M.C. Cadmus, L.G. Jayko, D.E. Hensley, H. Gasdorf and K.C. Smiley, Cereal. Chem., 43 (1966) 658-669.
114 M.S. Matyshevskaya, R.I. Gvozdyak, I.I. Maiko, O.A. Litvinchuk, L.A. Pasichnik, M.I. Lipkes, G.Y. Dedusenko and N.M. Kolodkova, Mikobiol. Zh (Kiev), 44 (1982) 36-40.
115 M. Charles and M.E. Radjai, Ref 1, pp. 27-39.
116 K.R. Stauffer and J.G. Leeder, J. Food. Science, 43 (1978) 756-758.
117 R.A. Moraine and P. Rogovin, Biotechnol. Bioeng., 13 (1971) 381-391.
118 U. Behrens, B. Heyne, R. Chop, W. Klipp, R. Kaiser, H.J. Mueller, W. Tarnow and M. Wagner, Rom.Pat.71, 875 (1982).
119 M.O. Bagby, I.A. Wolff and M.C. Cadmus, Tappi, 52 (1969) 442-445.
120 R. Bretschneider, J. Pridoha and V. Mikulkova, Czech. Pat. 161, 467 (1975).
121 H. Kanda, Japan Pat. 77 10,492 (1977).
122 P. Souw and A.L. Demain, J. Ferment. Technol., 58 (1980) 411-416.
123 P. Souw in C. Vezina and K. Singh (Eds.), Adv. Biotechnol., Vol. 3 (Proc. 6th Int. Ferment. Symp.), Pergamon, Toronto, 1981, pp. 447-453.
124 A.L. Demain and P. Souw, U.S. Pat. 4,245,046 (1981).
125 W.P. Weisrock, U.S. Pat. 4,301,247 (1981).
126 R.A. Moraine and P. Rogovin, Biotechnol. Bioeng., 8 (1966) 511-524.
127 G.W. Pace and S.D.J. Coote, Eur. Pat. 32, 293A (1981).
128 R.W. Silman and P. Rogovin, Biotechnol. Bioeng., 12 (1970) 75-83.
129 R.W. Silman and P. Rogovin, Biotechnol. Bioeng., 14 (1972) 23-31.
130 R.W. Silman and E.B. Bagley, Biotechnol. Bioeng., 21 (1979) 173-179.

131 D.C. Ellwood, I. Davidson, R.C. Righelato and C.E. Smith, Ger. Offen. 2,849,183 (1979).
132 U.K. Secretary of State for Defence, Neth. Appl. 76 12, 448 (1977).
133 W.P. Weisrock and H.S. Klein, U.S. Pat. 4,374,929 (1983).
134 T.R. Jarman and G.W. Pace, Eur. Pat. 66,961A (1982).
135 T.R. Jarman, Eur. Pat. 66,377A (1982).
136 W.P. Weisrock, Eur. Pat. 44,659A (1982).
137 B.D. Church, U.S. Pat. 4,218,538 (1980).
138 D.C. Orlowski and B.D. Church, U.S. 4,251,633 (1981).
139 C.J. Norton, D.O. Falk and W.E. Luetzelschwab, Soc. Pet. Eng. J., 21 (1981) 205-217.
140 J.F. Kennedy, P. Jones, S.A. Barker and G.T. Banks, Enzyme Microb. Technol. 4 (1982) 39-43.
141 A. Margaritis and J.E. Zajic, Biotechnol. Bioeng., 20 (1978), 939-1001.
142 N. Thomson and D.F. Ollis, Biotechnol. Bioeng., 22 (1980) 875-883.
143 H.T. Chang and D.F. Ollis, Biotechnol. Bioeng., 24 (1982) 2309-2318.
144 L.G. Maury, U.S. Pat. 4,352,882 (1982).
145 G. Chauveteau and N. Kohler, Soc. Pet. Eng. SPE 9295 (1980).
146 J.J. O'Connell, U.S. Pat. 3,355,447 (1967).
147 K.S. Kang and D.B. Burnett, G.B. Pat. 1,488,645 (1977).
148 J.T. Patton, U.S. Pat. 3,964,972 (1976).
149 G.T. Colegrove, U.S. Pat. 3,516,983 (1970).
150 G.T. Colegrove, U.S. Pat. 3,966,618 (1976).
151 K.S. Kang, G.B. Pat. 2,052,542A (1981).
152 R.A. Empey and D.J. Pettitt, G.B. Pat. 1,487,530 (1977).
153 F.E. Corley and J.B. Richmon, U.S. Pat. 4,135,979 (1979).
154 B.B. Buchanan and J.E. Cottle, U.S. Pat. 3,773,752 (1973).
155 M.K. Abdo, U.S. Pat. 3,711,462 (1973).
156 G.T. Colegrove, G.B. Pat. 1,443,507 (1975).
157 T.J. Holding and G.W. Pace, G.B. Pat. 2,065,689 (1981).
158 S.L. Wellington, U.S. Pat. 4,119,491 (1978).
159 M. Rinaudo, M. Milas and N. Kohler, Fr. Demande 2,491,494 (1982).
160 O.M. Stokke, U.S. Pat. 4,165,257 (1979).
161 W.E. Rader and J.C. Wong, Eur. Pat. 39,962A (1981).
162 C.W. Schroeck, U.S. Pat. 4,094,739 (1978).
163 F.J. Maske, U.S. Pat. 4,041,234 (1977).
164 G.W. Pace and R.C. Righelato, Adv. Biochem. Eng., 15 (1981) 41-70.
165 M.B. Inkson and C.K. Wilkinson, Eur. Pat. 28,446A (1981).
166 Societe des Usines Chimiques Rhone-Poulenc, Fr. Demande 2,251,620 (1975).
167 A. Bouniot, U.S. Pat. 3,988,313 (1976).
168 W.H. McNeely and J.J. O'Connell, U.S. Pat. 3,232,929 (1966).
169 J.E. Patton and W.E. Holman, U.S. Pat. 3,382,229 (1968).
170 G.A. Towle, U.S. Pat. 4,051,317 (1977).
171 I.H. Smith, Eur. Pat. 68,706A (1983).
172 W.J. Albrecht, S.P. Rogovin and E.L. Griffin, Nature, 194 (1962) 1279.
173 W.J. Albrecht, V.E. Sohns and S.P. Rogovin, Biotechnol. Bioeng, 5 (1963) 91-99.
174 S.P. Rogovin and W.J. Albrecht, U.S. Pat. 3,119,812 (1964).
175 G.P. Lindblom and J.T. Patton, U.S. Pat. 3,163,602 (1964).
176 J.W. Gill and P.G. Lim, U.S. Pat. 3,422,085 (1969).
177 J.F. Kennedy, S.A. Barker, I.J. Bradshaw and P. Jones, Carbohydr. Polymers 1 (1981) 55-66.
178 J.F. Kennedy and I.J. Bradshaw, J. Carbohydr. Nucleosides and Nucleotides, 8 (1981) 461-480.
179 J.F. Kennedy and I.J. Bradshaw, Brit. Polym. J., 15 (1983) 85-90.
180 C.W. Schroeck, U.S. Pat. 4,254,257 (1981).
181 J.F. Kennedy, S.A. Barker, P. Jones and I.J. Bradshaw, Carbohydr. Polymers 1 (1981) 85-99.
182 S.P. Rogovin, W.J. Albrecht and V.E. Sohns, Biotechnol. Bioeng., 7 (1965) 161-169.
183 H.J. Leder and G.M. Miescher, U.S. Pat. 3,316,241 (1967).

184 Kelco Division of Merck and Co., Inc., Xanthan Gum - a Natural Biopolysaccharide for Scientific Water Control, Kelco, San Diego.
185 R.S. Igoe, G.B. Pat. 2,039,712A (1980).
186 P.A. Sandford, J. Baird and I.W. Cottrell, Ref 11, pp. 31-41.
187 E. Unsal, J.L. Duda and E.E. Klaus, J. Can. Pet. Technol., 18 (1979) 17-19.
188 W.L. Griffith, L. Compere and J.M. Crenshaw, Dev. Ind. Microbiol., 21 (1980) 451-460.
189 W.L. Griffith, A.L. Compere and J.W. Holleman, G.B. Pat. 2,065,688A (1981).
190 J.L. Duda, E.E. Klaus and W.C. Leung, Preprints Amer. Chem. Soc. Div. Pet. Chem., 26 (1981) 146-155.
191 H.L. Lee, U.S. Pat. 4,299,825 (1981).
192 R.G. Schweiger, U.S. Pat. 3,236,831 (1966).
193 R.G. Schweiger, U.S. Pat. 3,349,077 (1967).
194 R.G. Schweiger, U.S. Pat. 3,244,695 (1966).
195 R.G. Schweiger, U.S. Pat. 3,256,271 (1966).
196 R.G. Schweiger, U.S. Pat. 3,446,796 (1969).
197 T.N. Tyler, G.B. Pat. 2,064,568A (1981).
198 D.J. Pettitt, U.S. Pat. 3,708,446 (1973).
199 I.W. Cottrell, J.L. Shim, G.H. Best and R.A. Empey, in R.G. Schweiger, J.L. Shim and G.H. Best (Eds.), A.C.S. Symp. Ser., Vol. 77 (Carbohydrate Sulphates), American Chemical Society, Washington D.C., 1978, pp. 193-202.
200 I.W. Sutherland, J. Appl. Bacteriol., 53 (1982) 385-393.
201 M.C. Cadmus, L.K. Jackson, K.A. Burton, R.D. Plattner and M.E. Slodki, Appl. Environ. Microbiol., 44 (1982) 5-11.
202 M.C. Cadmus, U.S. Pat. Appl. 345,512 (1982).
203 R.E. Cripps, H.J. Somerville and M.S. Holt, Eur. Pat. 30,393A (1981).
204 M. Rinaudo and M. Milas, Int. J. Biol. Macromol., 2 (1980) 45-48.
205 M. Glicksman, in M. Glicksman (Ed.), Food Hydrocolloids, Vol. 1, CRC, Boca Raton, Fla, 1982, pp. 3-18.
206 D.J. Robbins, J.E. Moulton and A.N. Booth, Fd. Cosmet. Toxicol., 2 (1964) 545-550.
207 G. Woodward, M.W. Woodward, W.H. McNeely, P. Kovacs and M.T.I. Cronin, Toxicol. Appl. Pharmacol., 24 (1973) 30-36.
208 Federal Register, 34 (1969) 5376 (March 19).
209 Federal Register, 36 (1971) 17333 (August 28).
210 Federal Register, 38 (1973) 1218 (January 10).
211 Federal Register, 38 (1973) 6883 (March 14).
212 Federal Register, 39 (1974) 4466 (February 4).
213 Federal Register, 36 (1971) 24217 (December 22).
214 FASEB Committee, Evaluation of GRAS Monographs (Scientific Literature Reviews), PB 80-203789, NTIS, U.S. Dept. of Commerce, Washington D.C., 1980.
215 Canada Gazette, Part II, 105 (1971) (January 27).
216 European Economic Community, Official Publication No. L.189/1, July 12, 1974.
217 M. Glicksman, in G.O. Phillips, D.J. Wedlock and P.A. Williams (Eds.), Progress in Food and Nutrition Science, Vol. 6 (Gums and Stabilisers for the Food Industry), Pergamon Press, Oxford, New York, Paris, Toronto, Frankfurt, and Sydney, 1982, pp. 299-321.
218 L.R. Stanislav and J.K. Sheets, Food Prod. Develop., 5 (1971) 52-56.
219 D. Wintersdorff, U.S. Pat. 3,784,715 (1972).
220 H.R. Schuppner, U.S. Pat. 3,413,125 (1968).
221 K. Kulp, F.N. Hepburn and T.A. Lehmann, Baker's Dig., 48 (1974) 34-7 and 58.
222 D.D. Christianson, Baker's Dig., 50 (1976) 34-36.
223 S. Atsumi, M. Sasaki and I. Kitamura, Can. Pat. 1,109,276 (1981).
224 G. Muschiolik and H.S.F. Schmandke, Ger (East) Pat. 123,157 (1975).
225 P. Kovacs and B.D. Titlow, Am. Dairy Rev., 38 (1976) 34J-34L.
226 H.R. Schuppner, U.S. Pat. 3,507,664 (1970).
227 H.R. Schuppner, U.S. Pat. 3,519,434 (1970).
228 H. Cheng, U.S. Pat. 4,219,582 (1980).
229 D.P.J. Moran and J.J. Hepburn, Eur. Pat. 11,344A (1980).
230 General Mill Chemicals Inc., Japan Pat. 77 114,054 (1977).

231 R.S. Igoe, G.B. Pat. 2,039,712A (1980).
232 K.R. Jackman, U.S. Pat. 4,163,807 (1978).
233 J.W. Bernotavicz, G.B. Pat. 1,537,209 (1978).
234 T.R. Andrew, Ref 1, pp. 231-241.
235 K.F. Gibsen and A.W. Saddington, U.S. Pat. 3,717,452 (1973).
236 K.F. Gibsen, U.S. Pat. 3,481,889 (1969).
237 P. Godet, Process Biochem., 8(1) (1973) 33-34.
238 W.H. McNeely, W.P. Fairchild and A.R. Hunter, Can. Pat. 727,071 (1966).
239 H. Ichikawa, H. Fukuda and M. Tanaka, Japan Pat. 76 82,739 (1976).
240 R.D. Carico, Soc. Pet. Eng. SPE 5870 (1976).
241 N.D. Lipton and D.B. Burnett, Soc. Pet. Eng. SPE 5872 (1976).
242 N.J. Iammarino, U.S. Pat. 3,736,275 (1973).
243 G.T. Colegrove, U.S. Pat. 3,326,733 (1967).
244 H. Van Pollen, Fundamentals of Enhanced Oil Recovery, Penn Well Books, Tulsa, 1980.
245 D. Lipton, Soc. Pet. Eng. SPE 5099 (1974).
246 D.B. Burnett, Soc. Pet. Eng. SPE 5372 (1975).
247 J.T. Patton, Soc. Pet. Eng. SPE 4670 (1973).
248 G.E. Tinker, R.W. Bowman and G.A. Pope, J. Pet. Tech., 28 (1976) 586-596.
249 N. Kohler and G. Chaveteau, Soc. Pet. Eng. SPE 7425 (1978).
250 P. Davison and E. Mentzer, Soc. Pet. Eng. SPE 9300 (1980).
251 S.L. Wellington, Soc. Pet. Eng. SPE 9296 (1980).
252 M.H. Akstinat, Soc. Pet. Eng. SPE 8979 (1980).
253 W.R. Hause, Pet. Eng. Int., 53 (November) (1981) 55-72.
254 I.W. Cottrell, K.S. Kang and P. Kovacs, in R.L. Davidson (Ed.), Handbook of Water-Soluble Gums and Resins, McGraw-Hill, New York and London, 1980, pp. 24.1-24.31.
255 R.H. McComb and R.H. McCready, Anal. Chem., 29 (1957) 819-821.
256 J.H. Sloneker and D.G. Orentas, Nature, 194 (1962) 478-479.
257 M. Duckworth and W. Yaphe, Chem. Ind., (1970) 747-748.
258 J.F. Kennedy, D.L. Stevenson, C.A. White, M.S. Tolley and I.J. Bradshaw, Brit. Polym. J., Accepted for Publication.
259 I.W. Sutherland, in H.J. Rehm and G. Reed (Eds.), Biotechnology, Vol. 3, Verlag Chemie, Weinheim, 1983, pp. 531-574.
260 P.A. Sandford, Adv. Carbohydr. Chem. Biochem., 36 (1979) 265-313.
261 D.J. Pettitt, Ref 2, pp. 263-282.
262 D.J. Pettitt, Ref 205, pp. 127-149.
263 K.S. Kang and I.W. Cottrell, in H.J. Peppler and D.Perlman (Eds.), Microbiol Technology, Vol. 1 (Microbial Processes), Academic Press, New York and London, 1979, pp. 418-481.
264 M.E. Slodki, and M.C. Cadmus, Adv. Appl. Microbiol., 26 (1979) 19-54.
265 I.H. Smith and G.W. Pace, J. Chem. Technol. Biotechnol., 32 (1982) 119-129.
266 A. Jeanes, J. Polym. Sci. Symp., 45 (1974) 209-227.
267 G.R. Sanderson, Brit. Polym. J., 13 (1981) 71-75.

PROGRESS IN SOY SAUCE AND RELATED FERMENTATIONS

B. J. B. WOOD
Applied Microbiology Division, Department of Bioscience and Biotechnology, University of Strathclyde, Glasgow, G1 1XW, (Scotland)

I. INTRODUCTION

Numerous reviews and accounts of the basic biochemistry and microbiology of soy sauce and related fermentations are now available in the English-language scientific literature (refs 1 to 9). This situation is in marked contrast with that prevailing ten years or so ago. For the purposes of the present review I propose to give a brief account of the principles of the fermentations concerned, then follow it with a review of the more recent publications pertaining to each.

To the occidental, the use of moulds as the source of hydrolytic enzymes is a particularly striking feature of these ancient and highly traditional fermentations. Microbiologists will tend to feel alarm at the possibilities for mycotoxin formation presented by fermentations involving moulds, carried out by methods unchanged since antiquity and of markedly non-aseptic nature. Despite this, there seems to be a concensus that these fermentations are wholesome and

TABLE 1

Glossary of terms used in soy sauce and similar fermentations

1. Koji basically this refers to the mass of raw materials covered in (or developing a covering of) mould. The mould itself may be called a koji-mould, while the inoculum or seed-koji is also called a tane-koji. Thus the word "koji" is both a noun and an adjective and may be used as a verb.
2. Miso Japanese word for fermented soy paste. Numerous similar preparations are made throughout S.E. Asia and are known by a bewildering variety of local names. Outwith S.E. Asia the consumer is most likely to come across miso described by Japanese varietal names. These include:- hatcho or mame; made from soybeans only. Some authors also quote soybeans and wheat. Mugi; made from soybeans and barley. At least two types, sweet and salty. Genmai; made from soybeans and brown rice. Kome; made from soybeans and polished rice. This is the commonest form of miso and there are many varieties differing in their ratios of salt, beans and rice, times and methods of cooking the ingredients and duration of fermentation and maturation.
3. Moromi the secondary (salty) stage of shoyu and miso fermentation. Also called the salt-mash. Moromi is used as both a noun and an adjective.
4. Shoyu Japanese word for soy sauce, strictly the fermented variety, as opposed to that produced by acid hydrolysis of soybeans. According to Yokotsuka the Japanese Agricultural Standard recognises 5 types of shoyu, each of which is subdivided into 3 grades (Special, Upper and Standard) depending on organoleptic and analytical evaluation (ref 20). The names of the types, their percentages of the Japanese market and distinguishing characteristics are;

   Kokiuchi (85%); made from equal parts beans and wheat, dark in colour and strong flavour.
   Usukuchi (12%); more wheat than beans in the koji, lighter in flavour, lower in nitrogen and alcohol.
   Tamari (2%); made entirely or mainly from beans, high in nitrogen, low in alcohol, strong flavour.
   Shiro (0.5%); very light in colour, made from a koji almost entirely of wheat and very low in soybeans.
   Saishikomi (0.5%); apparently an enzymic extract of wheat and soybeans in shoyu.
5. Tamari a form of soy sauce made with little or no wheat flour, ie mainly or entirely from soy beans. Name frequently misapplied to the more normal type (Kokiuchi, q.v.).
6. Teriyaki a sauce prepared by blending soy sauce, white wine (and/or wine vinegar), garlic, sugar and spices. It is used as a marinade for meat and poultry or can be thickened with starch for use on previously broiled or grilled fish (ref 10).

do not lead to the formation of toxins, although no convincing scientific explanation for this state of affairs has been advanced as yet.

II. SOY SAUCE

A. Introduction

Basically this is a two stage fermentation. In the first stage soy beans are soaked, cooked by boiling, either at atmospheric pressure or in an autoclave, cooled, then inoculated (after mixing with wheat flour in S.E. Asia and Japan, but not in classical Chinese practice) with a strain of *Aspergillus oryzae* or a closely related species. The inoculated material is then incubated in warm, humid conditions for 2, 3 or several days (depending on local practice) resulting in rapid growth of the mould with consequent heating up of the fermenting material or "koji" and a tendency for the surface of the koji to dry out.

At the conclusion of this stage the mould has released a variety of hydrolytic enzymes into the substrate; these include proteinases, amylases, cellulases, lipase and sucrase, and their action results in a degree of hydrolysis of the proteins, polysaccharides, etc. present in the substrate. This mature koji is then mixed with brine to give a mixture containing 18% w/v of salt. Spontaneously, or as the result of deliberate inoculation, a lactic fermentation develops dominated by *Lactobacillus casei* and/or *Pediococcus spp.* resistant to the high salt levels. This takes the pH down from the original value near neutral to around pH 4.5 to 5.0. This then permits the development of a halotolerant yeast flora; normally *Saccharomyces rouxii* is dominant. In S.E. Asia this moromi or "salt-mash" stage is carried out in earthenware vessels exposed to the heat of the tropical sun and lasts for about 3 months. In Japan the process traditionally lasted for 9 months to 3 years. Japan, unlike S.E. Asia, has very marked seasonal variations in temperature, and exposure of the moromi to these changes is traditionally considered to be important to the development of the finest flavour. Of recent years the tendency in Japan has been toward shortening the time for which this second fermentation is carried out. Since importance continues to be attached to the view that the seasonal variations in temperature contribute to the subtle organoleptic qualities of the sauce, it appears that the modern factory practice requires exposure to an artificial simulation of the seasonal cycle of temperature. Of course some producers continue to adhere strictly to traditional precepts, and to the educated Japanese palate different soy sauces exhibit subtle variations which are probably part of the pleasure derived from using them.

When the moromi fermentation is complete, the sauce must be separated from the insoluble residues of beans and wheat. In visits to soy sauce breweries in Thailand and Malaysia I saw this being done by straining the mash through woven banboo or rattan baskets, sometimes lined with cloth or hessian. The liquid portion is then heated to stop fermentation and to develop the full flavour. Most non-Japanese soy sauce comes blended with caramel and sweetners such as molasses, and these will be added prior to heating. The finished sauce is then packaged for distribution and sale. The solid residues are mixed with fresh brine and returned to the fermentation jars for a further period of a week or so, then the filtration is repeated, and so forth. The entire cycle is repeated several times, and it will be evident that the extract will gradually become weaker. It seems that the first extracts may be retailed without additions but the later ones are blended with molasses, etc. as described above. Additionally the extract may be placed in shallow bowls and exposed to sun and air; evaporation of the water concentrates the liquor and salt crystallises out; the concentrated liquid is filtered off for use. Thus numerous types and grades of soy sauce are produced and find different uses in cooking and serving food. The solid residues from extracting the sauce are not wasted, but are used as human or animal food.

Japanese practice seems to favour pressing the moromi in order to remove the sauce. Traditionally this is done by charging moromi into wooden frames lined with filter cloth, folding the cloth over the box, stacking boxes into a vertical press and applying controlled pressure for several days. More recently however some companies have introduced more modern presses. One major producer has a patented design of press in which moromi is continuously fed to a moving filter cloth which is subjected to high pressure in order to effect the filtration.

After filtration, the solid residues, which have been pressed to the point where they resemble moist hardboard, are used as animal food. The liquid is stored for a time to allow oil forced out from the beans to rise to the surface, whence it is skimmed off and used for industrial purposes; this does not seem to occur in the process described earlier, presumably the oil remains trapped in the solid residues.

Japanese practice attaches importance to the correct heat treatment of the raw soy sauce. It is claimed that the importance of heating as a means of stopping the fermentation process and stabilising the soy sauce was recognised in Japan some centuries before the discovery of Pasteurisation in the West. Correct heating is also essential to develop the full aroma and colour of the soy sauce and to precipitate heat-coagulable proteins. It is impossible to make completely general statements about a product with as many varieties as soy sauce exhibits, but Japanese soy sauce is often lighter in colour than are other varieties and hence clarity and brightness in appearance are important to a much greater extent than with the dark-coloured Chinese types of sauce. Thus careful filtration of the sauce after heat treatment is an important part of the production process. It is of interest to note that the Japanese exported sauce to the French court of Louis XIV (ref 10) after packing it into pottery vessels which were sealed and then heated to 80°C for a time in order to preserve their contents.

Soy sauce is used in many ways in cooking and presenting food, and is also the basis of other condiments such as the Japanese Teriyaki sauce, a blend of soy sauce, white wine, sugar and spices. A considerable part of the "soy sauce" sold in the West is made by acid hydrolysis of soy beans, but for the full benefit of soy sauce to be obtained it is essential that the fully brewed product be used.

A detailed survey of the history and technology of soy sauce is given in ref 4. This review is somewhat updated in ref 7, which also attempts an economic survey of world trade in soy sauce. My own group's studies on the biochemistry and microbiology of soy sauce are reported in a series of papers (refs 11 - 16) based on studies carried out under aseptic conditions and using selected pure cultures of micro-organisms typical of soy sauce fermentations. It is, of course, recognised that such studies have only limited validity in relation to events taking place in the more varied and variable environment of a soy sauce production facility, but it is argued that such limited studies are an essential preparation for the scientific scrutiny of the much more complex changes which take place in production fermentations. For the purposes of the present review current understanding of the soy sauce fermentation may be summarised as follows;-

1. The koji stage is dominated by the growth of moulds belonging to the *Aspergillus oryzae/sojae* group. Other moulds and also bacteria are undesirable. The technology evolved over many centuries in the Far East empirically selected conditions favourable to the development of the desirable moulds.

2. The mould produces extra-cellular enzymes. These are hydrolases active against various components of the soybeans and wheat. They include proteinases, amylases, phosphatase(s), sucrase, cellulase(s) and probably other enzymes as well.

3. Control of temperature and humidity are important to the success of the koji stage. This is equally true of the traditional tray fermentations and of the highly mechanised systems employed in the more modern Japanese, S.E. Asian and American factories. Regular and careful mixing and close control of ventilation are essential. Errors will lead to premature sporulation of the mould. The purpose of the koji stage is to obtain maximum yields of the hydrolytic emzymes and some degree of hydrolysis of the various components of the substrate. Premature sporulation will obviously reduce the yield of enzymes and the extent of hydrolysis, and it is also accompanied by the production of ammonia (by deamination of amino acids) and the development of a harsh smell in the koji, quite different from the mild, sweet smell of a koji in which the mould is in a state of active vegetative growth.

4. The moromi stage comprises a double fermentation. First lactic fermentation(s) by species of *Lactobacillus* and/or *Pediococcus* forms acid in the mash, leading to a reduction in pH to 5.0 or below. As the pH decreases and the lactic acid bacteria begin to die out, a yeast flora develops. *Saccharomyces rouxii* seems to be a consistent member of this flora. Reports on the occurrence and importance of *Torulopsis* species are less unambiguous, but it seems that *Torulopsis* is probably important for the production of the finest grades of Japanese soy sauce, although the development of excessive numbers of this yeast is said to be harmful to the soy sauce. Some hearsay reports from Japan convey

the impression that the only satisfactory growth of *Torulopsis* arises from chance contamination of the vats, and that attempts to achieve consistent results by deliberate inoculation lead to overgrowth of the organism.

5. The salt concentration in the moromi is very important, since it prevents the growth of undesirable putrefactive bacteria, while permitting the growth of the yeasts and lactic acid bacteria which are essential for the development of the aroma and flavour of a good soy sauce.

6. While the fermentations taking place in the moromi are essentially anaerobic in nature, it seems to be generally accepted that some introduction of oxygen from time to time is beneficial, apparently through the stimulation of yeast growth and by the development of pigments desirable for the preferred colour of soy sauce (the Maillard reaction).

7. The high salt concentration and low pH might be expected to impose severe constraints on the mould enzymes' ability to continue their hydrolytic actions in the salt mash, but in practice they seem to be fairly stable in the mash and, although their activity will be well below that achievable under optimal conditions, it seems that they may effect significant hydrolysis over the long periods of time for which the moromi stage is carried out (ref 13).

This is not an exhaustive account of the soy sauce fermentation, but will suffice to set the background for a discussion of more recent advances in the understanding of this complex fermentation. It must be appreciated that the rapid change typical of modern industry has already significantly affected soy sauce brewing practice in Japan and is beginning to influence makers in S.E. Asia, most notably Taiwan and South Korea, but increasingly also countries such as Malaysia and Thailand. South Kerea has an important export trade in soy sauce (ref 7), whereas the latter countries produce almost entirely for domestic consumption. Despite this they have substantial university research programmes on soy sauce, linked to the producers both through provision of technical advice and, for example, the supply to the manufacturers of pure cultures of moulds for the production of safe kojis of consistent quality. In Malaysia, a special conference on soy sauce organised by the Malaysian Institute of Food Technology in August 1976 was so successful that not all those who wished to attend it could be accommodated (ref 17).

Further evidence of the desire for high and consistent quality products is provided by the development of national standards for these products, and of standard methods of analysis for them. A selection culled from Food Science and Technology Abstracts (FSTA) is listed in Table 2, and includes Standards from Taiwan, Malaysia, Thailand and the U.S.A.

As might be expected, it is still very difficult to obtain any data on soy sauce production in China; one can only hope that the current policies of the Chinese government will make it easier in the future. With Japan the problem

TABLE 2

Recently published national standards for soy sauce, etc.

1. Methods of Test for Soy Sauce.
   Taiwan, National Bureau of Standards.
   No. CNS N6008, 4pp (1979)
   Cited in FSTA *14*, 3U196 (1982).

2. Methods of Test for Soy Bean Paste.
   Taiwan, National Bureau of Standards.
   No. CNS N6009, 1p (1979)
   Cited in FSTA *14*, 3U197 (1982).

3. Miso.
   Taiwan, National Bureau of Standards.
   No. CNS N5106, 1p (1980)
   Cited in FSTA *14*, 4U266 (1982).

4. Miso Preserved Vegetables (Packaged).
   Taiwan, National Bureau of Standards.
   No. CNS N5163, 1p (1980)
   Cited in FSTA *14*, 3U210 (1982).

5. Soy-sauce Preserved Vegetable.
   Taiwan, National Bureau of Standards.
   No. CNS N5165, 1p (1980)
   Cited in FSTA *14*, 3U212 (1982).

6. Standard for Soy Sauce.
   Thailand, Industrial Standards Institute.
   No. T1S 252, 15pp (1977)
   Cited in FSTA *13*, 4U152 (1981).

7. Specification for Caramel in the Manufacture of Soya Sauce.
   Malaysian Standard.
   No. MS 8 23:1977 40pp (1977)
   Cited in FSTA *12*, 12U968 (1980).

8. Soy Sauce.
   U.S.A. Federal specification, revision.
   No. EE-s-610G, 16pp (1978)
   Cited in FSTA *13*, 1U67 (1981).

is somewhat different; there is a great deal published on soy sauce, and much of the published material is indexed in standard sources such as FSTA, but not unreasonably, there is a tendency to publish in Japanese, and it is rather frustrating to be aware of the existence of an important paper, such as the reviews of work on shoyu and miso published by Mori in 1979 and 1980 and comprising 632 and 625 references respectively, while being unable to make use of them (refs 18, 19). An excellent review of the "Industrial Application of Proteinous Food Fermentation" in English by Tamotsu Yokotsuka, including upwards of 150 references, was presented at an International Symposium organised by the Association for Science Cooperation in Asia and held at Medan, Indonesia, in February, 1981. Unfortunately I have no information as to the plans (if any to publish the proceedings of this Symposium, and it must therefore be classed for the time being as a "personal communication" (ref 20).

B. Raw Materials and Koji

Soybeans and wheat are the most suitable raw materials for the production of soy sauce, not only because of their traditional use and the consequent familiarity which the manufacturer has with their use and properties, but also because the size of the world markets in these commodities means that their price is reasonably stable and fairly low; soybeans in particular are cheaper than any other generally available beans. Furthermore supplies of soybeans and wheat are reasonably reliable even in a year such as 1980, when the weather in the U.S. Midwest depressed harvest yields of soybeans. However, there will be a continuing interest in the possibilities offered by other raw materials, both for product improvement and to permit use of locally available crops, just as Robinson and Kao (ref 21) used chickpeas and horsebeans to make tempeh. Ooraikul *et al.* (ref 22) report on the production of an acceptable product comparable with shoyu by the fermentation of rapeseed meal with the soy sauce method and micro-organisms. A more radical departure is that of Nagase and Nishijima (ref 23), who used supplementation with an hydrolysate of maize protein, of interest because the protein has a high glutamic acid content. Benefit could be obtained by adding the hydrolysate to the moromi or to the finished soy sauce. A semi-chemical (acid hydrolysate, see ref 4) soy sauce could also be prepared from this material.

C. The Unami Taste

In the West we think of glutamic acid (in the form of its sodium salt monosodium glutamate or M.S.G.) as simply a flavour enhancer, slthough there is little that is simple about this concept. M.S.G. has received a rather unfavourable press of late due to adverse effects in some people which are said to be due to the ingestion of excessive amounts of M.S.G., the so-called "Chinese restaurant syndrome". The effect of glutamate in bringing out the full flavour of foods was probably first recognised as the result of work on shoyu and miso. Japanese flavour scientists contend that there are not just the 4 basic tastes recognised in Western food science, but there is also a fifth one of equal importance. This is referred to as "tastiness" (ref 10) or the "unami taste" (ref 24) and is seen as being associated with amino acids (especially glutamate) and nucleotides. This type of thinking offers a considerable challenge to the Western scientist; it cannot just be dismissed out of hand but demands careful examination as an explanation of the role of amino acids and nucleotides in food flavours. If correct, it justifies attempts to increase the flavour-enhancing properties of soy sauce by additions such as the maize protein described in ref 23. A somewhat different approach to reaching the same goal is described by Nakamura *et al.* (ref 25). In this study Antarctic krill (shrimp-like creatures found in Antarctic waters) was boiled then frozen soon after catching. On return to port the krill was ground up while still frozen, then added to the shoyu fermentation. It is not clear from the cited abstract as to the stage (koji or moromi) at which the ground krill was added. It seems that adding the krill gave a substantial increase in the lysine content of the shoyu, contributed "abundant taste components" and did not give rise to any observable unacceptable taste. The last claim is a little surprising since even freshly caught krill has a rather characteristic odour which, although not necessarily objectionable in itself, is distinctly different from anything normally associated with shoyu.

D. Defatted Soybean Meal

In preparing normal soybean and wheat koji there has been a steady increase in the use of defatted soybean meal ("grits" according to Yokotsuka, see ref 20) in place of whole beans in Japan. The defatted soybean is produced by solvent extraction of dehulled and milled beans with hexane. Yokotsuka (ref 20) details the advantages of the defatted material, mainly its slightly lower cost. He is doubtful of the claim that defatted meal is intrinsically more digestible than is the whole bean in the koji fermentation, and he points out that the koji stage is more difficult to operate with defatted meal than with whole beans; this is because the latter gives an easily-mixed koji with good air penetration between the larger particles represented by whole beans. The marked tendency of kojis made with defatted meal to form rather impenetrable porridge-like masses is now overcome by using mechanised koji troughs featuring forced aeration from beneath, plus control of temperature and moisture content of the air via feedback from probes buried in the koji, plus automatic mixing of the koji with powerful flails.

Earlier studies also indicated that shoyu made from defatted beans had lower concentrations of ethanol and glycerol than did whole-bean shoyu, but modern research indicates that under the conditions new employed the yeast fermentation gives rise to high concentrations of both ethanol and glycerol. The Kikkoman Shoyu Co Ltd have been at the forefront of work to improve the performance of defatted meal in the fermentation, and have fully documented their claim that the method of cooking the meal is crucial to successful fermentation. Work over a 10 year period enabled them to develop from a situation where conventional cooking gave 69% digestibility of the soy proteins during a one-year moromi fermentation in 18% brine, to the current situation of 95% digestibility. These discoveries are protected by Japanese Patents Nos 929,910(1966) and 794,915(1970) and U.S. Patent No 3,764,708(1971) assigned to the Kikkoman Co and its employees. In essence the process consists of a short time, high temperature (e.g. 4 to 8kg, cm^{-2} for 15 seconds) cooking in steam, followed by discharge to atmospheric pressure, leading to explosive release of the steam trapped in the meal, and to a puffing of the material, so increasing the surface area and the scope for access of enzymes to the bean proteins. The process also has other advantages, notably that the meal is thoroughly sterilised, so permitting the use as raw material of meal carrying a reasonable burden of bacterial spores. Continuous cookers have been constructed and give excellent results. Yokotsuka (ref 20), from whom the above information derives, points out that improvement of these methods still continues, citing references up to 1977.

The method for preparing the wheat is also the subject of study, with Japanese research indicating the many advantages of toasting whole wheat at temperatures up to 180°C then coarsely crushing it to yield a mixture of larger lumps and a fine powder. When working with whole beans this method of preparing the wheat has the advantage that the fine powder coats the beans so giving them a dry surface, while the lumps hold the beans apart so improving air penetration.

As was noted above, the Japanese contend that phenolic acids and alcohols formed in the toasting of the wheat make an important contribution to shoyu's flavour. This will be very important no matter if whole beans or defatted meal be used. It is probably also of advantage to have lumps present, even if defatted meal is used, in order to prevent the mixture becoming too set and porridge-like. The more advanced Japanese practice favours the use of what seems to be a fluidised bed of sand for the roasting; it is claimed that the hot sand grains impinging on the wheat grains effect a superior toasting effect.

E. Microbiological Considerations

Given the scale on which soy sauce making is carried out in the major producing countries, it will be appreciated that it is difficult to operate under totally aseptic conditions, but it will clearly be of advantage to the producer to obtain the sterilising effects afforded by the high temperature treatments described above. However the control of bacteria in the koji remains as an important problem for the manufacturer. Hayashi *et al.* (ref 26) have shown that adding small amounts of acetic acid (up to 0.8% w/w) to the koji exerts a substantial suppressing effect on contaminating bacteria, while actually benefiting the mould's growth and its production of protease. Further work on this approach to the inhibition of bacteria in koji is reported by Noda and Hayashi (ref 27).

Yokotsuka (ref 20) discusses the selection of mould strains for soy sauce production and points out that no evidence of aflatoxin production has been recorded among more than 200 Japanese shoyu moulds which his group had examined; this is in line with results by numerous other authors, including examinations of nearly 300 isolates from Malaysia by Ho and his group (personal communication). Yokotsuka's group did observe a very few cases of the production of non-toxic fluorescent compounds by koji-moulds. Rather more than a third of a smaller group of moulds which they examined for aspergillic acid production would yield the compound (29 out of 69 strains examined) but none of these gave the compound under koji conditions.

There is a considerable amount of variability in hydrolytic enzyme production between different strains of koji mould (ref 12), but hearsay evidence and the work of Yokotsuka's group described above tend to agree that the really important factor controlling the extent of protein digestion is the way in which the beans or defatted soybean meal are cooked, and the subsequent management of the koji stage. Yokotsuka (ref 20) discusses the problems met in selecting conditions which give good growth of mycelium and good yield of protease together with minimal growth of bacterial contaminants, the last two being favoured by rather lower temperatures than is the first. Close control of temperature is more conveniently achieved in the modern koji makers than in the traditional trays.

F. Moromi Stage of Fermentation

Traditionally this has been a slow fermentation taking months or years to achieve completion, and regulated principally by the combination of the high salt concentration provided by the brewer and the low pH provided by the growth of lactic acid bacteria, which together gave an environment unsuitable for the growth of putrefactive bacteria, while promoting the growth of beneficial yeasts whose activities do much for the production of the characteristic fragrance of a fully-brewed soy sauce. The temperature of the mash is important, both for its effects on microbial growth and for its influence on the activity and the denaturing of enzymes in the mash which derive from the koji stage. Yokotsuka (ref 20), in accordance with Japanese traditions of shoyu fermentation, stresses the need for a low initial temperature of the moromi, in order to permit the *Aspergillus* alkaline protease a reasonable time for activity before the decrease in pH consequent upon the lactic fermentation inactivates this important enzyme. S.E. Asian practice on the other hand is to expose the moromi in its earthenware vessels to the full heat of the tropical sun as soon as it has been mixed. Yokotsuka also stresses the importance for shortening the traditionally prolonged Japanese moromi fermentation of the increase in protein digestibility consequent upon the improvements in soybean cooking and koji management which were discussed above.

It is convenient to say something about salt concentration before detailed discussion of the microbes active in the moromi. Long experience of the fermentation has shown shoyu producers that a high salt concentration favours the satisfactory progress of the moromi stage. The demands which a hot and frequently humid climate makes on the body have predisposed the traditional users of soy sauces and fermented soy pastes towards a diet high in salt, a substantial part of which has been supplied by the fermented soy products (about one ounce of shoyu and a half ounce miso per head per day in the Japanese diet, for example). Changes toward a more sedentary life-style, greater use of air conditioning and the move away from traditional farming methods and other labour-intensive jobs and into factory and service-industry work are all reducing the need for salt in the diet. At the same time, concern about the role of high salt intakes in the incidence of hypertension and other clinical manifestations, has spread from Europe and North America to the Far East and Japan. In consequence there is a marked growth of interest in the production of shoyu and miso of reduced salt content. Leading equipment manufacturers the Fujiwara Brewing Machinery Co. Ltd., offer plant for concentrating soy sauce, removing the resulting deposit of salt crystals and then diluting as required. Careful control of the operating conditions reduces flavour and aroma loss to a very low level. Other workers have concentrated on developing modified fermentation conditions. Nishijima *et al.* (ref 28) describe a method using ethanol at 1.0 to 5.0% v/v, preferred concentration 2%, to replace part of the salt; it was claimed that this also gave a product of superior flavour and material utilisation to those obtained by the conventional fermentation. Chen (ref 29), reporting on "studies on the new manufacturing method of soy sauce", and using defatted soybean meal as well as the traditional whole beans, obtained reductions in the moromi fermentation time from 8 months to 1 to 2 months by using different salt concentrations. Improvements in bacterial counts and product quality are claimed with the new method over those found in the more traditional brewing processes.

Okada *et al.* (refs 30, 31, 32) give the latest in a series of reports (Nos. VIII, IX and X) of their studies on the reduction of salt concentration in fermented foods. No. VIII deals with quality improvement in low-salt, soybean plus rice misos, while IX and X deal with the preparation and quality of a low-salt tamari. Tamari of satisfactory quality could be made with as little as 9% (w/w) salt, provided that ethanol (up to 8% w/w was tested) was added to the moromi fermentation. Omission of the alcohol resulted in the development of an ethyl acetate aroma which is undesirable in soy sauce. We have observed the same off-aroma in moromis where an unusually light inoculum of yeast has been added. Addition of alcohol to the fermentation is also dealt with in a Japanese Patent issued to the Ajinomoto Co. Ltd. in 1981 (Table 3, No. 18). The Kikkoman Co. Ltd. have two patents on fermentations using salt at 13 to 15% concentration (Nos. 15 and 16 in Table 3) and they also claim (Table 3, No. 5) a process for anaerobic fermentation with yeast to give shoyu of high ethanol content. A fairly high level of ethanol (around 2% v/v) has long been regarded as a mark of a high quality shoyu in Japan, although this can produce problems with the Customs and Excise regulations in some countries (ref 7). The evidence presented here suggests that there is a considerable effort being made towards the production of fermented soy products of acceptable organoleptic quality in which part of the salt has been replaced by ethanol.

The bacteria present in healthy soy sauce moromi are drawn from the genera *Lactobacillus* and *Pediococcus* (ref 4). Yokotsuka (ref 20) reports that the *Pediococcus* formerly called *P. soyae* but now assigned to *P. halophylus* is the major species and that some workers have obtained good soy sauce by using it as a pure culture. On the other hand he quotes authorities who stress the desirability of having a diversity of lactic acid bacteria if the best aroma, pH, colour, organic acid content and levels of certain amino acids (arginine, histidine and aspartic acid) are to be secured. Onishi *et al.* (ref 33) have shown that bacteria isolated from soy sauce moromis, and those from the crude salt used in its production, belong to the "moderately halophilic" group on the scheme of classification used in their Paper; it will be interesting to see if prolonged operation in conditions of reduced salt concentration supplemented with ethanol will result in any change in the status of the microbes present in the mash.

The emphasis which Japanese authorities place on *Pediococcus spp.* casts some doubt on the significance of *Lactobacilli* in the fermentation although Hesseltine (ref 34) is quite clear as to the involvement of *L. casei (L. delbrueckii* according to some workers) in the fermentation.

The yeasts, which play such an important part in the development of the flavour and aroma of fully-brewed shoyu, are dominated by *Saccharomyces rouxii*, but *Torulopsis spp.* also have a significant contribution to make (refs 4, 11, 20). Mogi and Mogi (ref 35) have shown that, while *Sacch. rouxii* flourish then disappear in the first 3 to 6 months of the moromi stage, *Torulopsis* reaches maximum population after 6 months. Some Japanese workers are of the opinion that artificial inoculation with *Torulopsis* will give too much growth of this yeast. In the great shoyu brewing centres of Japan, the craft has gone on so long that the entire environment is endowed with the appropriate microbes, just as are the great wine centres of Europe with wine yeasts. In consequence they can perhaps rely with reasonable certainty upon chance inoculation to supply the *Torulopsis*.

Tachi *et al.* (ref 36) have examined the ability of 3 species of *Torulopsis* to participate in the moromi fermentation. All batches were inoculated with *Sacch. rouxii*; one was reserved as a control and the others were inoculated with one each of *T. etchellsii*, *T. versatilis* and *T. magnoliae*. Of these *T. etchellsii* gave the best improvement in quality, "due to its superior content of ethyl, isoamyl and isobutyl alcohols" (FSTA abstract). The *T. versatilis* batch had a low lactic acid concentration (0.13 to 0.21g per litre) in comparison with the other batches (5.7 to 6.8g per litre). No significant differences in amino acid composition were observed.

The contribution of lactic acid bacteria in depressing pH to the point at which the yeasts can grow in the high salinity has been firmly established (ref 11). However, Noda *et al.* (ref 37) report that the acids produced by lactic acid bacteria tend to depress alcohol formation by osmophilic yeasts. They have shown that this effect is primarily due to the acetic acid produced by the

Pediococcus halophilus, with lactic acid being only mildly inhibitory. Depending upon the concentrations involved, this work may have relevance for the studies on controlling bacterial growth in koji by adding acetic acid, which was discussed above. A recent study by Yong *et al.* (ref 38) has shown that *Saccharomyces rouxii* NRRLY-1096 produces acetic acid from glucose and also formic acid (apparently by fixation of carbon dioxide) when grown at neutral pH. This would explain his earlier observation (ref 11) that the pH of a moromi inoculated under aseptic conditions with *Sacch. touxii* alone, exhibits a fall in pH. In view of the earlier comments on the depressing effect of acetic acid on alcoholic fermentation by moromi yeasts, this would suggest that it is desirable to secure a reasonably low pH in the moromi before undertaking heavy inoculation with a *Sacch. rouxii* culture.

G. Refining of and Properties of Shoyu

The matured moromi must be filtered in order to remove the solid residues and give a clear liquid suitable for bottling. Subsequently the filtrate is pasteurised and this treatment also helps the clarification of the sauce. Sakurai *et al.* (ref 39) have used gel filtration to show that the undialysable nitrogenous compounds in raw, filtered shoyu contained 3 principal fractions, of which 2 disappeared on pasteurisation. Another study of this process (ref 40) showed that during the early part of pasteurisation the newly-formed particles of coagulum underwent a decrease in size, but that thereafter particle size and density increased as time passed at pasteurisation temperature, thereby improving the sediment's tendency to settle out of the soy sauce.

With the high compression developed by the filters used in modern Japanese practice, most of the oil present in whole soybeans will be forced out with the filtrate, requiring that the shoyu be held for a time in store, so as to permit the oil to float to the surface for subsequent removal. Using defatted soybean meal should remove this need. Work by Ono and Aoyama (ref 41) shows that sugars and polypeptides present in soy sauce act synergistically to "solubilise" (emulsify?) soybean lipids through mycelle formation, with the effect increasing sharply at total shoyu solids concentrations in excess of 35%. Thus some part of the oil from the beans and wheat is bound to remain in suspension in soy sauce, especially if it is made with whole beans, and miso will inevitable contain some plant lipids from the soybeans and grains used in its preparation. These lipids tend to be easily oxidisable under normal circumstances, with their oxidation giving rise to strong and disagreeable rancid smells and tastes. A clue as to why these undesirable qualities are not normally found in well-made soy sauce and miso is offered by the work of Yamaguchi *et al.* (ref 42) who have shown that shoyu and miso show anti-oxidative action for unsaturated fatty acids (linoleic acid was used for the study). The power to protect the fatty acid against oxidation correlated well with nitrogen content and colour (correlation coefficients 0.8) except for colour of shoyu (c.c. approximately 0.53), properties which in turn relate to quality in shoyu and miso, although shoyu colour is less significant because of the scope for altering it. The workers suggested that the chelation of metals, notably copper and iron, might be involved.

Colour of soy sauces is a fascinating matter with many cultural overtones. Chinese-type sauces tend to be very dark with added caramel and with molasses or other sweeteners also contributing to their colour. Some Japanese sauces are also very dark, while others have a much lighter red-brown colour. A review by Chiba (ref 43) deals with the production and properties of the caramel needed for shoyu production, the detection and determination of levulinic acid in shoyu and caramel, and the detection of caramel in shoyu. The interest in levulinic acid arises because it is absent from fully-brewed soy sauce whereas it is an unavoidable constitutent of "soy sauce" made by acid hydrolysis of beans, etc. (ref 4) and it is therefore assayed (preferably by H.P.L.C. which also shows the presence of the fermentation acids such as lactic acid) as part of routine quality assessment procedures in Japan.

H.P.L.C. can also be used in the simultaneous determination of saccharin and the two most usual preservatives (benzoic acid and p-hydroxybenzoic acid esters) in soy sauce samples (ref 44). The procedure is a simple one, consisting of extraction with ether, filtration of the extract and H.P.L.C. An alternative procedure for saccharin determination (ref 45) consists of extraction into an organic solvent as a complex with the tetrabutylammonium ion, followed by methylation with methyl iodide and gas liquid chromatography without further clean-up.

In any fermentation involving the participation of moulds there will be a proper concern to ensure that the product is free of mycotoxins. This will be especially true of foodstuffs for direct human consumption. Much concern has been expressed from time to time about the possibility that mycotoxins may develop in soy sauce fermentations, especially when carried out under primitive conditions. A literature too extensive for convenient review here has been built up as the result of many careful studes by the experts in Japan and S. E. Asia. The overwhelming conclusion of all these studies is that soy sauce samples prepared with reasonable care, even under the fairly primitive conditions frequently met with in the small S. E. Asian breweries, are consistently free of any significant (or indeed detectable) amount of soy sauce. That this result is not an artefact of the analytical procedures employed is shown by studies on the recovery of added aflatoxins from shoyu and miso (e.g. ref 46), which show that good recoveries are obtained. The cited study also demonstrates that the toxins are not outstandingly stable in soy sauce, the half-life for Aflatoxin B1 being 60 days, while that for G1 is 120 days.

The delicate and subtle flavour of shoyu has been the subject of many investigations. Yokotsuka points out (ref 20) that a correct harmony of the 5 basic tastes is essential. The nitrogenous constituents are most important in giving a desirable taste, follwed by the sugar constituents, "delicious" taste and the organic acids. The last-named components seem to be very significant in the expression of the shoyu taste even though it is not noticeably sour to the palate. Neutralisation with alkali seems to destroy much of the characteristic fragrant organoleptic quality, and it cannot be restored by re-acidifying the sample.

The fragrance is very complex; Japanese workers have identified nearly 300 components of the volatile aroma. Of these the phenolic compounds are the most important but, as Yokotsuka shows, the position is complicated. He points out the special part played by *Torulopsis sp.* (but not by *Sacc. rouxii*) in the provision of the desirable phenolics, especially 4-ethyl guaiacol which has the interesting and unique property of ameliorating the salty taste of shoyu. (Even to the naive western palate it is noticeable that the best quality soy sauces have a smootheness, free of the salty "bite" of inferior products which sometimes try to hide the saltiness with an excess of sweeteners). The preferred *Torulopsis* species for this purpose are *T. etchellsii* and *T. versatilis*.

Nunomura *et al.* (ref 47) identified 93 compounds when they analysed the acidic flavour (aroma?) components of shoyu, 36 of them for the first time. The conclude that methyl hydroxy furanones are chiefly responsible for the aroma of shoyu. Kihara (ref 48) reports a method for the separation and determination of "flazin", described as an important compound with a characteristic shoyu flavour. Unfortunately the abstract does not make it clear as to how much is known about the stucture of "flazin".

Two papers by workers at the Nestlé Company's Swiss Laboratories (refs 49, 50) report that the flavour compounds of brewed soy sauce contain 7 lactones, including 4 which are reported for the first time in soy hydrolysates. Also present were basic substances, including alkylpyrazines, alkylpyridines and amides; these were mainly products of metabolic activity during the koji stage of brewing.

Three papers (refs 51, 52, 53) report on the use of G.L.C. in objective evaluation of soy sauce aroma profiles. They show that statistical treatment of the G.L.C. profiles yields results which enabled the sauces to be classified into groups which corresponded with those which could be identified by sensory evaluation, so giving an objective assessment, which would not require the participation of highly trained and statistically large enough taste panels. Mori *et al.* (ref 54) employed statistical analysis of taste profiles and odour preferences of soy sauces. They concluded that for a more complete understanding of the complexities of shoyu flavour, taste quality information for each component, and chemical analytical information are needed if the relation between taste sensation and the components of soy sauce is to be further clarified.

H. Patents

Patents provide an interesting insight into current trends in an industry. Patents relating to soy sauce are listed in Table 3. No 1 seems to describe a procedure for effecting a second preparation of soy sauce from the residues of a normal fermentation, while No. 13 describes a similar-sounding process with the additional use of an enzyme (amylase) but starting from saké residues rather than shoyu residues. No. 20 uses the residues (the whey or solids or both, the abstract does not make this clear) from soy curd (presumably tofu) making. These are apparently added to the koji, so giving a high enzyme activity, presumably through stimulation of the koji mould. Depending on the amount of residues that could be so used, this might also represent a useful economy in raw materials. Nos. 24 and 25 use corn starch as raw material for a lactic fermentation, apparently carried out separately from the main fermentation; this, together with changes in the procedures for the moromi stage, is claimed to permit a very short fermentation.

The U.S. Patent listed at No. 27 relates to the operation of the koji fermentation; from the abstract it is difficult to ascertain how it differs from the koji methods used elsewhere in the U.S. for a number of years.

Patent No. 8, which describes the use of koji extracts as enzyme preparations for use in shoyu brewing, and No. 6 which covers the use of enzymic hydrolysis of the raw materials for shoyu brewing, seem to have something in common. No. 19 also relates to the use of an enzyme to improve the utilisation of the raw materials; here the enzyme is specified as polygalacturonase (polyglacturonidase?). No. 17 may again relate to the use of koji extracts possessing enzyme activity. It appears that Japanese producers are recently becoming significantly interested in the possible economic advantages of pure enzymes or enzymically active extracts of koji moulds as adjuncts or aids to more rapid and/or economical conduct of the koji stage of shoyu brewing. This is comparable to the use of enzymes in the mashing stage of beer brewing. The importance of obtaining a satisfactory concentration of glutamic acid, discussed above in connection with the "Unami" taste, is illustrated by No. 7 on the list of patents; this refers to the use of a specified strain of *Aspergillus* in order to give a shoyu of high glutamic acid content.

Numerous patents relating to stability and shelf life of soy sauce are given in Table 3. Nos. 9 and 10 relate to adding plant extracts. Nos. 11 and 11a cover using ethanol as a stabilising agent. More unusually they also refer to ethyl esters in this connection, although it might have been expected that

TABLE 3

Recent patents relating to shoyu and miso production and processing

Entry No.	Patent No.	Company or Patentee	FSTA reference	Nature of claims
	JAPANESE, 1980		SHOYU	
1	5,533,301	Kikkoman Co. Ltd.	13(1981)6J 851	Soy lees sprayed with water, heat sterilised, cultured with a mould, then salt extracted with water.
2	5,533,303	Kikkoman Co. Ltd.	13(1981)6J 852	Sediment formation in soy sauce prevented by 2-stage rapid heating followed by rapid cooling.
3	5,533,304	Kikkoman Co. Ltd.	13(1981)6T 292	Storage stability of shoyu improved by rapid heating to 65 to 80°, holding at that temperature for up to 20h, then bottling at not less than 45°.
4	5,536,305	Kikkoman Co. Ltd.	13(1981)7T 394	Shoyu or miso heated at 70-120° to give a material powdered seasoning with the flavour of the original.
5	5,539,307	Kikkoman Co. Ltd.	13(1981)7T 401	Process for the anaerobic fermentation of yeast in the production of soy sauce with a high ethanol content.
6	5,539,308	Kikkoman Co. Ltd.	13(1981)7H 1178	Enzymic hydrolysis of raw materials followed by lactic and yeast fermentations.
7	5,542,618	Kikkoman Co. Ltd.	13(1981)9T 503	Use of a specified *Aspergillus* strain to give shoyu with a high glutamic acid content.
8	5,542,619	Noda Institute of Scientific Research.	13(1981)9T 493	Extraction of a solid culture of the koji mould to give an enzyme preparation for use in brewing shoyu.
9	5,542,812	Nakano Su-Mise Co.	13(1981)9T 502	Improvement of shoyu storage stability by incorporation of extracts from plants of the genus *Hosta*.
10	5,542,813	Nakano Su-mise Co.	13(1981)9T 501	As in the previous patent, but using *Erythronium japonicum* extracts.

TABLE 3 (Contd.)

Entry No.	Patent No.	Company or Patentee	FSTA reference	Nature of claims
11 11a	5,542,815 5,542,816	Kikkoman Shoyu Co.	13(1981)9G 671, 670	Prevention of fungal growth on shoyu and miso incorporation of ethanol and ethyl lactate, malonate, succinate, maleate or citrate, together with heating.
12	5,544,590	Noda Institute of Scientific Research.	13(1981)9T 499	Improving the colour stability of shoyu by treatment with molecular sieves.
JAPANESE, 1981				
13	5,600,018	T. Ida.	13(1981)7T 414	Sake residues (lees) heated, dried, treated with amylase, aged with addition of moromi, thus yielding a clear soy sauce.
14	5,606,262	Chosi Shoyu Co. Ltd.	14(1982)2T 72	Moromi Fermentation with anaerobic lactic acid bacteria, using conditions to promote fermentation without increasing viscosity.
15	5,608,572	Kikkoman Co. Ltd.	14(1982)3T 155	Process for rapid fermentation of a concentrated shoyu in a mixture containing 13-15% salt.
16	5,608,573	Kikkoman Co. Ltd.	14(1982)3T 156	As above; koji + 5 to 20% protein + brine to salt concentration of 13 to 15%, then subjected to moromi fermentation.
17	5,609,101	Ajinomoto Co. Ltd.	14(1982)1T 48	A rapid brewing process for shoyu using a solid cultured product of Aspergillus or its autolysis product.
18	5,609,297	Ajinomoto Co. Ltd.	14(1982)1T 47	Flavour improved by adding 2 to 7% ethanol during an intermediate brewing stage or after brewing, plus heating of product.
19	5,609,903	Kikkoman Co. Ltd.	14(1982)2T 70	Addition of polygalacturonase to starting material for shoyu to give improved utilisation of the material.
20	5,609,904	Ajinomoto Co. Ltd.	14(1982)2T 71	Adding bean curd (tofu) waste to raw material for shoyu koji, so giving a high enzyme activity.

TABLE 3 (Contd.)

Entry No.	Patent No.	Company or Patentee	FSTA reference	Nature of claims
21	5,609,905	Mitsubishi Chemical Industries Ltd.	14(1982) 1T 46	Use of a fine-pored anion exchange resin to decolourise shoyu.
22	5,609,906	Tanabe Seiyaku Co. Ltd.	14(1982)iT 45	Preventing precipitate formation by treating shoyu with protease and (optionally) polygalacturonidase or hemi-cellulase before pasteurisation, in order to prevent precipitate formation.
23	5,644,695	Nihon Shokuhin Kako Co. Ltd.		Process to permit drum drying of shoyu, etc., without loss of flavour, etc., achieved by adding mixed dextrin and cyclodextrin.
U.S.A., 1979				
24	4,180,590	Shono Denpun Co. Ltd.	12(1980)6T 310	Corn starch saccharified with amylase, lactic fermentation, heated, koji added, fermented with yeast, salt concentration increased in step with ethanol concentration, thus fermentation completed and ripened in 25 to 35 days.
25	4,180,591	Shono Denpun Co. Ltd.	12(1980)6T 311	As above except that a koji exteact is used, and the salt is not added until the yeast fermentation has gone to completion.
U.S.A., 1980				
26	4,209,535	Beatrice Foods Co.	13(1981)2T 88	Method for producing koji in a bed through which air is forced under defined conditions for 2 to 10 days. The koji mixture has a moisture content of 18 to 40%.
27	4,241,095	Nakano Vinegar Co. Ltd.	13(1981)9T	Shoyu preserved by adding acetic acid and a salt of an organic acid selected from malic, acetic, tartaric, citric and lactic acids. The addition is made before or after the sauce is sterilised.

TABLE 3 (Contd.)

Entry No.	Patent No.	Company or Patentee	FSTA reference	Nature of claims
	JAPANESE, 1979		MISO	
28	5,417,036	Shimizu.	12(1980)2J 238	A miso is described of improved flavour, and containing "yeast-fermented pine leaves".
29	5,417,037	Seisan Kaihatsu Co. Ltd.	12(1980)2J 237	Drying miso to yield a product stable to storage; process includes incorporation of a porous crushed soybean powder.
30	5,418,357	H. Goto.	12(1980)2J 236	Improving miso's odour by treatment with an *Asp. oryzae* enzyme or water to remove odour before ageing.
31	5,432,079	Chokai Miso Co. Ltd.	12(1980)5M 635	Unpolished rice is immersed in boiling water, then autoclaved, then cultured with *Asp. oryzae* "as a leaven for miso production".
32	5,434,839	Nakano Su-mise Co. Ltd.	12(1980)5G 363	Miso is preserved by adding to it vinegar semi-neutralised with egg-shells and bicarbonate or caustic soda; addition is after fermentation is complete.
	JAPANESE, 1980			
33	5,533,299	Ichibiki Co. Ltd.	13(1981)6J 877	Ethanol addition, plus heat treatment, to sterilise miso and prevent gas formation during storage.
34	5,542,814	Kikkoman Shoyu Co. Ltd.	13(1981)10G 754	Cereal cooked under pressure, then discharged to swell them, prior to use in brewing miso, etc.
	JAPANESE, 1981			
35	5,639,178	Ishiyama Miso Shoyu Co. Ltd.	14(1982)8G 539	Use ethanol, miso yeast and 5 - 10% salt to give a healthy miso in a 2-stage fermentation, with improved taste and flavour of product.
			TEMPEH	
36	U.S.A. 1979 4,151,307	General Mills Inc.	12(1980)1G 2	Dough made from cereal, potato flour, starch, plus tempeh (5 - 60% dry wt); the mixture is shaped and deep-fried.

TABLE 3 (Contd.)

Entry No.	Patent No.	Company or Patentee	FSTA reference	Nature of claims
37	U.S.A. 1979 4,157,984	Z-L Ltd.	12(1980)3T 179	Extraction of an alcohol-soluble mixture from tempeh, purification, use as a food, oils, fats stabiliser.
38	U.S.A. 1980 4,218,489	Z-L Ltd.	13(1981)4T 168	Recovery of steroid-type anti-oxidant for lipids from tempeh.

they would affect the sauce's aroma, and it is difficult to imagine that organoleptically acceptable amounts of such esters would exert any significant effect on fungi growing on shoyu. No. 27 refers to use of acetic acid and salts of organic acids in a similar way. No. 12 uses molecular sieves to improve the sauce's colour stability, while No. 21 describes the use of ion exchange resin to decolourise the shoyu, although why this should be desirable is not clear from the cited abstract. Nos. 3 and 23 relate to the production of dried products retaining as much as possible of the flavour of the original liquid product. No. 22 recalls modern beer brewing practice in its use of enzymes to prevent the formation of precipitates.

I. Work Outwith Japan

The concentration on work in Japan which is inevitable in a review of this type, should not be taken to mean that there is no research programme in the other major soy sauce producing countries. In the present state of the industry's development however, it is inevitable that the bulk of the published work will come from Japan, as the other countries' research programmes have more immediate and local aims, whereas the Japanese industry has matured to the point where it can afford the commitment of time and effort needed to undertake long-term research and to publish its findings. The interesting Malaysian review has already been referred to (ref 17). Yokotsuka (ref 20) has some informative comments on the types and quantities of soy sauce produced in Korea, Taiwan and Southern China, although the lack of literature citations strongly suggests that he is drawing mainly upon his own extensive travel and observations. South Korea's soy sauce he characterises as being mainly similar to the Japanese kokiuchi type when produced industrially. He also describes a method for domestic production in Korea, wherein cooked soybeans are pulped, made into small balls and allowed to undergo a spontaneous fermentation dominated by *Aspergillus* and *Rhizopus spp*. This seemingly occupies the winter months, and the moromi stage takes place in spring. The solid residues which remain after extraction of the balls are not left in the fermenting mash but are filtered off and made into a form of miso. Taiwanese commercial production is probably also dominated by a koikuchi type of sauce but 10% is a special sauce called Inyu, which is made from black soybeans instead of the more usual yellow variety. It is possible to buy in London, another form of black soybean product, which seems to be well known to our S. E. Asian students; here the beans are entire, soft and salty; although I am not fully certain that they are fermented I consider that this is probable. A recent report (ref 55) from Taiwan deals with Inyo (a transliteration difference, the product is evidently the same). Standard practice uses a koji made of beans only. A strange feature is that the developed koji is washed in water before adding the brine. Wang *et al.*

(ref 55) described pre-incubating the washed koji at 45° for 4 hours before add-adding the brine. They found that this procedure gave a shorter ripening time (45 days rather than 90 days) and higher acidity that if it was omitted. After the pre-incubation, halophilic yeasts (4×10^3 per g.), *Lactobacilli* (2.8×10^8 per g.) and other bacteria (1.27×10^5 per g.) were present. After the pre-incubation the reducing sugar and formolnitrogen had each doubled in concentration, reaching 3.4 and 3.74mg/mℓ.

According to Yokotsuka, sauces "similar to Inyu and Tamari" are "still being produced in the southern part of China" (ref 20). He goes on to describe a method used in Peking and Shanghai in which a fairly standard koji is made from beans and wheat (6 : 4). The developed koji is mixed with sufficient salt water to give a moisture content of about 80% and a salt concentration of 6 to 7%. This he calls a "hard mash". It is incubated at 40° to 50° for about 3 weeks, then extracted first with brine then with hot plain water. He states that "there is no alcoholic fermentation of mash or pressing of mash as there is in the case of Japanese shoyu manufacture" but it would be surprising if the "hard mash", which must resemble miso, does not ferment. Use of defatted soybean meal, pressure cooked and then subjected to explosive decompression as described above, is reported to give a 75 to 80% yield of the nitrogen from the raw material in the finished liquid. The solid residues, being low in salt, can be used as animal feed.

Published reports on current work in Thailand are rather lacking at present, although that country has a substantial research effort devoted to improvement of their soy sauce industry. An interesting paper by Bhumiratana *et al.* (ref 56) reports on the isolation of moulds from Thai soy sauce kojis. They isolated an organism identified as *Aspergillus flavus var. columnaris*. This gave superior protease production, and there was no detectable production of aflatoxin when the mould was grown on various substrates. When tested in a soy sauce factory, its use gave an improvement in the quality of the product in comparison with the traditional process.

Reports on Vietnamese processes are very rare indeed and the appearance of such a paper is most welcome (ref 57). The report describes the examination of Vietnamese traditional rice-soy sauce (made from 75% rice, 25% soy beans) and soy sauce for the presence of aflatoxins. None of the 59 samples examined contained the toxins. The paper claims that the sauces are foods as well as seasonings because of their high protein content (6.1% and 15% respectively) although it is not clear from the cited abstract as to the daily consumption of these sauces by the Vietnamese people.

III. MISO

Miso is the generic term for an extensive family of fermented pastes made from soybeans alone or in combination with rice (brown or white), barley, occasionally wheat and sometimes other materials (Table 1). By variations in the ratio of grain to beans, in the amount of salt used and the time allowed for fermentation, many types of miso can be produced, with flavours ranging from very light and sweet, through gradations of saltiness and maturity to beans-only misos matured for years, whose flavour dominates whatever food it is used in, to an extent unattractive to most Europeans. Miso has recently been the subject of an extensive review (ref 9) and the reader is referred to that paper for detailed discussion of the history, nature, types and production of miso.

The production of miso has similarities to and differences from the production of soy sauce. A major difference is in the koji stage. Where grains (rice, barley, etc.) are used these are cooked and inoculated with the koji-mould; no beans are present in the koji. The selected mould is highly productive of amylases but is less productive of proteinases than is its soy sauce counterpart. The mould belongs to the same species (*Aspergillus oryzae)* as does the soy sauce mould. Traditionally the miso koji is only incubated for 55 hours or so, rather than the 70+ hours which was traditionally used for soy sauce koji. The koji is otherwise managed very much as for soy sauce, being spread in trays, incubated in a warm and humid atmosphere and mixed from time to time. After incubation, the mould-covered grain is mixed with salt, with soaked, cooked soy beans, and with additional water if needed. The mixture is seeded with a portion of a previous good mash and is packed into barrels. It is moved from time to time into different barrels to obtain mixing and aeration to promote good microbial growth. It thus matures for the appropriate time, during which enzymes from the koji hydrolyse proteins and polysaccharides, while yeasts and lactic acid bacteria grow on the resulting sugars, amino acids and peptides. When maturation is finished the miso is removed from the barrels, mixed and packaged for sale. It is normally still slowly fermenting and so it should not be placed in a tightly closed container, although pasteurised versions are now available. In Oriental practice, importance is attached to the fact that the miso is still living and fermenting, therefore it is usual to add it at the end of cooking the dish which it is to embellish. It is held that miso has great powers to restore and aid the digestive tract, and that these properties are destroyed by boiling. Rice miso in its various forms accounts for 80% of the miso used in Japan. Although it is frequently stated that the Japanese consume rougly equal weights of shoyu and miso each day, Yokotsuka (ref 20) says that only half as much miso as shoyu is used. He reports that Japanese use of miso is declining, but Western interest in it is increasing; it could develop a more important role in the Western diet than is possible for shoyu. Modern packaging, blending and pasteurising before packaging, judicious use of preservatives and ethanol to control fermentation and spoilage, all these will contribute to its market potential in the West.

On examining the citations to papers and patents on miso in Food Science and Technology Abstracts it soon becomes clear that less work is being published and patented on this product than is the case with shoyu. However, the basic concerns are much the same with both products.

The preparation of raw materials to optimise solubilisation of proteins is of less importance here than with shoyu since no filtration step is involved. However various workers are interested in puffing the grains and soybeans by explosive decompression as was described for shoyu. A patent (Table 3, citation No. 34) covers puffing grains for this purpose. Mogi *et al.* in a series of papers (refs 58 to 60) report that the use of puffed soybeans gave a miso of inferior quality. On the other hand puffing the rice to be used in the production of Shiro Miso had some advantages, they found. The chief problem was a smell of puffed rice but this could be controlled by using equal amounts of puffed and steamed rice. Puffing could give higher-than-usual glutamate contents if sufficient water was added to the moromi. The misos made with puffed rice were more intensely coloured than was the case with normal misos.

Three papers, from two different groups, report on the use of peanuts to replace soybeans in miso production. Miura *et al.* (ref 61) found that full-fat peanuts gave "miso" with reduced levels of hydrolysis of saccharides and proteins, with consequent reduction in free amino acid content. This could be corrected by using partially defatted substrate, but even so the "miso" was less acceptable than the conventional product, presumably because of the distinctive peanut taste. Shieh *et al.* (refs 62, 63) compared the fermentation of soybean and peanut misos, using two moulds (*Aspergillus oryzae* and *Rhizopus oligosporus*, alone or in combination) and two salt levels (6% and 12%). Free fatty acid contents of all misos increased rapidly during the first 4 days of fermentation. They found that their peanut "miso" had higher soluble nitrogen than did the corresponding soybean misos, but that the levels of free amino acids is comparable misos were similar.

Honma (ref 64) reviews the properties of miso made with mould only and compares the product with conventional miso. Unfortunately the cited abstract does not detail the findings of the work. On the other hand Kiosumi *et al.* (ref 65) are quoted as saying that lactic acid bacteria and yeasts are both essential for the preparation of good quality miso. A balance between the organisms is needed, and lactic acid bacteria in large numbers, while improving colour, saltiness and taste, gave a sourness to the product.

The need for control of dietary salt levels has been referred to already, and miso, being a paste, with consequent reduction in water activity over shoyu moromi, is an obvious candidate for this type of study. Miyasaka discusses the need for, and methods of producing, low-salt misos, but again no details are given in the cited abstract (ref 66). Tsukuhara *et al.* (ref 67) describe a complex-sounding study in which 25 different types of miso were prepared, featuring variations in salt level and the use of different isolates of yeasts and lactic acid bacteria. Even small reductions in salt level made it essential to inoculate with desirable yeasts and lactic acid bacteria, thus preventing acid putrefaction. *Saccharomyces rouxii* prevented putrefaction at salt concentrations at or above 6% (the bais, dry or wet weight, or weight/volume? of these figures is not made clear). Saké yeasts grew in miso with reduced salt content and suppressed the growth of lactic acid bacteria and the development of sour putrefaction. This suppression of lactic acid bacteria by yeasts was general, while the presence of lactic acid bacteria promoted yeast growth. They concur with Koizumi *et al.* (ref 65) in finding that both types of microbe are needed for best maturation of the miso. Yoshi (ref 68) reviews the question of water activity in miso and shoyu and discusses the question of the "relation between water activity and the growth of halophilic micro-organisms".

The preservative action of ethanol (Table 3, patent citation Nos. 33 and 35) may well be due in part to its effect on water activity. Its preservative action, along with that of sorbic acid, on two types of miso, is discussed by Kawanishi *et al.* (ref 69). The misos are described as "light yellow type" and "koji type"; the significance of the latter description is not clear. The study showed that 0.1% sorbic acid or 2 to 3% ethanol were very effective at prolonging the shelf-life of the light yellow miso, and afforded some protection to the seemingly more unstable "koji type" miso; these results even applied under the stringent test of storage at $30^{\circ}C$.

Three papers (refs 70, 71, 72) deal with the carbohydrate chemistry of miso. Pentose sugars and glucose play an important part in the browning reaction, necessary for the development of the proper miso colour. The free sugars of mature miso (type(s) not specified) included glucose (70 to 80%) iso-maltose, galacturonic acid and fructose. The free pentose concentration was very low. During the fermentation the starch content declined rapidly, being completely gone after 50 days at 30°. Free glucose showed a corresponding increase during the early stages of the fermentation, as did iso-maltose, but the glucose concentration later declined; galactose showed a similar pattern to that of glucose. Maltose and sucrose decreased rapidly from the beginning of the fermentation, while stachyose decreased more slowly. Mannose, perhaps surprisingly, increased in concentration through the fermentation. The importance of the part which polysaccharides, especially pectins, play because of their water-holding capacity, their effect on viscosity, and hence their effect on the texture of miso, is reported.

Ohara *et al.* (ref 73) report on the use of electrophoresis to separate and identify the proteins of soybean, rice and barley in extracts prepared from miso. The method can be applied to the quantitative estimation of the barley content of Mugi Miso.

Kiuchi *et al.* (refs 74, 75) discuss the fate of the di-unsaturated fatty acid linoleic acid in misos. The free acid, resulting from hydrolysis of lipids present in the soybeans etc. shows a decrease in concentration during the moromi stage of fermentation. That this is due to assimilation of the free fatty acid by *Saccharomyces rouxii* was confirmed by respirometer experiments with the yeast provided with the fatty acid as sole carbon source. Several different *Sacch. rouxii* isolates showed this ability. Since free polyunsaturated fatty acids are a source for the development of rancid off-flavours, the effect of this assimilatory activity of the yeast on the flavour of miso (and shoyu?) is of considerable interest. To enable further study of the matter, ref 75 describes the production of otherwise normal mutants of *Sacch. rouxii*, unable to oxidise linoleic acid.

The general question of miso flavour has been reviewed recently by Yasuhira (ref 76) citing 40 references and Ito (ref 77) citing 24 references. Ishima (ref 78) showed that sensory evaluation of miso samples was related to their chemical analyses in some respects, for example the concentrations of reducing sugars and salt. Colour also showed a correlation. These correlations held true for salty wheat miso and for soybean miso, although samples were not sufficiently numerous to permit the statistical assessment by Principal Component Analysis which was applied to light-coloured rice miso. Iwabuchi *et al.* (ref 79) discuss the production of volatile carbonyls during the heating of miso. The production of aldehydes and furfural were examined, and the importance to flavour of 3-deoxyglucozone was stressed. Kiyuchi *et al.*, whose work on linoleic acid was discussed above, have reported (ref 80) on miso lipids and free fatty acids and their effect on miso's quality. The lighter and more delicate the flavour of the miso, the more important it is that the extent of lipid hydrolysis should be small.

Ebine (ref 81) has prepared a review surveying the colour of miso, factors contributing to the colour, mechanisms of colour formation, and the preservation of desirable colour.

Hamano and Sugimoto, of the Kikkoman Research Laboratories, have prepared an English-language report on water sorption isotherms of powdered miso and shoyu (ref 82). These are important for applications of powders in cooking and for their packaging and handling under factory conditions. In general the powders resembled powdered orange juice and custard powder in their behaviour. Shoyu powder became damp and difficult to handle at Relative Humidities (R.H.) above 40% At R.H. between 20 - 33% it had a tendency to cake, but could easily be broken to the free-flowing condition. Miso powder would cake at R.H. 40 - 45% but could easily be broken to a free-flowing powder and it only became really difficult at R.H. above 50%. Clearly these results will be altered depending on how the powders are prepared, etc., but in general the results are a useful guide to the conditions for handling the powders. The workers suggest that adding fatty acids or hydrocarbons with 5 to 9 carbon atoms will improve the powders' resistance to humidity.

Rather little seems to have been published on miso-like products from outwith Japan. Ha (ref 83) has reported on Korean fermented soybean foods, which have a history spanning 1300 years. Kanjan is Korean shoyu, Tenjan is miso and Kochuman is a mustard rice made by mixing rice, pwdered miso, Korean mustard and kanjan.

IV. TEMPEH

Least familiar to most Europeans of the three fermented soybean products discussed in this paper, tempeh is used as a food in its own right in its country of origin, Indonesia. It is made by cooking soybeans, inoculating them with a mould, normally *Rhizopus oligosporus*, incubating for about a day in a partially closed container, then slicing the resulting cake of soybeans bound together by mould mycelium, and frying, roasting or otherwise cooking it. It may be improved as a food by making it from a mixture of soybeans and wheat or other grain; this gives a better balance of amino acids and a tastier product in the opinion of some, including the present reviewer. The really remarkable thing about tempeh is how very good it is to eat, even when served to people who are totally unfamiliar with it. Key references on its production, microbiology and biochemistry are refs 84 to 90. The major change to the beans during the fermentation is the partial hydrolysis of bean proteins by extracellular proteases liberated by the mould. A particularly important development in understanding of the nutritional value of tempeh is the demonstration by Steinkraus' group that bacteria which develop in the maturing tempeh contribute substantial amounts of vitamin B12, a nutrient which is normally rather lacking in vegetarian diets unless some animal products (e.g. milk, eggs) are consumed (ref 91).

A recent review of knowledge about tempeh is that by Dien and Hesseltine (ref 92), while Malaysian tempeh is discussed by Ismail (ref 93). As with other soy-based products, there is some interest in making tempeh with substrates other than soybeans. Robinson and Kao (ref 21) used chickpeas and horsebeans to make a palatable and generally acceptable "tempeh". The use of a mixture of equal parts of soybeans and peanuts is described in ref 94. The product was acceptable and of equal nutritional value to normal tempeh. The authors also describe and discuss the effect of supplementing the tempehs with amino acids.

Some interest in tempeh is now developing in the U.S.A. In Table 3, Patent No. 36 refers to a way of making a deep-fried product based on tempeh, while Nos. 37 and 38 discuss the production of antioxidants and food stabilisers of "exceptional" properties by methanol extraction of tempeh. The mixture contained two isoflavones and ergostadientriol. The last-named compound seems to have been the active anti-oxidant.

V. CONCLUSION

This paper has reviewed current work in the production, biochemistry and microbiology of fermented foodstuffs based on the ubiquitous soybean. Bau and Debry (ref 95) have recently discussed the nutritive values and deficiencies of soybeans and the basic principles underlying the production of both fermented and non-fermented soyfoods. They argue the case for the greater use of soyfoods in both developed and developing countries, with new, simple production methods being needed for the latter. Wang *et al* (ref 96) highlight a particular advantage of mould fermentation, the ability of many moulds isolated from Oriental food fermentations to hydrolyse phytic acid. All of the moulds tested, except *Actinomucor elegans* (from ontjom, a fermented peanut press cake) and *Mucor dispersus,* possessed phytase activity. The hydrolysis of phytic acid is important because it increases the nutritional availability of the important divalent metal ions such as calcium, which are strongly sequestered by phytic acid. Thus, in a variety of ways contributions of fermentation to the basic foods are made in addition to flavour improvement and a degree of pre-digestion of proteins and polysaccharides. It may reasonably be anticipated that further discoveries are waiting to be made. A case in point is the fermentation of castor beans (*Ricinus spp.*) by a rather uncontrolled method practiced at domestic level in Nigeria and which converts these normally toxic and purgative seeds into a valued condiment. So far no definitive work seems to have been carried out on this remarkable fermentation, described to me by visiting workers and postgraduate students coming to us from Nigeria.

Finally, it may be that some readers will want information on the culinary uses of some of the products mentioned in this survey, while others may be interested in the historical background, the Oriental philosophy attaching to the subject or the practical production of the various foodstuffs. The Kikkoman Way of Fine Eating (ref 10) has material of interest on all these topics as they relate to shoyu. Shurtleff and Aoyagi (refs 97-100) have a series of interesting books on miso and tempeh and a corresponding one on shoyu is currently in an advanced state of preparation. Their writings will not be to the taste of all scientists but they have much to say which is of interest. They must also take much of the credit for the establishment of the Soycrafters Association of North America and its journal Soycraft; these, plus the annual meeting which they organise, provide fascinating common ground for government and university scientists to meet and work with small and large scale producers and with the exotic wholefood, organic and mystical side of the soybean movement.

TABLE 4

Japanese journal titles and their English translations.

Eisei Shikenjo Hokoku.
    Bulletin of the National Institute of Hygenic Sciences.

Hakko Kogaku Zasshi.
    Journal of Fermentation Technology.

Miyazaki Daigaku Nogakubu Kenkyu Jiho.
    Bulletin of the Faculty of Agriculture, Miyazaki University.

Nihon Jozo Kyokai Zasshi.
    Journal of the Society of Brewing, Japan.

Nihon Nogei Kaggakai-shi.
    Journal of the Agricultural Chemical Society of Japan.

Nihon Suisan Gakkai-shi.
    Bulletin of the Japanese Society of Scientific Fisheries.

Nippon Shokuhin Kogyo Gakkaishi.
    Journal of Japanese Society of Food Science and Technology.

Nyu Fudo Indasutori.
    New Food Industry.

Shokuhin Eiseigaku Zasshi.
    Journal of the Food Hygenic Society of Japan.

Shokuryo Kenkyusho Kenkyu Hokoku.
    Report of the National Food Research Institute.

Tokyo Nogyo Daigaku Nogatu Shuho.
    Journal of Agricultural Science.

This list covers publications mainly or entirely in the Japanese language, cited in the list of references and whose papers are quoted via their citations in Food Science and Technology Abstracts. Transliterations and translations are as used in FSTA. Japanese journals published in English are cited by their English language titles.

VI. REFERENCES

(See Table 4 for a guide to the Japanese journals cited here)

1. C.W. Hesseltine and H.L. Wang, Biotechnol. Bioeng., 9(1967) 275-288.

2. T. Watanabe, N.M.I.D.O. Expert Group Meeting on Soybean Processing and Use, U.S.D.A., Peoria, Illinois, U.S.A., 1969.

3. C.W. Hessletine and H.L. Wang, in A.K. Smith and S.J. Circle (Eds), Soybean: Chemistry and Technology, Vol. I, Avi Publishing Corp., Westport, Conn., 1972, pp 389-419.

4. F.M. Yong and B.J.B. Wood, Adv. Appl. Microbiol., 17(1974) 157-194.

5. B.J.B. Wood and F.M. Yong, in J.E. Smith and D.R. Berry (Eds), Filamentous Fungi Vol. I, Industrial Mycology, Edward Arnold, London, 1974, pp.262-280.

6. B.J.B. Wood, in J.E. Smith and J.A. Pateman (Eds) Genetics and Physiology of *Aspergillus*, Academic Press, London, 1977, pp.481-498.

7. B.J.B. Wood, in A.H. Rose (Ed.) Economic Microbiology Vol. 7, Fermented Foods, Academic Press, London, 1982, pp.39-86.

8. D. Fukushima, J. Amer. Oil Chem. Soc., 58(1981) 346-354.

9. S.H. Abiose, M.C. Allan and B.J.B. Wood, Adv. Appl. Microbiol., 28(1982) 239-265.

10. S. Otsuka, in The Kikkoman Way of Fine Eating, Kikkoman Shoyu Co. Ltd., Tokyo, 1973, pp.10, 71.

11. F.M. Yong and B.J.B. Wood, J. Food Technol., 11(1976) 525-536.

12. F.M. Yong and B.J.B. Wood, Ibid., 12(1977) 163-175.

13. F.M. Yong and B.J.B. Wood, Ibid., 12(1977) 263-273.

14. S.K. Goel and B.J.B. Wood, Ibid., 13(1978) 243-247.

15. K.E. Aidoo, R. Hendry and B.J.B. Wood, Ibid., 18(1981) 543-548.

16. K.E. Aidoo, R. Hendry and B.J.B. Wood, Eur. J. Appl. Microbiol. Biotechnol., 12(1981) 6-9.

17. Malaysian Institute of Food Technology Bulletin, Symposium Supplement 1977, Soya Sauce, M.I.F.T., Petaling Jaya.

18. H. Mori, Nihon Jozo Kyokai Zasshi, 74(1979) 291-316.

19. H. Mori, Ibid., 75(1980) 384-410.
Cited in FSTA 13(1981) 10J 1578.

20. T. Yokotsuka, Proceedings of a Symposium of the Association for Science Cooperation in Asia, Medan, Indonesia, February 9-15, 1981. (peraonal communication).

21. R.J. Robinson and C. Kao, Cereal Chem., 54(1977) 1192-1197.

22. B. Ooraikul, H.M. Mei, S.K. Sarkar and H. Jackson, J. Food Sci., 45(1980) 200-203.

23. I. Nagase and H. Nishijima, Niyu Fudo Indasuteri, 23(1981) 50-57.
Cited in FSTA 14(1982) 8T 465.

24. S. Yamaguchi, Abstracts of Papers, Amer. Chem. Soc., 177(1979) A.G.F.D. 64.

25. H. Nakamura, Y. Mohri, I. Muraoka and K. Ito, Nihon Suisan Gakkai-shi, 45(1979) 1389-1393.
Cited in FSTA 13(1981) 2R 76.

26. K. Hayashi, U.S. Patent No. 4,028,470(1977).

27. F. Noda and K. Hayashi, Hakko Kogaku Zasshi, 58(1980) 345-351.
Cited in FSTA 14(1982) 1G 51.

28. H. Nishijima, S. Yamagishi, H. Tate, F. Yanagita and I. Nagasi, Nihon Jozo Kyokai Zasshi, 75(1980) 77-80.
Cited in FSTA 13(1981) 6T 345.

29. S.L. Chen, J. Chinese Agric. Chem. Soc., 18(1980) 42-52.
Cited in FSTA 14(1982) 5T 249.

30. Y. Okada, H. Yoshii, H. Kato and T. Takeuchi, Nippon Shokuhin Kogyo Gakkaishi, 28(1981) 194-200.
Cited in FSTA 14(1982) 10G 667.

31. Y. Okada, T. Amano, T. Takeuchi and H. Yoshi, Ibid., 28(1981) 201-207.
Cited in FSTA 14(1982) 10G 668.

32. Y. Okada, H. Yoshii and T. Takeuchi, Ibid., 28(1981) 208-215.
Cited in FSTA 14(1982) 10G 669.

33. H. Onishi, H. Fuchi, K. Konami, O. Hidako and M. Kamekura, Agric. Biol. Chem., 44(1980) 1253-1258.

34. C.W. Hesseltine, Mycologia, 57(1965) 149-197.

35. K. Mogi and K. Mogi, J. Agric. Chem. Soc. Japan, 42(1968) 466.
Cited in ref 20.

36. H. Tachi, I. Kamoi, I. Nagase and T. Obara, Tokyo Nogyo Daigaku Nogatu Shuho, 90th Anniversary Edition (1981) 239-250.
Cited in FSTA 14(1982) 4G 217.

37. F. Noda, K. Hayashi and T. Mizunuma, App. Environment. Microbiol., 40 (1980) 452-457.

38. F.M. Yong, K.H. Lee and H.A. Wong, J. Food Technol., 15(1980) 421-428.

39. H. Sakurai, H. Hashiba and A. Okuhara, Nihon Nogei Kaggakai-shi, 55(1981) 7-14.
Cited in FSTA 14(1982) 3T 137.

40. H. Matai, T. Ishiyama and E. Matsumoto, Ibid., 55(1981) 325-331.
Cited in FSTA 14(1982) 6G 430.

41. F. Ono and Y. Aoyama, Nippon Shokuhin Kogyo Gakkaishi, 24(1977) 230-235. Cited in FSTA 14(1982) 6T 324.

42. N. Yamaguchi, Y. Yokoo and M. Fujimaki, J. Jap. Soc. Food Sci. Technol. 26(1979) 71-75.
Cited in FSTA 13(1981) 4N 191.

43. H. Chiba, Nihon Jozo Kyokai Zasshi, 74(1979) 601-603.
Cited in FSTA 13(1981) 11T 639.

44. Y. Kitada, K. Tamase, M. Sasaki, Y. Nishikawa and K. Tanigawa, Shokuhin Eiseigaku Zasshi, 21(1980) 480-484.
Cited in FSTA 14(1982) 4T 204.

45. T. Takatsuki and K. Sakai, Ibid., 21(1980) 476-479.
Cited in FSTA 14(1982) 5L 354.

46. R.D. Wei, S.C. Chang and S.S. Lee, J. Assoc. Off. Anal. Chem., 63(1980) 1269-1274.

47. N. Nunomura, M. Sasaki and T. Yokotsuka, Agric. and Biol. Chem., 44(1980) 339-351.

48. K. Kihara, Abstracts of Papers, Amer. Chem. Soc., 177(1979) (1) AGFD 50.
Cited in FSTA 12(1980) 2T 123.

49. R. Liardon and G. Philippossian, Z. für Lebensmittel-Untersuchung und-Forschung, 167(1978) 180-185.

50. R. Liardon and S. Ledermann, Ibid., 170(1980) 208-213.

51. T. Aishima, Agr. Biol. Chem., 43(1979) 1905-1910.
Cited in FSTA 13(1980) 4T 207.

52. T. Aishima, Ibid., 43(1979) 1935-1943.
Cited in FSTA 13(1980) 4T 208.

53. T. Aishima, M. Nagasawa and D. Fukushima, J. Food Sci., 44(1979) 1723-1727, 1731.

54. S. Mori, T. Someya, A. Okuhara and N. Iguchi, Abstracts of Papers, Amer. Chem. Soc., 177(1979) AGFD66.
Cited in FSTA 12(1980) 3T 146.

55. I.H. Wang, S.K. Chen and W.H. Hsu, Research Report, Food Industry Research and Development Institute (Taiwan) No.2 (1980), 14pp.
Cited in FSTA 14(1982) LT 23.

56. A. Bhumiratana, T.W. Flegel, T. Glinsukon and W.Somporan, Appl. Envir. Microbiol., 39(1980) 430-435.

57. Le Van N'ong, Voprosy Pitaniya (1981) No. 5, 62-65.
Cited in FSTA 14(1982) 7T 374.

58. M. Mogi, S. Nakajima, T. Nakadai and S. Nasuno, Nihon Jozo Kyokai Zasshi, 74 (1979) 110-112.
Cited in FSTA 12(1980) 7G 525.

59. M. Mogi, S. Nakajima, T. Nakadai and S. Nasuno, Ibid., 74(1979) 113-115. Cited in FSTA 12(1980) 7G 526.

60. M. Mogi, S. Nakajima, T. Nakadai and S. Nasuno, Ibid., 75(1980) 750-752. Cited in FSTA 14(1982) 7G 478.

61. M. Miura, N. Ouchiyama and Y. Shinoda, Miyazaki Daigaku Nogakubu Kenkyu Jiho, 25(1978) 215-225. Cited in FSTA 12(1980) 5J 802.

62. Y.-S. C. Shieh and L.R. Beuchat, J. Food Sci. 47(1982) 518-522.

63. Y.-S. C. Shieh, L.R. Beuchat, R.E. Worthington and R.D. Phillips, Ibid., 47(1982) 523-529.

64. N. Honma, Nihon Jozo Kyokai Zasshi, 74(1979) 436-442. Cited in FSTA 13(1981) 1J 189.

65. Y. Koizumi, H. Hatori, F. Yanagida and A. Ito, Ibid., 76(1981) 206-210. Cited in FSTA 14(1982) 7G 440.

66. M. Miyasaka, Ibid., 75(1980) 960-965. Cited in FSTA 14(1982) 6G 363.

67. T. Tsukuhara, R. Nakazato and M. Takeda, Tokyo Nogyo Daigaku Nogatu Shuho, 90th Anniversary Edition (1981) 53-60. Cited in FSTA 14(1982) 4G 266.

68. H. Yoshi, Nihon Jozo Kyokai Zasshi, 74(1979) 213-218. Cited in FSTA 12(1980) 6G 423.

69. T. Kawanishi, A. Suzuki, H. Konuma, S. Takayama and K. Mizushima, Eisei Shikenjo Hokoku, 96(1978) 148-155. Cited in FSTA 12(1980) 3J 307.

70. S. Hondo and T. Mochizuki, Nippon Shokuhin Kogyo Gakkaishi, 26(1979) 461-468. Cited in FSTA 14(1982) 4G 241.

71 S. Hondo and T. Mochizuki, Ibid., 26(1979) 469-474. Cited in FSTA 14(1982) 4G 242.

72. S. Hondo and T. Mochizuki, Ibid., 26(1979) 509-513. Cited in FSTA 14(1982) 4G 243.

73. T. Ohara, T. Yoneyama, T. Ohike, T. Miyazaki and M. Negishi, Ibid., 26 (1979) 287-293. Cited in FSTA 14(1982) 4G 237.

74. K. Kiuchi, O. Suzuki, T. Obta, E. Sato and H. Ebine, Ibid., 25(1978) 508-514. Cited in FSTA 12(1980) 5G 389.

75. K. Kiuchi, M. Kakezawa and H. Ebine, Ibid., 27(1980) 60-67. Cited in FSTA 14(1982) 3G 157.

76. H. Yasuhira, Nihon Jozo Kyokai Zasshi, 75(1980) 506-515. Cited in FSTA 13(1981) 10G 741.

77. H. Ito, Ibid., 75(1980) 881-884.
Cited in FSTA 14(1982) 4A 304.

78. T. Ishima, Shokuryo Kenkyusho Kenkyu Hokoku, No. 34(1979) 252-257.
Cited in FSTA 12(1980) 2G 65.

79. S. Iwabuchi, M. Sato and K. Shibasaki, Nippon Shokuhin Kogyo Gakkaishi, 24(1977) 65-71.
Cited in FSTA 14(1982) 6G 399.

80. K. Kiuchi, T. Ohta, T. Ishima, T. Takabayashi and H. Ebine, Ibid., 24 (1977) 295-299.
Cited in FSTA 14(1982) 6G 400.

81. H. Ebine, Nihon Jozo Kyokai Zasshi, 75(1980) 145-148.
Cited in FSTA 13(1981) 6J 904.

82. M. Hamano and H. Sugimoto, J. Food Processing and Preservation, 2(1978) 185-196.
Cited in FSTA 12(1980) 5G 342.

83. T.M. Ha, Fermentation and Industry, 37(1979) 202-211.
Cited in FSTA 12(1980) 11G 781.

84. K.H. Steinkraus, Y.B. Hwa, J.P.V. Buren, M.I. Provvidenti and D.B. Hand, Fd. Res., 25(1960) 777-788.

85. K.H. Steinkraus, J.P.V. Buren and D.B. Hand, Studies on Tempeh, an Indonesian fermented food, in Publication 843, National Academy of Sciences, National Research Council, Washington, D.C., U.S.A. (1961).

86. K.S. Djian and C.W. Hesseltine, Soybean Digest, 22(1961) 14-15.

87. C.W. Hesseltine, M. Smith, B. Bradle and K.S. Djian, Devel. Indust. Microbiol., 4(1963) 275-287.

88. A.K. Smith, J.J. Rackis, C.W. Hesseltine, M. Smith, D.J. Robbins and A.N. Booth, Cereal Chem., 41(1964) 173-181.

89. A.K. Smith, J.J. Rackis, C.W. Hesseltine, D.J. Robbins and A.N. Booth, Cereal Sci. Today, 8(1963) 129.

90. A.M. Jurus and W.J. Sundberg, Appl. Environ. Microbiol., 32(1976) 284-287.

91. I.T.H. Liem, K.H. Steinkraus and T.C. Cronk, Ibid., 34(1977) 773-776.

92. Ko Swan Dien and C.W. Hesseltine, in A.H. Rose (Ed) Economic Microbiology, Vol. 4, Microbial Biomass, Academic Press, London, 1979, pp.115-140.

93. M. Ismail, Pertanika, 4(1981) 129-132.
Cited in FSTA 14(1982) 8J 1279.

94. R. Girija Bai, T.N. Ramachandra Rao, T.N. Prabha, V.P. Sreendharan and N. Sreedhara, Indian Food Packer, 33(1979) 26-33.

95. H.M. Bau and G. Debry, Cahiers de Nutrition et de Dietetique, 15(1980) 227-284.
Cited in FSTA 14(1982) 1J 39.

96. H.L. Wang, E.W. Swain and C.W. Hesseltine, J. Food Sci., 45(1980) 1262-1266.

97. W. Shurtleff and A. Aoyagi, The Book of Miso, Ballantyne Books, New York, 1976.

98. W. Shurtleff and A. Aoyagi, Miso Production, New-Age Foods Study Center, Lafayette, Califonria, 1980.

99. W. Shurtleff, and A. Aoyagi, The Book of Tempeh, Harper and Row, Publishers, New York, 1979.

100. W. Shurtleff and A. Aoyagi, Tempeh Production, New Age Foods Study Center, Lafayette, California, 1980.

ACKNOWLEDGMENTS

These are gratefully due to Mrs Jean Winter for her careful and laborious typing and correction of this paper, with its many and varied problems of spelling, unusual citations, etc.

The unknown abstractors working for Food Science and Technology Abstracts are thanked most sincerely for making the Japanese language papers known to me.

FERMENTATIONS INVOLVED IN THE PRODUCTION OF COCOA, COFFEE AND TEA

K.L. JONES* and S.E. JONES

Cadbury Schweppes Group Research, The Lord Zuckerman Research Centre, Whiteknights, Reading, England.

*Present address: National College of Food Technology, University of Reading, Food Studies Building, Whiteknights, Reading RG6 2AP.

CONTENTS

1. INTRODUCTION

There is no authentic record of the origin of cocoa, coffee or tea. Cocoa is believed to have originated from the Orinoco and Amazon valleys of South America (ref. 1) and following initial cultivation by the Aztecs has become established in America, the West Indies, Africa and Asia (refs. 2, 3). Coffee may well be native of Ethiopia and is now well established in Africa, South America and India (ref. 4). Tea is produced mainly in China, India, Ceylon, Japan, Taiwan and the Soviet Union, although it was originally cultivated in China (ref. 4). All three crops require tropical climates, the best tea and coffee flavours are produced at high altitudes.

The development of chocolate flavour (refs. 5, 6, 7, 8) requires the fermentation, drying and roasting of fresh seeds. In this instance fermentation can be regarded as having two purposes. Firstly, it removes the endocarp or pulp which surrounds the bean and secondly, it initiates the reactions leading to flavour development within the cotyledons of the beans. The role of coffee fermentation is purely to remove the pulp material surrounding the coffee cherries. Unlike cocoa and coffee fermentation tea fermentation can be regarded as an oxidative process which has no microbiological involvement.

Under optimum fermentation conditions all three crops will develop flavours of the desired standard. In practice fermentations are often carried out by traditional methods, often on small farms, resulting in wide variations in flavour. Attempts to improve the processes have been empirical, a task made difficult, particularly in the case of cocoa, by variations in fermentation practice. In order to standardise and improve fermentation it is necessary to understand the microbiological, biochemical and chemical aspects of all three processes. This review outlines the current understanding of these processes.

COCOA FERMENTATION

2. CLASSIFICATION AND DISTRIBUTION OF COCOA

The seed of the cocoa tree *Theobroma cacao* Linn consists of two cotyledons and a radicle surrounded by a testa or seed coat which is enveloped by a layer of mucilagenous endocarp termed the pulp. A mature cocoa pod contains thirty to forty seeds.

A newcomer to the field may find the classification of cocoa confusing. Consequently this aspect is covered in greater detail than may be necessary for reviewing cocoa fermentation. Two principal types of cocoa are recognised, Criollo and Forastero. Hart (ref. 9) proposed a further group, the Callabacillo group, which is no longer regarded as distinct enough to be separated from the Forastero type. Criollo and Forastero cocoas can be differentiated on the basis of cotyledon colour, the former having white or pale violet cotyledons, whereas the latter have red or purple cotyledons. The term Criollo (native) was first used in Western Venezuela for the type of cocoa that was regarded as native to that country. Usage has resulted in the application of the name to similar high quality cocoas from other countries. The term Forastero came into use at the same time (ca. 1825) to distinguish the native cocoa of Venezuela from cocoas introduced from Trinidad. Cheesman (ref. 10) regarded the cocoa introduced from Trinidad to be similar to the Trinidadian cocoa of today. This is thought to have risen from crosses between plants transferred to Trinidad from Eastern Venezuela, following destruction of the native cocoa population in 1727, and surviving Trinidadian plants. Forastero cocoa was regarded as being inferior in quality to Criollo and the term was extended to include the cocoas of Trinidad and Ecuador and later those of West Africa and Brazil. Botanically, the cocoas of Ecuador, Brazil and West Africa differ from the original Forasteros of Venezuela. Cocoa was introduced to the state of Bahia in Brazil from Belem on the Amazon (ref. 11). The seeds were purple in colour and fall into the Forastero class. Similar cocoa reached Ghana and Nigeria via Sao Tome and Fernando Po and is termed Amelonado (melon shaped).

The term Amazonian Forasteros is used to differentiate the older Forastero cocoas of West Africa and Brazil, and newer (post 1930) cocoas resulting from collection expedititions around the Amazon Basin. The expression Amazon hybrid is commonly used to denote crosses between Amazons and Amelonados or Trinitarios. Trinitarios, as already implied, are hybrid populations which resulted from crosses between surviving Criollo cocoa and imported varieties. The origin of the Criollo cocoa of Trinidad is unclear (ref. 10) although it is thought to have come from Venezuela. The plants introduced following the devastation of crops in 1727 also appear to have come from Eastern Venezuela and were either Amazonian Forastero types, similar to the yellow-podded Amelonado found wild in Surinam, or hybrids (ref. 10). The term Trinitario is now used not only for the cocoa of Trinidad but also for that of Sri Lanka, Indonesia and Papua-New Guinea, where crosses between Criollo and Forastero cocoas have taken place. Malaysian cocoas are genetically similar to the Amazon hybrid cultivars that have been approved for planting in West Africa (ref. 12).

There is no definitive classification system for cocoa. Cheesman (ref. 10) proposed that the division of Forastero cocoa on the basis of differences in pod shape into Angoleta, Cundeamor, Amelonado and Calabacillo varieties was unsatisfactory and proposed the following classification:-

Criollo	Central American Criollos
	South American Criollos
Forastero	Amazonian Forasteros
	Trinitarios

Urquhart (ref. 13) proposed a commercial classification system consisting of three groups:-

1. Criollos
2. Fine Forasteros including Ecuador and Trinidad cocoas and cocoas grown from Trinidad stock in Venezuela, Ceylon and Indonesia.
3. Ordinary Forasteros, grown in West Africa and Brazil.

The fine Forasteros correspond to the Trinitarios of Cheesman (ref. 10).

The bulk of world cocoa production is of the older Amelonado Forastero type from West Africa and Brazil, and the newer Amazonian Forastero types. Criollo cocoa, although of high quality, is now produced in negligible amount. Similarly,the Criollo Forastero hybrids (Trinitario types) found in Trinidad, Venezuela, Ceylon and Indonesia, produce high quality beans though production is not extensive.

3. HARVESTING

Harvesting practices and processing prior to fermentation differ markedly not only in different countries but also from farm to farm (refs. 1, 14). Although recommendations have been made (ref. 15) that harvesting should be carried out at intervals of three weeks, this practice is not always adopted. Attempts to save on labour costs in Nigeria for example have resulted in a wider spectrum of pod maturity than desired ideally (ref. 16). One important consequence of such action is variation in the level and composition of pulp surrounding the beans which, as will be described later, affects fermentation (ref. 17).

Following harvesting, the beans must be removed from the pods prior to fermentation. The stage at which the pods are broken open varies with locality. Usually, pods are transported intact although in Brazil beans may be transported to central fermentaries. Provided that the beans are received within twenty-four hours of removal from the pods the fermentation process is not affected (ref. 18). On small farms, where harvesting yields a limited number of pods, it is conceivable that pods may be stored for a number of days prior to fermentation.

4. FERMENTATION PRACTICE

The practice of fermenting cocoa beans for the specific purpose of developing chocolate flavour has only developed during the present century. Originally, fermentation resulting from the collection of beans into heaps was seen mainly as a method of removing pulp. The manner in which fermentation is carried out in different countries varies greatly (ref. 19) and the fermentation practice adopted depends, or at least should depend, on the type of cocoa grown. The Criollo type of cocoa, which is grown in relatively small amounts, requires a shorter period of fermentation than the Forastero type.

The two main types of fermentation processes are heap and box fermentation. As suggested by the terminology heap fermentation simply consists of collecting beans, varying in amounts from <10 - >1,800 kg (ref. 15), into a heap. Ideally, the heaps are mixed after 48 and 96 hours in order to avoid variation in flavour and quality, and to aid aeration. Early work indicated (ref. 15) that fermentations were carried out without mixing; unfortunately practices have not changed particularly on small farms where the amounts of beans available are not great. Heap fermentations are limited to the West African countries of Ghana, Nigeria and the Ivory Coast.

It is immediately apparent that cocoa fermentation is strictly empirical. The duration of fermentation should be shorter during wet weather and the length of fermentation should be varied according to the time between harvesting and breaking of the pods. The higher quality of Ghanaian cocoa compared with Nigerian or Ivorian underlines the importance of good farm practice and care during fermentation. It is generally believed that fermentation does not receive adequate attention in Nigeria and the Ivory Coast.

Box fermentation (refs. 1,20) is common in Trinidad, Brazil and Malaysia although the detailed process varies from country to country. In Trinidad and Malaysia boxes with slatted bottoms, to aid drainage and oxygenation, containing approximately 150 kg of wet beans are arranged in rows. Beans are tipped from the front box to the second, to aid aeration and mixing, after twenty-four hours and from the second to the third after forty-eight hours. The total fermentation time being four to nine days. It is generally believed that similar care and attention to detail is not common in Brazilian box fermentations.

Venezuela, Ecuador, Costa Rica and Mexico practice variations on the above schemes (ref. 1) although the cocoa produced is variable in quality.

4.1 Composition of the pulp - the substrate for fermentation

Published information regarding the composition of pulp is limited despite the fact that both initial microbial colonisation and subsequent microbial activity is largely dependent on the composition of the pulp.

Pulp is generally quoted as consisting of 80-90% water and 10-13% sugar (refs. 21, 22, 23, 24, 25) with about 1% pectin (ref. 22). Examination of pulp carbohydrate levels during growth and maturation of Trinidad cocoa (ref. 26), West African cocoa (ref. 27), Venezuelan cocoa (ref. 28) and Sri Lankan cocoa (ref. 29) showed the concentration of glucose, sucrose and fructose to be dependent on the age of the fruit examined. Further, the pulp to bean ratio can vary both during maturation (ref. 26) and between cultivar types (refs. 30, 31).

Volatile acids and alcohols are considered to be absent from cocoa pulp (ref. 20). Earley work (ref. 21) showed pulp to contain 0.2 - 0.4% non-volatile acids, citric acid being the main component (refs. 32, 33). More recent work has indicated a citric acid level of 2% (refs. 24, 25) and a malic acid level of 0.1 - 0.2% (ref. 30). Amino acids in the form of aspartic acid, glutamic acid and asparagine accounted for the nitrogen content of pulp (refs. 33, 34). The approximate composition of pulp is summarised in Table 1 (ref. 35).

TABLE 1

Composition of fresh cocoa pulp (ref. 35)

Component	Average content %
Water	80 - 90
Albuminoids, astringents, etc.	0.5 - 0.7
Glucose	8 - 13
Sucrose	0.4 - 1.0
Starch	Trace
Non-volatile acids (like tartaric)	0.2 - 0.4
Iron oxide	0.03
Salts (potash, soda, calcium and magnesium)	0.4 - 0.45
Volatile acids	Nil
Alcohol	Nil

4.2 Microbial ecology of fermentation

Characterisation of the microbiological changes taking place during fermentation (refs. 30, 36, 37, 38, 39) has demonstrated the involvement of three key microbial groups, namely - yeasts, lactic acid bacteria and acetic acid bacteria. Members of the genus *Bacillus* dominate the later stages of fermentation (ref. 40) and may be responsible for the production of the flavouring component tetramethylpyrazine (ref. 41). The numbers and growth rates of the microbial groups is dependent on the sampling position within the fermenting mass. Ostovar (ref. 40) attempted to analyse the frequency of occurrence of different microbial species during fermentation. It is unfortunate that the thesis contained neither critical evaluation of the methodology adopted nor any form of statistical analysis. The factors controlling the observed growth patterns will be discussed later. Sporulating bacilli found to be dominant in the latter stages of fermentation (ref. 37) were probably present as spores.

4.2.1 Yeasts

Following the initial colonisation of pulp, yeasts were found to proliferate rapidly forming *ca.* 90% of the microbial population after 24 hours. Characterisation of isolates (refs. 42, 43, 44) showed the presence of a range of species (Table 2). Roelofsen (ref. 23) isolated 24 strains of yeasts, all capable of growing on citric acid, from West African and Javanese cocoa. A succession pattern in the types of yeasts colonising fresh pulp has been noted (ref. 23); *Saccharomyces* species dominate the early phase of fermentation whereas pseudo-mycelial yeasts (e.g. *Candida krusei*) predominate in the later stages.

Similarly, Carr *et al.* (ref. 44) found members of the genus *Kloeckera* to be present earlier than *Saccharomyces* species but suggested the latter were the main fermenting organisms, as did Martelli and Dittmar (ref. 45).

It has been suggested (refs. 2, 46) that differences in yeast populations may contribute to differences in chocolate flavour. Although differences in yeast populations have been detected, as outlined above, the presence or absence of species may be a function of the isolation technique and may not reflect *in situ* activity. Further, the ubiquitous nature of genera such as *Trichosporon*, *Torulopsis* and *Candida* is well established (ref. 47).

TABLE 2

Yeast species isolated from fermenting cocoa

Species	Author
Candida krusei	(ref. 43)
Torulopsis rosei	(species listed
Pichia farinosa	in order of
Saccharomyces cerevisiae var *ellipsoides*	decreasing
Kloeckera apiculata	frequency)
Hansenula anomala	
Schizosaccharomyces pombe	
Pichia fermentans	
Trichosporon pullulans	
Pichia membranefaciens	
Candida catenulata	
Saccharomyces cerevisiae	(ref. 42)
S. cerevisiae var *ellipsoides*	
S.globosus	
S. oviformis	
Schizosaccharomyces pombe	
Pichia fermentans	
Hansenula anomala	
Rhodotorula aurantiaca	
Rh. mucilaginosa	
Torulopsis sp.	
Kloeckera apiculata	
Schizoblastosporion sp.	
Candida krusei	
C. krusei var. *vanlaeriana*	
Trichosporon sp.	
Hansenula sp.	(ref. 44)
Kloeckera sp.	
Torulopsis sp.	
Saccharomyces sp.	
Candida sp.	
Pichia sp.	
Schizosaccharomyces sp.	
Saccharomycopsis sp.	

4.2.2 Lactic acid bacteria

Roelofsen (ref. 23) identified lactic acid bacteria isolated from Javanese fermentations as heterofermentative whereas Carr et al. (ref. 44) identified heterofermenters such as Lactobacillus collinoides and L. fermentum as well as homofermenters such as L. plantarum and L. mali. The ecological significance of the species detected could not be assessed on the information available (ref. 44). Differences in the composition of the Lactobacillus populations of Ghanaian and Malaysian cocoa could not be detected and population trends were also similar in both countries (ref. 44). Numbers of lactic acid bacteria did however show variation with time and sampling location within the fermenting mass (ref. 30). Microbial numbers also show variation depending on the time of harvesting (ref. 48) due to differences in pulp volume affecting the aeration of the fermentation.

4.2.3 Acetic acid bacteria

With the continuance of fermentation, loss of pulp material by "sweating" renders the fermenting mass permeable to oxygen so producing conditions suitable for the growth of acetic acid bacteria (ref. 49) which form 80-90% of the microbial population 48 hours after initiation of fermentation (refs. 23, 42). The distribution of acetic acid bacteria within a fermenting heap follows the trends outlined by Carr et al. (ref. 30). Numbers of acetic acid bacteria are influenced by the level of pulp surrounding the beans (ref. 48).

The characterisation of isolates from Java and Trinidad (ref. 23) and Ghana and Malaysia (ref. 44) showed the predominant species to be Acetobacter rancens, A. ascendes and A. xylinum (now considered to be a strain of A. aceti). Carr et al. (ref.44) found a novel species with some similarity to A. rancens in Malaysia. Strains of Gluconobacter were detected in both Ghana and Malaysia but only during the initial stages of fermentation. The amount of information available is not sufficient to correlate the activity of acetic acid bacteria with changes in numbers or distribution of different species.

4.3 Changes in pulp chemistry resulting from microbial activity during fermentation

Changes in sugar, ethanol, acetic acid and lactic acid levels during fermentation have been recorded by a number of workers (refs. 20,48,50,51,52,53,54,55, 56,57,58) and are best exemplified by the work of Carr et al., (ref. 30). The data of Carr et al. (ref. 30) although it indicates trends, does not characterise the changes required for the optimum fermentation of cocoa beans since acid tainted chocolate was produced from all the beans fermented. Comparative experiments on Amelonada and Amazon cocoa in Ghana (ref. 30) indicated acid production to be greater during fermentation of Amazon cocoa.

Citric and malic acids were catabolised during fermentation (refs. 23, 30) and pulp pH values increased from *ca.* 3.8 to 4.5 (refs. 30,37,48,57,59).

During fermentation the temperature increased to 40-45°C (refs. 30,32,37,48, 51,59,60,61). The maximum temperature reached during fermentation varie over the harvesting season (ref. 63) possibly due to differences in pulp carbohydrate levels (ref. 64). The degree of maturity of cocoa also influences temperature changes during fermentation (refs. 29, 63).

As will be discussed later, the degree of aeration of fermenting pulp is a key factor in controlling microbial activity. The effect of aeration has been discussed in the early literature (refs. 19,24,25,65,66) and is influenced by whether or not the beans are turned. Increased oxygenation resulted in higher maximum temperatures during fermentation due to the exothermic nature of microbial oxidation of ethanol and acetate.

4.4 Rationalisation of the microbiology of cocoa fermentation

Prior to fermentation, harvested cocoa pods are opened by traditional methods (ref. 1) resulting in the hitherto sterile pulp being infected with a wide variety of micro-organisms from the surrounding environment (refs. 30,37, 39,48), 65% of which are soil or air contaminants which take no further part in fermentation (ref. 37). The low pH (3.5) of fresh pulp is selective for yeasts whose facultative metabolism allows them to utilise pulp carbohydrates (refs. 30, 57) under both aerobic and anaerobic conditions. The initial anaerobic state of a mass of beans results in fermentation of pulp glucose, fructose and sucrose (refs. 67, 68) and the accumulation of ethanol (refs. 30, 57, 58). Pectinase activity (refs. 23, 69, 70, 71) during the initial stages of fermentation results in the breakdown of the cellular structure of pulp which leads to an ingress of oxygen. Roelofsen (ref. 23) suggested pulp oxygen tension to be important in controlling yeast succession patterns. In terms of microbial energetics, fermentation, as compared with oxidative metabolism, is a wasteful process due to the accumulation of ethanol and other by-products. Consequently, whenever conditions permit, oxidative metabolism will prevail. High carbohydrate concentrations inhibit oxidative metabolism in yeasts via inhibition of respiratory enzyme synthesis and mitochondria formation (refs. 67, 68).

Increases in pulp pH during fermentation are attributed to assimilation of citric acid (ref. 20) and 24 strains of yeasts isolated from fermenting cocoa showed such activity (ref. 23). Citric acid metabolism could only be initiated following establishment of aerobic conditions. Acetic acid supports some growth of *Candida utilis* although yields are lower than on sugars and *Saccharomyces cerevisiae* has been shown to respire acetic acid (ref. 72).

Influx of oxygen and increases in pH due to the metabolism of citric acid renders pulp suitable for the growth of the strictly aerobic bacteria belonging to the genera Gluconobacter and Acetobacter. In addition to hexoses and glycerol, some Acetobacters oxidise ethanol firstly to acetate and secondly to CO_2. The so-called over-oxidation of ethanol to acetate and CO_2 (ref. 49) occurs most markedly at low concentrations of acetic acid with strains tolerant to low pH, the reaction being limited to a narrow species-specific pH range. Over-oxidation cannot therefore be regarded as a major pathway of acetate removal during fermentation. Lactate is similarly metabolised. Sugar and ethanol oxidation in the Gluconobacters proceeds only as far as acetate.

The pH and temperature requirements of Acetobacters and Gluconobacters are summarised in Table 3. The initial low growth rates of acetic acid bacteria in fermenting pulp (refs. 20, 30, 37) can be attributed to low oxygenation and sub-optimal pH. Cessation of growth correlates similarly with increases in pulp temperature above 40° and nutrient depletion (ref. 30). The final decline of the acetic acid bacterial population could be rationalised in terms of temperature inhibition of maintenance metabolism.

TABLE 3.
pH and temperature requirements of the genera Acetobacter and Gluconobacter. (Many strains will not grow above 37°C)

	Acetobacter	Gluconobacter
Temperature range (^{o}O)	5-42	7-41
Optimum growth Temperature	30	25-30
pH range	3-8	4.0-6.5 (unpublished reports of growth at pH 3.2-3.5)
Optimum pH for growth	5.4-6.3	5.5-6.0

The lactic acid bacteria form a group of fermentative, anaerobic, microaerophilic or facultative organisms which are highly saccharoclastic (ref. 73). Two physiological subgroups are identified on the basis of glucose fermentation. The homofermentative lactic acid bacteria convert glucose quantitatively to lactic acid. The heterofermentative lactic acid bacteria convert glucose to a mixture of lactic acid, ethanol and CO_2. The differences in end products result from the mechanism of sugar breakdown (Embden-Meyerhoff pathway or hexose monophosphate shunt). Homofermentative and heterofermentative lactic acid bacteria also differ with respect to fructose fermentation. The heterofermenters reduce fructose to mannitol coupled with the formation of acetic acid:

3 fructose ⟶ lactic acid + acetic acid + CO_2 + 2 mannitol

Heterofermentation of sucrose therefore leads to the accumulation of both lactic and acetic acids.

Lactic acid bacteria such as L. plantarum, L. delbrukii and L. bovis can perform limited oxidation of substrates such as glucose and pyruvate. Although lactobacilli are catalase negative, H_2O_2 resulting from direct oxidation of metabolisable substrates is removed by a "pseudo catalase" (see refs. 74,75).

The faster growth rate and higher population levels of lactic acid bacteria in the anaerobic parts of fermentation heaps and boxes is difficult to rationalise in terms of the physiology of lactic acid bacteria. It can only be suggested that anaerobic conditions permit the growth of all lactic acid bacteria whereas aerobic conditions inhibit the growth of the anaerobic forms, possibly due to the accumulation of hydrogen peroxide. It is also likely that the production of inhibitory compounds by lactic acid bacteria is involved in these processes.

4.5 Effect of microbial metabolism on temperature during fermentation.

Quesnel (refs. 24, 25, 65) drew attention to the exothermic nature of ethanol and acetic acid oxidation and the correlation between temperature changes and known microbial succession. The role of aeration in cocoa fermentation has attracted much discussion (refs. 76,77,78,79,80,81,82). Calculations based on the ethanol and acetic acid content of fermenting pulp showed 700 litres of oxygen to be consumed per 100 kg of cocoa during a 7 day fermentation (ref. 65) such a value being 40 times less than the figure calculated by Roelofsen (ref. 23). The differences could be due to the over-estimation of the volume of air space within the fermentation by Roelofsen (ref. 65). The temperature increases observed during fermentation can be accounted for by microbial fermentation and oxidation reactions within the pulp (refs. 65, 83). Further, the potential energy of pulp sugar was enough to heat cocoa to a much higher temperature than that observed.

4.6 Control of fermentation

As already outlined, conditions within a fermenting mass of beans are not uniform and attempts have been made to remedy such problems, for example, fermenting beans in shallow trays (refs. 1,60,61). Central to the control of fermentation lies the control of aeration (ref. 24) which, as already discussed, determines the relative activity of the organisms involved in fermentation. Several factors influence the aeration of fermentations: pulp water content, size and shape of the fermenting mass, number of times the mass is turned, type of fermentation box, if any, and the type of covering (refs.24,25). Although variables such as the frequency of turning and size of the fermentation have been investigated the fermentations appeared to be sub-optimal. Chong et al. (ref. 12) found that removal of pulp prior to conventional box fermentation resulted in acid production. Lopez (ref. 59) however showed removal of pulp by pressing to accelerate acetic acid production.

5. EFFECT OF FERMENTATION ON THE CYTOLOGY, BIOCHEMISTRY AND CHEMISTRY OF COCOA BEAN COTYLEDONS

Fermentation results in the disruption of the cellular structure of cocoa bean cotyledons, due to the ingress of fermentation products, and in the production of flavour precursors.

5.1 Ingress of fermentation products

Lactic and acetic acids accumulate in cotyledons during fermentation (refs. 51,52,54,55). Concomitant changes in pH were reported. Changes in lactate, acetate and pH were dependant on the position within the fermenting mass.

Forsyth and Rombouts (ref. 76) and Forsyth (ref. 79) suggested heat to be the agent responsible for the death of beans during fermentation. The claims of Roelofsen and Giesberger (ref. 48) that acetic acid was the primary cause of death with temperature and alcohol being contributory factors were partly confirmed by Quesnel (ref. 62); temperature was found to be unimportant.

5.2 Changes in cotyledon oxygenation

Forsyth and Rombouts (ref. 76) suggested that cotyledons were anaerobic during the initial phase of fermentation based on the evidence that the pulp of fermenting cocoa decolorised methylene blue, and loss of epicatechin* from the beans resulted from exudation through the testa rather than by oxidation. Maravalhas (ref. 84) also stressed the importance of this anaerobic, hydrolytic phase. Observations regarding the oxygen requirement for fermentation (refs. 77,78,60,85) led Powell (ref. 80) to suggest incorrectly (according to Quesnel, refs. 81, 86) that anaerobic cotyledons did not exist during fermentation. Quesnel (refs. 81, 86) discussed in detail the evidence for anaerobic cotyledons and concluded (ref. 86) that oxygen does penetrate cotyledons, but only to a limited extent, after four days of fermentation (ref. 87).

It is during these initial four days of fermentation that major biochemical activity takes place (ref. 86).

5.3 Composition of unfermented cotyledons

Information regarding the sugar composition of unfermented cocoa beans is sparse. Rohan and Stewart (ref. 88) found Nigerian and Ghanaian Amelonado cocoa to have the composition outlined in Table 4. Forsyth (ref. 33) and Humphries (ref. 34) detected D-fructose, D-glucose and sucrose. L-arabinose,

* Epicatechin (3,37,47,5,7-pentahydroxyflavan) can be regarded as a condensed tannin being built up by polymerisation of C_{15} flavan units.

D-galactose, and D-xylose and the polysaccharides cellulose, starch, pectic substances, araban, galactan and xylan have also been detected. Seiki (ref. 57) showed glucose, maltose and fructose to be present in order of decreasing concentrations in Nigerian cocoa. The determinations were not, however, strictly quantitative.

TABLE 4
Sugar content of Nigerian and Ghanaian Amelonado cocoa beans on a % dry weight basis (ref. 88).

Reducing Sugars (% of cotyledons)		Total Sugars (% of cotyledons)	
Ghana	Nigeria	Ghana	Nigeria
0.12	0.33	1.44	2.20

Starch formed 7% of cotyledon dry weight (ref. 89) and was isolated in granular form (ref. 90).

Weissberger _et al._, (ref. 91) found unfermented Trinidadian cocoas to contain lactic, oxalic, succinic, malic, and citric acid at levels of 0.15, 0.25, 0.06, 0.11 and 0.98 g/100g beans (apparently wet weight).

The total nitrogen content of Amelonado beans was found to vary between 27 and 30 mg/bean (refs. 92, 93, 94). Variability in bean composition within a seedling cocoa population was too great to allow for examination of the effect of environmental factors on nitrogen content (ref. 92). Miranda (ref. 95) found Forestero and Amelonado cocoa to contain similar nitrogen levels.

The amino acid content of unfermented Amelonado beans was as indicated in Table 5 (refs. 57, 96). Because of differences in the analytical methods used by different authors, comparisons are difficult.

Cocoa contains high levels of purines (theobromine and caffeine) and polyphenols. Levels of theobromine in unfermented cocoa were quoted as 2.8% by Humphries (ref. 97), 0.7% (dry weight) by Humphries (ref. 34), 1.7% by Forsyth and Quesnel (ref. 20) and 3% by Obiakor and Nwanko (ref. 98). Caffeine levels were 0.4% according to Fincke (ref. 99) and 0.1% according to Forsyth and Quesnel (ref. 20). In addition to purine bases approximately 30% (dry weight) of fat free cotyledons consists of flavonoid material (ref. 100).

TABLE 5
Amino acid composition of unfermented Amelonado cocoa beans (refs. 57, 96).

Amino Acid	mg Amino Acid/ Bean (ref. 57)	g Amino Acids/ 100g Extract (ref. 96)
Lysine	11	0.56
Histidine	10	0.036
Arginine	3	0.35
Aspartic acid	5	-
Threonine	5	0.84
Serine	24	1.99
Glutamic acid	72	1.77
Proline	10	1.97
Alanine	38	3.61
Valine	18	2.60
Methionine	3	-
Iso-leucine	19	1.68
Leucine	24	4.75
Tyrosine	24	1.27
Phenylalanine	24	3.36
Glycine	-	0.35

The phenolic compounds of plant materials which are grouped together under the term 'flavonoid' are based on the flavan structure (2-phenylbenzpyran) (ref. 101). The main differences between the various classes of phenolic compounds are due to differences in the oxidation states of the pyran ring. The cocoa polyphenols are formed by polymerisation of the C_{15} flavan units. The simplest group of cocoa polyphenols are the catechins (3.3'.4'.7-pentahydroxyflavan). Catechin and epicatechin differ in the spatial arrangement of the 2-phenyl and 3-hydroxyl groups about the pyran ring. Other members of this group, gallocatechin and epigallocatechin (3.3'.4'.5.5'hexahydroxyflavan) also form part of cocoa polyphenols. The anthocyanins are glycosylated derivatives of anthocyanidins (which are one of the oxidised forms of flavan). The leucoanthocyanidins are substances which yield anthocyanidins on treatment with hot acids and may be present in cocoa beans as mono or oligomeric forms (see ref. 102). Other phenols, which are not flavonoid compounds such as chlorogenic acid (3-caffoylquininc acid) and p-coumarylquinic acid (presumably an ester of courmaryl alcohol and quininc acid) have been detected in trace amounts. The levels of the various polyhydroxyphenols present are summarised in Table 6 (refs. 20, 103, 104).

TABLE 6

Polyphenols of cocoa beans

Catechins	37%
Anthocyanins	4%
(Cyanidin 3-galactoside, Cyanidin 3-arabinoside)	
Leucoanthocyanidins	58%
Chlorogenic acid	Trace
p-Coumarylquinic acid	Trace

The major groups of polyphenols in cocoa beans, catechins and leucoanthocyanidins (refs. 104,105,106,107) have tanning properties, i.e. are capable of forming stable cross links with proteins. Tannins are high molecular weight compounds (500-5,000) containing sufficient phenolic hydroxyl groups to permit the formation of stable cross-links with proteins. Catechols are effective tannins in monomeric form whereas leucoanthocyanidins can act as tannins in polymeric forms. The structure of the L_1 - L_4 leucoanthocyanidins have been investigated by Forsyth and Roberts (ref. 102), Weinges et al. (ref. 108) and Quesnel (ref. 104). Epicatechin forms 90% of the catechin fraction of cocoa beans (ref. 105). The anthocyanin fraction, although present in relatively low amounts, is responsible for the purple colour of unfermented beans and consists structurally of arabinosides and galactosides of cyanidin (refs. 18, 19).

The protein content of unfermented Bahian cocoa beans was as indicated in Table 7 (ref. 109) and was shown to decrease progressively during ripening. It was further shown that the spectrum of the proteins produced changed during ripening (ref. 109).

TABLE 7

Protein changes in cocoa beans during ripening (ref. 109)

	Days after pollination			
	135	143	150	160
Defatted cocoa bean mass				
Total protein	23.6	21.6	21.4	17.6
Extractable protein	14.2	13.2	12.5	11.5

5.4 Effect of fermentation on cotyledon cytology

Cocoa bean cotyledons consist of two types of cells: parenchyma cells containing starch granules, aleurone grains and fat droplets (ref. 23) and larger, vacuolated, polyphenol cells which form 11-13% of the cotyledon tissue. Parenchyma cell constituents are compartmentalised by membranes which separate enzymes and substrates (refs. 110,111,112). The conclusion that the larger vacuolated cells contained polyphenols but no enzymes (refs. 20,18,19,113) based on the separation technique of Brown (ref. 114) was apparently incorrect since the technique isolated vacuoles and not complete cells (ref. 100). As a result of infusion with fermentation products, pectinase and polygalacturonase activity (ref. 115) compartmentalisation of enzymes and substrates is destroyed (refs. 100,116). It has been suggested (refs. 100, 117) that the rate of cellular and organelle destruction and fusion of lipid vacuoles results in the isolation of cytoplasm and the separation of enzymes and substrates.

5.5 Changes in bean composition as a result of fermentation

Rohan and Stewart (ref. 88) and Seiki (ref. 57) showed the sucrose content of fresh Ghanaian and Nigerian cocoa beans to be hydrolysed to glucose and fructose during fermentation and Rohan (ref. 118,119,120) and Reineccius et al (ref. 121) found that reducing sugar levels increased from 0.1 - 1.0% due to hydrolysis of disaccharides. Analysis of various commercial cocoa beans indicated differences in sugar content (ref. 88). Analysis of fermented beans by Cerbulis (ref. 122,123) indicated the presence of D-fructose, D-glucose, sucrose,melobiose, raffinose, manninotriose, stachyose, planteose, verbascotetraose, verbascose, glycerol, meso-inositol and a range of unidentified sugars and sugar alcohols. In addition to the sugars listed, mannitol and sorbitol were identified by Reineccius et al (ref. 121) in beans from Ghana, Trinidad and Brazil.

A survey of acids and salts in fermented cocoa beans (ref. 124 - Table 8) indicated wide variation according to source. Acetate was the principal acid present.

TABLE 8

Volatile and non-volatile acids in commercial cocoas (ref. 124).

Sample	Volatile acids % dry weight estimated as acetic acid			Non-volatile acids % dry weight based on a molecular weight of 60		
	free	combined	total	free	combined	total
Accra	0.27	0.13	0.40	0.91	1.07	1.98
Arriba	0.37	0.54	0.90	0.54	0.50	1.04
Bahia	0.43	0.10	0.53	0.48	1.48	1.96
Carupano	0.23	0.91	1.14	4.43	0.82	5.25
Grenada	0.17	0.17	0.34	0.64	1.18	1.82
New Guinea	0.88	0.21	1.09	1.44	1.23	2.67
Puerto Cabello	0.16	0.17	0.33	0.87	1.33	2.20
Sanchez	0.22	0.15	0.37	0.80	0.79	1.59

Cocoas containing high levels of free volatile acids contained low levels of combined volatile acids. Citrate was the principal non-volatile acid in fermented cocoa (refs. 91, 124, see Table 9) with lower levels of tartaric and lactic acid. Increases in lactate levels during fermentation were accompanied by decreases in citrate levels (ref. 91, Table 10).

TABLE 9

Concentrations of several organic acids in cocoa beans, extracted with water and measured as methyl esters (ref. 91).

Type of Bean	Acid concentration, g/100g beans				
	Lactic	Oxalic	Succinic	Malic	Citric
Sanchez	0.14	-	0.03	0.10	1.32
Sanchez	0.12	0.24	0.02	0.05	0.67
Sanchez	0.14	-	0.03	0.10	1.04
Tabasco	0.11	0.28	0.03	0.05	0.68
Tabasco	0.11	-	0.04	0.07	0.58
Arriba	0.43	0.39	0.04	0.10	0.79
Trinidad	0.71	0.29	0.05	0.04	0.62
Trinidad	0.55	0.35	0.05	0.05	0.68
Ghana	0.31	0.36	0.04	0.06	0.59
Ghana	0.22	0.43	0.03	0.05	0.62
Bahia	0.42	0.38	0.04	0.04	0.56
Bahia	0.29	0.41	0.07	0.02	0.64
Bahia	0.13	-	0.03	0.03	0.63
Bahia	0.19	-	0.03	0.04	0.57
Samoa	0.66	-	0.05	0.10	1.05

- = Not Determined

TABLE 10

Concentrations of several organic acids in unfermented and fermented beans from Trinidad (ref. 91).

Acid	Unfermented (g/100g Beans)	Fermented (g/100g Beans)
Lactic	0.15	2.34
Oxalic	0.25	0.28
Succinic	0.06	0.03
Malic	0.11	0.09
Citric	0.98	0.42

In addition to the acids indicated in the Tables provided, butyric (ref. 125) and gluconic acids (ref. 91) have also been detected. The accumulation of C_3-C_5 volatile fatty acids during fermentation was quantified by Lopez and Quesnel (ref. 54).

Changes in total nitrogen levels were described by Bracco _et al_. (ref. 94) and Rohan and Stewart (ref. 93). Total nitrogen decreased from 2.2-1.8% (dry weight) and soluble nitrogen levels increased.

Changes in amino acid profiles during fermentation have received considerable attention (refs. 57,84,93,126,127,128,129) due to the involvement of amino acids in flavour development. Although Rohan and Stewart (ref. 129) found the distribution of amino acids to differ in commercially fermented, unroasted beans from different countries, changes in amino acid profiles during fermentation follow the trends indicated in Table 11 (ref. 57). During small scale fermentation a steady loss of protein bound nitrogen was accompanied by increases in amino acid and peptide nitrogen (refs. 130,131,132). Increases in extractable amino acid levels resulted from protein hydrolysis and little protein remained at the end of the fermentation period (ref. 128). Fermentation was also accompanied by changes in the electrophoretic patterns of extractable proteins (ref. 109).

TABLE 11.

Changes in the amino acid content of Amelonado cocoa beans during heap fermentation (ref. 57).

Amino Acid	Fermentation Time (Hours) 24	48	72	96	168
Lysine	18	28	42	46	43
Histidine	11	19	12	15	9
Arginine	13	20	34	36	37
Aspartic acid	5	6	10	11	10
Threonine	6	9	10	12	12
Serine	27	29	36	40	33
Glutamic acid	27	34	32	41	27
Proline	13	16	12	14	11
Alanine	31	41	38	55	37
Valine	24	28	27	45	28
Methionine	5	7	11	14	13
Iso-Leucine	21	24	24	31	22
Leucine	46	63	90	107	88
Tyrosine	40	49	50	64	46
Phenylalanine	41	59	68	92	68
Total	328	432	496	623	484

Unfermented Forastero cocoa beans contain 3-β-D galactosidyl and 3-α-L arabinosidyl cyanidins (refs. 18,19) which are rendered colourless by hydrolytic reactions during the anaerobic phase of fermentation (refs. 18,51, 106). Residual purple pigment has been regarded as indicative of inferior fermentation (ref. 133) and more than 10% residual anthocyanidin results in poor flavour development (ref. 85). Decreases in levels of epicatechin and simple leucocyanidins occurs by exudation and non-enzymic destruction (refs. 20,106,134). Decreases in purine levels also result from exudation through the testa (ref. 98). Tanning reactions between leucocyanidins and proteins lead to formation of insoluble complexes (refs. 18,19).

5.6 Chocolate flavour and flavour precursors

Despite numerous publications on the flavour components of chcolate (refs. 5,6,7,76,118,119,120,121,135,136,137,138,139,140,141,142) the exact composition is unresolved due to its complex nature (refs. 20,143). Synthetic mixtures of the volatile components detected have not reproduced chocolate flavour (refs. 121,144). Flavonoids (refs. 76,118,145,146) and amino acids and sugars (refs. 118,120,147) have been cited as flavour precursors. Aqueous and methanol extracts of raw cocoa were mostly composed of sugars and amino acids and yielded typical chcolate flavour on roasting (refs. 76,118,119,120,140,141,144,148). The role of flavonoids as flavour precursors may therefore be regarded as unresolved. (refs. 118,119,120). Of established importance, (refs.

polyphenols, quinones can condense spontaneously to form phlobaphenes (products of condensation of polyphenolics which contain flavan rings). Under acid to neutral pH values quinones react preferentially with amines and SH groups. Amino acids resulting from protease activity (ref. 158) react with quinones both by nucleophilic substitution and by addition/condensation reactions to form quinonimine. The adduct product resulting from nucleophilic substitution can take up a second amino acid molecule forming a quinonimine from which phenolimine can be produced by rearrangement.

The suggested reaction pathways of Purr (ref. 158) indicate a metabolic link between tanning and the tricarboxylic acid cycle via glyoxylate. Equally important in the proposed system are transamination and decarboxylation reactions.

Although the reactions cited present a logical basis for research into fermentation biochemistry, they are far from being well established: Maravalhas (ref. 84) concluded that an anaerobic phase was not essential for the formation of flavour precursors although, as already outlined, this would lead to excessive quinone tanning and inhibition of proteolysis. Further, it was concluded that polyphenol oxidase was not essential for flavour development, the system for flavour precursor development being active at pH 4.5 (the lower limit of activity of polyphenol oxidase). These results would suggest polyphenols to be unimportant in flavour development. The presence of such compounds would lead to astringency (ref. 156). Biehl (ref. 100) stressed the importance of quinone tanning in flavour precursor development, but only if preceded by adequate proteolysis. Maravalhas (ref. 84) concluded that polyphenol oxidase was not essential to flavour development, since the system for flavour precursor development was active at pH 4.5 (lower than the lower limit of polyphenol oxidase activity, refs. 159, 163), and that anthocyanin pigments were required for flavour development. Such results would suggest that only the reducing sugar moiety of the pigments is necessary for flavour development. Rohan found glucose and epicatechin to be essential to flavour development but suggested the most important function of the aerobic phase to be the reduction of astringency and bitterness via polyphenol oxidase (also see refs. 20, 23). Polyphenols could therefore be regarded as playing a negative role in flavour development.

7,149)are the Maillard reactions between specific amino acids (and peptides) and sugars at roasting temperatures (ref. 150) and the Strecker degradation (refs. 88,149,151,152,153,154) under similar conditions.

5.7 Enzymology and chemistry of flavour precursor development

A wide range of enzyme activities have been detected in cocoa bean cotyledons during fermentation:- amylase, protease, lipase, pectinase, polygalacturonase, glucosidase, decarboxylase and coenzyme A (refs. 20,155, 156,157,158). Although both amylases and lipases were detected in cotyledons, starch and fat levels were unchanged during fermentation (ref. 100). Also present within the cotyledons are the enzymes of primary metabolism, which, under conditions suitable for germination, would lead to the development of a new plant. The biochemistry of cocoa fermentation can therefore be regarded as an ad hoc process resulting from cell disruption. The principal enzyme activities involved in fermentation decline during the fermentation period (ref. 159). The loss of activity was due to progressive deactivation resulting from the tanning action of polyphenols (ref. 115) and the general hydrolysis of proteins (ref. 100).

Two types of tanning reactions have been identified (ref. 100). Tanning can result firstly from hydrogen bridging of proanthocyanidin dimers and proteins and secondly from reactions between quinones and protein. Quinone tanning can only take place in the presence of oxygen and completely inhibits proteolysis (refs. 160,161) due to the formation of stable compounds. Artificially controlled fermentation indicated premature oxygenation of the fermenting mass to render protein insoluble by quinone tanning (ref. 160) and resulted in low levels of soluble condensates (ref. 100). Tannin tanning was not shown to inactivate proteases (ref. 159), because of the reversible nature of hydrogen bonding.

During the initial anaerobic phase of fermentation, reactions are generally lytic and non-specific. Anthocyanins were split into sugars and aglycones under the action of glycosidases and were the only polyphenols broken down enzymatically (ref. 100). β-glucosidase activity results in the hydrolysis of disaccharides and the generation of reducing sugars (ref. 115). It was further shown (ref. 115) that amylase, β-glucosidase, catalase and peroxidase activities were low after 40 hours of fermentation whereas polyphenoloxidase activity was not similarly affected (refs. 18, 159).

Polyphenol oxidase, penolase or cacao oxidase (ref. 162) non-specifically oxidise O-diphenols to quinones (ref. 163). Due to enzyme inactivation during fermentation residual polyphenol oxidase activity after 48 hours was low. Quesnel and Jugmohunsingh (ref. 163) demonstrated that residual polyphenol oxidase activity differed from polyphenol oxidase activity in unfermented beans both as regards solubility and optimum pH. Following oxidation of

6. CLASSIFICATION AND DISTRIBUTION OF COFFEE

Although more than 40 species of the genus Coffea exist only a few are cultivated commercially for coffee production (ref. 164). The most commonly grown varieties are Coffea arabica, which accounts for three-quarters of world production (ref. 165) and Coffea robusta Linden, a form of the species Coffea canephora Pierre ex. Froehner (ref. 166) accounting for one quarter of the world consumption (ref. 165). Small quantities of Coffea liberica are grown for the African home market (ref. 4). New varieties of cultivated coffee have been obtained from hybrids of canephora and arabica species and crosses resulting from backcrossing with arabica strains have been shown to be promising as regards coffee quality, growth and disease resistance (ref. 167). The arabica varieties produce the best, most flavourful and expensive coffees which exhibit fine aroma and a desirable degree of acidity (ref. 167) whereas the robusta varieties have less acidity but impart more body to the beverage. The robusta varieties are often used in Europe as low-priced ground coffees.

Geographically, coffee is well dispersed. Arabica and robusta coffees are cultivated extensively in many South American countries such as Brazil (ref. 168), Colombia (ref. 169), Guatemala and Costa Rica (ref. 170), African countries, particularly Kenya (ref. 171), India (ref. 172) and Hawaii (ref. 173). The best quality arabica coffees are cultivated in high altitude regions of Africa and South America whereas robusta coffees are generally grown in lower-lying areas (refs. 170,174).

The coffee tree is a perennial plant restricted to tropical and sub-tropical regions. The coffee fruits are known as cherries and usually contain two beans (more correctly termed seed)(ref. 165). The ripe seeds are covered with an outer layer of pulp and an inner sticky mesocarp (the mucilage) which is usually 0.5-2mm thick depending on species and altitude (ref. 170). There is no evidence of cellular structure within the mucilage (ref. 175). The sole object of fermentation is the removal of mucilage which adheres closely to the parchment layer or outer covering of the seed. The colour of the green, or unroasted, coffee seed after fermentation may be blue, green, yellow or brown (ref. 171).

The best quality beans are blue whereas yellow or brown colouration is indicative of poor quality. As already mentioned the initial indicator of coffee quality is species type, arabica hybrid varieties producing better beverages than robusta varieties (refs. 4,165,171,174). Coffees are further classified by the quality, colour, size and shape of the bean after roasting (grading is dictated by region and climate, cultivation practices, harvesting, processing and storage conditions). Since classification with respect to beverage quality is important in describing coffee, some mention must be made of the different quality descriptors. Classification of coffee quality varies with country of origin and because of the subjective nature of assessments, blenders' descriptors seldom indicate the cause of taints. Thus South American classifications grade coffee from best to worst in the following way:- soft (mild), almost soft, hard (astringent), slightly Rio and Rio (medicinal) (ref. 165). Best Kenyan arabica coffee is described as having a desirable "Solai" flavour (ref. 171), probably equivalent to a soft grade of Brazilian arabica coffee. Ratings for Indian coffees are described somewhat elusively as fine, fair and falling off (ref. 176).

7. CHEMICAL COMPOSITION OF GREEN (UNROASTED) COFFEE BEANS

The green coffee bean was found to be composed of about 60% carbohydrate, 14% protein and about 13% lipid, together with smaller quantities of non-volatile acids, trigonelline, caffeine and theobromine (ref. 4). Genetic differences, time of harvest, degree of ripeness of the fruit and the geographical location of the tree all produce variation in the chemical composition of coffee beans. The approximate levels of constituents are presented in Table 12.

TABLE 12.

Chemical composition of green coffee beans (approximate composition based on dry weight) (ref. 4)

Classes and Components	Water Solubility	Percentage of Green Coffee Total	Compound or Group	Soluble
Carbohydrates		60		
Reducing sugars	Soluble		1.0	
Sucrose	Soluble		7.0	
Pectins	Soluble		2.0	
			10.0	10
Starch	Easily solubilised		10.0	
	Easily solubilised		5.0	
			15.0	-
Hemicelluloses	Hydrolyzable		15.0	-
Holocelluloses	Non-hydrolyzable fiber		18.0	
Lignin	Non-hydrolyzable fiber		2.0	
			35.0	
Oils	Insoluble	13		-
Protein (N x 6.25)	Depends on amount denatured	13		4
Ash as oxide	Depends on amount hydrolyzed	4		2
Non-volatile acids				
Chlorogenic	Soluble		7.0	
Oxalic	Soluble		0.2	
Malic	Soluble		0.3	
Citric	Soluble		0.3	
Tartaric	Soluble		0.4	
			8.2	8
Trigonelline	Soluble	1		1
Caffeine (Arabica 1.0%, Robusta 2.0%)	Soluble	1		1
		92		26

Polysaccharides were found to form the major part of the carbohydrate fraction (ref. 177), the constituent sugars being D-mannose with lower levels of L-arabinose, D-galactose and D-glucose (ref. 4). The principal polysaccharide of green coffee beans is an insoluble (1-4)-β-D-mannan (ref. 178). Most of the L-arabinose and D-galactose components are in the form of a water soluble arabinogalactan (ref. 4). Cellulose is present at a level of 5% (ref. 178). Small quantities of sugars such as sucrose and the galactosides raffinose and stachyose have been demonstrated in green beans (ref. 4). Levels of these oligosaccharides have been shown to be species dependant (ref. 179).

The predominant proteins of green coffee were demonstrated, using sodium dodecyl sulphate gel electrophoresis and permeation chromatography, to be high molecular weight species (ref. 180) which correlated with coffee quality. Amorim & Josephson (ref. 181) suggested monomeric forms of water soluble proteins to be few in number and of relatively low molecular weight. Further, it was suggested that disulphide bonding may contribute significantly to coffee protein structure. Good quality Brazilian beans contain high molecular weight proteins which are absent in poor quality beans (ref. 181). Free amino acids (ref. 4) and low molecular weight peptides (ref. 165) are also found in green coffee beans. Percentage levels of free amino acids are listed in Table 13.

Considerable attention has been paid to the fat composition of coffee because of the role of lipids as carriers of coffee aroma. The lipid content of green Colombian coffee beans has been reported as phytosterine esters, hydrocarbons and waxes 2.0%, triglycerides 81.3%, diterpene fatty acid esters 15.9%, free sterols 0.39% and diterpenes 0.15% (ref. 182).

TABLE 13.

Levels of free amino acids in green coffee

Amino Acids	% of Bean
Aspartic acid	0.33
Serine	0.12
Asparagine	0.3
Glutamic acid	0.49
Proline	0.14
Glycine	0.02
Alanine	0.24
Valine	0.02
Isoleucine	0.03
Leucine	0.03
Tyrosine	0.04
Phenylalanine	0.81
γ-aminobutyric acid	0.3
Lysine	0.04
Histidine	0.04
Arginine	0.04

Chlorogenic acids, trigonelline, caffeine and theobromine are considered to be important components of coffee and contribute to final coffee flavour (ref. 4). The chlorogenic acids consist of chlorogenic acid, neochlorogenic acid and the isochlorogenic group of acids (dicaffeolquinic acids) (ref. 183). Trigonelline has a pyridine structure which may be transformed into nicotinic acid during roasting. The alkaloid caffeine forms 1% of arabica coffee and 2% of robusta coffee (ref. 4).

8. CONSTITUENTS OF COFFEE FLAVOUR AND AROMA

Suboptimal fermentation can result in defects in coffee flavour and aroma. The taste and aroma of coffee (refs. 184, 185) develops only during roasting and begins to form when the bean reaches 180-190°C (refs. 186, 187). The principal flavour precursors are trigonelline, sucrose, fructose, glucose, free amino acids and peptides (refs. 186,188). Both the ratio as well as the total amount of amino acids appears to be important in the development of coffee flavour. The amino acid composition and relative heat stabilities of coffee proteins will be important in flavour development (ref. 189). Although an aroma reminiscent of coffee was produced on heating the water soluble fraction of green coffee containing trigonelline, sucrose, glucose and amino acids (ref. 188) more than 500 compounds have been found in coffee aroma (ref. 171). Many of the important flavour components are present at low concentrations being produced under the pyrolytic conditions of roasting (ref. 171).

9. COFFEE PROCESSING

Two basic methods have evolved for the production of coffee from coffee cherries, the dry and wet processes.

9.1 The Dry Process

The dry process is non-microbial, produces poor quality beverages and will only be mentioned briefly. In the dry method of coffee preparation the overripe fruits are left on the tree until partially dry, collected, dried to a moisture content of 10-11% and hulled (refs. 4, 190). This process is cheaper than the wet process which involves a microbial fermentation.

9.2 The Wet Process

The wet process is adopted for the production of high quality coffee particularly from arabica beans (refs. 171,174). In South America, processing is either carried out on the farm in more remote regions or, where economic forms of transport are available, in central processing plants (ref. 170). Essentially, the process involves mechanical removal of the outer pulp leaving the mucilage layer adhering to the bean. The mucilage may be removed by natural fermentation, by the addition of enzymes or alkali or by mechanical force. For the successful completion of these processes it is essential that the coffee cherries are ripe otherwise difficulties will arise with mucilage removal.

9.3 Composition of the Mucilage

As outlined in the context of cocoa fermentation the composition of the mucilage layer governs the course of fermentation. Consideration of the chemical composition of the mucilage is therefore pertinent.

The polysaccharide component of coffee mucilage has been found to form 77.6% of the dry weight (ref. 191) being composed mostly of pectin and pectic acids. Pectic acid is a polymer of galacturonic acid associated with low levels of arabinose, xylose and rhamnose. Pectin is composed mostly of α1-4 linked galacturonic acid with smaller amounts of arabinans and galactans (refs. 192, 193, 194). The pectin present in coffee was found to have a degree of esterification of 70% and chain length varies from 140 units for robusta varieties to 250 units for arabica varieties (ref. 195). Reducing sugars, sucrose, caffeine, chlorogenic acid (ref. 196) and amino acids (ref. 197) were also found in mucilage.

Mucilage is devoid of cellular structure (ref. 175) and varies in thickness between 0.5 and 2.0 mm, the level being dependant on variety, ripeness of fruit and environmental conditions (ref. 170). Several hydrolytic and oxidative enzymes such as galacturonase, α-galactosidase, peroxidase and polyphenol oxidase have also been detected (ref. 165). Pectinolytic enzymes were reported to be absent from coffee mucilage by Arakji and Yang (ref. 198). Conflicting results were noted by Rolz et al (ref. 199). The discrepancy may have been due to the presence of plant derived enzymes in the latter case. Mucilage is not thought to be removed by pectinases of plant origin since it has been shown that the breakdown of pectin does not take place in a sterile environment (ref. 165). Further surface sterilisation of cherries prevents mucilage breakdown (ref. 200). Thus, there is a requirement for microbial enzymes to completely catalyse the solubilisation of mucilage although conclusive proof is lacking.

9.4 Microbial groups associated with fermentation

It was originally believed that the removal of the mucilagenous layer adhering to the parchment was achieved firstly by endogenous mucilage enzymes, followed by bacteria, filamentous fungal and yeast attack (ref. 201). Since the mucilage consists predominantly of pectic substances (ref. 191) the ability to elaborate enzymes broadly classed as pectinases is a prerequisite to successful colonization and utilization of the mesocarp. Various workers have investigated the microflora of fermenting coffee and have attributed the breakdown of pectin to different microbial groups (refs.173,174,195,202,203).

Penicillium, Fusarium, Cladosporium and Aspergillus species isolated from Brazilian coffee fruits metabolised both pectic acid and galacto araban. Pectinolytic Fusarium, Aspergillus and Penicillium species were also isolated by Vaughn and co-workers (ref. 203). Major pectinolytic activities were attributed to Gram-negative, non sporeforming, coliform-like bacteria since the numbers of these organisms increased significantly during fermentation from an initial surface population of 10^2-10^3 bacteria per gram to 10^9 bacteria per gram after 24 hours of fermentation. Pectinolytic Bacillus species were also detected. Very low numbers of yeasts were isolated, suggesting little involvement in mucilage breakdown.

Studies on fermenting Hawaian coffee (ref. 173) demonstrated that the predominant microbial flora were Gram-negative bacteria which fermented lactose rapidly. The cultures were identified as species of Erwinia, Paracolobactrum (now reclassified amongst other members of the Enterobacteriaceae) and Escherichia. Of the 168 strains tested, 44 strains all identified as Erwinia dissolvens and were capable of removing mucilage from depulped cherries. Erwinia dissolvens has been reclassified as a Klebsiella species (ref. 173). Low numbers of yeasts were present, none of which were pectinolytic. Thus the work of Frank et al (ref. 173) would suggest that bacteria, particularly Klebsiella-like members of the Enterobacteriaceae, play an important role in coffee fermentation. However, during the course of this work only lactose fermenting isolates were tested for "mucilaginase" activity on the basis of an earlier report (ref. 203), that fermentation of pectic material in Brazilian coffee cherries was carried out by coliform-like bacteria resembling Aerobacter and Escherichia. Since lactose is not a component of mucilage, the use of this criterion alone for isolating organisms responsible for the degradation of pectin may have produced misleading results. A Klebsiella strain has also been reported to be the sole isolate exhibiting pectinolytic activity in fermenting African coffee cherries (ref. 174). However investigations into the properties of the pectinases produced by a Klebsiella strain (ref. 174, 195)which preferentially attack pectic acid, indicated exopectate lyase to be the only form elaborated. The pectinases of this particular strain would not therefore degrade the highly polymerised form of pectin found in coffee mucilage (ref. 195). Pectin depolymerising enzymes and possibly pectin methyl esterases would also be required. Although there are conflicting reports concerning the presence of plant derived pectinases in coffee mucilage (ref. 198,199), it is generally held that the mucilage enzymes do not play a role in pectin breakdown (ref. 174). It was thus concluded (ref. 174) that in addition to Klebsiella species pectinolytic organisms possessing pectin methyl esterase and/or pectin esterase were also involved in the removal of mucilage from robusta coffee.

Agate and Bhat (ref. 200) studied the microflora of fermenting Indian robusta coffee and showed mucilage breakdown to be due to the activities of pectinolytic yeasts. In this study yeast species formed the largest part of the microbial population (47-52%) on both the cherry surface and the mucilage. The bacterial flora was also found to be different to the flora isolated by Vaughn et al. (ref. 203). and Van Pee and Castelein (ref. 174), and was composed of streptococci, pseudomonads, Flavobacterium and Proteus. The predominant pectinolytic yeasts were Saccharomyces marxianus, Sac. bayanus, Sac. cerevisiae var ellipsoides and Schizosaccharomyces spp. The Saccharomyces spp. produced more pectinolysis than the Schizosaccharomyces spp. and none of the bacteria, with the exception of Flavobacterium, were pectinolytic. Demucilaging experiments showed that greatest mucilage removal was obtained when either an inoculum from fermenting coffee or a mixture of three Saccharomyces spp. (Sac. marxianus, Sac. bayanus and Sac. cerevisiae var ellipsoides) were used. Of particular interest was the observation that different yeast species produced different fermentations, producing changes in colour, odour and degree of pectinolysis of the ferment. The results of this study are summarised in Table 14.

TABLE 14.
Fermentation of Indian coffee cherries by pure cultures of yeasts (ref. 200).

		Criteria observed			
No.	Yeasts tested	External appearance	Odour of ferment	Disappearance of stickiness	Pectin decomposition %
1	Sac. marxianus	Greyish-brown	Pungent	Almost complete	86.4
2	Sac. bayanus	Cream to yellow	Vinegar-like	Partial	59.0
3	Sac. cerevisiae var elliposoides	White till 48 hours, black after 70 hours	Cheesey	Nil	27.6
4	Mixture of 1 and 2	Yellowish-brown	Vinegar-like	Almost complete	70.8
5	Mixture of 2 and 3	Brownish-black	Vinegar-like	Nil	38.5
6	Mixture of 1 and 3	Greyish	Pungent	Partial	53.5
7	Mixture of 1, 2 and 3	Blackish	Vinegar-like	Complete	94.6
8	Natural process	Original colour retained with brown tinge	Vinegar-like	Complete	98.2

The natural fermentation process produced the best results since the mucilage was completely removed and discolouration of the beans did not occur. A pooled inoculum of three yeasts brought about almost complete elimination of pectic substances after 7-8 hours at room temperature although the beans became discoloured. A single strain inoculum could not produce complete mucilage breakdown. It is concluded therefore that the full array of pectinase enzymes required to catabolise the highly polymerised and esterified form of pectin found in coffee cherries is not possessed by any of the yeast species examined. A mixed culture inoculum is therefore essential for optimum fermentation.

Functionally, the fermentations reported by Agate and Bhat (ref. 200) and Van Pee and Castelein (ref. 174) are similar in that the process results in complete pectin breakdown and production of good quality coffee beans. The fermentations differed in the types of micro-organisms involved. Although the same coffee variety was examined, the Indian coffee fermentation was dominated by yeast species, whereas in the African fermentation Klebsiella spp. were predominant. The reasons for these differences are not immediately apparent. Although the climatic conditions are similar in both countries, geographical variations could well result in differences in the mesocarp. Because of the differences in the type of pectinases produced by the organisms mentioned above it may be surmised that the degree of polymerisation of the pectin component may be affected which would influence the selection of the dominant microbial group.

Given that the pectin in coffee mucilage is a highly esterified polymer, it would be expected that the organisms deemed responsible for its breakdown would elaborate depolymerising enzymes and esterases. Highly esterified polygalacturonic acid can be split directly only by pectin lyase (poly-(methoxygalacturonide) lyase, EC 4.2.2.10). Pectin lyases are produced by fungi such as Aspergillus, Sclerotinia and Penicillium spp. but have not been reported in higher plants and bacteria. All known pectin lyases are endo-enzymes which specifically attack glycosidic linkages located next to a methyl ester group, causing a rapid fall in viscosity. Apart from the fungal isolates, none of the predominant microbial groups considered to be responsible for mucilage degradation are able to elaborate this particular lyase. Bacterial pectinases are generally of the pectate lyase type which act on partially or fully de-esterified pectin, low methoxyl pectin, pectic and polygalacturonic acid.

The yeasts and Erwinia spp. isolated from fermenting coffee are ubiquitous members of epiphytic communities. It has been suggested (refs. 174, 200) that fermentation is the result of inoculation from the cherry surface rather than from the air or water used in pulp removal. Frank et al. (ref. 173) found the soil surrounding coffee trees to contain a high proportion of the genera found in fermenting coffee. Inoculation of the mesocarp is believed to occur through insect damage and pulp removal (ref. 173).

9.5 Changes in mucilage chemistry resulting from microbial activity during fermentation.

The sole aim of fermentation is the removal of pectinaceous mucilage. The most obvious change that takes place during fermentation is a decrease in mucilage viscosity after two to three hours due to pectinase activity. The mucilage was completely degraded after seven to eight hours (ref. 200). Mesocarp sugar levels also fall during fermentation (refs. 199, 205) and pH decreases from an initial value of 6.4 to 5.4 (ref. 200) or 3.7 (ref. 170). The increase in acidity is due predominantly to the exposure of pectin carboxyl groups during pectinolysis (ref. 175) and also to the production of microbial acids (refs. 170,206). Numerous workers (refs. 175,197,201,206,207) have reported the production of low levels of ethanol during fermentation. Quantitative data was not published until 1973 however, when Menchu and Rolz (ref. 170) fermented Guatemalan coffee under both aerobic and anaerobic conditions. Ethanol production during anaerobic fermentation increased from an initial level of 180mg litre^{-1} to 220mg litre^{-1} during the first 6 hours of fermentation and then declined. During aerobic fermentation ethanol levels fell steadily from 160mg litre^{-1} after one hour to 40mg litre^{-1} at 10 hours.

Fermentations are normally carried out in deep tanks open to the atmosphere. Thus, in practice, only the upper layers of a fermentation are aerobic. Work carried out on aerobic and anaerobic fermentations (ref. 170) has indicated that the aerobic process enhanced mucilage removal accompanied by a more rapid fall in pH. The anaerobic fermentation produces a higher ethanol content than the aerobic process. In the latter case, ethanol levels drop steadily over the first 10 hours of fermentation as a result of evaporation due to aeration. It is unlikely that disappearance of alcohol was due to microbial oxidation since ethanol oxidation was only observed when carbohydrate levels fell below 0.01% (ref. 210). In the early stages of fermentation (i.e. up to 10 hours) carbohydrate concentrations would exceed this level. Organoleptic tests indicated the anaerobic process to produce the best quality coffee. A control fermentation with an end point midway between the aerobic and anaerobic process produced a beverage of intermediate quality (ref. 170).

Acetic, propionic and butyric acids were produced during the course of fermentation (ref. 170,206). Acetic acid was detected early in the fermentation, propionic acid levels increased after 23 hours with the concomitant appearance of butyric acid. The quantitative data of Menchu & Rolz (ref. 170) indicated acid levels, expressed as percentage of total volatiles, to range from 5-15%. Evaporation of volatile components during fermentation may have led to erroneous results. Thus, increases in levels of organic acids may have been artifacts resulting from evaporation of ethanol.

Propionic (ref. 211) and butyric acids (ref. 170) have been associated with formation of off flavours. Acetaldehyde was detected in increasing amounts during the latter stages of fermentation (ref. 212). It has been suggested that the presence of acetaldehyde may be indicative of spoilage of Kona (Hawaiian) coffee.

9.6 Rationalisation of mucilage chemistry and fermentation

As already outlined ethanol production during coffee fermentation varies. Further, the final pH values attained have been found to vary (ref. 200). The work of Menchu and Rolz (ref. 170) indicated pH to fall by three units compared with a fall of one unit recorded by Agate and Bhat (ref. 200). These anomalies may be due to the involvement of different microbial groups. The work of Menchu and Rolz (ref. 170) did not include a microbiological investigation. Where microbiological investigations have been carried out, different microbial groups have been associated with the breakdown of pectin:-

a. Filamentous fungi - *Aspergillus*, *Penicillium*, *Cladosporium* and *Fusarium* (ref. 202).

b. Members of the Enterobacteriaceae, particularly *Klebsiella* spp. (refs. 173,203).

c. Yeasts - *Sac. marxianus*, *Sac. bayanus*, *Sac. cerevisiae var ellipsoides*, *Schizosac* spp. (ref. 200).

It is therefore suggested that yeasts may be the dominant group where ethanol levels are high. Lower levels of ethanol could be produced by members of the Enterobacteriaceae by mixed acid or butanediol fermentation.

Lactic acid reported in the latter stages of fermentation (ref. 206) could also be produced by *Leuconostoc mesenteroides*, *Lactobacillus* spp. or *Streptococcus faecalis* which were isolated from coffee cherry samples transported by air from Mexico to Geneva (ref. 213). These organisms may well be responsible for acidogenesis.

Traces of acetaldehyde (ref. 212) are likely to be the product of either mixed acid fermentation, butanediol fermentation or butyric acid fermentation. Although butyric acid has been detected in fermenting coffee (ref. 170) saccharoclastic *Clostridia* have not been isolated. Similarly, although propionic acid was detected (ref. 170) *Propionibacterium* spp. have not been isolated.

It is apparent from the preceeding remarks that more definitive work is required to clarify the process of coffee fermentation.

9.7 Effect of altitude on coffee fermentation

It is generally known that high grown coffees harvested during the dry season are of better quality than low grown coffee picked during the rainy season. A negative correlation between total mucilage content and altitude has been demonstrated (refs. 170,199), low grown coffees having a thicker mucilage coat than high grown coffee due to higher moisture content. Physico-chemical differences between the mucilage of different coffees could influence microbial interactions and the course of fermentation. A cursory analysis of aerobic or anaerobic fermentations of beans grown at different altitudes (ref. 170) indicated the effect of altitude to be quantifiable (Table 15).

TABLE 15.
Fermentation time and final pH of coffee cherries from plantations located at different altitudes (ref. 170).

Sample	Altitude above sea level (m)	Aerobic Hours	Aerobic pH	Anaerobic Hours	Anaerobic pH
1	365	25	3.7	28	3.7
2	915	24	3.3	53	3.7
3	1370	26	-	68	-
4	1550	51	3.7	215	3.7
5	1830	120	3.2	250	3.4

\- Information not provided

The fermentation time required to remove mucilage increased with altitude and pH dropped to lower values at the higher altitudes.

Fermentation of low grown coffee produced twenty times as much ethanol as fermentation of high grown coffee (ref. 170). No attempt was made to explain the results although it is reasonable to surmise that different microbial groups may have been involved. Anaerobic fermentation would select facultative or anaerobic groups such as yeasts or members of the Enterobacteriaceae whereas the aerobic process would enhance the growth of filamentous fungi. It is also quite likely that altitude affects mucilage composition and thus influences microbial colonisation.

9.8 Assessment of fermentation end point

Following fermentation remaining mucilage is removed by washing. The duration of the fermentation period has been found to depend on coffee variety, climate and regional factors, degree of anaerobiosis of the ferment and presumably microbial groups effecting degradation of the mucilage. Fermentation periods of 16-24 hours are generally considered to be optimal (Refs. 167,214). Prolonged fermentation results in the accumulation of microbial products which are associated with off flavours in the beans and final beverage (refs. 170, 212). Exposure of coffee to fermentation liquor for too long causes discolouration of the silver skin and the production of taints due to ingress of products such as end products of microbial fermentation and chlorogenic acid. Oxidation of chlorogenic acid due to polyphenol oxidase and peroxidase activity has been shown to result in browning reactions (ref. 215). Because of the importance of free amino acids in coffee aroma, the reaction of oxidation products with amino acids and peptides cause flavour alterations (ref. 188). To avoid browning reactions Wootten (ref. 206) proposed the addition of 1% sodium sulphite. Fermentation time would be inhibited however due to the effect of sulphite on microbial growth. Long fermentations decrease polyphenol oxidase in dried beans (ref. 214) and produce a depreciation of beverage quality.

Fermentation is regarded as being complete when residual mucilage may be washed away from the bean (ref. 216). Washing is essential if secondary fermentation is to be avoided. Residual mucilage adhering to the bean will also impede correct drying.

In practice fermentation periods are often longer than the recommended 24 hours (ref. 190), particularly if under-ripe beans are used (ref. 216). To overcome problems of quality depreciation due to increased fermentation times arabica-canephora hybrids have been developed which may be fermented for 48 hours without adverse effects on beverage quality (ref. 167).

In certain countries such as Kenya fermented beans are soaked in water for 24-48 hours in order to enhance flavour (ref. 171).

9.9 Non-microbiological removal of mucilage

Ideally, the fermentation phase of coffee production should last for no more than 24-48 hours (ref. 167) to obviate microbial damage to the bean and resultant formation of off-flavours and odours. Unfortunately, the natural process may often take much longer. Robusta varieties have thicker mesocarp layers than arabica coffees and a fermentation time greater than 72 hours may not be sufficient to remove all the adherent mucilage (ref. 190). The data of Menchu and Rolz (ref. 170) indicated fermentation time to increase with plantation altitude. Further, it has been demonstrated that the period required to remove mucilage from coffees fermented under identical conditions varies between 20 and 84 hours (ref. 217). Seasonal variation, coffee species, climatic conditions during growth and processing, standard of picking, fertilisers, minor alterations in processing techniques and other factors which have not been fully investigated, such as the level of pollution in the water used, contribute towards fluctuations in fermentation time (ref. 218).

Variation in fermentation requirements represents serious threats to the continuity of coffee production. It may therefore be necessary for a producer to change or omit parts of the process with a resultant deterioration in coffee quality. Alternatively, spontaneous fermentation may be accelerated to within 5-6 hours by the use of pectinases or the process may be replaced by alkali washing. Both types of treatment have been found to produce a good quality drink (ref. 190).

9.10 Enzymic treatment of coffee.

This was first reported by Johnston and Foote (ref. 219) and was marketed under the name of Benefac as part of a process for curing coffee. Further pectinolytic preparations, Ultrazyme, Pectozyme and Cofepec, have been available and tested in many countries (refs. 165, 220, 221). Coffee quality appears not to be impaired.

The most effective method of enzyme application was found to be by spraying (ref. 220). A dosage of 10 g of Ultrazyme 100 was found to be sufficient to process one tonne of beans (ref. 220). It was also discovered that by recycling the decanted liquid the preparation was suitable for three treatments (ref. 170). Browning of the silver skin during fermentation was avoided by addition of 250 ppm metabisulphite to 1 kg of coffee (ref. 170, 212). Microbial growth was also suppressed. The application of enzyme reduces the fermentation period. In East Africa a fermentation period of 40-80 hours was reduced to 16 hours (ref. 220).

Although enzyme preparations have been available for thirty years, cost prohibits their use on a large scale. The use of inocula prepared from over-fermented liquors provides a more economical alternative (refs. 218, 222). Inocula were prepared by fermenting waste mucilage and parchments in heaps or tanks for 4-6 days. Following overfermentation the liquid was washed and filtered. Butty (ref. 222) found 10 kg of this crude enzyme preparation to be sufficient to treat 1,000 kg of washed ripe cherries. Fermentation time for African coffees was reduced to 12 hours. Similarly, Agate and Bhat (ref. 200) found the addition of yeast enzymes to reduce fermentation time to 6-7 hours.

9.11 Alkali removal of mucilage

This is feasible since fermentation plays no part in flavour development. Dilute sodium hydroxide (10%) effectively removed mucilage from arabica coffees in 30 minutes and robusta coffee in 45 minutes (ref. 190). Differences between naturally fermented and NaOH treated coffee cherries could not be detected by organoleptic procedures.

9.12 Mechanical removal of mucilage

This is feasible but produces very poor quality coffee (ref. 190).

10. WASHING, SOAKING AND DRYING

Following fermentation residual mucilage is removed by washing. The washing process is followed by a period of soaking which improves coffee quality. During this period changes in the carbohydrate components of the beans have been noted (ref. 179). Levels of oligosaccharides such as raffinose, stachyose and sucrose decreased while monosaccharide levels increased. Activities of β-galactosidase increased correspondingly during soaking. These chemical changes may improve the quality of the final beverage. Greater galactosidase activity and low oligosaccharide levels in high quality coffee (ref. 179) supports this hypothesis.

Soaking is followed by drying which reduces the moisture content of the bean to 10-11%. In India, beans are commonly surface dried on trays supported on wiremesh trollies about three feet above floor level (ref. 190), for 24-48 hours and then transferred to the floor to a depth of 3-4 inches to complete the process. Elsewhere, weather permitting, beans are dried on the ground with regular turning to reduce variability.

Coffee quality and drying temperature appears to be related (ref. 223) and sun-drying appears to produce the best results. The optimum drying temperature was shown to be 45°C, temperatures of 50-60°C produced poor quality drinks whereas temperatures of 60-90°C produced reasonable beverage.

The relationship between the polyphenol oxidase activity of the dried bean and temperature have also been investigated (ref. 224). Generally, activity decreases as temperature increases except at 40°C where activity is lower than at 30 or 50°C. Since high polyphenol levels have been equated with good coffee quality (ref. 165) this result is significant. In some coffee growing regions, harvesting of the crop must be carried out during the rainy season (refs. 170, 216) and cannot therefore be sun dried. Wet coffee is highly susceptible to microbial attack and the production of off flavours due to overfermentation. Prolonged soaking produces a dulling in colour of the bean, induces mould and bacterial growth, off-odours and flavour, and a weakening of the parchment layer which may split. In an attempt to overcome these problems methods of artificial drying have been investigated (ref. 216). Conditions for optimum drying are critical, however; too high a drying temperature produced deleterious flavour changes and discolouration of the beans. Discolouration was due to oxidation of chlorophyll break-down products present in the silver skin which would normally be bleached during sun drying.

11. CONSEQUENCES OF OVER-FERMENTATION

If coffee beans are fermented for extended periods, such that microbial degradation of the beans occurs, very poor quality coffee is produced which characteristically has an onion-like flavour (ref. 216). Over-fermented beans are commonly termed "stinker" beans (ref. 225), which are produced at the death of the embryo. The embryo is killed by the rises in temperature concurrent with extended fermentation and ingress of end-products of microbial fermentation: methanol, ethanol and acetic acids (ref. 225). Stinker beans may also arise from anaerobic conditions of fermentation created when river water with a high biological oxygen demand is used. Inorganic pollutants may also increase pH and anaerobic conditions producing an undesirable microbial flora (ref. 171). Increased levels of acetoin are also found in stinker beans (ref. 225). Low numbers of stinker beans will taint a much larger volume of healthy beans. Fluorescence detection has been found to be effective in sorting defective and healthy beans (ref. 226), since cellular damage is exhibited by intense white fluorescence under ultraviolet light, whereas healthy beans do not fluoresce.

TEA FERMENTATION

Although tea fermentation is often grouped with cocoa and coffee fermentation the former, unlike the latter two, is purely an oxidative process initiated by the enzymic oxidation of catechin and gallocatechin and does not require any microbiological involvement (refs. 227, 228, 229). A brief description of tea fermentation is therefore included purely for comparative purposes since the initial enzyme oxidations are somewhat reminiscent of the microbiologically induced oxidations which occur during cocoa fermentation.

12. TEA PREPARATION AND FERMENTATION

Tea is commercially available in one of three forms, black tea which is fully fermented, oolong tea which is partially fermented and unfermented green tea.

12.1 Black tea

The production of black tea involves four basic steps - withering, rolling, fermenting and frying. Once plucked from the tea bushes the leaves are spread thinly on hessian or wire-netting trays and arranged in a manner which allows for free access of air. In certain factories the rate of withering may be controlled by changing temperature and humidity levels whereas in other systems control may only be exerted by varying the depth of the tea leaves (ref. 230). It is essential that withering is carried out slowly and evenly, the process usually taking approximately 18 hours. During this period leaf moisture content is reduced from 75-80% to 58-68% although exact values depend on the growing region (refs. 4, 230, 231). Prolonged withering may result in microbially induced changes in the tea leaves (ref. 230).

On completion of withering the tea leaves are subjected to a rolling process which damages the leaf structure and initiaties "fermentation". In order for rolling to be successful it is vital that the correct moisture content is achieved during withering (ref. 231). The induction of enzymic oxidation of tea polyphenols by rolling is reminiscent of the induction of enzymic oxidation of cocoa polyphenols by microbial fermentation.

The fermentation process is carried out at a relative humidity of 100% for a period of approximately 3 hours. During this period the colour of the tea leaves changes from green to a coppery-red colour and the odour of tea develops. The duration of fermentation depends on the time of harvesting (ref. 232). Fermentation results in loss of astringency due to the oxidation of polyphenols and the darkening of the leaves due to the type of tanning reactions already described in the context of cocoa fermentation (refs. 4,227, 228,229,231).

The final frying process terminates fermentation and reduces the moisture content of the leaves to approximately 3%. Roasting temperatures of 82-95°C result in the loss of volatile flavour components and coagulation of leaf proteins (ref. 4).

12.2 Oolong tea

The withering period for the production of Oolong tea is reduced so as to induce a limited form of fermentation. The tea leaves are laid in bamboo baskets for a period of 5-6 hours at 28.3-29.4^{o}C (refs. 4,231). The fermentation is terminated by drying with agitation for 10 minutes at 204°C. The leaves are then fired firstly for 3 hours and secondly for 5-12 hours at a temperature of 100°C. Long firing times produce high quality tea.

12.3 Green tea

Green tea is produced by steaming or heating the plucked leaves to inactivate oxidative enzymes and rolling and drying to a final moisture content of 2%.

13. COMPOSITION OF FRESH TEA LEAVES

The composition of fresh and fermented tea leaves has been summarised by Sanderson (ref. 228). Polyphenols form approximately 25-30% of the total weight of fresh tea, 20-25% of which are oxidised during fermentation. Protein, amino acids and carbohydrates form 15, 4 and 7% of fresh tea leaves and show little change after fermentation. Flavour development is therefore largely associated with the oxidation of simple polyphenols. The tea polyphenols may be classified as flavanols (catechins), flavonols and flavonol glycosides, leucoanthocyanins and acids and depsides. The flavanols include (-)-epicatechin, (-)-epicatechin gallate, (-)-epigallocatechin, (-)-epigallocatechin gallate, (-)-catechin, (+)-gallocatechin. Flavonols and flavonol glucosides comprise quercetin, kaempferol, quercetin 3-rhamnoglucoside, kaempferol, quercetin 3-rhamnoglucoside, kaempferol 3-rhamnoglucoside, quercertin 3-rhamnodiglucoside and kaempferol 3-rhamnodiglucoside. The acids and depsides include gallic acid, chlorogenic acids, p-coumarylquinic acids, theogallin and ellagic acid.

A range of enzyme activities have been detected in tea leaves (refs. 233, 234,235,236,237,238) including polyphenol or catechol oxidase, peroxidase, pectin methyl esterase, proteinase, phosphatase and 5-dehydroshitimate reductase. Catechol oxidase and polyphenol oxidase are probably the most important components from the point of view of tea production.

14. CHANGES IN TEA LEAF CHEMISTRY RESULTING FROM FERMENTATION

The fermentation process results firstly in the enzymic oxidation of catechin and gallocatechin (refs. 239,240,241,242) and secondly in subsequent chemical oxidations to produce theaflavins and thearubigins (ref. 235). Theaflavins are pigmented components formed by oxidation and dimerisation of catechins (refs. 243, 244). Also formed during fermentation are epitheaflavic acids (ref. 245). These compounds belong to a heterogeneous group of pigments termed thearubigins (ref. 240). Epitheaflavic acids may undergo further oxidation during fermentation to form other thearubigins (ref. 245).

15. CONCLUDING REMARKS

The fermentation of cocoa, coffee and tea may be regarded as an art rather than a science. The cost of raw materials and a desire for product uniformity are heralding changes however. As outlined in the preceding account the factors controlling cocoa and coffee fermentations, and their effect on the biochemistry of flavour development, are poorly understood despite the considerable volume of literature available on these topics. Fundamental work on the biochemistry and chemistry of flavour development is therefore necessary in order to establish the criteria for the development of controlled fermentation systems. The application of cloning techniques to cocoa and coffee allied to developments in fermentation technology could result in the optimum fermentation of beans and cherries from plants specifically tailored to particular flavour requirements, soil type or climate.

The authors wish to acknowledge the help and guidance provided by colleagues and are grateful to Cadbury Schweppes both for their support and permission to produce this review.

16. REFERENCES

1 T.A. Rohan, The Processing of Cocoa. Food and Agriculture Organisation of the United Nations, Rome, 1960.
2 E.M. Chatt, Cocoa: Cultivation, Processing, Analysis. Interscience, New York and London, 1953.
3 V.C. Quesnel, Esso Cultivar, Special Cacao issue (1966) 4-6.
4 F.A. Lee, Basic Food Chemistry, Avi, Westport, 1975, pp.282-327.
5 T.A. Rohan, Gordian, 9 (1969) 443-590.
6 T.A. Rohan, Gordian, 12 (1969) 587-590.
7 T.A. Rohan, Food Process. Market., 38 (1969) 12.
8 O.G. Vitzthum, P. Werkhoff and P. Hubert, J. Food Sci., 40 (1975) 911-916.
9 H.J. Hart, Cacoa, Duckwork, London, 1911.
10 E.E. Cheesman, Tropic. Agric., 21 (1944) 144-159.
11 D.B. Murray, in G.A.R. Wood (ed.), Cocoa, Longman, London, 1975.
12 C.F. Chong, R. Shepherd and Y.C. Poon, Proc. Int. Conf. on Cocoa and Cocoanuts, Kuala Lumpur, Malaysian Agricultural Research and Development Institute, 1978, pp. 387-414.
13 D.H. Urquhart, Cocoa, Longmans, Green and Co., London,1955.
14 W.G.C. Forsyth and V.C. Quesnel, J. Sci.Food Agric., 9 (1957) 505-509.
15 P.S. Hammond, Rept. Cocoa Conf., Cocoa Choc. and Confect. Alliance, London, 1953, pp. 29-32.
16 A.V. Gibberd, Rept. Cocoa Conf., Cocoa Choc. and Confect. Alliance, London, 1953, pp. 26-29.
17 A.W. Knapp, J.Soc.Chem.Ind., 45, (1926) 140-142.
18 G.R. Howat, B.D. Powell and G.A.R. Wood, J. Sci. Food Agric., 8 (1957) 65-72.
19 W.G.C. Forsyth and V.C. Quesnel, Biochem. J., 65 (1957) 177-179.
20 W.G.C. Forsyth and V.C. Quesnel, Advan. Enzymol. 25 (1963) 457-492.
21 L. Nicholls, in H.H. Smith (Ed.), The Fermentation of Cacoa, Bale, Sons and Danielsson, London, 1913.
22 H.F.K. Dittmar, Gordian, 56 (1956) 16-18.
23 P.A. Roelofsen, Adv. Food Res., 8 (1958) 225-296.
24 V.C. Quesnel, J.Agric. Soc. Trinidad Tobago, 67 (1967) 41-50.
25 V.C. Quesnel, J.Agric. Soc. Trinidad Tobago, 67 (1967) 527-545.
26 E.C. Humphries, Ann. Bot., 8 (1943) 45-61.
27 J.A. Maclean and R. Wickens, Rept. Cocoa Conf. Cocoa Choc. and Confect. Alliance, London, 1953, pp.34-39.
28 K. Saposhnikova, Agron. Tropic, 2 (1952) 185-195.
29 E.V. Packiyasothy, E.R. Jansz, U.M. Senanayaka, R.C. Wijesundara and P. Wickremasinghe, J.Sci. Food Agric., 32 (1981) 873-876.
30 J.G. Carr, P.A. Davies and J. Dougan, Research Rept., University of Bristol Research Centre, Long Ashton, Bristol, 1979.
31 H.T.L. Liau, Proc. Seminar on Cocoa and Cocoanuts, Shaba, 1976, pp. 184-196.
32 F. Hardy, J. Soc. Chem. Ind., 44 (1925) 305-306.
33 W.G.C. Forsyth, Nature, 164 (1949) 25.
34 E.C. Humphries, Tropic. Agric., 21 (1944) 166-169.
35 H.H. Smith, The Fermentation of Cacoa, Bale, Sons and Danielsson, London, 1913.
36 P.A. Roelofsen and G. Giesberger, Arch. Koffiecult. Nederland., 16 (1947) 1-159.
37 J.E. Rombouts, Proc. Soc. Appl. Bact., 15 (1952) 103-111.
38 R.L. Camargo and A. Martinelli, Food Tech., 17 (1963) 116-118.
39 K. Ostovar and P.G. Keeney, J. Sci. Food Agric., 38 (1973) 611-617.
40 K. Ostovar, Ph.D. Thesis, Pennsylvania State University, University Park, 1971.
41 D.L. Zak, K. Ostovar and P.G. Keeney, J. Food Sci., 37 (1972) 967-968.
42 J.E. Rombouts, Tropic. Agric., 30 (1953) 34-41.
43 J.E.Rombouts, Private Communication cited by P.A. Roelofsen (1958), 1955.
44 J.G. Carr, P.A. Davies and J. Dougan, Research Rept. University of Bristol Research Centre, Long Ashton, Bristol, 1980.
45 H.L. Martelli and H.F.K. Dittmar, Appl. Micro., 9 (1961) 370-371.
46 V.C. Quesnel, Rept. Cocoa Conf., Grosvenor House, London, 1957.

47 R.R. Davenport, in F.A. Skinner, S.M. Passmore and R.R. Davenport (Ed.), Biology and Activities of Yeasts, Soc. Appl. Bact. Symp. Ser. 9, Academic Press, London and New York, 1980, pp. 1-27.
48 P.A. Roelofsen and G. Giesberger, Arch. Koffiecult, Nederland., 16 (1947) 1-159.
49 T. Asai, Acetic Acid Bacteria Classification and Biochemical Activities, Univ. of Tokyo Press and University Park Press, 1968.
50 W.G.C. Forsyth, Rept. Cocoa Conf., Cocoa Chocolate and Confect. Alliance, London, 1953, pp. 32-34.
51 L.A. Griffiths, Rept. Cocoa Res., Imp. Coll. Tropic Agric., St. Augustin, Trinidad, 1961, pp.42-46.
52 B. Biehl, Zeitschrift fur Lebensmittel-Untersuchung und Forschung, 136 (1968) 76.
53 H.F.K. Dittmar, Rev. Brazil. Farm., 50 (1969) 347-350.
54 A. Lopez and V.C. Quesnel, J. Sci. Food Agric., 24 (1973) 319-326.
55 A. Lopez and V.C. Quesnel, Ann. Rep. on Cocoa Res., Univ. West Indies, 1973, pp. 39-50.
56 K.S. Morinago, Rev. Int. Choclat., 26 (1973) 19-24.
57 K. Seiki, Rev. Int. Chocolat., 28 (1973) 38-42.
58 V.C. Quesnel and A. Lopez, Tropic. Agric., 52 (1975) 309-316.
59 A.S. Lopez, Rev. Theobrom., 9 (1979) 25-37.
60 H.W.S. Allison and T.A. Rohan, Tropic. Agric., 35 (1958) 279-288.
61 H.W.S. Allison and R.H. Kenten, Tropic.Agric., 40 (1963) 15-24.
62 V.C. Quesnel, J. Sci. Food Agric., 16 (1965) 441-447.
63 H.W.S. Allison and R.H. Kenten, Tropic. Agric., 40 (1963) 15-24.
64 J.A.R. Maclean and R. Wickens, Rept. Cocoa Conf., Cocoa Choc. and Confect. Alliance, London, 1951, 116-123.
65 V.C. Quesnel, Turrialba, 18 (1968) 110-114.
66 V.C. Quesnel, Rept. Cocoa Res., Imp. Coll. Tropic. Agric., St. Augustin, Trinidad, 1968, pp. 59-60.
67 H. Suomalainen and E. Oura, in A.H. Rose and J.S. Harrison (Ed.), The Yeasts, Vol. II, Yeast Nutrition and Solute Uptake, Academic Press, London and New York, 1971, 271-307.
68 A. Sols, C. Gancedo and G. Delafuente, in A.H. Rose and J.S. Harrison (Ed.), The Yeasts, Vol. II, Energy-Yielding Metabolisms in Yeast, Academic Press, London and New York, 1971, 271-307.
69 P.A. Roelofsen, Biochim. Biophys. Acta, 10 (1953) 410-413.
70 W. Gamble, Experiment., 29 (1973) 421-422.
71 B. Gauthier, J. Guirand, J.C. Vincent, J.P. Parvais and P. Galzy, Rev. Ferment. Ind. Aliment., 32 (1977) 160-163.
72 H.A. Krebs, Biochem. J., 29 (1935) 1620-1644.
73 R.E. Buchanan and N.E. Gibbons, Bergey's Manual of Determinative Bacteriology, Williams and Wilkins, Baltimore, 1974.
74 F. Gotz, B. Sedewitz and E.F. Elstner, Archiv. Microbiol., 125 (1980)209-214.
75 F. Gotz, E.F. Elstner, B. Sedewitz, and E. Lengfelder, Archiv. Microbiol., 125 (1980) 215-220.
76 W.G.C. Forsyth and J.E. Rombouts, Rept. Cocoa Conf., Cocoa Chocolate and Confect. Alliance, London, 1951, pp. 73-78.
77 L.A. Brigland and R.J. Friend, Rept. Cocoa Conf., Cocoa Chocolate and Confect. Alliance Ltd., London, 1958, pp. 177-190.
78 G.R. Howat, B.D. Powell and G.A.R. Wood, Tropic. Agric., 34 (1957) 249-259.
79 W.G.C. Forsyth, Rept. Cocoa Conf., Cocoa Chocolate and Confect. Alliance, London, 1958, pp. 145-148.
80 B.D. Powell, Chem. Ind., 31 (1959) 991-992.
81 V.C. Quesnel, Chem. Ind., 4 (1960) 101-102.
82 J.B. Roberts, Chem. Ind., 45 (1959) 1410-1411.
83 R.H. Kenten and B.D. Powell, J. Sci. Food Agric., 11 (1960) 396-400.
84 N. Maravalhas, Rev. Int. Chocolat., 27 (1972) 22-23.
85 T.A. Rohan, J. Sci. Food Agric., 9 (1958) 542-551.
86 V.C. Quesnel, Conferencia Internacional de Pesquisas em Cacau, Ilheus, BA, CEPLAC, Brazil, 1969, pp. 503-509.

87 B. Biehl, 1st International Congress on Cocoa and Chocolate Research, 1974, pp. 55-70.
88 T.A. Rohan and T. Stewart, J. Food Sci., 32 (1967) 399-402.
89 H. Fincke, Handbuch der Kakaoerzeugnisse, 2nd edn., Springer Verlag, Berlin, 1965.
90 I. Geilinger, R. Amado and H. Neukom, Starch, 33 (1981) 76-79.
91 W. Weissberger, T.E. Kavanagh and P.G. Keeney, J. Food Sci., 36 (1971) 877-879.
92 F. Haworth, Rept. Cocoa Res. Imp. Coll. Tropic. Agric., St. Augustin, Trinidad, 1953, pp. 92-97.
93 T.A. Rohan and T. Stewart, J. Food Sci., 32 (1967) 395-398.
94 U. Bracco, N. Grailhe, W. Rostagno and R.H. Egli, J. Sci. Food Agric., 20 (1969) 713-717.
95 S.S. Miranda, Reuniao do Gomite Tecnico Interamericano de Cacau, Salvador, Brazil, ICB, 1956, pp. 151-155.
96 T.A. Rohan, J. Food Sci., 29 (1964) 456-459.
97 E.C. Humphries, Rept. Cocoa Res. Imp. Coll. Tropic. Agric., St. Augustin, Trinidad, 1939, pp. 227-266.
98 E.K. Obiakor and S.I.N. Nwanko, Proc. 5th Int. Cocoa Res. Confer., Ibadan, CRI, 1977, pp. 568-573.
99 H. Fincke, Zeitschrift fur Untersuchung der Lebensmittel, 55 (1928) 559-568.
100 B. Biehl, Ann. Technol. Agric., 21 (1972) 435-455.
101 T. Swain, British Food Manufacturing Industries Research Association, Scientific and Technical Survey No. 23, 1954.
102 W.G.C. Forsyth and J.B. Roberts, Biochem. J., 74 (1960) 374-378.
103 G.F. Schubiger, E. Roesh and R.H. Egli, Fette Seifen Astrichmittel, 59 (1957) 631-636.
104 V.C. Quesnel, Rept. Cocoa Res., Imp. Coll. Tropic. Agric., St. Augustin Trinidad, 1966, pp. 57-66.
105 W.G.C. Forsyth, Biochem. J., 51 (1952) 511-516.
106 W.G.C. Forsyth, Biochem. J., 51 (1952) 516-520.
107 J.A.R. Maclean, Chem. Ind., 49 (1957) 1597-1598.
108 K.W. Weinges, W. Bahr, K. Ebert, K. Gorlitz and H.D. Marx, in L. Zechmeister (Ed.)., Progress in the Chemistry of Organic Natural Products, Vol. 27, Konstitution, Entstehung und Bedentung der Flavonoid - Gerbstoffe, Springer Verlag, New York, 1969, pp. 159-260.
109 D.L. Zak and P.G. Keeney, J. Agric. Food Chem., 24 (1976) 483-486.
110 P. Matile, J.P. Balz, E. Semadeni and M. Jost, Zeitschrift fur Naturforschung 20 (1965) 693-698.
111 A. Hofer, Planta, 65 (1964) 137-159.
112 T. Takeo, Agric. Biol. Chem., 29 (1965) 558-563.
113 W.G.C. Forsyth, Biochem. J., 60 (1955) 108-111.
114 H.B. Brown, Nature, 173 (1954) 492.
115 M. Holden, J. Sci. Food Agric., 10 (1959) 691-700.
116 B. Biehl, Zeitschrift fur Lebensmittle-Untersuchung und Forschung, 153 (1973) 137-150.
117 B. Biehl, U. Passern and D. Passern, J. Sci. Food Agric. 28 (1977) 41-52.
118 T.A. Rohan, Gordian, 70 (1970) 5-8.
119 T.A. Rohan, Gordian, 70 (1970) 59-62.
120 T.A. Rohan, Gordian, 70 (1970) 158-160.
121 G.A. Reineccius, D.A. Andersen, T.E. Kavanagh and P.G. Keeneg, J. Agric. Food Chem., 20 (1972) 199-202.
122 J. Cerbulis, Archiv. Biochem. Biophys., 49 (1954) 442-450.
123 J. Cerbulis, Archiv. Biochem. Biophys., 58 (1955) 406-413.
124 T.A. Rohan and T. Stewart, Rev. Int. Choclat., 19 (1964) 502-506.
125 H.T.L. Liau, International Conference on Cocoa and Coconuts, Kuala Lumpur, 1978, paper 24.
126 H.F. Birch, Rept. Cocoa Res., Imp. Coll. Tropic. Agric., St. Augustin, Trinidad, 1941, pp. 23-33.
127 B. Biehl, Separate de Gordian, 1444 (1961) 14-20.
128 B. Biehl, Gordian, 61 (1961) 62-68.

129 T.A. Rohan and T. Stewart, J. Food Sci. 30 (1965) 416-419.
130 K.W. De Witt, Tropic. Agric., 30 (1953) 228-236.
131 K.W. De Witt, Rept. Cacao Res., Imp. Coll. Tropic. Agric., St. Augustin, Trinidad, 1954, pp. 54-57.
132 K.W. De Witt, Rept. Cacoa Res. Imp. Coll. Tropic. Agric., St. Augustin, Trinidad, 1961
133 R.V. Wadsworth, Rept. Cocoa Conf., Cocoa Chocolate and Confect. Alliance Ltd., London, 1955, pp. 131-137.
134 D.W. Duthie, Rept. Cocoa Res., Imp. Coll. Tropic. Agric., St. Augustin, Trinidad, 1937, pp. 47-51
135 H.B. Brown, Rept. Cocoa Conf., Cocoa Chocolate and Confect. Alliance, London, 1958, pp. 165-170.
136 E.N. Boyd, P.G. Keeney and S. Patton, J. Food Sci., 30 (1965) 854-859.
137 M. Pragg, S.S. Herbert and M.S. Tibbetts, J. Agric. Food Chem., 16 (1968) 1005-1008.
138 D.A. Forss, Gordian, 11 (1971) 331-332.
139 B. Wal, D.K. Kettenes, J. Stoffelsma, J. Sipma and A.J. Semper, J. Agric. Food Chem., 19 (1971) 276-280.
140 I.I. Rusoff, US Patent 2,835,590, 1958.
141 T.A. Rohan, Food Technol., 19 (1965) 122.
142 A.S. Lopez, Ph.D. Thesis, University of the West Indies, 1974.
143 T.A. Rohan, J. Sci. Food Agric. 14 (1963) 799-805.
144 W. Mohr, M. Rohrle and T. Severin, Fette Seifen Anstrichmittel, 73 (1971) 515-521.
145 H.B. Brown, Chem. Ind., 6 (1957) 164-166.
146 T.A. Rohan, J. Sci. Food Agric., 14 (1963) 799-805.
147 J.B. Roberts, BFMIA Science and Technical Survey Number 38, 1961.
148 T.A. Rohan, J. Food Sci., 29 (1964) 456-459.
149 W.J. Herz and R.S. Shallenberger, Food Res., 25 (1960) 491-494.
150 R. Rapp, Susswaren, 4 (1981) 126-132.
151 T.A. Rohan and T. Stewart, J. Food Sci., 31 (1966) 202-205.
152 T.A. Rohan and T. Stewart, J. Food Sci. 31 (1966) 206-209
153 R.R. Darsley and V.C. Quesnel, Rept. Cacao Res., Imp. Coll. Tropic. Agric., St. Augustin, Trinidad, 1970
154 R.R. Darsley and V.C. Quesnel, J. Sci. Food Agric., 23 (1972) 215-225.
155 A. Purr, A. Helfenberger and J. Nadal, Rev. Int. Choclat., 17 (1962) 534-548.
156 A. Purr, A. Helfenberger and J. Nadal, Rev. Int. Choclat., 18 (1963) 194-207.
157 A. Purr, R. Springer and H. Morcinek, Rev. Int. Choclat., 28 (1965) 38-42.
158 A. Purr, Susswaren, 16 (1972) 543-550.
159 W.G.C. Forsyth, V.C. Quesnel and J.B. Roberts, J. Sci. Food Agric., 9 (1958) 181-184.
160 B. Biehl, Zeitschrift fur Lebensmittel Untersuchung und Forschung, 119 (1963) 105-108.
161 B. Biehl, Zeitschrift fur Lebensmittel Untersuchung und Forschung, 133 (1967) 145-158.
162 R.G. Moores, D.M. Greninger and I.I. Rusoff, J. Amer. Chem. Soc., 74 (1952) 928-930.
163 V.C. Quesnel and K. Jugmohunsingh, J. Sci. Food Agric., 21 (1970) 537-541.
164 R.H. Chenney, Coffee: A Monograph of the Economic Species of the Genus Coffea L, University Press, New York, 1925.
165 H.V. Amorim and V.L. Amorim, Amer. Chem. Soc. Symp. Ser., 47 (1977) 27-56.
166 J.D. Tothill and A.S. Thomas, Agriculture in Uganda, Oxford Univ. Press London, 1940.
167 K. Suryakantha Raju, S. Vishveshwara and C.S. Srinivasan, Indian Coffee, 42 (1978) 195-197.
168 H.V. Amorim, M.G. Legendre, V.L. Amorim, A. St. Angelo and R.L. Ory, Turrialba, 26 (1976) 193-195.
169 V.H. Calle, Cenicafe (Colombia), 14 (1963) 187-194.

170 J.F. Menchu and C. Rolz, Cafe, Cacao, The, 17 (1973) 53-61.
171 G.W. Kulaba, Kenya Coffee, 46 (1981) 351-360.
172 N.H. Gopal, D. Venkataramana and N.G.N. Ratna, Indian Coffee, 40 (1976) 29-33.
173 H.A. Frank, N.A. Lum and A.S. Dela Cruz, Appl. Microbiol., 13 (1965) 201-207.
174 W. Van Pee and J.M. Castelein, J. Food Sci., 37 (1972) 171-174.
175 R.J. Carbonnel and T.M. Vilanova, El cafe de El Savador, 22 (1952) 411-556.
176 Anonymous, Indian Coffee, 44 (1980) 28.
177 M.L. Wolfrom and D.L. Patin, J. Org. Chem., 30 (1965) 4060-4063.
178 M.L. Wolfrom and D.L. Patin, J. Agr. Food Chem., 12 (1964) 376-377.
179 M. Shadaksharaswamy and G. Ramachandra, Enzymologia, 35 (1968) 93-99.
180 H. Thaler and W. Arneth, Zeitschrift fur Lebensmittel-Untersuchung und Forschung, 140 (1969) 101-109.
181 H.V. Armorim and R.V. Josephson, J. Food Sci., 40 (1975) 1179-1184.
182 H.P. Kaufmann and R. Schickel, Fette Seifen Anstrichmittel, 67 (1965) 115-120.
183 J. Corse, L.L. Layton and D.C. Patterson, J. Sci. Food Agric., 21 (1970) 164-168.
184 H.L. Weidemann and W. Mohr, Lebensmittel-Wiss, V. Technol., 3 (1970) 23-32.
185 O.G. Vitzthum and P. Wekhoff, Annals Falsif. Expert. Chim., 69 (1976) 725-735.
186 W. Baltes, 7th International Colloquium on the Chemistry of Coffee, Hamburg, 1976, pp. 91-107.
187 D. Reymond, Abst. 171st Amer. Chem. Soc. Meeting, New York, 1976, pp. AGFD 23.
188 H. Russwurm, 4eme Colloque International sur la Chemie du Cafe Vert, Torrefies et leur Derives, Amsterdam, 1969, pp. 103-109.
189 J.C. Oliveira, 5eme Colloque International sur la Chemie des Cafes, Lisbon, 1971, pp. 115-119.
190 Anonymous, Indian Coffee, 40 (1976), 226-270.
191 R.J. Coleman, J.F. Lenney, A.T. Coscia and F.J. Di Carlo, Arch. Biochem. Biophys., 59 (1955) 157-164.
192 J.B.C. Correa and J.D. Fontana, An. Acad. bras. Cienc., 43 (1971) 803-812.
193 J.B.C. Correa, S. Odebrecht and J.D. Fontana, An. Acad. bras. Cienc., 46 (1974) 357-360.
194 J.B.C. Correa, E.O. Coelho and J.D. Fontana, An. Acad. bras. Cienc., 46 (1974) 357-360.
195 J.M. Castelein and W. Pilnik, Lebensmittel-Wiss, V. Technol., 9 (1976) 277-283.
196 C.M. Franco, J. Agronomia (Brasil), 2 (1939) 131-139.
197 M. Sivetz and H.E. Foote, Coffee Processing Technology, Vol. I, Avi, Westport, 1963, pp. 48-99.
198 O.A. Arakji and H.Y. Yang, J. Food Sci., 34 (1969) 340-342.
199 C. Rolz, J.F. Menchu, R. Espinosa and A. Garcia-Prendes, 5eme Colloque International sur la Chemie des Cafes Verts, Torrefies et leur Derives, Lisbon, 1971, pp. 259-268.
200 A.D. Agate and J.V. Bhat, Appl. Microbiol. 14 (1966) 256-260.
201 V.H. Calle, Cenicafe (Colombia), 16 (1965) 3-11.
202 G. Wosiaki and G.T. Zancan, Arq. Biol. Tecnol., 16 (1973) 129-134.
203 R.H. Vaughn, R. De Camargo, H. Falanghe, G. Mello-Ayres and A. Serzedello, Food Technol. Suppl. 4, 12 (1958) pp.57
204 W. Pilnik and F.M. Rombouts, Int. Congress Food Sci., Technol.-Abs., 1978, pp.42.
205 H.A. Frank and A.S. Dela Cruz, J. Food Sci., 29 (1964) 850-853.
206 A.E. Wootton, Report C.R.12, East African Research Organization, 1963.
207 R. Wilbaux, Les Cafeiers au Congo Belge, Publication de la Direction de l'Agriculture, Bruxelles, 1956.
208 Y.B. Goto and E.R. Fukunaga, Hawaii Agr. Exp. Station, Circular 359, Hawaii, 1956.
209 R. Coste, Le cafeier, Maisonneuve et Larose, Paris, 1968, pp.310.

210 J.C. Lievense and H.C. Lim, in G.T. Tsao (Ed.)., Annual Reports on Fermentation Processes, Vol. 5, The growth and dynamics of Saccharomyces cerevisiae, Academic Press, New York, 1982, Ch.6, pp. 211.
211 A.E. Wootton, Report C.R.11, East African Research Organization, 1960.
212 D.B. Rodriguez, H.A. Frank and H.Y. Yamamoto, J. Sci. Fd. Agric., 20 (1969) 15-19.
213 C.S. Pederson and R.S. Breed, Food Res., 11 (1946) 99-106.
214 O.B. Sanint and G.A. Valencia, Cenicafe (Colombia), 21 (1970) 59-71.
215 M. Melo and H.V. Amorim, Turrialba, 25 (1975) 243-248.
216 J.F. McLoy, Kenya Coffee, 44 (1979) 13-26.
217 A.E. Wootton, Report C.R. 14, East African Research Organization, 1971.
218 M. Butty, Kenya Coffee, July (1973) 214-224.
219 W.R. Johnston and H.E. Foote, Food Technol., 5 (1951) 464-467.
220 R.O. Arunga, Kenya Coffee, 38 (1973) 354-357.
221 B. Kupferschmied, Revista Cafetalera, 42 (1975)
222 M. Butty, Annual Report East African Industrial Research Organisation, (1971) pp. 79-92.
223 M.B. Ferraz and A.A. Veiga, Bol. Supta. Serv. Cafe, 29 (1954) 5-6.
224 P.J. Arcila and A.G. Valencia, Cenicafe (Colombia), 26 (1975) 55-71.
225 M. Barel, F. Challot and J.-C. Vincent, Cafe Cacao The, 20 (1976) 129-134.
226 A. Gibson and M. Butty, 7th International Colloquium on the Chemistry of Coffee, Hamburg, 9-14 June 1975, pp. 141-152.
227 D.J. Millin and D.W. Rustidge, Process Biochem., June (1967) 1-5.
228 G.W. Sanderson, Rec. Adv. Phytochem, 5 (1972) 247-316.
229 G.V. Stagg, Rep. Prog. Appl. Chem., 59 (1974) 451-465.
230 T. Eden, Tea. 2nd edn., Longmans, London, 1965.
231 N.D. Pintauro, Tea and soluble tea products manufacture, Noyes Data Corporation, Park Ridge, 1977.
232 N.M. Abdul-Gaffar, C. Kandappah, T. Alawattegama and S. Samarasingham, Tea Q, 49 (1980) 58-64.
233 H.B. Sreerangachar, Curr. Sci., 8 (1939) 13-14.
234 H.B. Sreerangachar, Biochem. J., 37 (1943) 661-667.
235 E.A.H. Roberts, J. Sci. Food Agric. 9 (1958) 212-216.
236 G.W. Sanderson, Biochim. Biophys. Acta, 92 (1964) 622-624.
237 G.W. Sanderson, J. Sci. Food Agric. 15 (1964) 634-639
238 M.A. Dix, C.J. Fairley, D.J. Millin and D. Swaine, J. Sci. Food Agric., 32 (1981) 920-932.
239 A.R. Pierce, H.N. Graham, S. Glassner, H. Madlin and J.G. Gonzalez, Anal. Chem., 41 (1969) 298-302.
240 E.A.H. Roberts, in T.A. Geissman (Ed.), The Chemistry of Flavonoid Compounds, Economic Importance of Flavonoid Substances: Tea Fermentation, London, Pergamon, 1962.
241 K.M. Dzhemukhadze, G.A. Shalneva and L.F. Mileshko, Biokhimiya, 22 (1957) 836-640.
242 K.M. Dzhemukhadze, G.A. Shalneva and L.F. Mileshko, Biokhimiya, 29 (1964) 882-888.
243 Y. Takino, H. Imagawa, H. Horikawa and A. Tanaka, Agr. Biol. Chem., 28 (1964) 64-71.
244 Y. Takino and H. Imagawa, Agr. Biol. Chem., 28 (1964) 255-256.
245 J.E. Berkowitz, P. Coggon and G.W. Sanderson, Phytochem., 10 (1971) 2271-2278.

Subject index